Edited by
Lutz H. Gade and Peter Hofmann

Molecular Catalysts

Related Titles

Cornils, B., Herrmann, W.A., Beller, M., Paciello, R. (eds.)

Applied Homogeneous Catalysis with Organometallic Compounds

A Comprehensive Handbook in Three Volumes

3 Edition

2015

Print ISBN: 978-3-527-32897-0, also available in digital formats

Wilson, K., Lee, A.F. (eds.)

Heterogeneous Catalysts for Clean Technology

Spectroscopy, Design, and Monitoring

2014

Print ISBN: 978-3-527-33213-7, also available in digital formats

Kamer, P.P., van Leeuwen, P.P. (eds.)

Phosphorus(III) Ligands in Homogeneous Catalysis – Design and Synthesis

2012

Print ISBN: 978-0-470-66627-2, also available in digital formats

Hashmi, A.S., Toste, F.D. (eds.)

Modern Gold Catalyzed Synthesis

2012

Print ISBN: 978-3-527-31952-7, also available in digital formats

Che, M., Vedrine, J.C. (eds.)

Characterization of Solid Materials and Heterogeneous Catalysts

From Structure to Surface Reactivity

2012

Print ISBN: 978-3-527-32687-7, also available in digital formats

Beller, M., Renken, A., van Santen, R.A. (eds.)

Catalysis

From Principles to Applications

2012

Print ISBN: 978-3-527-32349-4

Itsuno, S. (ed.)

Polymeric Chiral Catalyst Design and Chiral Polymer Synthesis

2011

Print ISBN: 978-0-470-56820-0, also available in digital formats

van Leeuwen, P.W., Chadwick, J.C.

Homogeneous Catalysts

Activity – Stability – Deactivation

2011

Print ISBN: 978-3-527-32329-6, also available in digital formats

Zhou, Q. (ed.)

Privileged Chiral Ligands and Catalysts

2011

Print ISBN: 978-3-527-32704-1, also available in digital formats

Edited by Lutz H. Gade and Peter Hofmann

Molecular Catalysts

Structure and Functional Design

WILEY-VCH
Verlag GmbH & Co. KGaA

The Editors

Prof. Dr. Lutz H. Gade
Ruprecht-Karls-Universität Heidelberg
Anorganisch-Chem. Institut (ACI)
Im Neuenheimer Feld 270
69120 Heidelberg
Germany

Prof. Dr. Peter Hofmann
Ruprecht-Karls-Universität Heidelberg
Organisch-Chemisches Institut (OCI)
Im Neuenheimer Feld 270
69120 Heidelberg
Germany

Cover picture
Source: BASF Pressefoto

Library of Congress Card No.: applied for

British Library Cataloguing-in-Publication Data
A catalogue record for this book is available from the British Library.

Bibliographic information published by the Deutsche Nationalbibliothek
The Deutsche Nationalbibliothek lists this publication in the Deutsche Nationalbibliografie; detailed bibliographic data are available on the Internet at <http://dnb.d-nb.de>.

Print ISBN: 978-3-527-33521-3
ePDF ISBN: 978-3-527-67330-8
ePub ISBN: 978-3-527-67329-2
Mobi ISBN: 978-3-527-67328-5
oBook ISBN: 978-3-527-67327-8

Cover Design Adam-Design, Weinheim, Germany
Typesetting Laserwords Private Limited, Chennai, India
Printing and Binding Markono Print Media Pte Ltd., Singapore

Printed on acid-free paper

Contents

List of Contributors

Marcel Brill
Ruprecht-Karls-Universität
Heidelberg
Organisch-Chemisches Institut
(OCI)
Im Neuenheimer Feld 270
69120 Heidelberg
Germany

Dominik Brox
Ruprecht-Karls-Universität
Heidelberg
Physikalisch-Chemisches Institut
(PCI)
Im Neuenheimer Feld 229
69120 Heidelberg
Germany

Peter Comba
Ruprecht-Karls-Universität
Heidelberg
Anorganisch-Chemisches
Institut (ACI)
Im Neuenheimer Feld 270
69120 Heidelberg
Germany

Thomas Debnar
Rheinische Friedrich-
Wilhelms-University of Bonn
Department of Organic
Chemistry and Biochemistry
Gerhard-Domagk-Street 1
53121 Bonn
Germany

Markus Enders
Ruprecht-Karls Universität
Heidelberg
Anorganisch-Chemisches
Institut (ACI)
Im Neuenheimer Feld 270
69120 Heidelberg
Germany

Stefan Fischer
Ruprecht-Karls University of
Heidelberg
Interdisciplinary Center for
Scientific Computing (IWR)
Computational Biochemistry
Im Neuenheimerfeld-368
69120 Heidelberg
Germany

Lutz H. Gade
Ruprecht-Karls-Universität
Heidelberg
Anorganisch-Chemisches
Institut (ACI)
Im Neuenheimer Feld 270
69120 Heidelberg
Germany

A. Stephen K. Hashmi
Ruprecht-Karls-Universität
Heidelberg
Organisch-Chemisches Institut
(OCI)
Im Neuenheimer Feld 270
69120 Heidelberg
Germany

Achim Häußermann
Ruprecht-Karls-Universität
Heidelberg
Organisch-Chemisches Institut
(OCI)
Im Neuenheimer Feld 270
D-69120 Heidelberg
Germany

Katja Heinze
Johannes Gutenberg-University
Institute of Inorganic and
Analytical Chemistry
Duesbergweg 10-14
D-55128 Mainz
Germany

Günter Helmchen
Ruprecht-Karls-Universität
Heidelberg
Organisch-Chemisches Institut
(OCI)
Im Neuenheimer Feld 270
69120 Heidelberg
Germany

Dirk-Peter Herten
Ruprecht-Karls-Universität
Heidelberg
Physikalisch-Chemisches Institut
(PCI)
Im Neuenheimer Feld 229
69120 Heidelberg
Germany

Hans-Jörg Himmel
Ruprecht-Karls-Universität
Heidelberg
Anorganisch-Chemisches
Institut (ACI)
Im Neuenheimer Feld 270
D-69120 Heidelberg
Germany

Peter Hofmann
Ruprecht-Karls-Universität
Heidelberg
Organisch-Chemisches Institut
(OCI)
Im Neuenheimer Feld 270
69120 Heidelberg
Germany

Olaf Hübner
Ruprecht-Karls-Universität
Heidelberg
Anorganisch-Chemisches
Institut (ACI)
Im Neuenheimer Feld 270
D-69120 Heidelberg
Germany

Andres Jäschke
Ruprecht-Karls-Universität
Heidelberg
Institut für Pharmazie und
Molekulare Biotechnologie
(IPMB)
Im Neuenheimer Feld 364
69120 Heidelberg
Germany

Gregor Jung
Universität des Saarlandes
Biophysikalische Chemie
Campus B2 2
Postfach 15 11 50
66041 Saarbrücken
Germany

Eva Jürgens
Eberhard Karls Universität
Tübingen
Institut für Anorganische
Chemie
Auf der Morgenstelle 18
72076 Tübingen
Germany

Farooq Ahmad Kiani
Ruprecht-Karls University of
Heidelberg
Interdisciplinary Center for
Scientific Computing (IWR)
Computational Biochemistry
Im Neuenheimerfeld-368
69120 Heidelberg
Germany

Günter Klatt
Ruprecht-Karls-Universität
Heidelberg
Theoretische Chemie
Physikalisch-Chemisches Institut
(PCI)
Im Neuenheimer Feld 229
69120 Heidelberg
Germany

Horst Köppel
Ruprecht-Karls-Universität
Heidelberg
Theoretische Chemie
Physikalisch-Chemisches
Institut (PCI)
Im Neuenheimer Feld 229
69120 Heidelberg
Germany

Roland Krämer
Ruprecht-Karls-Universität
Heidelberg
Physikalisch-Chemisches Institut
(PCI)
Im Neuenheimer Feld 229
69120 Heidelberg
Germany

Doris Kunz
Eberhard Karls Universität
Tübingen
Institut für Anorganische
Chemie
Auf der Morgenstelle 18
72076 Tübingen
Germany

Dirk Menche
Rheinische Friedrich-
Wilhelms-University of Bonn
Department of Organic
Chemistry and Biochemistry
Gerhard-Domagk-Street 1
53121 Bonn
Germany

Thomas J. J. Müller
Heinrich-Heine-Universität
Düsseldorf
Institut für Organische Chemie
und Makromolekulare Chemie
Universitätsstr. 1
40225 Düsseldorf
Germany

Arina Rybina
Ruprecht-Karls-Universität Heidelberg
Physikalisch-Chemisches Institut (PCI)
Im Neuenheimer Feld 229
69120 Heidelberg
Germany

Ilme Schlichting
Max-Planck-Institut für Medizinische Forschung
Jahnstraße 29
69120 Heidelberg
Germany

Solveig A. Scholl
Ruprecht-Karls-Universität Heidelberg
Anorganisch-Chemisches Institut (ACI)
Im Neuenheimer Feld 270
69120 Heidelberg
Germany

Bernd F. Straub
Ruprecht-Karls-Universität Heidelberg
Organisch-Chemisches Institut (OCI)
Im Neuenheimer Feld 270
D-69120 Heidelberg
Germany

Michael E. Tauchert
Ruprecht-Karls-Universität Heidelberg
Organisch-Chemisches Institut (OCI)
Im Neuenheimer Feld 270
D-69120 Heidelberg
Germany

Oliver Trapp
Ruprecht-Karls-Universität Heidelberg
Organisch-Chemisches Institut (OCI)
Im Neuenheimer Feld 270
69120 Heidelberg
Germany

Johannes Troendlin
Ruprecht-Karls-Universität Heidelberg
Organisch-Chemisches Institut (OCI)
Im Neuenheimer Feld 270
69120 Heidelberg
Germany

Marcel Wirtz
Universität des Saarlandes
Biophysikalische Chemie
Campus B2 2 Postfach 15 11 50
66041 Saarbrücken
Germany

Foreword

1
The Magic Force That Became a Science

In Heidelberg Castle, among many wonders, one finds the German Pharmacy Museum (*Deutsches Apotheken-Museum*). As one wanders through it, one comes to the object illustrated below (Figure 1). This is Döbereiner's lighter or igniter, *Döbereinersches Feuerzeug*. For about 40 years in the nineteenth century, it was an object of industrial and middle class utility. Chemical light years behind the wonders of catalysis described in this book, it was, nevertheless, in its time the first example of a practical designed application of catalysis. I would like to tell you its story; as I do I will call it the Feuerzeug, because the German word is more poetic than its English equivalent.

The story I will tell you is of a catalyst for one of the simplest chemical reactions, the combustion of hydrogen: $2H_2 + O_2 \rightarrow 2H_2O$. It is also a story of chemistry in culture, of a Russian–German geopolitical tie that, in the 1820s, helped a Jena professor to invent a new way of lighting fires using, of all metals, platinum. So, this is a story of fire too. And one of modern surface chemistry.

2
It Should Go Off, Shouldn't It?

Is there a problem with the above reaction, hydrogen burning? The hydrogen-filled balloon set off by a taper is the chemistry lecturer's favorite demonstration. You can vary the effect by adjusting the mixture of hydrogen and oxygen in the balloon – pure hydrogen gas (H_2) will give you a respectable pop and a neat flame, just a little H_2 will simply not go off. The most bang for the buck comes from a mixture of hydrogen and oxygen gas (O_2). I remember waking up sleeping dogs in my class, not to mention students, with this demonstration.

The reaction is highly exothermic: the change in free energy for all gaseous components under standard conditions is a very respectable 242 kJ per mole of H_2. And all it takes is a lighted taper or match to set it off.

Figure 1 Döbereiner's Feuerzeug, from the German Pharmacy Museum in Heidelberg. This one is from around 1840. (Reproduced with permission of the Deutschen Apotheken-Museum Heidelberg.)

" … a match to set it off." That's just the point: the radicals in the flame and the heat of the match initiate the reaction, after which it indeed proceeds, posthaste. The mixture of hydrogen and oxygen, in the absence of that match or of a catalyst, would just sit there—the activation energy for the uncatalyzed reaction is very high.

Hydrogen was first well identified by Cavendish in 1766. Its burning to water and the parallel and more difficult decomposition of water to H_2 and O_2 were cornerstones of Lavoisier's chemical revolution. The reaction was just as reluctant to go in the 1780s as it is today. There were no safety matches until 1855. So Lavoisier set it off with an electric spark. And within 50 years, a German chemist, Johann Wolfgang Döbereiner, used the same H_2 and O_2 reaction, now catalyzed, as a ready source of fire, replacing other sources of fire in home and laboratory.

3
Incendiary Acts

How were fires lit before? To begin with, from other fires, of course. Two further techniques evolved around the world – the first generated heat by rubbing wood rapidly against wood, the second created sparks by striking hard stones against stone or metal. In both cases, the heat or spark had to be "caught" by a flammable material nearby. Tinder could be mostly anything organic, but certain dried mushrooms were particularly valued. The ingenuity in the construction of bows for rubbing wood sticks or the compact steel-silex-tinder kits of seventeenth-century soldiers is remarkable.

In the seventeenth and eighteenth centuries, as optics evolved, the burning glass or mirror became an important fire source, albeit a fair weather one. Ehrenfried Walther von Tschirnhaus (1651 – 1708), a natural philosopher and polymath with a practical bent, built the best of these, capable of reaching the highest temperatures then possible on earth. Sparks also came from the newly discovered electricity. Still another source was invented in 1770 and has a fascinating connection to chemistry and physics. This is the pneumatic lighter, in which the heat generated in a rapidly compressed gas is sufficient to inflame tinder.

4
Döbereiner's Feuerzeug

Johann Wolfgang Döbereiner was born in Hof an der Saale in 1780. His beginnings were simple. He was largely self-educated, the son of a coachman. But Döbereiner's talents were recognized, and in 1810 he was appointed to a professorship in Jena. This town was in the Grand Duchy of Saxe-Weimar-Eisenach, a princely state under the administration at just that time of another Johann Wolfgang, namely Goethe. Goethe and Döbereiner had an extensive correspondence, *inter alia* dealing with the tarnishing of silver spoons in red cabbage and the composition of Madame de Pompadour's toothpaste. Goethe went to Jena to study analytical chemistry with Döbereiner [1][1)]. A contemporary analog would be if the present French Prime Minister Manuel Valls took off a few weeks to learn about supramolecular chemistry with Jean-Marie Lehn at the University of Strasbourg. It would be good for Valls, but ...

Döbereiner did much interesting chemistry. For instance, he was responsible for noting an important regularity in the chemistry of the elements, that of triads, one of the forerunners of Mendeleev's periodic table. And Döbereiner observed in 1823 that when platinum metal (in a finely dispersed form called platinum (Pt) sponge) was exposed to hydrogen, much heat was generated. The platinum in fact glowed red hot to white hot, and if more hydrogen were supplied, the hydrogen burst into a hot but nearly colorless flame.

Döbereiner writes to his Prime Minister Goethe, as follows:

> *Ich erlaube mir, Eurer Exzellenz von einer Entdeckung Nachricht zu geben, welche [...] im hohen Maße wichtig erscheint. Ich finde [...], dass das rein metallische staubfeine Platin die höchst merkwürdige Eigenschaft hat, das Wasserstoffgas durch bloße Berührung [...] zu bestimmen, dass es sich mit*

1) I thank Dr. Linke for introducing me to the Döbereiner story.

> *Sauerstoffgas zu Wasser verbindet, wobei eine bis zum Entglühen des Platins gesteigerte Summe von Wärme erregt wird.* [2, 3]
> *[I allow myself, your Excellency, to give you news of a discovery which appears to most important. I found that finely divided metallic platinum has the most remarkable property, to induce hydrogen to combine with oxygen to give water, on simple contact. In the process heat is released, sufficient to make the platinum glow.]*

And Berzelius, in his Annual Reports on the Progress of Chemistry, writes the year after:

> *From any point of view the most important, and, if I may use the expression, the most brilliant discovery of last year is, without doubt, that fine platinum powder has the ability to unite oxygen and hydrogen even at low temperatures.*

Within days, Döbereiner turned this beautiful observation into a practical igniter. Figure 2 shows the design. One has a bottle that can be tightly sealed. Inside a glass cylinder in that bottle hangs a piece of zinc (*d*). the bottle is filled with sulfuric acid (typically 25% sulfuric acid (H_2SO_4)). There is a controlled outlet from the glass bottle, the stopcock (*e*). The zinc (Zn) reacts with sulfuric acid, generating hydrogen gas *in situ*:

$$Zn + H_2SO_4 \rightarrow Zn^{2+} + SO_4{}^{2-} + H_2$$

When the stopcock is opened, the H_2 is directed through a thin tube (*f*) onto a bit of platinum sponge (*g*). A flame lights, essentially instantaneously. When the stopcock is closed, the flame goes out. More H_2 is generated, but comes to a stop as gas pressure build.

5
Geopolitics and Science

Döbereiner continued his research with the catalytic properties of platinum. Actually Humphry Davy had discovered the phenomenon 6 years ago, heating a fine platinum wire above a coal gas flame. Davy compared copper, silver, gold, palladium, and iron in this effect, and found that only Pt and Pd initiated exothermic burning; he was thus the first to focus on selectivity, a matter of some importance in this book. Döbereiner actually made a supported catalyst (a mainstay of industrial catalysis and automotive catalytic converters today) by shaping small balls of potter's clay impregnated with platinum.

The word catalysis came from further north, from the authoritative pen of Jöns Jacob Berzelius in 1835. In his highly influential Annual Reports (translated into German and many other languages; but it is good to see a language besides English and German in chemistry ...), he wrote the following:

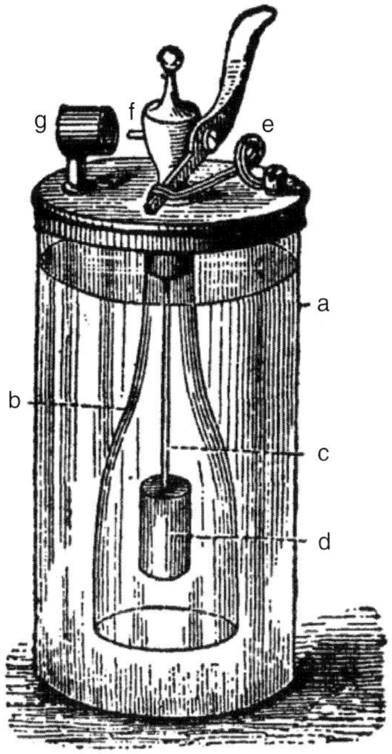

Figure 2 Schematic of an early Döbereiner lighter. (Adapted from Nordisk familjebok/ Uggleupplagan. Vol. 7. Egyptologi - Feinschmecker, p. 163–164, 1907. From Wikipedia.)

> *Jag skall derföre … kalla den kroppars* katalytiska kraft, *sönderdelning genom denna kraft* katalys, *likasom vi med ordet analys beteckna åtskiljandet af kroppars beståndsdelar medelst den vanliga kemiska frändskapen. [emphasis in original]* [4]
> *[I shall, therefore … call it that body's catalytic force, the decomposition of other bodies by this force catalysis, just as we signify by the word analysis the separation of the constituents of bodies by the usual chemical affinities.]*

Berzelius viewed catalysis as a special force, to some not that different from the heritage of the Philosopher's Stone [3]; it took 60 years and the great science and intuition of Wilhelm Ostwald to point us in another direction.

For his work, Döbereiner needed great supplies of the precious metal, and a geopolitical note is in order here: platinum originally came from Spanish colonial mines in the New World, and that is presumably Döbereiner's original source. Around 1824, major deposits were discovered in the Urals. How could Döbereiner, who was struggling desperately in his laboratory finances, get the precious white metal? Well, the Empress of Russia, Catherine the Great (1729–1796), was a German princess from the Duchy of Anhalt-Zerbst. There were close Russian–German ties throughout this period, and they continued until the World War I. In Döbereiner's principality, the wife of Carl Friedrich, the then heir to the Grand Duke, was Maria Pavlovna, the daughter of Czar Paul I of Russia. Platinum from the Urals came easily to Jena; he may have had a kilogram of it.

6
Chemistry in Culture

Döbereiner's lamp became a common way to light fires in industrial settings in the first half of the nineteenth century. Within 5 years of its discovery, 20,000 lamps were in use in Germany and England. It entered the middle-class home as well. (Nothing like this could happen today; imagine the horror of today's risk-avoiding society at the thought of filling a lamp with sulfuric acid!) And if a utilitarian technology is to be accepted into society, it must be culturally processed. What I mean is that it is clothed according to the prevalent esthetics of the time. In 1829, a Berlin manufacturer could offer "… as a pleasant and useful Christmas present a lighting machine, outfitted with platinum, elegant, clean, and sturdily constructed, with Chinese and other decoration, insensitive to wetness and cold … "

In time the safety match, the cerium frictional spark source (see Primo Levi's Ce chapter in his *Periodic Table*), the cigarette lighter and the gas stove electronic lighter put Döbereiner's Feuerzeug into the museum.

Look at the Pharmacy Museum Feuerzeug again. The image on the housing is more than a genre painting. Two young women appear to be playing a game: one watching and the other active. There is a basin with water (see the jug in front to replenish the water). And in the basin are two floating "boats," perhaps made of wood or paper. In each is a candle. The boats look circular, but their action will be more pronounced if they were not, or if a piece of the rim of one is cut out a bit. Try it yourself, please. The young women are doing a scientific experiment. And one which is fun too.

7
How Does Hydrogen Burn?

One hundred and seventy-two years after Döbereiner's discovery, in 1995, Laurens K. Verheij and Markus B. Hugenschmidt write:

> *In recent years many studies on the reaction between hydrogen and oxygen on metal surfaces have been reported. Although this reaction is expected to be one of the simplest oxidation reactions, rather complex phenomena are observed which make a determination of the reaction mechanism difficult. Even for the water formation reaction on Pt(111), the system which has been most widely studied, an understanding of the reaction process seems only just emerging* [5].

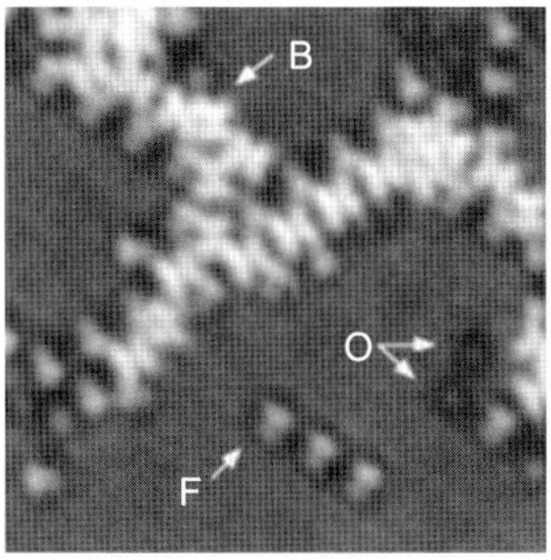

Figure 3 Schematic of O_2 molecules and O atoms (gray) adsorbed on a platinum surface (left). The image generated by the scanning tunneling microscope reveals two shapes of oxygen molecules on the platinum surface (right). Molecules can appear to form "clover-leaf" (B) or "pear-like" (F) shapes. In some places, one sees the molecules dissociated to adsorbed oxygen atoms (dark circles, marked "o"). (Photograph courtesy of Wilson Ho.)

In the next 20 years, understanding did emerge, as incomplete as it is. It came through the beautiful work of Gerhard Ertl [6] and of others [7]. The sequence of events is not simple, not easily revealed, despite the availability of tools unimaginable to either Döbereiner or Ostwald, such as scanning tunneling microscopy. I show a teaser, an image relevant to the very first steps of the reaction, when oxygen molecules impinge, as they must, on a platinum surface (Figure 3).

At low temperatures, way below room temperature, O_2 bonds to the surface first as a molecule and in several different ways. The clover-leaf pattern is for an O_2 lying across two Pt atoms. Do you know why one "sees" clover-leafs and not dumbbells? The pear-like features are likely O_2 molecules bound at an angle to the surface, one atom further up than the other. As one heats up the surface, the diatomic (O_2) ruptures into individual oxygen atoms, which sit bonded to triangles of platinums. At ambient temperatures, it is not likely that an O_2 coming onto the surface survives very long before it breaks apart. The hydrogen molecules break apart even more readily on the same surface.

And what happens after that? How do the chemisorbed H and O atoms find each other, form water? Good questions, the subject of current research. Something as simple a burning hydrogen is … not simple.

8
Knowing without Seeing

The stories of the book before you are different from Döbereiner's, for they are, for the most part, tales of catalysis in solution, so-called homogeneous catalysis. This type of seeming magic is also old, even if we do not consider the many instances of catalysis that nature has wrought, and limit ourself to human–crafted interventions. So, in 1860, the aging Liebig found an aldehyde-catalyzed transformation of some utility from cyanogen to oxamide, shown below [8]:

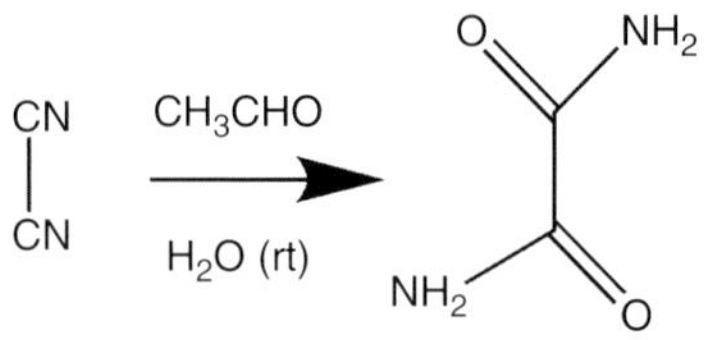

On surfaces or in solution (and the distinction is being undermined everyday now by the catalysis on nanoparticles), modern catalysis operates within the framework of Wilhelm Ostwald's transformation of catalysis from magic to rate change (as often slowing down as enhancing, but that is another story). Here is how Ostwald put it:

> *Ein Katalysator ist jeder Stoff, der, ohne im Endprodukt einer chemischen Reaktion zu erscheinen, ihre Geschwindigkeit verändert* [9]
> *A catalyst is a substance that changes the rate of a chemical reaction without appearing as a product of it.*

It took the next hundred years after Ostwald to shift the emphasis from the observation of a changed rate to the search for the partial establishment of the underlying mechanism of chemical reactions, both organic and inorganic. These mechanisms, incredibly rich in their variety, are a *Leitmotif* of the book before you. The tools for "seeing" in solution are not yet developed. And I like it so, for it leaves room for what has always been in my mind the strength of chemistry, its great intellectual achievement – the marshaling of chemical and physical fragmentary evidence into a knowing *without* seeing. People do know, putting together pieces from kinetics, and from signals from within. The latter detected by the lovely tools we have invented of spectroscopic attention to those signals, and in turn elicited by small electromagnetic perturbations. And now theory too. It is absolutely amazing how much we may know, without seeing *per se*. That we do that – understand without direct sight – I think should make sometimes arrogant scientists more tolerant of the ways of knowing that artists and writers and social scientists have. They also listen, create, and understand.

Roald Hoffmann

References

1. Linke, D. (1981) *ZeitschriftfiirChemie*, **21**, 309–319.
2. Schiff, J. (1914) *Briefwechsel zwischen Goethe und Johann Wolfgang Döbereiner (1810–1830)*, Hermann Böhlhaus, Weimar, p. 78.
3. I was led to this quotation, and to much other information in this essay by the wonderful article of Ertl, E. and Gloyna, T. (2003) *Z. Phys. Chem.*, **217**, 1207–1219.
4. (a) Berzelius, J.J. (1835) *Årsberättelsen om framsteg i fysik och kemi*, Royal Swedish Academy of Sciences, Stockholm; (b) For an earlier use of the word by Libavius in 1597(!), see Lindström, B. and Pettersson, L.J. (2003) *CATTECH*, **7**, 130–138.

5. Verheij, L.K. and Hugenschmidt, M.B. (1995) *Surf. Sci.*, **324**, 185–201.
6. Zambelli, T., Barth, J.V., Wintterlin, J., and Ertl, G. (1997) *Nature*, **390**, 495–497.
7. Stipe, B.C., Rezaei, M.A., and Ho, W. (1997) *J. Chem. Phys.*, **107**, 6443–6447.
8. (a) Von Liebig, J. (1860) *Justus Liebigs Ann. Chem.*, **113**, 246–247; (b) See also Dalko, P. (ed.) (2007) *Enantioselective Organocatalysis: Reactions and Experimental Procedures*, Wiley-VCH, Weinheim, Chapter 1, pp. 1–17.
9. Ostwald, W. (1902) *Phys. Z.*, **3**, 313–322.

Introduction

Lutz H. Gade and Peter Hofmann

> *Fragt man [...], welche Stellung insbesondere das wissenschaftliche Gebiet der Katalyse [...] einnimmt, so erkennt man alsbald, dass es sich noch in den ersten Stadien seiner Entwicklung befindet. Die Hauptarbeit besteht zurzeit noch darin, die verschiedenen Fälle von Katalyse überhaupt erst ausfindig zu machen und wissenschaftlich festzustellen.*
>
> (Annalen der Naturphilosophie 1910, 9, 1).[1]

I

In his Nobel lecture, entitled "On Catalysis" and delivered on 12 December 1909, Wilhelm Ostwald gave an overview of the development of a field of research which was still in its infancy. At that stage, knowledge about catalysis was purely phenomenological and deeper insight into its mechanisms was lacking. A century later, it is a highly developed field of research, in which insight into the detailed processes underlying catalytic transformations is obtained by exploitation of the whole armory of modern analytical methods.

From the early stages of research in all areas of catalysis, dating back to the beginning of the twentieth century, fundamental research at universities and the more applied research programs in industry have been intimately interlinked. This combination of fundamental research and commercial development is almost unique in the history of science and technology.

The importance of catalysis for the development of chemistry subsequent to the foundational work of the nineteenth century has been recognized several times by the Nobel Committee. Following Ostwald's award (1909), early important contributions to the field led to Nobel Prizes to Paul Sabatier (1912), Fritz Haber (1918), and Carl Bosch and Friedrich Bergius (1931). Subsequently and

1) "[...] it will immediately be realized on inquiry into the particular position occupied [...] by the scientific field of *catalysis* that it is in the *first* stages of its development. At present, the main task is still essentially to discover and scientifically to establish the various cases of catalysis." (English Translation: Wilhelm Ostwald – Nobel Lecture: On Catalysis. *Nobelprize.org.* Nobel Media AB 2013. Web. 2 December 2013 *http://www.nobelprize.org/nobel_prizes/chemistry/laureates/1909/ostwald-lecture.html*)

throughout the past decades, the elucidation of enzyme mechanisms has been recognized by several Nobel Prizes. Finally, since the turn of the millennium, four Nobel Prizes in Chemistry have been awarded to 10 laureates for their contributions to catalysis.[2)]

Although the principles of catalysis are general and applicable to all forms of catalysts, several sub-disciplines have emerged over time that focus on the type of catalyst and are characterized by the experimental methods employed to generate and study catalysts. The production of many basic chemicals on a large scale involves heterogeneous catalysts for which the catalytic transformations occur at the inner and/or outer surfaces of solids that are in contact with the fluid reactant phase(s). On the other hand, enzymes – rarely in free form but rather trapped in living cells or immobilized on artificial supports – are widely used for the synthesis of intermediates and targets relevant to the production of organic compounds. They compete with small-molecule, homogeneous catalysts that have been used for the synthesis of both basic chemicals and structurally complex targets, such as aroma and flavoring substances, herbicides and pesticides, diagnostics, and drugs.

II

Molecular catalysis has thus emerged as a key methodology in modern chemical synthesis. Ideally, it allows the energy- and atom-efficient generation of complex target compounds with high selectivity. The development of new catalytic systems, including new catalytic phases that allow novel chemical transformations and/or render known catalytic reactions more efficient and selective, has been identified as a key scientific and technological goal of chemical research in the decades to come. In the twenty-first century, catalyst discovery, catalyst design, and catalyst improvement will contribute to a sustainable, both economically and ecologically optimized, use of mankind's natural environment and resources in view of the growing global energy demand.

Admittedly, truly novel catalysts still tend to be discovered rather than developed by design. However, an ever-increasing insight into the mechanisms of the more or less complicated reaction networks, frequently referred to simplistically as *catalytic cycles*, will provide the foundation for more rational approaches. A combination of detailed studies into the reaction kinetics, isotope labeling, and the identification of key intermediates by *in situ* spectroscopy, along with the theoretical modeling of individual reaction steps or even complete reaction networks, constitutes the methodological state of the art in this field.

To cover recent developments in the field of molecular catalysis alone would require a multivolume opus of encyclopedic proportions; in fact, several highly developed and frequently used catalytic transformations themselves might

2) 2001: W. S. Knowles, R. Noyori, K. B. Sharpless (enantioselective catalysis); 2005: Y. Chauvin, R. H. Grubbs, R. R. Schrock (alkene metathesis); 2007: G. Ertl (chem. processes on solid surfaces/mechanisms of heterogeneous catalysts); 2010: R. F. Heck, E. Negishi, A. Suzuki (Pd-catalyzed cross coupling reactions).

demand that level of coverage. This research-oriented book therefore focuses on individual case studies. It is characterized by its unusual combination of organic, organometallic, inorganic, theoretical, bio-, and physical chemistry, with simultaneous emphasis on synthesis and mechanistic studies.

III

This monograph highlights several key aspects of molecular catalysis, ranging from the in-depth analysis of an individual key reaction step to mechanistic studies of selected biocatalysts in living cells, and from studies of the impact of catalyst immobilization to large-scale industrial catalysis for commodity chemicals. The individual chapters emphasize the state of the art in the different areas and provide a general frame of reference. The case histories that are presented to illustrate their more subtle aspects focus on results obtained in Heidelberg over more than a decade, involving more than 30 closely interacting research groups of a "Collaborative Research Centre" working together in the field of molecular catalysis. This local research network (Sonderforschungsbereich (SFB) 623, entitled "Molecular Catalysts: Structure and Functional Design" of the German Research Foundation, DFG) began to operate in Heidelberg in July 2002, based on an initiative by one of the editors, which had been originally launched in 1999 within the Chemistry Department (PH acting as SFB Chairman from 2002 until 2009, until LHG took over during 2009–2013).

Looking back, it is interesting to note that it was also in 1999 that Ryoji Noyori, who 2 years later became one of the three Nobel Laureates of 2001 in the field of enantioselective catalysis, was awarded the "Georg Wittig Lectureship" founded jointly by the University of Heidelberg and BASF. When, in the spring of 2002, a panel consisting of prominent DFG reviewers came to Heidelberg in order to conduct the mandatory on-site evaluation of the funding proposal, it was by pure accident that Professor Roald Hoffmann from Cornell University, USA, Nobel Laureate of 1981, Honorary Senator of our university, and former postdoctoral research advisor of one of the editors (PH), happened to visit Heidelberg on that same day. Roald Hoffmann, as an accidental witness of the very beginning of Heidelberg's Collaborative Research Centre on Molecular Catalysis, followed its development on the occasion of various subsequent visits and through contacts with individual researchers over the years. The editors were delighted when he agreed to write the foreword.

As detailed below, the chapters of this book, based upon various SFB research projects, intend to deepen our insight into a range of catalytic transformations, to provide rational concepts for catalyst optimization, and to develop synthetic procedures employing molecular catalysis. Part I focusing on *Mechanisms of elementary reactions in catalytic processes* highlights both theoretical and spectroscopic methods for the investigation of the dynamics of individual reaction steps. This includes the structural identification of frequently labile and thus transient intermediates. Case histories illustrating the interplay between structural and mechanistic insight and actual catalyst development are presented in Part II entitled *New catalysts – new and old reactions*, which is followed up by a wide range of examples of *Catalysts in synthesis* (Part III). The underlying mechanisms of enzyme or ribozyme catalysis offer higher levels of complexity. This is reflected in the experimental and theoretical studies highlighted in Part IV on *Structures and mechanisms in biocatalysis.* Finally, catalyst recycling is an important practical objective in the development of supported molecular catalysts on solids or well-defined macromolecular carriers such as dendrimers. Moreover, catalyst support may be crucial for the application of high-throughput screening for the identification of novel catalyst lead structures. Such issues will be addressed in the chapter entitled *Studies of immobilized catalysts* (Part V).

IV

The book targets scientists who are active in molecular catalysis both in academic/research institutions and industry. It focuses on a deeper understanding of reaction mechanisms and catalytic cycles at the molecular level. This level of insight is required for a "semi-rational" development of novel catalysts for many key transformations. It also provides the basis for rational improvements of existing catalysts in order to boost their performance.

Utilizing the conceptual framework and the synthetic and analytical methodology of modern molecular chemistry, including the complete toolkit of quantum chemistry and modeling, examples drawn from catalysis lend themselves to high-level and research-oriented teaching, especially at the graduate student and post-doctoral level. From the student perspective, the book will therefore be of interest as a collection of case histories that illustrate how to identify and tackle challenging and important research problems.

Part I
Mechanisms of Elementary Reactions in Catalytic Processes

Lutz H. Gade

Catalytic reactions involve a series of chemical transformations resulting from substrate–catalyst interactions, which taken together are referred to as the *mechanism of the catalytic reaction*. Frequently mentioned in textbooks and research papers alike as "catalytic cycles," they should be more appropriately referred to as *catalytic reaction networks* comprising multiple bifurcation points and parallel reaction pathways. Detailed knowledge of the mechanism is a prerequisite for a rational approach in catalyst development, and a great deal of effort goes into their systematic investigation.

In adopting the reductionist approach underlying much of modern scientific research, the reaction networks of catalytic transformations can be broken down to elementary reaction steps. These may involve the formation or cleavage of a single chemical bond but also complex rearrangements representing themselves cascades of addition or dissociation and thus bond-breaking and bond-forming steps. A closer look at such elementary reactions may uncover more complex reaction pathways involving "hidden" intermediates associated with shallow minima in the free-energy landscape, and this complexity then adds onto the overall mechanism of the catalytic reaction.

Our understanding of reaction mechanisms, therefore, begins with a detailed study of the elementary reaction steps and the nature of intermediate species, corresponding to local minima in free enthalpy. This is frequently complicated by the rapidity of the processes and the short lifetimes of the intermediate species and is, consequently, a challenge for the application of analytical methods appropriate to the relevant reaction phase and/or medium. Once such data are available, the theoretical modeling of the process represents the next stage in its understanding. This involves not only modeling the "free-energy landscape" but also the dynamics of the process itself.

A quantum dynamical modeling of olefin insertion into a metal–hydride bond and its reverse reaction is the topic of Chapter 1 by Klatt and Köppel. They employ the wave packet methodology, which is based on the time-dependent Schrödinger equation and allows the description of coherence and tunneling effects as well as

Molecular Catalysts: Structure and Functional Design, First Edition. Edited by Lutz H. Gade and Peter Hofmann.

the vibrational structure of electronic spectra and thus goes far beyond transition-state theory. The results described in this chapter are relevant to a deeper understanding of the electronic factors that govern, *inter alia*, olefin polymerizations.

The ligands in molecular metal catalysts may be viewed as moderators of their chemical reactivity, shielding the active site from nonspecific reactive interactions with the surrounding reaction medium and orientating the substrates in their transformation within the coordination sphere. Ligated metal atoms and clusters are thus "passivated" reagents, a notion that is all too obvious whenever chemical reactions of the bare atoms and clusters are studied. This has been achieved, for example, in the collision chambers of mass spectrometers where chemical reactions are observed of which more traditionally minded coordination chemists may dream. An alternative to gas-phase studies is provided by low-temperature matrix isolation in inert gas matrices, the origins of which date back to more than half a century. This approach has been taken by Himmel and Hübner, who describe their studies on bare metal clusters that bind and activate small molecules such as H_2, N_2, and O_2, in Chapter 2. Such studies provide insight into processes that are relevant to both molecular and heterogeneous catalysis.

Most analytical methods employed to study elementary reactions and reaction intermediates provide information about ensembles of reactive molecules. However, for more than a decade now, it has been possible to monitor transformations of single molecules by means of fluorescence microscopy. This involves the labeling of at least one of the reactants with a fluorophore, which in the case of a catalytic reaction would be either the catalyst or the substrate. The visualization of a chemical reaction then relies on changes of the fluorescence induced by a transformation at the reactive center of a catalyst. Herten and coworkers present several case histories. The use of single-molecule spectroscopy and the way in which the methods developed in this field may give rise to potential "single-molecule catalysis" are described in Chapter 3.

The two final chapters in this first part go beyond the elementary reaction steps and focus on reactivity issues in two important areas of molecular catalysis. Molecular gold catalysis has developed rapidly during the past decade and has provided new tools for organic synthesis. Key steps involve the π-coordination and electrophilic activation of an unsaturated substrate. An understanding of the complex subsequent transformations and reaction cascades involved in gold-catalyzed reactions requires a deeper knowledge of the relevant intermediates and their reactive potential. Moreover, the unambiguous assignment of the reactive gold species remains another challenge for many recently discovered reactions. Hashmi discusses some of these issues in Chapter 4.

Olefin metathesis has completely transformed the conceptual framework of organic synthesis and retro-synthesis. It is appropriate to state that its advent in organic synthesis has had an impact that is comparable to the addition of a new move to a chess game. Disconnections of target molecules, which were deemed impossible previously, are feasible because of this type of transformation. The foundations were laid by the groups of Schrock and Grubbs, but continuous development and modification of their catalysts has significantly enhanced the scope of the method. As Straub points out in Chapter 5, the issue of stereoselectivity in olefin metathesis, which limited the early applications in organic synthesis, has received a lot of attention in recent years, and major advances in developing stereoselective metathesis catalysts have given the field a boost. This remains a rich field for ligand design!

Understanding the mechanism of a catalytic reaction involves many levels of study and analysis, as illustrated by the case histories presented in this section. They are representative examples of the type of research efforts carried out in many laboratories worldwide. While (serendipitous) catalyst discovery will continue to fuel progress in this field, a deeper understanding of the mechanisms of the elementary reactions involved will provide the foundation for more rational approach – possibly even catalyst design.

1
Quantum Dynamics of Molecular Elementary Processes in Catalytic Transformations

Günter Klatt and Horst Köppel

1.1
Introduction

Olefin insertion into an M–H bond and the reverse process, β-hydrogen elimination (BHE), represent elementary reaction steps that are of fundamental importance to many transition-metal (TM) catalyzed processes, such as hydrogenation, hydroformylation, and olefin polymerization, to mention a few. Correspondingly, there has been much theoretical effort toward characterizing the structural parameters and energetics of the intermediates and transition states [1–8], which complements the large amount of related experimental investigations (see, for example, Refs [9–12], and references therein).

Usually, the insertion proceeds in two steps, as is indicated in Scheme 1.1.

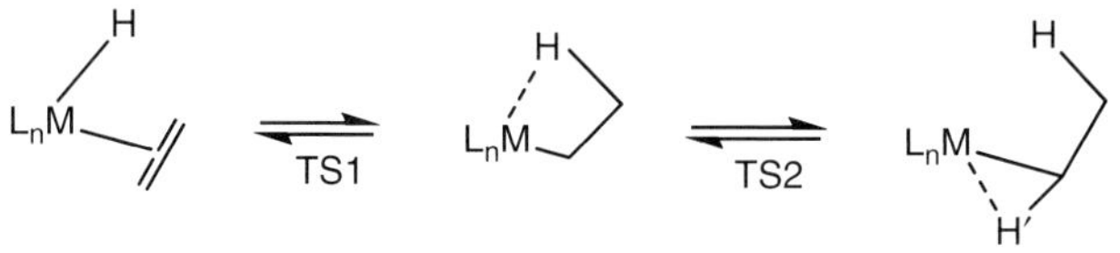

1 2 3

Scheme 1.1

Initially, the hydrogen atom migrates from the metal atom to which it is covalently bound in the olefin hydride structure **1** to the ethylene moiety to form the β-agostic structure **2**.

A further rearrangement yields the α-agostic complex **3**. These reaction steps are associated with the transition states TS1 and TS2 and are equilibrium processes, which are indicated by the harpoon arrows. For early TMs, such as titanium, zirconium, or chromium, the equilibrium favors the insertion products **2** and **3**. Catalytic ethylene homo- and copolymerization has been largely dominated by early TM catalysts [13–25], while late-metal catalysts generally show reduced activity and a strong tendency to form oligomers because the termination reactions involving the transfer of β-hydrogen atoms compete with chain

Molecular Catalysts: Structure and Functional Design, First Edition. Edited by Lutz H. Gade and Peter Hofmann.

growth [26]. In recent years, these disadvantages have been overcome by a new generation of late TM polymerization catalysts, based on the 1995 discovery of diimine complexes of Ni and Pd as active catalysts in ethylene polymerization by Brookhart and coworkers [27]. Compared to metallocene catalysts, these late TM catalysts are characterized by facile BHE, resulting in highly branched polymers with new physical properties. They also exhibit less oxophilicity and therefore a greater tolerance toward polar reagents. Much attention has since been paid to olefin polymerization using late TM catalysts (see, for example, Ref. [12], and references therein).

This chapter is organized as follows. First, the structural and energetic aspects of the migratory insertion/BHE in complexes of early (Cr) and late (Co, Rh) TMs will be discussed. These results are based on the application of density functional theory (DFT) to complexes of early and late TMs, in particular the isomers of Scheme 1.1. During the last decade, DFT has established itself as a useful and reliable method to study the potential energy surfaces (PESs) and reaction mechanisms of compounds containing TMs and heavy elements [28]. The subsequent section is dedicated to the application of quantum dynamical methods to TM-catalyzed olefin polymerization. The results presented here have been obtained using the wave packet methodology, which is based on the solution of the time-dependent Schrödinger equation and is well established and widely used for different kinds of systems (see, for example, Refs [29–33]). Unlike previously performed calculations at the Car–Parrinello molecular dynamics (CPMD) level [34, 35], this approach uses a full quantum treatment to describe the nuclear dynamics, which is of importance for describing coherence and tunneling effects, as well as the vibrational structure of electronic spectra. It describes the reaction path following a vertical Franck–Condon transition and allows the deduction of oscillation periods and lifetimes for the systems under study. Experimentally, the real-time monitoring of bond formation and bond cleavage can be studied by femtosecond time-resolved pump–probe spectroscopy [36–38], which relies on a quantum description of the nuclear motion. It also brings the system into a highly nonequilibrium initial state by the very nature of the excitation process.

1.2
Structural and Energetic Aspects

While the migratory insertion/BHE of the $[CpM(L)(H)(C_2H_4)]^+$ complexes is characterized by the five stationary points depicted in Scheme 1.1, additional isomers may be relevant for the insertion into a metal–alkyl bond. For example, for M = Cr it is known that the initial insertion product of the ethylene–alkyl complex I is a γ-agostic complex IIγ, which subsequently isomerizes to β- and α-agostic species IIβ and IIα [26], as shown in Scheme 1.2.

For this system, the probable catalytic cycle and possible termination reactions for the polymerization of ethylene have been investigated for model donor ligands containing phosphorus or nitrogen ($L = PH_3$, PMe_3, NMe_3) [39]. The starting

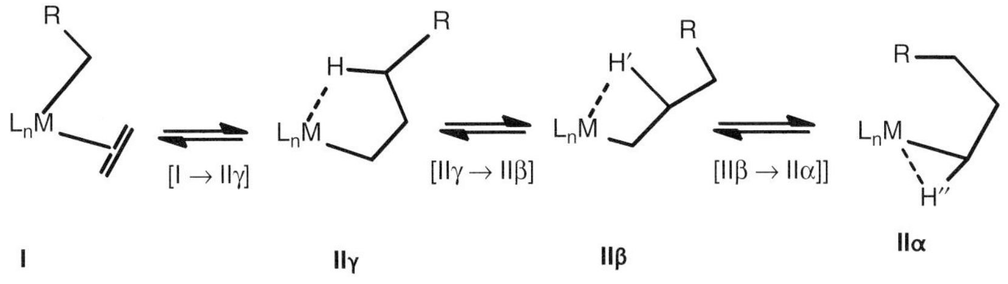

Scheme 1.2

point for these calculations was the cation $[CpCr(L)C_3H_7]^+$. The propagation and termination reactions are shown in Figure 1.1, starting with the insertion of ethylene into the Cr–C bond. Here, minima are designated by Roman numerals (I–VI), which in the case of an agostic structure Cr–H–C are followed by a Greek letter identifying the carbon atom involved. Transition states are labeled by the connected reactant and product states separated by an arrow and enclosed in square brackets. While the cationic Cr(III) complexes considered here usually have a quartet spin ground state, a possible termination process by β-hydrogen transfer (BHT) to a monomer involves a change of spin to the doublet state; in the following, the molecular term symbol is provided in parentheses where necessary (high-spin quartet configuration ^{4}A or low-spin doublet configuration ^{2}A).

The catalytic cycle shown at the top of Figure 1.1 consists of four intermediates, which differ by the kind of agostic interaction displayed (none, α-, β-, or γ-agostic). The insertion step always leads to the γ-agostic isomer as the initial product, except for the insertion of ethylene into the Cr–H bond, as there is no γ-H atom in the resulting complex [39]. Furthermore, two alternative mechanisms for the termination of the polymerization process exist:

- BHE involving the transfer of a hydrogen atom to the metal center and subsequent dissociation of the resulting olefin;
- BHT, in which the hydrogen is transferred to a monomer unit also coordinated to the metal, and which is thought to involve a change in spin multiplicity from the quartet to the doublet state.

The relative energies E and free energies G_{298} for the first two insertion steps with ligands L = (PH_3, PMe_3, NMe_3) are given in Figure 1.2. The findings confirm the rate-determining character of the insertion of ethylene into the chromium–alkyl bond, with an insertion barrier $\Delta E^{\ddagger} = 8.7-10.1\ \mathrm{kcal\,mol^{-1}}$ for the first insertion step. Unlike Jensen *et al.* [26], we find that the coordination of ethylene proceeds from the β-agostic isomer IIβ, which represents the global minimum and resting state of the catalytic cycle. First, an "outer" van der Waals complex $\mathrm{II^{vdW}}\beta$ is formed, in which the ethylene is loosely bound to IIβ, which then rearranges via a transition state $[\mathrm{II^{vdW}}\beta \rightarrow \mathrm{I}']$ to the ethylene complex I′.[1)]

1) The primed (′) complexes contain an additional ethylene unit in comparison to the preceding complexes.

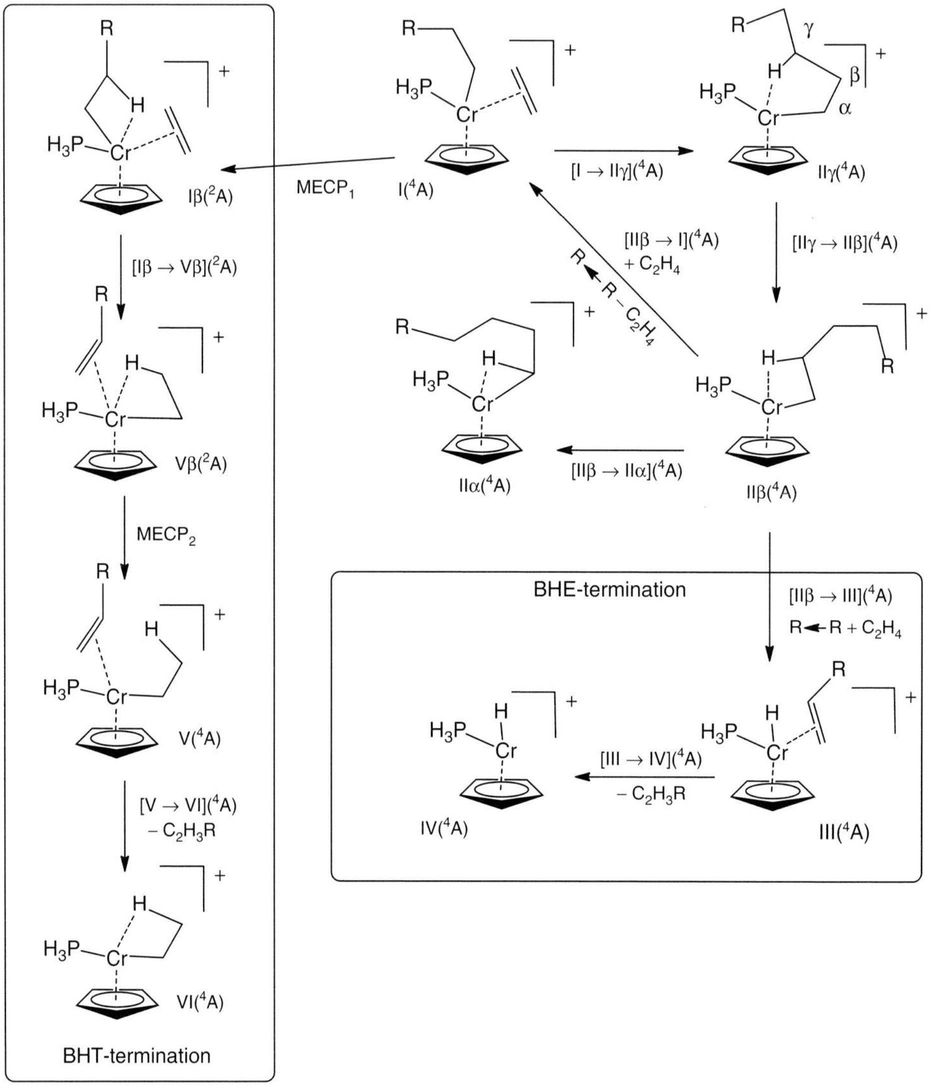

Figure 1.1 Catalytic cycle for the propagation reaction of $[Cp(PH_3)Cr\text{-}C_2H_4R]^+$ with ethylene (top), and the corresponding termination reactions by β-hydrogen elimination (BHE, bottom) or β-hydrogen transfer (BHT, left); method DFT (BP86/6-31G**/SDD(Cr)), R = H. Note that R becomes $R{-}C_2H_4$ for the process indicated by the diagonal arrow.

This result was confirmed by intrinsic reaction coordinate (IRC) calculations and is indicated by the diagonal arrow in Figure 1.1. The overall reaction is strongly exothermic ($\Delta E > 30\,\text{kcal mol}^{-1}$), with only minor differences existing for the three ligands L. The insertion barriers [I → IIγ] are comparable in energy and lower by only about $1\,\text{kcal mol}^{-1}$ for the more compact NMe_3 ligand in the first insertion step; this effect becomes more pronounced for the second ethylene insertion because of the increased steric demand of the propyl residue. Comparing the values for E and G_{298}, it is noticed that the reaction is exergonic by only about 20 kcal mol^{-1} because of the decrease in particle number and hence entropy during propagation. The connectivities given in Figure 1.2 are not intended to suggest the existence of a single sequential reaction path but have been selected for illustrative purposes; the actual reaction path is branched because of the existence of interconnected pathways between all agostic isomers ($\alpha \leftrightarrow \beta$, $\alpha \leftrightarrow \gamma$, $\beta \leftrightarrow \gamma$).

The termination of the polymerization reaction by BHE to the chromium center and subsequent dissociation of the resulting olefin is found to require about 25 kcal mol^{-1} and to be thermodynamically much less feasible than the alternative termination process by BHT to a monomer. The latter process involves spin change; two minimum-energy crossing points (MECPs) and further transition states and intermediates have been identified [39]. The termination reaction may also be controlled by the use of polymerization additives [40].

A different picture is obtained for the migratory insertion into the M–H bond in late TMs according to Scheme 1.1. From inspection of the stationary points of the insertion in the $[CpRh(PH_3)(H)(C_2H_4)]^+$ complex in Figure 1.3, it follows that

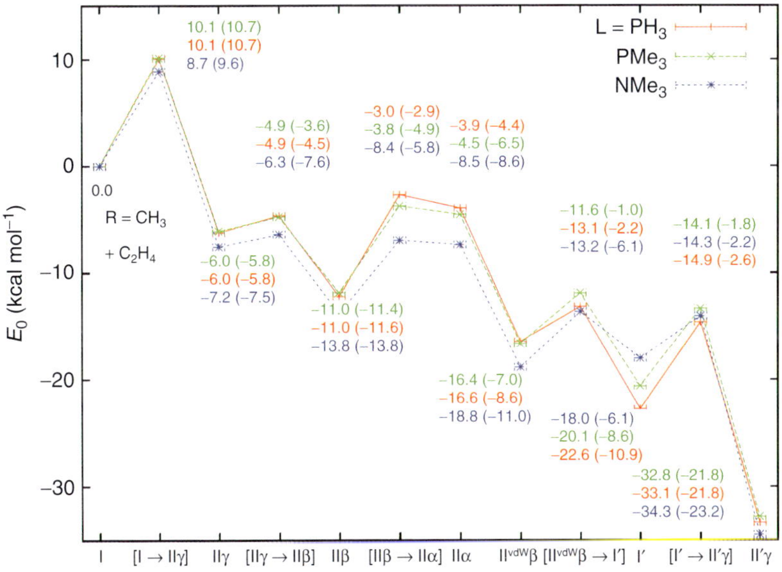

Figure 1.2 Energy profile for the first two insertion steps of the catalytic cycle for the complex $[Cp(L)Cr(C_2H_4)R]^+$ for $R = CH_3$ and $L = PH_3$, PMe_3, or NMe_3. All energies are corrected for zero-point vibrational energy and include the energy of separate C_2H_4 where required. Free energy values G_{298} are added in parentheses. For labels see Figure 1.1. The primed (′) complexes contain an additional ethylene unit in comparison to the preceding complexes, while for complex $II^{vdW}\beta$ the ethylene is attached to IIβ by van der Waals interactions only.

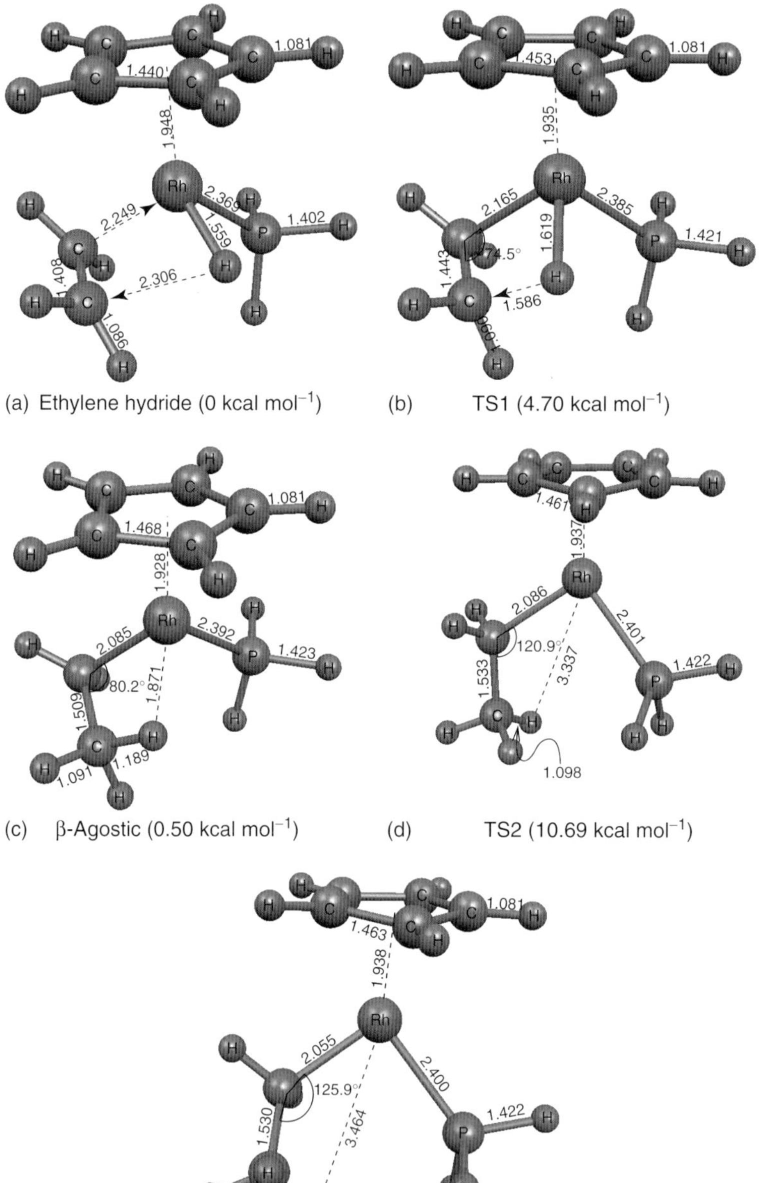

Figure 1.3 (a–e) Stationary points for the migratory insertion process. B3LYP/SDD geometries and energies of $[CpRh(PH_3)(C_2H_4)]^+$.

the overall insertion process is composed of two distinct steps, which are defined by two characteristic angles [41]:

1) The migration of the hydrogen atom from the TM to the ethylene to form the β-agostic structure (c) is described by the angle $\alpha(C_b-M-H)$; C_b is the carbon atom to which the migrating H will be attached.
2) The opening of the "chain" formed by the TM and the ethylene/ethyl moiety during the transformation of the β-agostic to the α-agostic structure (e) is described by the angle $\beta(M-C_a-C_b)$.

The four atoms involved here are thus considered key to describing the reaction.

A comparison of the energies of the stationary points for Cp-donor-type complexes of the late TMs Co and Rh is presented in Figure 1.4. Depending on the TM under consideration, either the ethylene hydride (Rh) or the β-agostic isomers (Co) represent the global minima, while the α-agostic species and their interconnecting transition states TS2 are rather high in energy. It is seen that the insertion barriers TS1 are negligible for the Co complexes, while the barriers to β-elimination given by $\Delta E = E_{TS1} - E_{\beta\text{-agostic}}$ are more than twice as high for the Co complexes than for any of their Rh analogs. Cobalt polymerization catalysts are therefore expected to exhibit less β-elimination and thus generate polymer of higher molecular weight and with less branching. Indeed, the complex $[Cp^*Co(P(OMe)_3)(Pr)]^+$ is an active polymerization catalyst [42]. If the phosphine ligands for the Co complexes are ordered according to barrier height ΔE, the following series is obtained:

$$PMe_3 \approx P(OMe)_3 < PH_3 < PF_3$$

Transition state theory thus predicts that the importance of β-elimination should diminish in the same order and identifies PF_3 as the ligand conferring

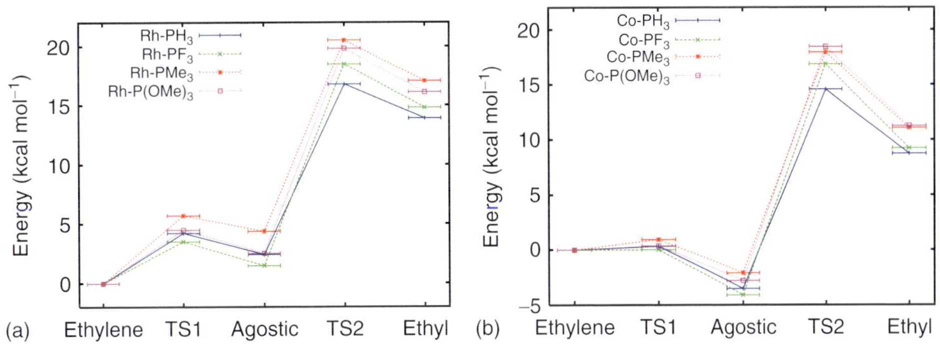

Figure 1.4 (a,b) Relative electronic energies (without zero-point energy) for the $[CpM(L)(H)C_2H_4]^+$ complexes. The energy of the ethylene structure is always set to 0. The other energies are defined in relation to the ethylene structure.

the greatest stability to the agostic intermediate. The relative energy barriers for the β-elimination step of the Co complexes increase approximately in the same order as the electron acceptor strengths of the ligands. For the Rh complexes, the analogous ligand order is found to be $PMe_3 < PH_3 < PF_3 < P(OMe)_3$.

The correlation with the ligand donor/acceptor strength is less clear in this case, but it is noteworthy nevertheless that the lowest ΔE values for both metals occur for the strong donor PMe_3 [43].

In addition to migratory insertion/BHE, the aforementioned complexes are subject to further dynamical processes. Four nuclear (C/H) scrambling processes have been examined theoretically; their computed activation energies agree very well with data from NMR studies [44].

1.3 Quantum Dynamical Calculations

1.3.1 Reaction Path Energy Profiles

Quantum dynamics is the propagation of a wave packet on a (in general) multidimensional PES according to the time-dependent Schrödinger equation. While Figure 1.4 provides an overview of the five stationary points of the PES for the insertion/elimination process, more detailed information about the energy variation in the course of the reaction is required. For a one-dimensional analysis confined to the reaction coordinate only, this can be obtained from either *ab initio* or DFT calculations by the IRC approach [45–48]. Zero-point effects may then be accounted for by a linear interpolation scheme. The reaction path energy curves for insertion/elimination in the $[CpM(L)(H)(C_2H_4)]^+$ complexes (M = Co, Rh; L = PMe_3, $P(OMe)_3$, PH_3, PF_3) are shown in Figure 1.5.

Inspection of the IRC reaction paths reveals that not all the five stationary points mentioned previously can be identified on all curves. Disregarding the complexes with the electron-rich PMe_3 ligand, we observe that for the Rh complexes TS1 corresponds to a narrow barrier, while for the Co complexes this barrier completely disappears after the zero-point energy (ZPE) correction is applied. This indicates that, from an energetic point of view, the first part of the migratory insertion (from the ethylene to the β-agostic conformation) is very fast for these complexes and should be easily observed at room temperature. Interestingly, the IRC curve for the Rh–PMe_3 complex is characterized by the absence of an agostic minimum, which would indicate that this compound might be the least effective as a polymerization catalyst. In contrast, the analogous Co complex is the only one for which a small TS1 insertion barrier is observed. For the second transition state TS2, we find a much broader and higher barrier for all complexes of both TMs.

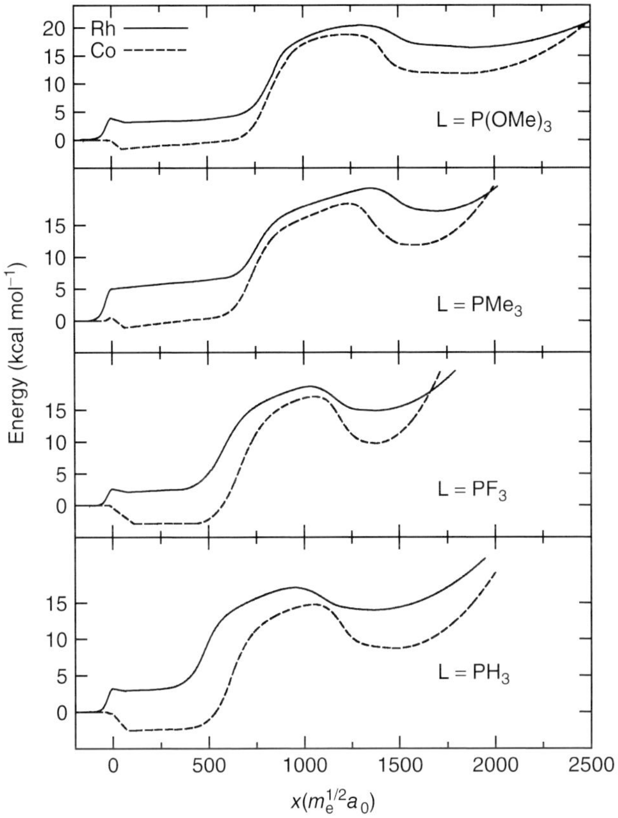

Figure 1.5 Zero-point energy-corrected potential energy curves along the reaction path in mass-weighted intrinsic coordinates for the complexes $[CpM(L)(H)C_2H_4]^+$.

1.3.2
Wave Packet Propagation for Late-Transition-Metal Complexes

If the wave packet maximum is traced along the reaction coordinate as a function of time, system-specific oscillation periods T may be determined. Figure 1.6 portrays the oscillation of the wave packet representing the $[CpRh(PH_3)(H)(C_2H_4)]^+$ complex within the agostic minimum for a start at TS1.

It is seen that, after 12 fs, the wave packet splits into two parts, moving to the left and to the right, respectively. The part on the left-hand side represents the molecules that isomerize to the ethylene complex. These parts are absorbed by a complex absorbing potential (CAP) that simulates decay of the complex, and they disappear for $t > 80$ fs. The part that moves along the right-hand side of the barrier runs into the local minimum of the β-agostic structure. It evolves further to the TS2 barrier, where it is reflected at 144 fs. Subsequently, the wave packet returns to the TS1 barrier, where a part of it overcomes the barrier and is thus

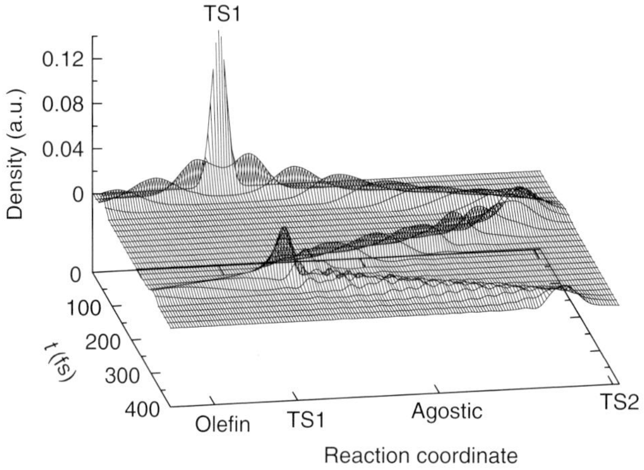

Figure 1.6 Oscillation of the time-dependent wave function within the β-agostic minimum for an initial wave packet located at TS1.

absorbed by the CAP. The rest of the packet continues to oscillate in the β-agostic minimum. During each oscillation period, a part of the wave packet is absorbed by the CAP.

Quantum dynamical simulations involving 2D and 3D PESs have also been performed. In the former case, the reaction is described by the reaction coordinates γ and θ, which are closely related to the angles α and β describing the two steps of the insertion process that have been defined in the previous section, while the latter additionally includes the metal–hydrogen bond length R_H, which was fixed to its value for the β-agostic minimum for the 2D PES. Figure 1.7 shows the 2D dynamics of a wave packet evolving from a saddle point on the $V(\gamma,\theta)$ surface which was visually identified to correspond to TS1 on the 3D PES after its center had been placed at this "quasi-TS1" at $t = 0$.

The reaction path is given in this figure by the black line that connects the stationary points (black dots). When the wave packet propagates, it splits up in two parts. One part propagates toward the ethylene minimum located near the top of the panels and is absorbed by the CAP. The other part of the wave packet moves into the agostic minimum where it oscillates (see panels for 32 fs and later). The potential close to the agostic minimum can be compared with a 2D harmonic potential. Each time the packet reaches TS1, a part of the packet evolves toward the ethylene minimum and will be absorbed by the CAP. As time increases, more and more molecules undergo the elimination process, and the wave packet gradually decays.

Figure 1.8 shows the time-dependent wave function resulting from the 3D calculation as a contour plot. In the first 5 fs, the wave packet is moving in the R_H direction because of the small reduced mass μ_H determining this motion. After 10 fs, the wave packet concentrates in the β-agostic minimum. Its subsequent oscillation and broadening in the β-agostic minimum is followed by a splitting

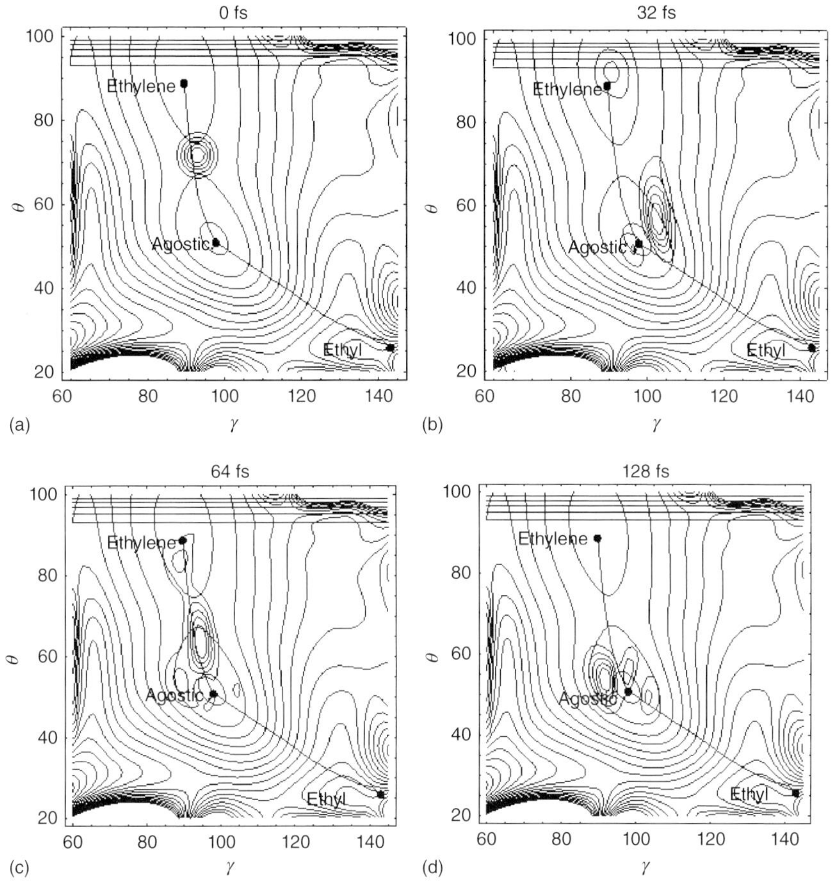

Figure 1.7 (a–d) Snapshots of the 2D time evolution of a wave packet started at quasi-TS1 on the agostic PES. Angles are in degrees.

into two parts. One part oscillates in the β-agostic minimum, while the other part moves toward the ethylene minimum, where it is absorbed by the CAP (20 fs and later). This process goes on periodically, and qualitatively in the same way as in the 1D and 2D calculations [49].

1.3.3
Norm Decay and Lifetimes

The overall progress of a wave packet propagation can be monitored by plotting the squared norm $N^2(t)$ of the wave packet as a function of time. Figure 1.9 presents the time evolutions of these survival probabilities of respective wave packets starting at TS1/ES (ethylene structure) and TS2 for all Co and Rh TM

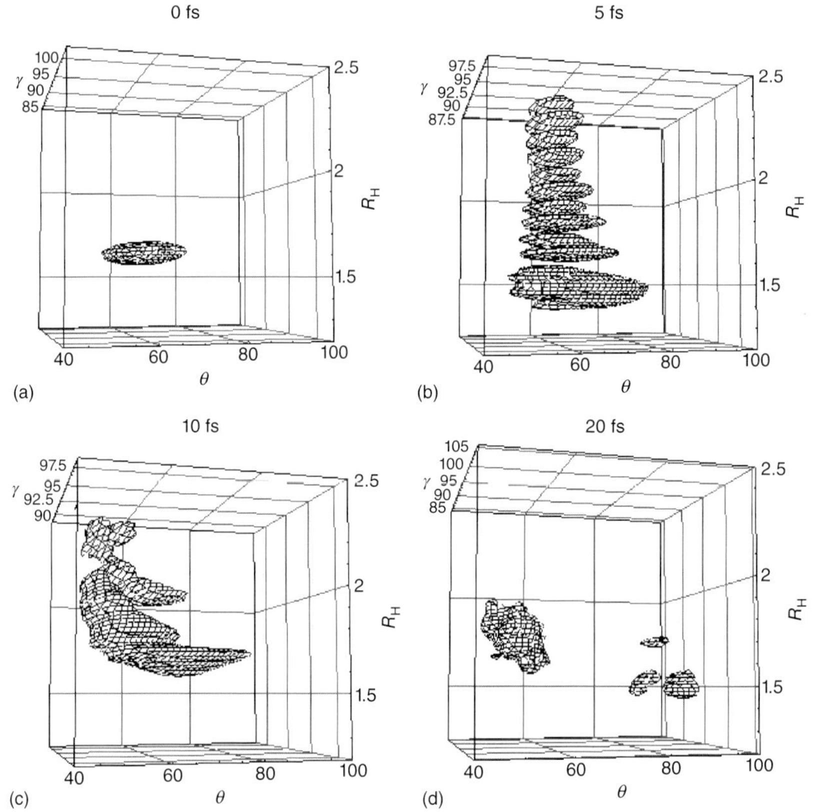

Figure 1.8 (a–d) Snapshots of the 3D time evolution of a wave packet started at TS1. Angles are given in degrees, R_H in angstrom. The contours correspond to a probability of $|\Psi(t)|^2 = 10^{-4}$.

complexes investigated.[2)] An exponential fit of $N^2(t)$ is also plotted. The squared norm decreases in a stepwise manner over time. The steps are nearly equidistant, with the distance corresponding to an oscillation period T of the wave packet moving in the respective minimum region (agostic for the initial wave packet located at TS1/ES, ethyl for the initial wave packet located at TS2). The steep sections of $N^2(t)$ correspond to the absorption of that part of the wave packet that has penetrated the region of the TS1 barrier and moved into the CAP region. Consequently, T can be obtained by inspecting the time evolution of $N^2(t)$. For instance, for a wave packet originating at TS2 for the ethyl minimum of the Rh–PF_3 complex, it is thus estimated to be 430 fs, and this value is close to that found by tracing the wave packet (454 fs). For each ligand L, the T_L for the Rh complex are longer than for the corresponding Co complex. This is because the

2) For all but one of the Co complexes, the TS1 barrier disappears after zero point correction. Wave packets are started from the ES in these cases.

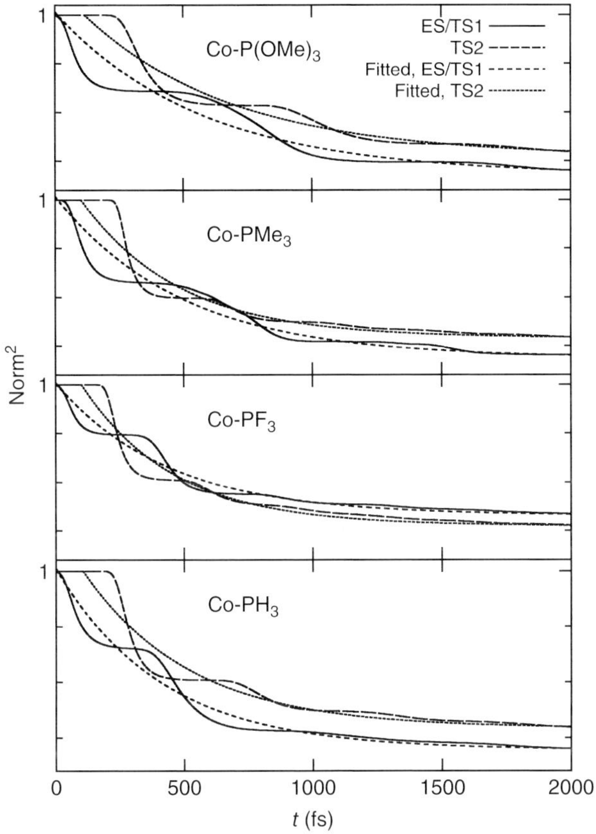

Figure 1.9 Time evolution of the squared norm (survival probabilities) for one-dimensional wave packets started at TS1 or TS2 for the $[CpCo(L)H(C_2H_4)]^+$ complexes. The fitted norm is based on $y = (1 - c)e^{-t/\tau} + c$.

higher ΔE values for the Co complex will result in faster propagation in the agostic potential well (due to its anharmonicity) and hence a shorter oscillation period. Since the ΔE values do not drastically differ among the Co complexes, we find that their T_L values increase in line with the ligand mass. Except for $L = PMe_3$, this correlation also holds for the Rh complexes. The short T for the trimethylphosphine ligand is due to its lack of an agostic minimum; instead of a genuine oscillation one therefore observes a complete decay of the norm after two absorption steps for this system. It is seen that the decay of $N^2(t)$ is almost complete after the first few oscillation periods. This process can be ascribed to the decay of the respective complex due to H-elimination following broadband excitation to TS1/ES or TS2. The lifetime τ of the system can be determined by fitting an exponential decay function to $N^2(t)$ and can be considered an indicator for the stability of the catalyst's β-agostic state against β-elimination. When comparing the τ_L for the same TM, it is found that they increase in parallel with

T_L: that is, a ligand with a longer oscillation time will also increase the lifetime of a particular complex. However, for identical ligands L, the lifetimes for the Co complexes are longer than for the Rh complexes. This is a consequence of the considerably lower energies (by over 2 kcal mol^{-1}) of the wave packet starting points at ES/TS1 for M = Co, resulting in a slower propagation in the region of the ES and the CAP. Our quantum dynamical results demonstrate that the broadband excitation lifetimes and hence the stability against β-elimination are influenced by the dynamics in both the agostic region and the ethylene region. For any particular ligand, cobalt catalysts are expected to generate a polymer that is characterized by higher molecular weight and less branching. Comparing the four ligands, we expect catalysts with L = P(OMe)$_3$ to lead to a polymer with the highest molecular weight. Indeed, the catalyst [Cp*Co(P(OMe)$_3$)(Pr)]$^+$ is used in the living polymerization of ethylene [50]. Data comparing the molecular mass and degree of branching for a polymer synthesized by a variety of late TM catalysts could corroborate our model once they become available.

1.3.4
Quantum Dynamics of Ethylene Insertion in Chromium Complexes

To contrast our findings for late TMs with data from a metal situated earlier in the d-block, we return to the Cp-donor-based chromium–alkyl complexes already addressed in Section 1.2.

The reaction path energy curve for insertion in the [CpCr(PH$_3$)CH$_3$]$^+$ complex in Figure 1.10 consists of five sections. The middle three parts correspond to IRC calculations starting from the transition states [II* → I], [I → IIγ], and [IIγ → IIα],

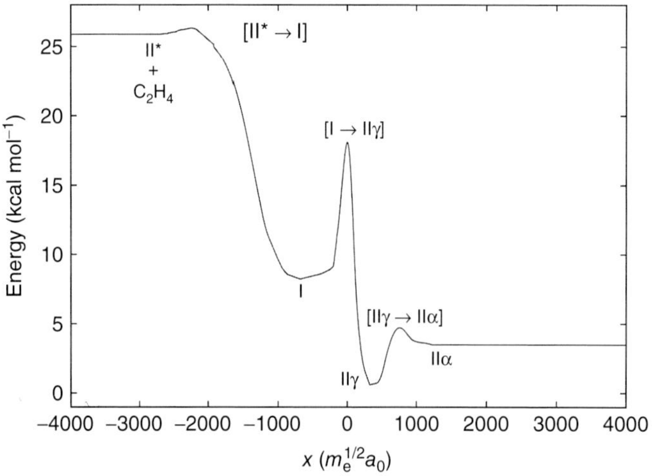

Figure 1.10 Zero-point energy-corrected potential energy curve along the reaction path in mass-weighted intrinsic coordinate for the complex [CpCr(PH$_3$)CH$_3$]$^+$. II* designates the structure of the methyl complex before the coordination of ethylene.

respectively, where II* is formed from I by dissociation of the coordinated ethylene. The two outermost parts serve to provide an extension of the grid range to the left and right, beyond structures IIα and II*, in order to enable the study of reactions by adding a CAP. As already mentioned, the actual reaction path is branched because the IIγ isomer may convert into either the IIα or the IIβ isomer; the IIα isomer has been selected here for illustrative purposes.

The IRC curves provide a direct and convenient way to probe the change of the energy and the structure of the complex along the reaction path. Consequently, the reaction sequence can be deduced from an analysis of the IRC curves. It is seen that the dissociated ES II* is about 18 kcal mol^{-1} above the ethylene–alkyl complex I, and the transition state [II* → I] for this dissociation still about 7 kcal mol^{-1} higher than the transition state [I → IIγ]. We therefore do not expect the dissociation of monomers from the complex to occur on the same time scale as the chain growth and subsequent isomerization of the insertion product. Indeed, the wave packet did not advance to the region of II* during our simulations. In comparison with the profiles for late TMs in Figure 1.5, the [I → IIγ] barrier (corresponding to TS1) is relatively high (10.1 kcal mol^{-1}), because here insertion into a metal–alkyl bond is considered. Furthermore, the agostic isomers are now considerably more stable than the ES, which is reflected in the diminished role of β-elimination in these systems. The energy differences of the agostic isomers and the barriers separating them (corresponding to TS2) are relatively small; their rapid interconversion is to be expected.

Inspection of Figure 1.11a displaying the time-dependent wave function with starting point [I → IIγ] confirms the above scenario. After starting its propagation, the wave packet separates into two parts in a rather short time. One part

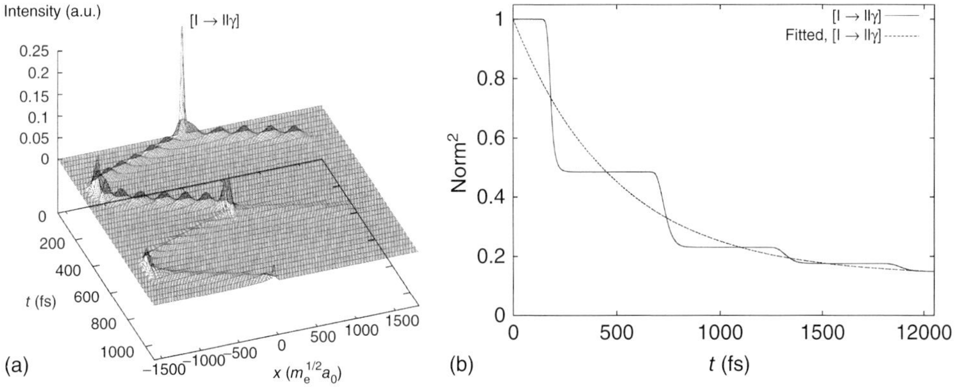

Figure 1.11 (a) Oscillation of the time-dependent wave function along the reaction coordinate for $[CpCr(PH_3)CH_3]^+$. The transition state $[I \rightarrow II\gamma]^‡(^4A)$ was selected as the starting point for the propagation. (b) Time evolution of the squared norm for a one-dimensional wave packet starting at $[I \rightarrow II\gamma](^4A)$ for $[CpCr(PH_3)CH_3]^+$. The fitted squared norm is based on $y = (1 - c)e^{-t/\tau} + c$.

propagates to the right of the barrier formed by [I → IIγ], moving into the potential energy well of the γ-agostic minimum, and further over the [IIγ → IIα] barrier to the region of the α-agostic minimum. It will be absorbed by the CAP after it has passed through this minimum (i.e., for $x > 1226\ m_e^{1/2} a_0$). Therefore, this part of the wave packet describes the migratory insertion of the complex.

The other part of the wave packet propagates toward the left-hand side of the barrier formed by [I → IIγ], where it starts to oscillate. This part of the packet represents the molecules isomerizing to the ethylene complex I.

During each oscillation period, this wave packet evolves from the [I → IIγ] barrier toward the region of [II* → I], where it is reflected. It then proceeds back to the [I → IIγ] barrier, whereupon a part of the packet penetrates the barrier and subsequently evolves toward the α-agostic minimum and is absorbed by the CAP. The rest of the packet continues to oscillate in the ethylene minimum. Each time the packet completes an oscillation and reaches the [I → IIγ] barrier, a part of the packet overcomes the barrier and is absorbed by the CAP. As time increases, a larger and larger part of the wave packet is absorbed. This reflects the increasing probability for the complex to undergo migratory insertion. It is noteworthy that the region below $x = -2000$ is never reached by the wave packet during the simulation, which confirms the longer time scale and lower significance of monomer dissociation for our model catalyst.

The overall progress of a wave packet propagation can be monitored by plotting the square of the norm $N^2(t)$ of the wave function as a function of time. Figure 1.11b presents the time evolution of the squared norm starting at [I → IIγ] for $[CpCr(PH_3)CH_3]^+$. An exponential fit of the squared norm is also plotted, from which the lifetime τ of the system can be determined. In this case, τ measures the stability of the catalyst against migratory insertion, while for the late TM complexes described above it refers to stability against β-elimination, due to the different positioning of the CAP. Hence, the dynamics data derived for the two situations, for example, from the norm decays given in Figures 1.9 and 1.11b, are not directly comparable.

As shown in Figure 1.11b, the squared norm of the wave packet at the beginning of the propagation is set to 1. It then decreases in a stepwise fashion over time. The steps are nearly equidistant, with the distance corresponding to an oscillation period of the wave packet moving in the ethylene minimum region. The sections of steep decay of $N^2(t)$ correspond to the absorption of that part of the wave packet that has penetrated the region of the [I → IIγ] barrier and moved into the CAP region. By tracing the position of the wave packet maximum during the propagation, we obtained a value of $T = 590$ fs for the aforementioned complex. From Figure 1.11b, it is seen that the squared norm of a wave packet has decayed to about 20% after the first few oscillation periods. This corresponds to the migratory insertion reaction of the respective complex. The lifetime obtained for $[CpCr(PH_3)CH_3]^+$ is $\tau = 500$ fs.

1.4 Summary and Outlook

In this chapter, we have described a quantal approach to investigate the dynamics of elementary catalytic reaction steps, especially BHE from an agostic TM complex. Several reduced-dimensionality (1D–3D) treatments could be set up, and many new spectroscopic data and time constants (vibrational periods and lifetimes) were obtained, which shed new light on the species treated and on the process in general. Further improvement of the dynamical approach would consist in including additional degrees of freedom; however, the latter would require a more approximate treatment, such as a locally harmonic approximation, to save computational effort in generating the DFT PES. As a further step in the overall line of investigation, these studies are to be extended to the computation of thermal rate constants. (Remember that the present study focused on a highly nonequilibrium process, corresponding to a Franck–Condon excitation to the transition states of the system.) This is possible, for example, by utilizing the above techniques and results within the flux–flux correlation function formalism of Miller, Manthe *et al.* [51, 52]. Of particular interest in this context are rate calculations for hydrogen transfer reactions, because of the importance of tunneling and barrier-recrossing effects in these systems. A preliminary study of the isotope effect on migratory insertion confirmed the expected deviation from Arrhenius-type dynamics at low temperatures. It is hoped that these studies will provide more insight into the electronic factors governing, for example, the outcome of olefin polymerization, and thus help in its control. More generally, we point out the relevance of our reduced-dimensionality treatment to processes other than those investigated here. The above-mentioned four atoms explicitly retained in our description are the atoms between which bonds are broken and formed in the course of the reaction. The present analysis establishes that it is mostly their structural parameters that change significantly, whereas the others remain nearly constant. If this holds true also for other systems, it might open a route to extend this approach to a large variety of different reactive processes, catalytic and others.

Acknowledgments

This work was financially supported by the German Research Council (DFG) through SFB623 "Molekulare Katalysatoren: Struktur und Funktionsdesign." The authors are indebted to Markus Enders for helpful discussions.

References

1. Koga, K. and Morokuma, K. (1991) *Chem. Rev.*, **91**, 823.
2. Musaev, D.G. and Morokuma, K. (1996) *Adv. Chem. Phys.*, **95**, 61.
3. Ziegler, T. (1991) *Chem. Rev.*, **91**, 651.
4. Salahub, D.R., Castro, M., Fournier, R., Calaminici, P., Godbout, N., Goursot, A., Jamorski, C., Kobayashi, H., Martinez,

A., Papai, I., Proynow, E., Russo, N., Sirois, S., Ushio, J., and Vela, A. (1995) in *Theoretical and Computational Approaches to Interface Phenomena* (eds H. Sellers and J. Olab), Plenum Press, New York, p. 817.
5. Siegbahn, P.E.M. (1996) *Adv. Chem. Phys.*, **93**, 333.
6. Dedieu, A. (ed.) (1992) *Transition Metal Hydrides*, VCH Publishers, New York.
7. van Leeuwen, P.W., van Lenthe, J.H., and Morokuma, K. (eds) (1994) *Theoretical Aspects of Homogeneous Catalysts. Applications of ab initio Molecular Orbital Theory*, Kluwer Academic Publishers, Dordrecht.
8. Yoshida, S., Sakaki, S., and Kobayashi, H. (1992) *Electronic Processes in Catalysis*, VCH Publishers, New York.
9. Brookhart, M., Green, M.L.H., and Wong, L.-L. (1988) *Prog. Inorg. Chem.*, **36**, 1.
10. Brookhart, M., Lincoln, D.M., Volpe, A.F., and Schmidt, G.F. (1989) *Organometallics*, **8**, 1212.
11. Brookhart, M., Hauptman, E., and Lincoln, D.M. (1992) *J. Am. Chem. Soc.*, **114**, 10394.
12. Ittel, S.D., Johnson, L.K., and Brookhart, M. (2000) *Chem. Rev.*, **100**, 1169.
13. Ziegler, K., Holzkamp, E., Breil, H., and Martin, H. (1955) *Angew. Chem.*, **67**, 541.
14. Ziegler, K. (1964) *Angew. Chem.*, **76**, 545.
15. Natta, G. and Corradini, P. (1955) *Atti. Accad. Naz. Lincei Mem. CI. Sci. Fis. Mat. Nat. Sez. II*, **5**, 73.
16. Natta, G. (1956) *Angew. Chem.*, **68**, 393.
17. Natta, G. (1964) *Angew. Chem.*, **76**, 553.
18. Seymour, R.B. and Cheng, T. (eds) (1986) *History of Polyolefins*, Reidel, Dordrecht.
19. Eisch, J.J. (1983) *J. Chem. Educ.*, **60**, 1009.
20. Boor, J. (1979) *Ziegler-Natta Catalysts and Polymerisations*, Academic Press, New York.
21. Sinn, H. and Kaminsky, W. (1980) *Adv. Organomet. Chem.*, **18**, 99.
22. Pino, P. and Mülhaupt, R. (1980) *Angew. Chem., Int. Ed. Engl.*, **19**, 857.
23. Hogan, J.P. and Banks, R.L. (1955) (Phillips Petroleum Co.) Belg Patent 530,617; U.S. Patent 2825,721, 1958.
24. Hogan, J.P. (1970) *J. Polym. Sci. A*, **8**, 2637.
25. Groppo, E., Lamberti, C., Bordiga, S., Spoto, G., and Zecchina, A. (2005) *Chem. Rev.*, **105**, 115.
26. Döhring, A., Jensen, V.R., Jolly, P.W., Thiel, W., and Weber, J.C. (2001) *Organometallics*, **20**, 2234.
27. Johnson, L.K., Killian, C.M., and Brookhart, M. (1995) *J. Am. Chem. Soc.*, **117**, 6414.
28. Ziegler, T. and Autschbach, J. (2005) *Chem. Rev.*, **105**, 2695.
29. Schork, R. and Köppel, H. (2001) *J. Chem. Phys.*, **115**, 7907.
30. Mahapatra, S., Köppel, H., and Cederbaum, L.S. (2001) *J. Phys. Chem.*, **105**, 2321.
31. Somers, M.F., Kingma, S.M., Pijper, E., Kroes, G.J., and Lemoine, D. (2002) *Chem. Phys. Lett.*, **360**, 390.
32. Kulander, K.C. (1991) *Time-dependent Methods for Quantum Dynamics*, North-Holland Publishing Company.
33. Zhang, J.Z. (1999) *Theory and Application of Quantum Molecular Dynamics*, World Scientific, River Edge, NJ.
34. Senn, H.M., Blöchel, P., and Togni, A. (2000) *J. Am. Chem. Soc.*, **122**, 4098.
35. Meier, R.J., van Doremaele, G.H., Ialori, S., and Buda, F. (1994) *J. Am. Chem. Soc.*, **116**, 7274.
36. Zewail, A.H. (1999) Atomic-Scale Dynamics of the Chemical Bond Using Ultrafast Lasers. Nobel Lecture.
37. Zewail, A.H. (1994) *Femtochemistry*, vol. 1 and 2, World Scientific, Singapore.
38. Manz, J. and Wöste, L. (1995) *Femtosecond Chemistry*, Verlag Chemie, Weinheim.
39. Xu, R., Klatt, G., Enders, M., and Köppel, H. (2012) *J. Phys. Chem. A*, **116**, 1077.
40. Mark, S., Kurek, A., Mülhaupt, R., Xu, R., Klatt, G., Köppel, H., and Enders, M. (2010) *Angew. Chem. Int. Ed.*, **49**, 8751.
41. Bittner, M. and Köppel, H. (2004) *J. Phys. Chem. A*, **108**, 11116.
42. Daugulis, O., Brookhart, M., and White, P.S. (2003) *Organometallics*, **22**, 4699.

43. Xu, R., Bittner, M., Klatt, G., and Köppel, H. (2008) *J. Phys. Chem. A*, **112**, 13139.
44. Xu, R., Klatt, G., Wadepohl, H., and Köppel, H. (2010) *Inorg. Chem.*, **49**, 3289.
45. Fukui, K., Sato, S., and Fujimoto, H. (1975) *J. Am. Chem. Soc.*, **97**, 1.
46. Fukui, K. (1981) *Adv. Chem. Phys.*, **14**, 363.
47. Gonzalez, C. and Schlegel, H.B. (1989) *J. Phys. Chem.*, **90**, 2154.
48. Gonzalez, C. and Schlegel, H.B. (1990) *J. Phys. Chem.*, **94**, 5523.
49. Bittner, M., Köppel, H., and Gatti, F. (2007) *J. Phys. Chem. A*, **111**, 2407.
50. Tanner, M.J., Brookhart, M., and DeSimone, J.M. (1997) *J. Am. Chem. Soc.*, **119**, 7617.
51. Miller, W., Schwartz, S., and Tromp, J. (1983) *J. Chem. Phys.*, **79**, 4889.
52. Manthe, U. (2002) *J. Theor. Comput. Chem.*, **1**, 153.

2
Activation of Small Molecules with Metal and Metal Oxide Clusters in Inert Gas Matrixes

Hans-Jörg Himmel and Olaf Hübner

2.1
Introduction

Many heterogeneous catalytic reactions involve metals or metal oxides as catalysts. However, the catalytic mechanism is seldom fully understood, a rare exception being the Haber–Bosch process [1]. It has been shown in many studies that reactions on the surface occur at steps or other defects. Clusters are often used to model the reactivity of bulk surfaces, and this approach has been named *molecular surface science* [2]. This might be an oversimplification. However, the reactivities of clusters and surfaces are indeed often comparable. For example, copper clusters and copper surfaces react with O_2 and CO but not with H_2 [3]. On the other hand, it was shown in many studies that the reactivity of clusters largely varies with their size. This could have several reasons, and the structures and electronic situations of clusters of different sizes could vary to a large extend. One recent example should illustrate this point. Clusters anions with an even number of aluminum atoms, for example, Al_{14}^-, were found to react much faster with dioxygen in its triplet electronic ground state than clusters anions with an odd atom number, for example, Al_{13}^- [4]. For comparison, Al surfaces were also found to react slowly with triplet O_2. On the basis of thermodynamic and kinetic considerations, as well as the likely structure of Al_{13}^-, it was suggested that specifically the Al_{13}^- cluster might provide a suitable model for some aspects of the surface of bulk Al [4, 5]. Switching between the potential energy surfaces of different spins in the course of a reaction does not necessarily slow down a reaction. In a spin-accelerated reaction, the barrier for thermal reaction is very high (so that the reaction cannot proceed although it is spin-allowed), but spin inversion provides the means to switch in the course of the reaction to an energetically lower hypersurface (and again back) to generate the product species [6].

For experimentalists, it is not an easy task to generate clusters with a narrow size distribution. In one approach, small metal and metal oxide particles were grown on oxide bulk single-crystal surfaces, and the adsorption and activation of small molecules on these particles was studied with surface science techniques [7]. In some studies, cluster ions were first generated by laser evaporation techniques

Molecular Catalysts: Structure and Functional Design, First Edition. Edited by Lutz H. Gade and Peter Hofmann.

in the gas phase, then mass-selected by using a quadrupole mass spectrometer or (to obtain higher currents at the cost of resolution) a Wien velocity filter, and subsequently deposited on a substrate surface [8]. In many of these studies, oxides and especially MgO serve as substrates. There are several problems and limitations of this technique. For example, "soft-landing" of the clusters at the surfaces has to be achieved. This is challenging because the kinetic energy of the mass-selected clusters is quite high (generally higher than the metal–metal bond energies). Even when all technical problems are solved, the chemical interaction between the cluster and the surface atoms could be very high and lead to cluster decomposition and adsorption of single atoms. For example, the stability of two Pd atoms bound to the oxygen atoms of the MgO surface is larger than the Pd_2 metal–metal bond (1.3 eV) [9].

Clusters could also be synthesized by reactions in solutions [10], but in this case they exhibit stabilizing ligands which prevent further agglomeration. These ligand shells prohibit or at least influence catalytic reactions. An example is the $Au_{55}(PPh_3)_{12}Cl_6$ cluster first synthesized by Schmid [11, 12]. It has been shown that the ligand shell of this cluster can be removed [13], and therefore, in this special case, studies of the reactivity of ligand-free Au_{55} clusters on substrate surfaces become possible. Oxidation experiments show that even atomic oxygen is not capable of oxidizing this cluster, although under similar conditions other gold clusters (e.g., the Au_{13} cluster which also could be obtained from $Au_{55}(PPh_3)_{12}Cl_6$) and also bulk gold are oxidized quantitatively to Au_2O_3 [13].

In a further approach, clusters are prepared and investigated in the gas phase. This allows the study of clusters without the influence of a supporting surface and without ligands. It is possible to investigate neutral species, but more often ions are prepared and investigated because they can better be controlled, for example, by electric and magnetic fields. By ion-cyclotron resonance mass spectrometry, it is also possible to store the ions for longer times and perform series of several experiments on their reactivity.

Furthermore, clusters can be investigated by the matrix isolation technique. This allows the accumulation of somewhat larger amounts of clusters and is more suitable to study neutral systems. Although the noble gases usually used as matrixes are unreactive, some matrix effects have to be considered. These studies could also be used to obtain information about the reactivity of these species, either in their electronic ground state or in excited electronic states.

Research on small transition-metal clusters was reviewed in a comprehensive way in 1986 [14] and again in 2002 (taking then in also lanthanide clusters) [15]. Moreover, small metal oxides and, less detailed, metal oxide clusters of the transition metals were recently reviewed [16]. Herein we restrict the discussion to some examples, most of them involving the matrix isolation technique; this article is not a comprehensive overview of research in this field. Different from the three aforementioned review articles, our examples will also include main-group element compounds.

A detailed understanding of the electronic structure of metal atom dimers and clusters and also small metal oxide clusters is necessary to anticipate their special reactivity. However, this is not at all an easy task, and requires the interplay of sophisticated experimental techniques and modern quantum chemical calculations. Experiments or theory alone cannot lead to a detailed characterization. For an understanding of the special reactivity of these species, not only the electronic ground state but also energetically low-lying excited states have to be considered [17]. It might be tempting to argue that quantum chemical calculations for molecules with a small number of atoms should be no problem. However, this is definitely not the case, at least not for compounds involving transition-metal atoms. Concerning wave function-based calculations, the often large number of low-lying electronic states and the requirement to treat the correlation of nd and $(n+1)$s electrons in a balanced way are challenges for theoreticians. Because of the open d shell, transition metals usually require a multiconfiguration treatment, and the large active spaces needed for systems involving more than one transition metal or open shell ligands quickly touch the technical boundary behind which calculations cannot be reasonably performed. Therefore, in a recent example (one of a number of studies on Ti_2), wherein the electronic structures of Ti_2 and Ti_2^+ were analyzed by different types of calculations including multireference calculations [coupled cluster calculations including all single and double excitations and a perturbative estimate of the triple excitations (CCSD(T)), Multireference configuration interaction (MRCI), Multireference averaged coupled pair functional (ACPF)] [18], the authors conclude: "… the almost insurmountable computational and conceptual difficulties of the 3d-M_2 systems are due to the extremely dense manifold of the Λ-Σ states"[18]. Density functional theory (DFT) in principle provides a less demanding approach, but severely suffers from the approximations of current functionals. In some studies, it was claimed that DFT functionals describe acceptably the electronic structure of compounds including the electron correlation due to the self-interaction error (SIE) of the exchange functionals mimicking typical correlation effects, in particular also nondynamic correlation [19]. However, they only show that the changes in the density due to the SIE resemble those induced by the inclusion of correlation, but they even do not show that these changes in density also lead to changes in energies that are similar to those from the inclusion of correlation. The error cancellation may be or is very likely the reason for the surprisingly good performance of DFT for organic molecules. But, if the performance of the current density functional methods so heavily depends on the cancellation of errors, there will be no guarantee that this will work universally, and the observed failure for systems including transition metals points to the fact that current functionals are not well suited for transition-metal systems. For example, for the cyclic $Ni(O_2)$ molecule, with both gradient-corrected and hybrid functional DFT methods a triplet ground state was found with further states at energies higher by about 30 kJ mol^{-1}. Recent multireference calculations, however, clearly show that $Ni(O_2)$ exhibits a $^1A_1$ electronic ground state. The next terms (in energetic order) are the quintet terms $^5B_2$ and $^5A_1$ (with energies higher by 56 and 60 kJ mol^{-1}), whereas triplet terms do not follow below 70 kJ mol^{-1} [20]. Hence, whereas wave function-based methods in principle work but quickly touch the border of technical applicability, DFT-based methods with current functionals are unfortunately often unreliable for transition-metal systems.

2.2 The Matrix Isolation Technique – Advantages and Limitations

The matrix isolation technique is now well established for the isolation and characterization of metal atom dimers and clusters and metal oxide clusters. The technique has already been described in several seminal books and review articles [21–27]. Studies in matrixes have some advantages and some disadvantages compared to gas-phase studies; both techniques together could provide detailed information. Using Fourier transform ion cyclotron resonance (FT-ICR) mass spectrometric techniques, the group of Schwarz and other groups were able to establish complete catalytic cycles for charged atoms or clusters as well as metal oxides in the gas phase [28–32]. Unfortunately, it is not possible to establish catalytic cycles in such a way using the matrix isolation technique. Nevertheless, some key steps of catalytic cycles, for example, oxidative addition, can be analyzed. In matrix isolation experiments, the reactivities of *neutral* metal atom and metal oxide clusters (as well as charged species) can be analyzed, and in this sense the matrix isolation results are complementary to the gas-phase results on charged species (see the discussion below). Moreover, a strict control of the electronic state of the metal or metal oxide cluster is possible in matrix isolation experiments. At the low temperature of the matrix (4 K), all species are generally in their electronic ground state. With the modern solid-state laser sources, it is possible to excite the species very precisely into one specific electronic state even if many electronic states lie close by. Only in very special cases (if two terms are energetically extremely close by) is the electronic state influenced by the interactions with the matrix gas. For example, Ni atoms exist in the $^3F_4$ state (configuration $3d^84s^2$) in Ne matrixes (and also in the gas phase), but in both the $^3F_4$ and the $^3D_3$ (configuration $3d^94s^1$) in Ar, Kr, or Xe matrixes [33]. Another example is the change of the electronic state of the UO_2 molecules in Ne and Ar matrixes (corresponding to a change from a (5f)(7s) configuration to a $(5f)^2$ one at the uranium atom) [34].

Different from the gas phase, the matrix host dissipates the energy released in exothermic reactions (by excitation of matrix phonon modes). In the gas phase, the energy remains on the product [31] and could lead to rapid fragmentation or other reactions (if it is not dissipated by collision with the walls or other collision partners). For example, Andersson *et al.* [3d] studied the gas-phase reactions between Co_n clusters and O_2 and obtained reaction probabilities S in a collision. For larger clusters ($n > 20$), S could be quite reliably measured and came out to be almost independent of n ($S \approx 0.7$). However, as the authors summarize, the apparently lower reactivity of the smaller clusters is presumably due to the fast fragmentation of the product clusters of these highly exothermic oxidation reactions, which prohibits their detection. For reaction between B atoms and O_2, linear OBO is the major product observed in Ar matrixes [35] whereas BO is detected in gas-phase studies [36]. The reaction energy of OBO formation from B and O_2 is $-860\,\mathrm{kJ\,mol^{-1}}$, and exceeds significantly the energy needed to cleave one of the B–O bonds with formation of BO and O ($+560\,\mathrm{kJ\,mol^{-1}}$). In addition to the dissipation of the reaction heat, a matrix cage effect could prevent the escape of a fragment formed in a decomposition step. This cage effect largely depends on the matrix host material. Hence the diffusion of small species is more efficient in Ne compared to Ar matrixes. One example in which presumably both energy dissipation and the matrix cage effect are in play is the reaction of Co atoms with O_2. In Ar matrixes, only CoO_2 is formed, but in Ne matrixes, CoO is also formed in significant amounts [37].

Finally, it should be mentioned that a matrix shift is observed between the frequencies of the vibrational modes in the matrix and in the gas phase. This shift is similar to the solvent shift in solution, and can be modeled by an electrostatic model (the dielectric constant is then the host-dependent parameter). However, in some cases the matrix host could also chemically interact with the isolated species. Such an interaction is observed not only for Xe (which one generally avoids as the matrix host because of its chemical reactivity) but in some cases also for other noble gas matrixes, for example, Ar. An example are the late 3d metal oxides in noble gas matrixes, where the spectra point to a weakly bound complex NgMO (Ng = Ar, Kr, Xe) [38].

2.2.1
Thermal Evaporation Versus Laser Ablation

Metal atoms could be generated in various ways (see the discussion in Ref. [27]), the most frequently used being thermal evaporation and laser ablation. Laser ablation techniques are extremely important in the context of gas-phase studies of cluster ions which are not easy to generate thermally in large enough quantities. The most important advantages of laser ablation are, (i) the sample is heated up only at a very small spot and (ii) metals with a very low vapor pressure at very high temperatures can be brought into the gas phase [24]. On the other hand, the laser ablation technique also has some disadvantages. Hence a significant number of atoms is ionized by the laser beam (e.g., more than 10% for calcium) [39]. Moreover, laser irradiation leads to electronic excitation of the metal atoms. Finally, the reactant might also be irradiated, leading to electronically excited reaction partners or even fragmentation (e.g., oxygen atoms in dioxygen or nitrogen atoms in dinitrogen). An illustrative example is the detection of GaN when laser-ablated Ga atoms are co-condensed with N_2 [40]. In the case of thermal evaporation, only a weak $Ga{\cdot}N_2$ complex is formed [41]. Since all species (neutral and charged) are embedded in the matrix, and their separation is not possible, the matrix is very often rich in species, and this is a problem for the complete assignment of the vibrational modes. If one wants to probe the reactivities of metal atom clusters or metal oxide clusters, thermal evaporation is, in our view, the superior method.

2.2.2
Metal or CsI Substrates for the Matrix

In principle, two different kinds of setup are in use for light absorption and emission measurements, which are illustrated in Figure 2.1. For electron paramagnetic resonance (EPR) studies, the setup has to be modified considerably. In the setup shown on Figure 2.1a, the matrix is deposited onto a CsI window as substrate, and the spectroscopic measurements are carried out in transmission. In the setup shown on Figure 2.1b, the matrix is deposited onto a Cu surface, and the measurements are carried out in reflectance. If absorption measurements in the visible region should also be made, it is advisable to coat the Cu surface with a white metal layer, for example, Rh, which exhibits no absorption in the visible region. The Rh layer should be formed by electrolytic deposition from solution rather than through vapor deposition of the metal to guarantee a stable layer in a wide temperature region (4–300 K). This setup has several advantages. Most importantly, it allows measurements in a very broad spectroscopic window. For studies on metal dimers and clusters, it is of importance to record not only absorption spectra but also Raman (and fluorescence) spectra for the same matrix sample. Moreover, the absorption measurements should be carried out in a wide energy window, including the FIR (far-infrared) (where modes of heavier elements show), MIR (middle-infrared), and NIR (near-infrared) regions, as well as the visible light region (for detection of electronic transitions). Such measurements are possible with the setup shown on Figure 2.1b, but not for that on Figure 2.1a. Moreover, the IR light passes twice through the matrix and therefore the signal-to-noise ratio is better. On the other hand, vibrations of the Cu block have to be excluded strictly. This is not so easy since the cryostat head of a closed-cycle refrigerator includes a

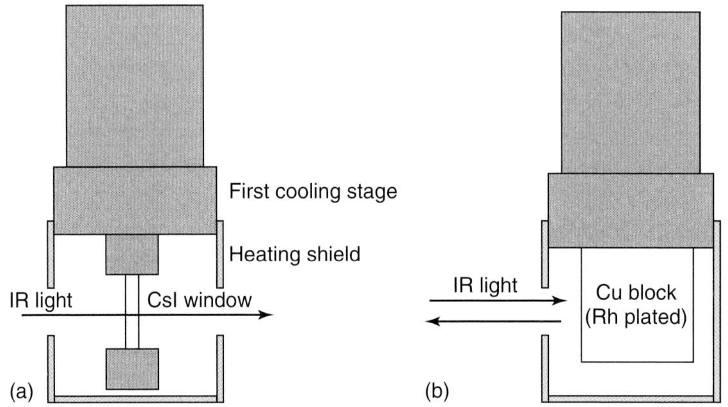

Figure 2.1 Possible matrix setups (only the cryostat head carrying the matrix substrate is shown). (a) Measurements in transmission through a CsI window. (b) Measurements in reflectance from a Cu block (ideally coated with a thin Rh layer).

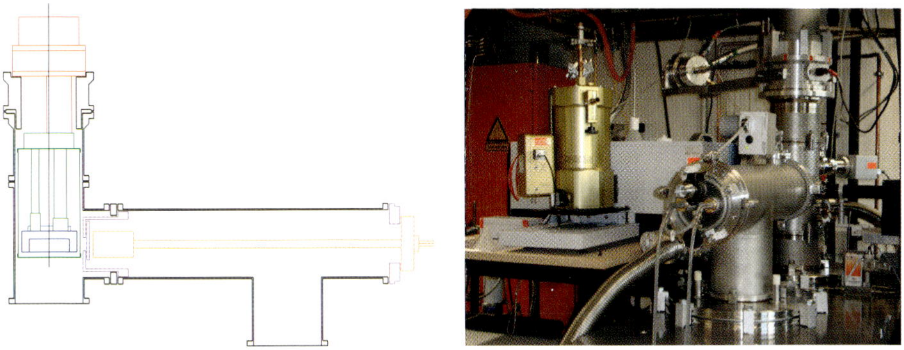

(a) (b)

Figure 2.2 (a) Schematic drawing and (b) photograph of the Heidelberg matrix apparatus.

valve which is constantly opened and closed. The matrix has to be kept in a high-vacuum chamber to avoid deposition of impurities. Owing to the differences in vacuum, the matrix setup has to be separated from the IR spectrometer.

The Heidelberg matrix setup (see Figure 2.2) consists of a vacuum chamber (matrix chamber) equipped with a rhodium-plated Cu block that has eight polished mirror faces for the deposition of the matrixes. The block is rotatable to allow the positioning of the mirror faces in front of a shutter which separates the chamber from a second vacuum chamber (preparation chamber) that contains the metal evaporator or in front of several windows that provide the connection to the IR spectrometer, the Raman spectrometer, or an excitation laser. The Cu block is cooled down to 4.3 K by a closed-cycle He pulse tube refrigerator (Vericold). To reduce radiative heating, the Cu block is surrounded by a Cu shield with eight window holes in front of the mirror faces. This shield is cooled to about 50 K, which is the temperature of the first stage of the refrigerator. The Cu block can be kept at higher temperatures by an electric heating device mounted on the Cu block. The second chamber is equipped with the thermal metal evaporator together with a quartz microbalance to monitor the deposition of the metal vapor. The metal is evaporated by heating a tungsten filament that is coated with the metal to be evaporated. Within the evaporator, the tungsten filament is mounted between two metal tubes which serve as contacts to electrically heat the filament and also allow water flow for cooling. The distance from the heater filament to the mirror faces is about 4 cm. The matrix gas is introduced into the matrix chamber by a small metal tube that ends near the shutter between the two chambers in such a way that the gas flow is directed to the mirror placed in front of the shutter. The flux can be regulated by a mass flow controller (Bronkhorst). The vacuum chambers are pumped by diffusion pumps to obtain a base pressure of 10^{-7} mbar. Spectra in the IR and visible region can be recorded by a Vertex 80v (IR) spectrometer (Bruker). By means of different detectors (Si diode, InSb detector, HgCdTe detector, bolometer) and radiation sources (W lamp, Globar, Hg vapor lamp), absorption spectra can be recorded between 20 and 25 000 cm^{-1}.

The (IR) spectrometer is separated from the matrix chamber by a rotatable disk that contains three windows, CaF_2, CsI, and high-density polyethylene, which can be chosen according to the desired spectral region. Raman and luminescence spectra can be recorded by a T64000 Raman spectrometer (Horiba Jobin Yvon).

2.3
Formation and Characterization of Metal Atom Dimers and Clusters

There have been numerous gas-phase studies of metal atom dimers and clusters. Bondybey [42] developed laser vaporization techniques and recorded laser-induced fluorescence spectra for metal dimers and small clusters. In their setup, the metals were first vaporized by a pulsed YAG (yttrium-aluminum garnet) laser, and then the gaseous species were excited with a pulsed dye laser. Especially, the dimer Be_2 was studied in a number of works, supplemented by quantum chemical calculations, also on small Be_n clusters ($n = 3-6$) [43, 44]. In matrix isolation studies, the number of metal atom dimers or clusters could be increased by diffusion of the metal atoms in the matrix host at higher temperatures. For example, potassium clusters were formed by tempering of adamantane matrixes [45]. In this context, the chosen matrix gas is of importance. The diffusion properties of Ne matrixes at temperatures below 10 K are better than those of Ar matrixes near 30 K. A problem, especially for matrix isolation studies, is that one normally obtains a mixture of different cluster sizes. In the past, some apparatuses were built in which the clusters were first charged and then mass-selected with a quadrupole mass spectrometer or a Wien velocity filter before they were brought into the matrix. Immediately before entering into the matrix, the clusters have to be neutralized with electrons, usually emitted from a tungsten filament. The different apparatuses were home-built, and an overview of some earlier machines could be found in Ref. [24]. Size-selected carbon chains and clusters were probably the most intensively studied species using these methods [46, 47]. Examples for mass-selected metal atom dimers and clusters characterized in inert gas matrixes include Mo_2 [48], Fe_3 [49], Ag_3 [49], and Ag_7 [50]. Raman spectra are extremely informative, especially if resonance Raman conditions (leading to a dramatic intensity increase) could be found; in the case of metal atom dimers, the detection of vibrational progressions allows the estimation of the anharmonicity constant and, if a sufficient number of overtones can be measured, the dissociation energy (in this case, a model for the potential energy curve has to be assumed). The harmonic frequency derived for a metal dimer could be used to calculate the force constant. In this context, it should be mentioned that simple models were developed to estimate the stretching force constants in small clusters by a "bond-connectivity" force field (which exclusively comprises primary and interaction bond-stretching force constants). For a cluster of n atoms, the force constant f_n is connected with the force constant of the dimer, f_2, by the relation $f_n = \beta_n \cdot f_2$, where β_n is $2/a_n$ and a_n is the sum of the number of the atoms to which each atom in the considered bond is connected (e.g., for a trinuclear cluster, $a_n = 4$ for a D_{3h} symmetric triangle and $a_n = 3$ for a $D_{\infty h}$ linear geometry) [51]. The formula implies that the force constant decreases monotonically with increasing number of atoms in the cluster, a trend that has been verified for several metals. This simple relation, of course, is applicable only if the bonds are equal and the type of bonding does not change (as is, for example, the case if multiple center bonding comes into play).

Two examples for matrix-isolation studies on metal dimers are Ga_2 and Ti_2. They are discussed at this place because their remarkable reactivities were studied in some detail (see next section). Resonance Raman spectra, excited with the 514 nm line of an Ar^+ ion laser, were obtained for the Ga_2 dimer (see Figure 2.3) [52]. The $^{69}Ga/^{71}Ga$ isotopic pattern could be used as a proof that all detected signals belong to this species [53]. The detection of a series of overtones in this case allowed the determination of the anharmonicity constant. A harmonic frequency of 178.2 cm^{-1} was then obtained for the isotopomer $^{69}Ga^{69}Ga$, and a force constant f(Ga–Ga) of 64.8 ± 0.3 N m^{-1}. An upper limit of 145 kJ mol^{-1} was estimated (on the assumption of a Morse potential) for the dissociation energy, a value which is slightly higher than estimates from gas-phase studies [54]. The Raman signature of larger clusters was also recorded, but a definite assignment was not possible. The knowledge of the structures of larger group 13 metal clusters has to rely in most parts on quantum chemical calculations. Owing to computational restrictions, only DFT methods could be applied (with the problems discussed above). Nevertheless, they yield interesting results, for example, for large aluminum clusters [55].

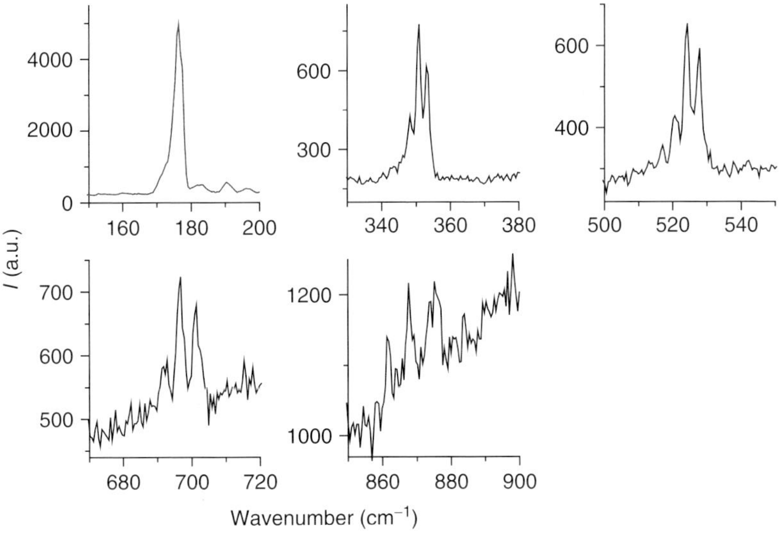

Figure 2.3 Resonance Raman spectra of Ga_2 in an Ar matrix (measured with an Ar^+-ion laser at the excitation wavelength of $\lambda = 514$ nm). The blue Raman signal is assigned to the fundamental vibrational transition.

Resonance Raman spectra were also obtained for matrix-isolated Ti_2 [56, 57]. In this case, it was possible to measure signals from the fundamental and seven overtones. Again, the isotopic pattern clearly proved that all signals belonged to Ti_2. The spectra recorded in different noble gas matrixes indicated the formation of a complex between Ti_2 and Xe, and also a weak dependence of the Ti–Ti force constant on the matrix host. In an Ar matrix, a force constant of 233 N m^{-1} was derived [57]. The dissociation energy was estimated with the aid of a formalism introduced by Le Roy, Bernstein, and Lam [58], and on the assumption of dominating quadrupole–quadrupole interactions near the dissociation limit. A value of 114 kJ mol^{-1} resulted [57].

Not only the vibrational spectra but also the electronic absorption and in some cases also emission spectra of clusters were intensively studied. In many of these studies, quite broad and unstructured bands in the UV/vis region were, often tentatively, assigned to clusters. In the case of matrix-isolated Pt_2, electronic absorption [59] and luminescence spectra [60] were recorded. Matrix-isolated Co_2 dimers show a number of electronic absorptions, with the lowest energetic one appearing already at 551.0 cm^{-1} [61]. Unfortunately, it was not possible to identify the excited electronic states involved in these transitions. Electronically excited states of Ti_2 energetically close to the ground state were intensively studied. Using a combination of experimental and quantum chemical results, detailed information about some excited states was obtained [62]. IR spectra reveal several energetically low-lying electronic transitions. Within the region 4000–6000 cm^{-1}, transitions into two different electronic states were recorded. In combination with multireference configuration interaction calculations, they could be identified as $^3\Pi_u \leftarrow {}^3\Delta_g$ and $^3\Phi_u \leftarrow {}^3\Delta_g$ excitations. A Frank–Condon analysis (assuming Morse potentials) yields for the $^3\Pi_u$ and $^3\Phi_u$ terms changes of the distances of 9 and 10 pm, respectively, with respect to the ground state, in good agreement with the calculated values (1.954, 2.127, and 2.136 Å for the $^3\Delta_g$, $^3\Pi_u$, and $^3\Phi_u$ terms). In the region 7000–9000 cm^{-1}, another two transitions could be assigned to further $^3\Pi_u \leftarrow {}^3\Delta_g$ and $^3\Phi_u \leftarrow {}^3\Delta_g$ excitations. Then an intense transition between 9000 and 10 000 cm^{-1} was assigned to a $^3\Delta_u \leftarrow {}^3\Delta_g$ excitation. Hence, between 4000 and 10 000 cm^{-1} altogether five electronic transitions could be analyzed and assigned. The investigation of the vibrational progressions also allowed the determination of the harmonic frequencies and the anharmonicity constants of the excited states [62].

Although there have been already a number of theoretical studies on Ti_2, Ti_2 and larger clusters are still the topic of actual theoretical work [18, 63]. Other metal atom dimers and clusters also are currently being studied. Whereas by DF methods comprehensive studies of a whole period of transition metals exist (but usually treating only a small number of states), wave function-based studies mostly are confined to one species only (see, for another example, work on V_2 [64]). This emphasizes the high demand of such calculations. Owing to its weak metal–metal bond (dispersion binding), Mn_2 is especially challenging. Accurate results have been obtained by combining MRCI results for the states with lower spin multiplicity with accurate coupled-cluster calculations (CCSD(T)) for the high-spin state,

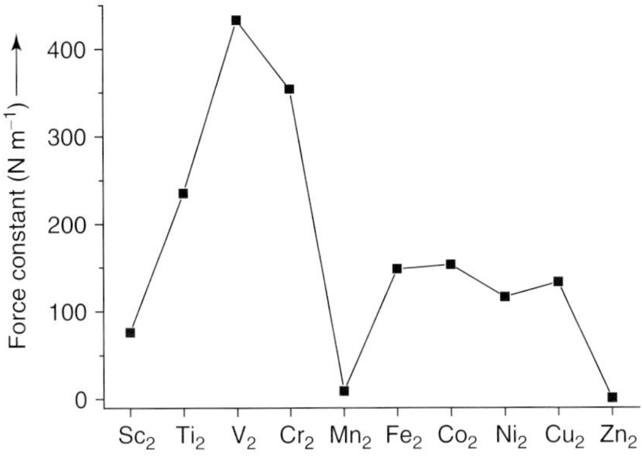

Figure 2.4 Trend of the force constants of 3d metal dimers.

which is a single-reference state [65]. Also, in the example of Mn_2, the notable failure of the (computationally less demanding) complete active space perturbation theory (CASPT2) methods for such systems has been shown [65, 66]. Discussions on the trends in the properties of metal atom dimers are limited. In Figure 2.4, the force constants of the transition-metal dimers of the 3d metals are plotted (the values were taken from Ref. [15]). Factors that are of importance to understand the trends are the atom sizes and the metal atom configurations. Many metal atoms exhibit d^ns^2 ground-state configurations with a filled s valence orbital, which are unsuitable for covalent bonding. For bonding, at least one of the atoms has to be promoted to the $d^{n+1}s^1$ configuration [15]. The possibility to correlate the force constant with the dissociation energy is discussed in Ref. [15].

2.4 Reactions of Atom Dimers or Clusters

Gas-phase reactions of transition-metal dimers and clusters with small molecules were reviewed by Knickelbein [2]. For example, silver clusters were reacted with ethylene, and IR photodissociation spectra of the resulting ethylene complexes were recorded. The ionization potentials of the clusters decreased monotonically with the number of ethylene molecules in $Ag_n(C_2H_4)_m$ ($n=3-7$, $m=1-4$) [67]. FT-ICR mass spectrometry was used extensively to study reactivities of charged metal atom dimers and clusters in the gas phase. Especially, Böhme and Schwarz [32] established complete catalytic cycles for charged atoms or clusters. For example, the reactivities of Rh_n^+ clusters (reaction with CH_4) [68], Pt_n^+ clusters (reaction with CH_4 and NH_3) [69], and anionic Au_n clusters (reaction with CO) [70] were studied. Some studies suggest that "the charge has only minor influence, and that in most cases the charged species exhibit very similar reaction channels as the corresponding neutrals" [71]. However, other studies indeed found differences. For example, V_n^+, Nb_n^+, and Ta_n^+ clusters with $n \leq 4$ are almost unreactive toward CO, but for $n > 4$ the formation of a carbonyl complex is abruptly observed. In contrast, the negatively charged clusters V_n^-, Nb_n^-, and Ta_n^- show little reactivity up to $n = 14-15$, and only larger clusters are much more reactive [71]. Unfortunately, only a few studies exist on the chemistry and its size dependence of mass-selected and neutralized clusters isolated in matrixes, so that a comprehensive comparison between gas-phase and matrix reactivities, especially a comparison between charged and neutral species, has not really been possible to date [72]. A few published studies deal with reactions between small metal atom clusters and CO (e.g., Fe_4 and CO) [73] or ethylene [72].

Interestingly, gas-phase studies and matrix isolation studies showed that metal atom dimers, charged or neutral, exhibit a special reactivity. Without ligands, Rh_2^+ is suitable for the dehydrogenation of CH_4. Using Pt_2^+ but not Pt^+ or other cationic Pt_n clusters, the dehydrogenation of NH_3 succeeds [69]. In the following, we discuss in more detail the striking reactivity of Ga_2 and Ti_2 as revealed in matrix isolation studies. The matrix-isolated metal atom dimer Ga_2 ($^3\Pi_u$ ground electronic state) [52] adds H_2 to give the cyclic Ga(I) compound Ga(μ-H)$_2$Ga (D_{2h} symmetry, see Scheme 2.1) [74–76]. The reaction barrier was estimated from experiments to be about 30 kJ mol^{-1}. This result stimulated theoretical studies on the thermodynamics of hydrogen addition to E–E bonds (E = main-group elements) [77]. Work on a preparative scale showed that digermynes [78] with bulky substituents (Ar′GeGeAr′, Ar′ = C_6H_3-2,6(C_6H_3-2,6-iPr$_2$)$_2$) indeed react already at room temperature with H_2, leading to a mixture of hydrogenation products (Ar′(H)GeGe(H)Ar′, Ar′H_2GeGeH_2Ar′, and Ar′GeH_3), the relative concentrations of which can be controlled by the H_2 offer. Distannynes were also shown to add H_2 [79], and the structure of the product is governed by the size of the ligands [80]. Further matrix-isolation studies revealed that Ga_2 also reacts with SiH_4 to give Ga(μ-SiH_3)GaH [81], and with SnH_4 to give a nido-cluster compound Ga_2Sn(μ-H)$_4$ (see Scheme 2.1) [82].

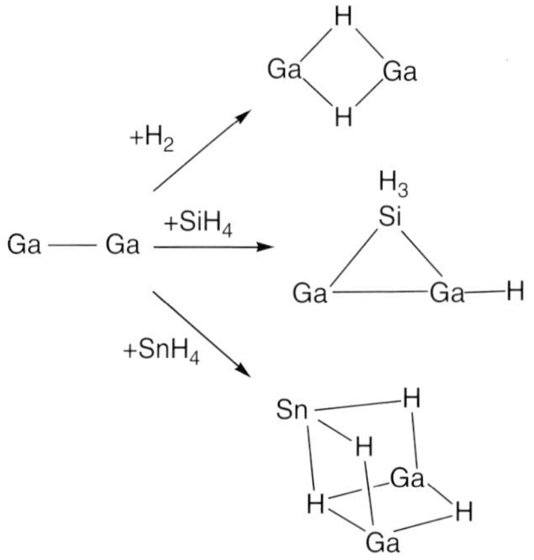

Scheme 2.1 Reactions of Ga_2 in matrixes.

Ti_2 dimers show an even more striking reactivity. They cleave the strong NN triple bond of dinitrogen in a concerted manner practically with no activation barrier to give the dinitrido species $Ti(\mu\text{-}N)_2Ti$ [83, 84]. By contrast, Ti atoms in their electronic ground state (^3F) interact repulsively with N_2. The product is a D_{2h} symmetric $Ti(\mu\text{-}N)_2Ti$ molecule which can be considered as the smallest building block of solid TiN. The finding of combination bands in addition to the three IR active fundamental vibrations allowed the determination of all fundamental frequencies of this symmetric molecule. Using data measured for the different isotopomers, the N–Ti–N bond angle was estimated to be 93°, in good agreement with the value of 89° from quantum chemical calculations (MRCI). The molecule possesses a singlet electronic ground state (^1A_g) with two unpaired electrons located at the Ti atoms. The first triplet state, $^3B_{1u}$, has an energy higher by 0.26 eV. Hence, the two electrons are coupled antiferromagnetically via the N bridges. Also, transitions to excited electronic states of $Ti(\mu\text{-}N)_2Ti$ were observed in the absorption spectrum. In the region 7500–9500 cm^{-1}, transitions with vibrational progressions occurred [84], which could be assigned with the help of quantum chemical calculations. The dinitrido compound $Ti(\mu\text{-}N)_2Ti$ reacts further with N_2. With increasing N_2 concentrations in Ne matrixes, several new species were detected with different degrees of N_2 activation. Finally, for very high N_2 concentrations or in pure N_2 matrixes, the nitrogen-rich product $[\{(N_2)_4Ti\}_2(\mu\text{-}N)_2(\mu\text{-}\eta^2\text{:}\eta^2\text{-}N_2)]$ is formed [85]. For comparison, the anionic complex $[\{[(Me_3Si)_2N]_2Ti\}_2(\mu\text{-}\eta^2\text{:}\eta^2\text{-}N_2)_2]^-$ (from single-crystal X-ray diffraction) [86] features two $\mu\text{-}\eta^2\text{:}\eta^2$-coordinated N_2 units rather than one in $[\{(N_2)_4Ti\}_2(\mu\text{-}N)_2(\mu\text{-}\eta^2\text{:}\eta^2\text{-}N_2)]$. These results motivate the use of titanium clusters stabilized on surfaces or inside nanopores as catalysts for ammonia formation. Titanium-coated carbon nanotubes and fullerenes were already discussed as possible hydrogen storage materials [87]. In theoretical studies, the possible anchoring of Ti_2 to C_{60} was studied [88]. Moreover, Kiwi and coworkers [89] studied the reaction of Ti_2 with C_{20} and calculated several possible product structures. Conversion of N_2 and H_2 into NH_3 with Ti_2C_{60} as catalyst was also analyzed [88].

The reaction of Ti_2 with CO was studied by Manceron *et al.* [90]. Reaction of ground-state Ti_2 at 9 K proceeds first to the carbonyl complex $Ti_2(CO)$. However, this complex rearranges during deposition or following electronic excitation into the dititanium oxycarbide OTi_2C. This is an extremely unusual reaction, and was not observed for any other metal atom dimer. Hence, both Fe_2CO [91] and Co_2CO [92], which were formed by matrix reaction of the metal atom dimers with CO, are carbonyl complexes that show no sign of rearrangement under cleavage of the strong CO bond [93]. In this context, it is worth mentioning that very recent quantum chemical calculations suggest that the trimer Nb_3 can cleave the CO bond (exothermic reaction and relatively low reaction barriers of the steps leading from the CO complex to the product) [94]. Scheme 2.2 summarizes the results of the matrix experiments on Ti_2, showing that this dimer is capable of cleaving the strongest chemical bonds virtually without any activation energy.

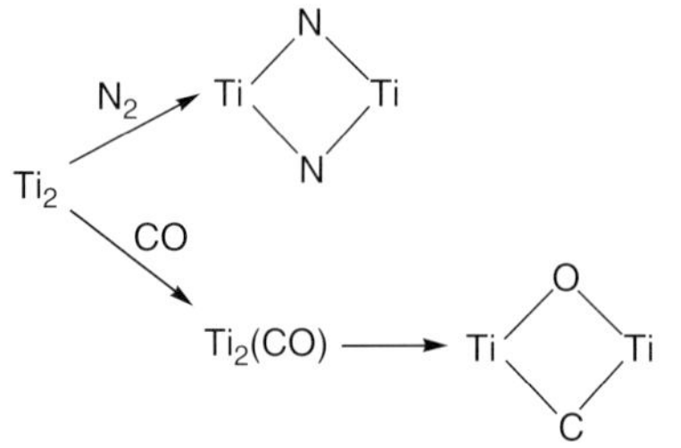

Scheme 2.2 Reactions of Ti_2 in matrixes.

The density of electronic states with energies close to that of the electronic ground state is generally the reason for the low activation energy and selectivity of metal cluster reactions. By contrast, high symmetry barriers often prevent metal atom reactions. This was shown explicitly for the reaction between Ga atoms and H_2 to give the bent radical hydride GaH_2. Experiments have shown that this reaction does not proceed thermally but only after photolytic excitation of the Ga atom ($^2S \leftarrow {}^2P$ or $^2D \leftarrow {}^2P$ excitation) [95]. According to MP2 calculations [96], the reaction is associated with an energy of $-13\,kJ\,mol^{-1}$, and therefore a significant reaction barrier for reaction from the ground state of Ga (^2P) must account for the experimental findings. Indeed, calculations have found a massive symmetry barrier, and suggested that a two-step formation pathway leading first to GaH and H atoms, which then recombine to the GaH_2 product, was preferred for the thermal reaction [96]. In the case of the reaction of Ga_2 ($^3\Pi_u$) with H_2 to give Ga_2H_2 ($^1A_1$ ground state), an intersystem crossing has to occur during the approach from a triplet energy hypersurface to a singlet one. Nevertheless, the thermal barrier for this reaction is small. Calculations using MRCI methods predict a barrier height of about $50\,kJ\,mol^{-1}$ [96]. The reaction mechanism was reinvestigated by Moc [97] using coupled-cluster methods, which showed a slightly lower activation barrier which is now in excellent agreement with the experimental estimate. One could assume in a rough approximation that the Ga_2 dimer is promoted during the $Ga_2 \cdots H_2$ approach from the $^3\Pi_u$ ground state to the $^1\Sigma_g{}^+$ state, which lies only slightly higher in energy [96]. The $^1\Sigma_g{}^+$ state then reacts with only a very small barrier with H_2. Of course, in the case of transition metals, the density of states is already high for the atom (see, e.g., the discussion in Ref. [98]). But still, it is higher for the dimers, as for example, shown in the case of Ti_2 [62].

2.5 Formation and Characterization of Metal Oxides

Metal oxides were also already intensively studied with the aid of gas-phase and matrix-isolation techniques. A comprehensive review article dealing with small transition-metal oxides and also their reactivity appeared in 2009 [16]. In this review article, both matrix-isolation and gas-phase studies were considered.

Studies of oxides containing only one main-group element atom were reviewed in 2002 [21]. The electronic structure of the diatomic MO molecules was reviewed in 2000 [98]. Finally, in a review article in 2001, the bonding in neutral and cationic transition-metal monoxides was compared [99]. Most matrix-isolation studies concentrate on small molecules containing one or two metal atoms. The reason for this restriction is that small oxides can be better generated in larger amounts. In matrix studies, the oxides are often the products of a reaction between metal atoms or dimers and dioxygen or other oxygen sources (O atoms generated in a microwave discharge, O_3, or N_2O). In very special cases, larger oxide clusters could be brought into the gas phase and then isolated in matrixes. An example is V_4O_{10} [100]. In the case of compounds with only one metal atom of the formula MO_n, those with a large number of oxygens (high formal oxidation number of the metal) were the focus of research in the past years. Generally, these species turned out to be side-on- or end-on-bonded O_2 complexes. An exception is the compound IrO_4, which could be described as an iridium tetraoxide, in which an extremely high oxidation number of Ir, namely +VIII, is realized [101]. Two antisymmetric Ir–O stretches of this molecule were observed at 870.5 and 859.5 cm^{-1}. In the case of cobalt, the series of compounds CoO_2 [102], CoO_4 [103], and CoO_6 [104] was characterized as products of the barrierless reaction of cobalt atoms in their ground state with one, two, or three dixoxygen molecules. CoO_6 can be described as a disuperoxo cobalt dioxide, $CoO_2(\eta^1\text{-}O_2)_2$ [104]. In these experiments, Co atoms were generated thermally. In contrast to IrO_4, RhO_4 does not form a rhodium tetraoxide. Instead, IR bands from the three species $[(\eta^1\text{-}O_2)RhO_2]$, $[(\eta^2\text{-}O_2)RhO_2]$, and $[Rh(\eta^2\text{-}O_2)_2]$ were sighted [105]. CrO_6 was identified as a side-on-bonded disuperoxo chromium dioxide, $CrO_2(\eta^2\text{-}O_2)_2$ [106], and CuO_6 as a trisuperoxo complex $Cu(\eta^2\text{-}O_2)(\eta^1\text{-}O_2)_2$, featuring both side-on- and end-on-bonded O_2 units [107]. Al atoms could bind up to three O_2 molecules. The vibrational spectra and quantum chemical calculations indicate that AlO_6 can be described as a tris-superoxo complex $Al(O_2)_3$ featuring three side-on-bonded O_2 units (D_3 symmetry) [108].

Not all metal atoms react in their ground state with O_2, and if they react, the structure of the product varies [16]. An interesting reaction is that between Ni atoms and O_2 leading to NiO_2. Two isomers of this product are formed, namely a cyclic and a linear one [20, 109]. The ratio between the two products depends on the matrix gas. Hence, in Ne matrixes linear ONiO dominates, while in Ar or CH_4 matrixes the cyclic form, which could be described as a side-on-bonded superoxo complex, $Ni(\eta^2\text{-}O_2)$, is the leading species (see Figure 2.5). Quantum chemical calculations (MRCI) yield for the linear ONiO molecule a $^1\Sigma_g^+$ ground term and, furthermore, $^3\Pi_g$, $^1\Pi_g$, and $^5\Pi_u$ excited states at 0.55, 0.95, and 1.21 eV. For the cyclic $Ni(O_2)$, a $^1A_1$ ground term is found, and there are $^5B_2$, $^5A_1$, $^3B_2$ excited states at 0.58, 0.62, and 0.73 eV. The $Ni(O_2)$ ground term is located 1.53 eV above the linear ONiO. Both isomers are separated by a barrier of 2.76 eV with respect to ONiO, and for the separated systems Ni + O_2, an energy just 0.08 eV below the barrier is obtained. Hence, ONiO is the lower lying isomer and, from a thermodynamic point of view, should essentially be formed. Nevertheless, according to

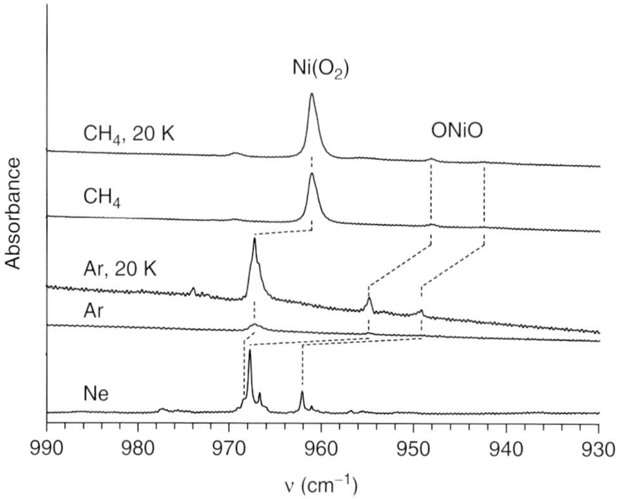

Figure 2.5 IR spectra of NiO_2 isomers in different matrixes (recorded at 4 K in Ne, and both at 4 K and after annealing to 20 K in Ar and CH_4). (For details see Ref. [20].)

the calculations, there is a small barrier (0.08 eV) to be surmounted to generate ONiO from the separated subsystems. But since the correlation treatment within the calculations is unfortunately incomplete, it is likely that a more complete correlation treatment will raise the energy of the separated systems with respect to the ONiO and $Ni(O_2)$ molecules and the transition structure. Therefore, both isomers in principle could be formed, and one should expect the lower lying ONiO isomer to be formed predominantly. This straightforwardly explains the finding within the Ne matrix. The fact that within the Ar and CH_4 matrixes the main product is not ONiO but $Ni(O_2)$ can be understood on the basis of the larger mass of Ar or the internal degrees of freedom of CH_4, because these properties allow the matrix host to more easily absorb the excess internal energy of the initially generated $Ni(O_2)$ and thereby to stabilize the higher lying isomer.

Another example, this time from main-group element chemistry, is the reaction between Ga or In atoms and O_2. This reaction again leads first to the cyclic superoxo complexes $M(O_2)$ with a $^2A_2$ electronic ground state [110]. The $M(O_2)$ complexes can be photoisomerized to the linear oxides OMO ($^3\Pi_g$), which are valence-isoelectronic to the known linear cation CO_2^+. Quantum chemical calculations indicate that the linear isomer is more stable than the superoxo complex [110]. However, multireference calculations are required; Hartree–Fock calculations give a significantly different picture. In contrast to the situation for NiO_2, the thermal barrier for isomerization from the superoxo complex to the linear dioxide is very high (about 300 kJ mol^{-1}), in line with the experimental observation that the linear molecule is formed only upon photolysis of the superoxo complex. A similar situation was observed in the case of TlO_2 [111]. In contrast to GaO_2, InO_2, and TlO_2, BO_2 is exclusively formed as linear dioxide OBO, when B atoms and O_2 are reacted together [35]. Interest in OBO and its homolog OAlO was spurred by their high electron affinities which exceed those of the halogen atoms. Therefore OBO and OAlO were denoted “superhalogens” [112]. Linear GaO_2 reacts further with O_2 to give the dioxygen complex $GaO_2(O_2)$ [113]. The structure of this complex is still not clear [114].

Alternatively to the reaction with O_2, other oxygen sources could be used. For example, oxygen atoms could be generated by microwave discharge. One of the best studied simple main-group element oxides is silicon monoxide. It can be produced in the gas phase by heating quartz to temperatures above 1500 K. However, larger quantities are obtained in the matrix by reaction of silicon atoms with oxygen atoms produced by microwave discharge. SiO shows a rich chemistry, and reacts further with O atoms to linear SiO_2 [115, 116] and with O_2 to give $OSi(O_2)$ with C_{2v} symmetry [117]. Finally, dimerization yields Si_2O_2 with a cyclic structure; trimerization was also observed [118, 119]. Alternatively, oxygen atoms could be produced from a precursor, especially N_2O, which liberates oxygen atoms in collision experiments in the gas phase or under the action of irradiation (see the equation below). For example, Bondybey *et al.* [120] studied the gas-phase reaction between N_2O and Pt_n clusters yielding different oxides.

$$M + N_2O \rightarrow MO + N_2$$

Oxides with two metal atoms are formed either in dimerization reactions, for example, between two MO molecules, or as products of matrix reactions between metal atom dimers and dioxygen. Such reactions could proceed either from the electronic ground state of the metal atom dimers or from their excited electronic states. It is often difficult to find out whether the ground state or an excited state of the dimer is involved, since the electronic states could be energetically close by. In the case of the generation of Co_2O_2, the reaction from the electronic ground state of Co_2 is presumably spin-forbidden, and experiments indicate that excited states of Co_2 are involved [121]. The product was described as the dioxo compound $Co(\mu\text{-}O)_2Co$, but this species could be reversibly photoconverted into two metastable Co_2O_2 species. The electronic structure of $Co(\mu\text{-}O)_2Co$ was studied in detail with the help of multireference methods [122], and a series of electronic states were found within an energy window of only $1000\ cm^{-1}$.

It is of interest to compare the bonding in neutral and cationic metal oxides. For the metal monoxides MO, such a comparison is possible because they were characterized in a number of studies. Figure 2.6 compares the bond dissociation energies for neutral and cationic metal monoxides of the first transition-metal period. It can be seen that, although the overall trend is similar, there are some significant differences. Except for ScO, TiO, and ZnO, the cations exhibit lower dissociation energies than their neutral counterparts. Especially large differences were found for NiO and CuO. Hence, the dissociation energy of CuO is 2.75 ± 0.22 eV, while that of CuO^+ is only 1.62 ± 0.15 eV. The lower dissociation energy of the cationic species might translate into a higher reactivity, in line with results discussed in the next section.

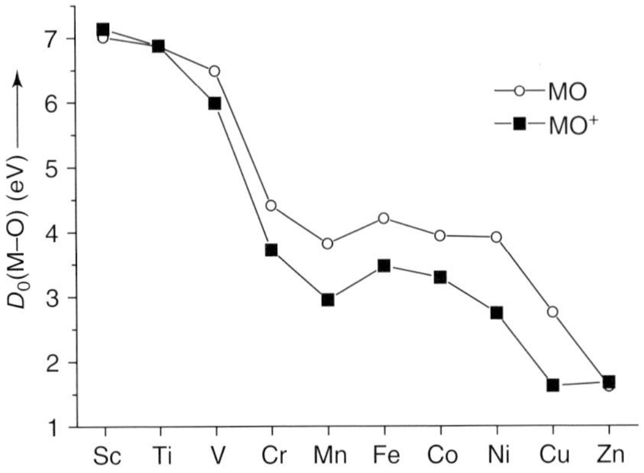

Figure 2.6 Comparison between the 0 K bond dissociation energies D_0(M−O) for neutral and cationic metal monoxides. (Data from Tables 5 and 6 in Ref. [99].)

For the example of Ni_2O_2, a detailed comparison of the neutral and cationic systems has been made on the basis of multireference configuration interaction calculations [123]. Neutral Ni_2O_2 has a ^1A_g electronic ground term. At energies lower than 1 eV, there are at least 21 further terms of singlet, triplet, and quintet multiplicity (see Figure 2.7). These terms can be grouped into Heisenberg spin ladders with three terms (singlet, triplet, and quintet) according to the spin coupling of

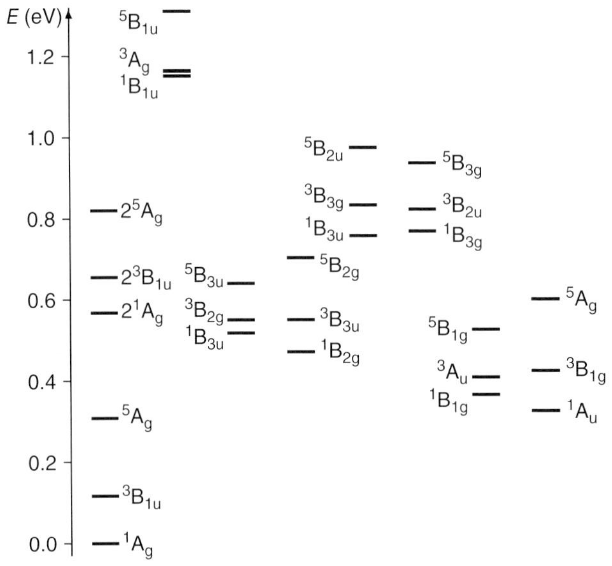

Figure 2.7 Relative term energies for Ni_2O_2 by Davidson corrected MRCI calculations at the structure of the ^1A_g ground term.

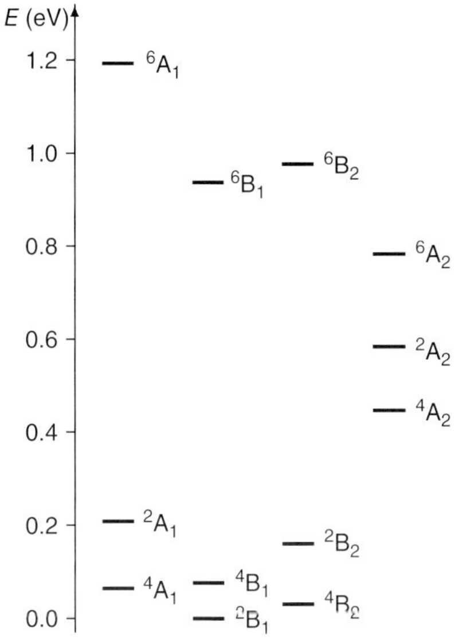

Figure 2.8 Relative term energies for $Ni_2O_2^+$ by Davidson corrected MRCI calculations at the structure of the $^2B_1$ ground term.

two $S = 1$ subsystems. The calculations clearly point to a system composed of two high-spin Ni atoms in the +II oxidation state and two O atoms in the −II oxidation state. The different spin ladders are just a consequence of different occupation patterns for the Ni 3d orbitals. The structure of Ni_2O_2 has D_{2h} symmetry with Ni–O distances of 1.789 Å and a Ni–Ni distance of 2.289 Å. $Ni_2O_2^+$ has a $^2B_1$ electronic ground term (see Figure 2.8). Low-lying excited states have doublet and quartet multiplicities; there are no sextet terms below about 0.6 eV. Its structure has only C_{2v} symmetry with two nonequivalent Ni atoms. The Ni–O distances are 1.706 and 1.826 Å. Concerning $Ni_2O_2^+$, one could think that compared to Ni_2O_2 one electron is removed from the metal centers and both Ni^{II} and Ni^{III} are present. Although this is formally correct concerning the assignment of oxidation states, namely by imposing −II to the oxygen atoms as the more electronegative atoms, the inspection of the orbitals within the actives space leads to a different conclusion. There are two orbitals (with occupation numbers of 1.7 and 0.3) that are bonding and antibonding combinations of O 2p orbitals and one Ni 3d orbital at one of the centers. Hence, in principle, one electron is removed from one of the O 2p orbitals, and this orbital then forms a bonding orbital with one Ni 3d orbital. The substantial occupation of the corresponding antibonding orbital indicates that there is radical character also at the O atoms. Therefore, comparing Ni_2O_2 and $Ni_2O_2^+$, for the cation a higher electronic flexibility and higher reactivity has to be expected, as a result of the removal of one electron out of the O 2p orbitals and a partial radical character at the O atoms. We will discuss this point further in the next section on the reactivity.

2.6 Reactions Involving Metal Oxides

The reactivity of metal oxides was predominantly assessed in the gas phase. Many of these studies are devoted to reactions with H_2O, H_2, or hydrocarbons. Not only small oxides but also larger clusters, in particular the cation $V_4O_{10}^+$ [124], were studied. The cations $[MO_2H_2]^+$ can be regarded as the products of reaction between MO^+ and H_2O. DFT calculations suggest that for M = Fe the dihydroxide $[Fe(OH)_2]^+$ is more stable than $[OFe(H_2O)]^+$, whereas in the case of M = Ni the $[ONi(H_2O)]^+$ isomer is preferred over $[Ni(OH)_2]^+$ [125]. Similar experiments and calculations are devoted to metal oxide anions. For example, gas-phase reactions between $[Co_nO]^-$ ($n = 1-3$) and H_2O yielded the dihydroxide anions $[Co_n(OH)_2]^-$ [126]. The reaction between FeO^+ and H_2 to give Fe^+ and H_2O is one of the most thoroughly studied examples of gas-phase reactions involving oxides [30]. The reaction is highly exothermic, but nevertheless proceeds only slowly. Mechanistic studies show that the barrier for thermal reaction is very high (although the reactants and product have the same spin multiplicity), and the reaction can proceed only by changing, during the course of the reaction, to an energetically lower lying hypersurface of different spin ("spin-accelerated reaction") [127]. Interestingly, gas-phase kinetic studies indicate that the neutral FeO is unreactive toward H_2 [128]. So far, only a few matrix-isolation experiments on the reactivity of metal oxides have been carried out. A few recent studies probe the possibility to cleave dihydrogen heterolytically by neutral metal oxides. TaO_4 reacts with H_2 to give the hydride complex $HTaO(OH)(\eta^2\text{-}O_2)$ [129]. In another work, group V metal dioxides were reacted with dihydrogen [130]. One of the results is that VO_2 in its electronic ground state ($^2A_1$) reacts without significant barrier with H_2 to give HVO(OH) whereas D_2 forms only a complex $VO_2(\eta^2\text{-}D_2)$. The product DVO(OD) could be formed only upon photolysis. There are only very few matrix-isolation studies dealing with the reaction of metal oxides with alkanes. Matrix-isolated thorium oxide, ThO, reacts with methane upon UV irradiation to give $CH_3Th(O)H$ [131]. In addition, methanol is formed. The results might suggest that methane could be converted to methanol by thorium monoxide.

Dehydrogenation of hydrocarbons by cationic metal oxide clusters was studied in several experimental gas-phase and theoretical works. Rozanska and Sauer [132] used DFT methods to establish the mechanism of the oxidative dehydrogenation of hydrocarbons by $V_3O_7^+$ (the smallest monocationic polynuclear closed-shell V^V species). Schwarz *et al.* [133] studied the gas-phase dehydrogenation of alkanes with $Ni_2O_2^+$ (see reaction below).

$$Ni_2O_2^+ + C_nH_{2n+2} \rightarrow Ni_2O_2H_2^+ + C_nH_{2n}$$

In a theoretical work, we compared cationic $Ni_2O_2H_2^+$ [134] with $Ni_2O_2H_2$ [135], which are the (potential) products of the reaction of $Ni_2O_2^+$ and Ni_2O_2 with H_2 or with alkanes. They also could be regarded as products of the $NiO/NiO^+ + H_2O$ reaction. The analysis of these systems completes the comparison of neutral and cationic nickel oxides. The lowest lying isomer of $Ni_2O_2H_2^+$ has a rhombic

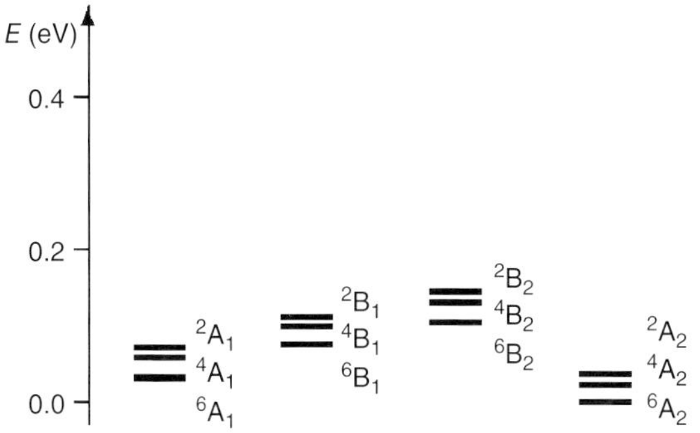

Figure 2.9 Relative energies of different terms of $Ni_2(OH)_2^+$ at the structure of the $^6A_2$ ground terms.

shape with two OH groups that bridge the Ni atoms, denoted in the following as $Ni_2(OH)_2^+$. At a small relative energy of 0.29 eV, there is a chainlike isomer, $NiONi(OH_2)^+$, with an O atom bridging two Ni atoms and a water ligand. Other isomers with the H atoms bound to the Ni atoms, $(NiH)_2O_2^+$, have considerably higher energies of more than 4 eV with respect to the ground state. Figure 2.9 shows the relative energies of the lowest lying states of $Ni_2(OH)_2^+$. The ground state is a $^6A_2$ term with a C_{2v} structure with nonequivalent Ni atoms; the Ni–O distances are 1.816 and 2.029 Å. Within 0.2 eV of the ground state energy, there are at least 11 further terms of sextet, quartet, and doublet multiplicity. These terms form Heisenberg spin ladders, with the lowest states being sextet states, corresponding to the ferromagnetic coupling of local $S = 1$ and $S = 3/2$ systems. The inspection of the natural orbitals reveals that, besides singly occupied 3d orbitals, there is one singly occupied 4s orbital, localized at the Ni atom with larger Ni–O distances. Hence, very clearly the system is composed of one Ni in the oxidation state +I and one Ni in the oxidation state +II.

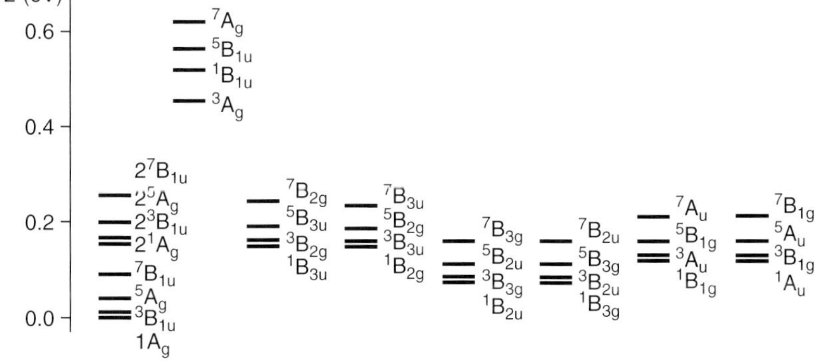

Figure 2.10 Relative energies of different terms of $Ni_2(OH)_2$ at the structure of the ^1A_g ground term.

The lowest lying isomer of neutral $Ni_2O_2H_2$ again has a rhombic shape with two bridging OH groups, $Ni_2(OH)_2$. Then there is also a chain-like isomer with a ligated water molecule, $NiONi(OH_2)$, now at an energy higher by 1.06 eV. Similarly, an isomer with the H atoms bound to Ni has a considerably higher energy of 2.93 eV. Figure 2.10 shows the relative energies of the low-lying terms of $Ni_2(OH)_2$. The ground term is a ^1A_g term. Its structure has D_{2h} symmetry, with Ni–O distances of 1.939 Å. Up to an energy of 0.26 eV, there are at least 32 terms, namely singlet, triplet, quintet, and septet terms, that form Heisenberg spin ladders with four levels corresponding to the coupling of two subsystems with $S = 3/2$. This is in agreement with the presence of Ni within the Ni^I oxidation state with occupied 4s orbitals.

Hence, both neutral and cationic $Ni_2(OH)_2$ are characterized by many very low lying electronic states including different spin multiplicities that form Heisenberg spin ladders. But, for the cations the two lowest-lying isomers, ($Ni_2(OH)_2^+$ and $NiONi(OH_2)^+$), are quite close in energy, 0.3 eV, whereas in case of the neutrals the difference is about 1.1 eV. Thus, for the cation it is more likely that different isomers play a role. In particular, one could think of larger structural flexibility. Of course, this would essentially be determined not by the relative energies of the isomers but by the barriers between them; maybe with the lower relative energy of $NiONi(OH_2)^+$ there is also a lower barrier for the conversion. Furthermore, for $NiONi(OH_2)^+$ a relatively small 3d–4s excitation energy within the molecule was found, that could point to a larger electronic flexibility.

A system composed of Ni_2O_2 and H_2 with a large distance between the subsystems has been considered to get estimates for the reaction energy. For the reactions $Ni_2O_2^{0/+} + H_2 \rightarrow Ni_2(OH)_2^{0/+}$, estimates of about 2 eV for the reaction of neutral Ni_2O_2 and 3.5 eV for the reaction of the cation have been obtained. These values very clearly support the assumption of a higher reactivity of the cation, inferred already from the electronic structure.

2.7 Concluding Remarks

Much more work, experimental (in the gas phase and using matrix isolation) as well as theoretical, is necessary to obtain a detailed picture of the electronic structures and reactivities of small metal atom clusters and metal oxides. Such work will assist in the understanding of the mechanism of the reactions taking place at the surfaces of heterogeneous catalysts ("molecular surface science"). However, the reactivities of the clusters greatly depend on their size, for various reasons, which we briefly discussed in this chapter. Metal atom dimers exhibit special reactivity; they are much more reactive than metal atoms. The access to energetically low-lying excited states is generally the reason for the high reactivity and selectivity. Therefore, a detailed analysis of the electronic structure not only of the electronic ground state but also of excited states is essential to anticipate the chemistry of metal atom dimers and clusters. Such analysis is possible only by a combination of experimental and calculational work. The restrictions of theoretical studies were briefly discussed. A comparison between the reactivities of charged and neutral species on the basis of a broad dataset is still not possible. Nevertheless, several studies indicate that the charge has a significant effect, which is stronger for smaller clusters. Work by us and others suggests that metal oxide cations are more reactive than their neutral counterparts.

Acknowledgments

We gratefully acknowledge the financial support for this project by the DFG (Deutsche Forschungsgemeinschaft) though the Collaborative Research Center SFB 623.

References

1. (a) Ertl, G. (1983) *J. Vac. Sci. Technol. A*, **1**, 1247–1253.(b) Schlögl, R. (2003), *Angew. Chem.*, **115**, 2050–2055, *Angew. Chem. Int. Ed.*, **42**, 2004–2008.
2. Knickelbein, M.B. (1999) *Annu. Rev. Phys. Chem.*, **50**, 79–115.
3. (a) Morse, M.D., Geusic, M.E., Heath, J.R., and Smalley, R.E. (1985) *J. Chem. Phys.*, **83**, 2293–2304; (b) Leuchtner, R.E., Harms, A.C., and Castleman, A.W. Jr., (1990) *J. Chem. Phys.*, **92**, 6527–6537; (c) Winter, B.J., Parks, E.K., and Riley, S.J. (1991) *J. Chem. Phys.*, **94**, 8618–8621; (d) Andersson, M., Persson, J.L., and Rosén, A. (1996) *J. Phys. Chem.*, **100**, 12222–12234.
4. Burgert, R., Schnöckel, H., Grubisic, A., Li, X., Stokes, S.T., Bowen, K.H., Ganteför, G.F., Kiran, B., and Jena, P. (2008) *Science*, **319**, 438–442.
5. Schnöckel, H. (2010) *Chem. Rev.*, **110**, 4125–4163.
6. (a) Schwarz, H. (2004) *Int. J. Mass Spectrom.*, **237**, 75–105; (b) Poli, R. and Harvey, J.N. (2003) *Chem. Soc. Rev.*, **32**, 1–8.
7. Freund, H.-J., Ernst, N., Risse, T., Hamann, H., and Rupprechter, G. (2001) *Phys. Stat. Sol. (A)*, **1**, 257–274.
8. Heiz, U. and Schneider, W.-D. (2000) *J. Phys. D: Appl. Phys.*, **33**, R85–R102, and references given therein.
9. Ferrari, A.M., Xiao, C., Neyman, K.M., Pacchioni, G., and Rösch, N. (1999) *Phys. Chem. Chem. Phys.*, **1**, 4655–4661.
10. (a) Siehe, Z.B. (1994) in *Cluster and Colloids* (ed. G. Schmid), VCH Publisher, Weinheim, New York; (b) Braunstein, P., Oro, L.A., and Raithby, P.R. (1999) *Metal Clusters in Chemistry*, VCH Publisher, Weinheim, New York; (c) Tran, N.T., Powell, D.R., and Dahl, L.F. (2000) *Angew. Chem.*, **112**, 4287–4291; *Angew. Chem. Int. Ed.*, (2000), **39**, 4121–4125; (d) Schnepf, A. and Schnöckel, H. (2002) *Angew. Chem.*, **114**, 3682–3704; *Angew. Chem. Int. Ed.*, (2002), **41**, 3532–3554; (e) Schnöckel, H., Schnepf, A., Whetten, R.L., Schenk, C., and Henke, P. (2011) *Z. Anorg. Allg. Chem.*, **637**, 15–23.
11. Schmid, G., Boese, R., Pfeil, R., Bandermann, F., Mayer, S., Calis, G.H.M., and van der Velden, J.W.A. (1981) *Chem. Ber.*, **114**, 3634–3642.
12. Schmid, G. (2008) *Chem. Soc. Rev.*, **37**, 1909–1930.
13. Boyen, H.-G., Kästle, G., Weigl, F., Koslowski, B., Dietrich, C., Ziemann, P., Spatz, J.P., Riethmüller, S., Hartmann, C., Möller, M., Schmid, G., Garnier, M.G., and Oelhafen, P. (2002) *Science*, **297**, 1533–1536.
14. Morse, M.D. (1986) *Chem. Rev.*, **86**, 1049–1109.
15. Lombardi, J.R. and Davis, B. (2002) *Chem. Rev.*, **102**, 2431–2460.

16. Gong, Y., Zhou, M., and Andrews, L. (2009) *Chem. Rev.*, **109**, 6765–6808.
17. Himmel, H.-J. (2005) *Inorganic Chemistry in Focus II*, Wiley-VCH Verlag GmbH, Weinheim, pp. 1–13.
18. Kalemos, A. and Mavridis, A. (2011) *J. Chem. Phys.*, **135**, 134302-1–134302-8.
19. Polo, V., Kraka, E., and Cremer, D. (2002) *Mol. Phys.*, **100**, 1771–1790.
20. Hübner, O. and Himmel, H.-J. (2012) *J. Phys. Chem. A*, **116**, 9181–9188.
21. Almond, M.J. and Downs, A.J. (1989) *Adv. Spectrosc.*, **17**, 1–12.
22. Andrews, L. and Moskovits, M. (eds) (1989) *Chemistry and Physics of Matrix-Isolated Species*, North-Holland Publishing Company, Amsterdam.
23. Clark, R.J.H. and Hester, R.E. (eds) (1989) *Spectroscopy of Matrix Isolated Species*, John Wiley & Sons, Ltd, Chichester, New York.
24. Bondybey, V.E., Smith, A.M., and Agreiter, J. (1996) *Chem. Rev.*, **96**, 2113–2134.
25. Dunkin, I.R. (1998) *Matrix-Isolation Techniques: A Practical Approach*, Oxford University Press, Oxford.
26. Bondybey, V.E., Räsänen, M., and Lammers, A. (1999) *Annu. Rep. Prog. Chem. Sect. C*, **95**, 331–372.
27. Himmel, H.-J., Downs, A.J., and Greene, T.M. (2002) *Chem. Rev.*, **102**, 4191–4241.
28. Eller, K. and Schwarz, H. (1991) *Chem. Rev.*, **91**, 1121–1177.
29. Parent, D.C. and Anderson, S.L. (1992) *Chem. Rev.*, **92**, 1541–1565.
30. Schröder, D. and Schwarz, H. (1995) *Angew. Chem.*, **107**, 2126–2150; *Angew. Chem., Int. Ed. Engl.*, (1995), **34**, 1973–1995.
31. Bondybey, V.E. and Beyer, M.K. (2001) *J. Phys. Chem. A*, **105**, 951–960.
32. Böhme, D.K. and Schwarz, H. (2005) *Angew. Chem.*, **117**, 2388–2406; *Angew. Chem. Int. Ed.*, (2005), **44**, 2336–2354.
33. Schroeder, W., Grinter, R., Schrittenlacher, W., Rotermund, H.H., and Kolb, D.M. (1985) *J. Chem. Phys.*, **82**, 1623–1630.
34. Infante, I., Andrews, L., Wang, X., and Gagliardi, L. (2010) *Chem. Eur. J.*, **16**, 12804–12807.
35. Burkholder, T.R. and Andrews, L. (1991) *J. Chem. Phys.*, **95**, 8697–8709.
36. Hanner, A.W. and Gole, J.L. (1980) *J. Chem. Phys.*, **73**, 5025–5039.
37. Danset, D. and Manceron, L. (2003) *J. Phys. Chem. A*, **107**, 11324–11330.
38. Zhao, Y., Gong, Y., and Zhou, M. (2006) *J. Phys. Chem. A*, **110**, 10777–10782.
39. Bondybey, V.E. and English, J.H. (1984) *Chem. Phys. Lett.*, **111**, 195–197.
40. Zhou, M. and Andrews, L. (2000) *J. Phys. Chem. A*, **104**, 1648–1655.
41. Himmel, H.-J. and Hebben, N. (2005) *Chem. Eur. J.*, **11**, 4096–4102.
42. The technique was described in: Bondybey, V.E. (1985) *Science*, **227**, 125–131.
43. Merritt, J.M., Bondybey, V.E., and Heaven, M.C. (2009) *Science*, **324**, 1548–1551.
44. Heaven, M.C., Merritt, J.M., and Bondybey, V.E. (2011) *Annu. Rev. Phys. Chem.*, **62**, 375–393.
45. Kornath, A., Ludwig, R., and Zoermer, A. (1998) *Angew. Chem.*, **110**, 1620–1622; *Angew. Chem. Int. Ed.*, (1998), **37**, 1575–1577.
46. Maier, J.P. (1998) *J. Phys. Chem. A*, **102**, 3462–3469, and references given therein.
47. Rechtsteiner, G.A., Felix, C., Ott, A.K., Hampe, O., Van Duyne, R.P., Jarrold, M.F., and Raghavachari, K. (2001) *J. Phys. Chem. A*, **105**, 3029–3033.
48. Kraus, D., Lorenz, M., and Bondybey, V.E. (2001) *PhysChemComm*, **4**, 44–48.
49. Haslett, T.L., Bosnick, K.A., Fedrigo, S., and Moskovits, M. (1999) *J. Chem. Phys.*, **111**, 6456–6461.
50. Bosnick, K.A., Haslett, T.L., Fedrigo, S., and Moskovits, M. (1999) *J. Chem. Phys.*, **111**, 8867–8870.
51. Ozin, G.A. and McIntosh, D.F. (1986) *J. Phys. Chem.*, **90**, 5756–5761.
52. Himmel, H.-J. and Gaertner, B. (2004) *Chem. Eur. J.*, **10**, 5936–5941.
53. Due to a lower resolution of the Raman machine, such a pattern was not visible in earlier studies: Froben, F.W., Schulze, W., and Kloss, U. (1983) *Chem. Phys. Lett.*, **99**, 500–502.
54. (a) Shim, I., Mandix, K., and Gingerich, K.A. (1991) *J. Phys. Chem.*, **95**,

5435–5442; (b) Tan, X. and Dagdigian, P.J. (2003) *J. Phys. Chem. A*, **107**, 2642–2649.
55. Ahlrichs, R. and Elliott, S.D. (1999) *Phys. Chem. Chem. Phys.*, **1**, 13–21.
56. Siehe, z.B., Cosse, C., Fouassier, M., Mejeau, T., Tranquille, M., DiLella, D.P., and Moskovits, M. (1980) *J. Chem. Phys.*, **73**, 6076–6085.
57. Himmel, H.-J. and Bihlmeier, A. (2004) *Chem. Eur. J.*, **10**, 627–633.
58. See, for example (a:) Le Roy, R.J. and Lam, W.-H. (1980) *Chem. Phys. Lett.*, **71**, 544–547. (b) LeRoy, R.J. and Bernstein, R.B. (1970) *J. Chem. Phys.*, **52**, 3869–3878.
59. Jansson, K. and Scullman, R. (1976) *J. Mol. Spectrosc.*, **61**, 299–305.
60. Wang, H., Liu, Y., Haouari, H., Craig, R., Lombardi, J.R., and Lindsay, D.M. (1997) *J. Phys. Chem. A*, **101**, 7036–7038.
61. Danset, D. and Manceron, L. (2004) *Phys. Chem. Chem. Phys.*, **6**, 3928–3932.
62. Hübner, O., Himmel, H.-J., Manceron, L., and Klopper, W. (2004) *J. Chem. Phys.*, **121**, 7195–7206.
63. For very recent calculations, see for example: (a) Joswig, J.-O. and Springborg, M. (2007) *J. Phys. Condens. Matter*, **19**, 106207-1–106207-18; (b) Medina, J., de Coss, R., Tapia, A., and Canto, G. (2010) *Eur. Phys. J. B*, **76**, 427–433; (c) Xu, X. and Truhlar, D.G. (2012) *J. Chem. Theory Comput.*, **8**, 80–90; (d) Kamal, C., Banerjee, A., Ghanty, T.K., and Chakrabarti, A. (2012) *Int. J. Quantum Chem.*, **112**, 1097–1106.
64. O'Brien, T.A., Albert, K., and Zerner, M.C. (2000) *J. Chem. Phys.*, **112**, 3192–3200.
65. Buchachenko, A.A., Chalasinski, G., and Szszesniak, M.M. (2010) *J. Chem. Phys.*, **132**, 024312-1–024312-10.
66. Camacho, C., Witek, H.A., and Yamamoto, S. (2009) *J. Comput. Chem.*, **20**, 468–478.
67. Koretsky, G.M. and Knickelbein, M.B. (1997) *J. Chem. Phys.*, **107**, 10555–10566.
68. Albert, G., Berg, C., Beyer, M., Achatz, U., Joos, S., Nieder-Schatteburg, G., and Bondybey, V.E. (1997) *Chem. Phys. Lett.*, **268**, 235–241.
69. Koszinowski, K., Schröder, D., and Schwarz, H. (2003) *J. Phys. Chem. A*, **107**, 4999–5006.
70. Balteanu, I., Balaj, O.P., Fox, B.S., Beyer, M.K., Bastl, Z., and Bonybey, V.E. (2003) *Phys. Chem. Chem. Phys.*, **5**, 1213–1218.
71. Balteanu, I., Achatz, U., Balaj, O.P., Fox, B.S., Beyer, M.K., and Bondybey, V.E. (2003) *Int. J. Mass Spectr.*, **229**, 61–65.
72. Siehe zum Beispiel: Mile, B., Howard, J.A., Histed, M., Morris, H., and Hampson, C.A. (1992) *Faraday Discuss.* **92**, 129–145.
73. Fedrigo, S., Haslett, T.L., and Moskovits, M. (1999) *Chem. Phys. Lett.*, **307**, 333–336.
74. Himmel, H.-J., Manceron, L., Downs, A.J., and Pullumbi, P. (2002) *J. Am. Chem. Soc.*, **124**, 4448–4457.
75. Himmel, H.-J., Manceron, L., Downs, A.J., and Pullumbi, P. (2002) *Angew. Chem.*, **114**, 829–832; *Angew. Chem. Int. Ed.*, (2002), **41**, 796–799.
76. Himmel, H.-J. (2005) *Z. Anorg. Allg. Chem.*, **631**, 1551–1564.
77. Himmel, H.-J. and Schnöckel, H. (2002) *Chem. Eur. J.*, **8**, 2397–2405.
78. Spikes, G.H., Fettinger, J.C., and Power, P.P. (2005) *J. Am. Chem. Soc.*, **127**, 12232–12233.
79. Peng, Y., Brynda, M., Ellis, B.D., Fettinger, J.C., Rivard, E., and Power, P.P. (2008) *Chem. Commun.*, 6042–6044.
80. Rivard, E., Fischer, R.C., Wolf, R., Peng, Y., Merrill, W.A., Schley, N.D., Zhu, Z., Pu, L., Fettinger, J.C., Teat, S.J., Nowik, I., Herber, R.H., Takagi, N., Nagase, S., and Power, P.P. (2007) *J. Am. Chem. Soc.*, **129**, 16197–16208.
81. Gaertner, B., Himmel, H.-J., Macrae, V.A., Downs, A.J., and Greene, T.M. (2004) *Chem. Eur. J.*, **10**, 3430–3443.
82. Gaertner, B., Himmel, H.-J., Macrae, V.A., Pardoe, J.A.J., Randall, P.G., and Downs, A.J. (2004) *Chem. Eur. J.*, **10**, 5836–5847.
83. Himmel, H.-J., Hübner, O., Klopper, W., and Manceron, L. (2006) *Angew. Chem.*, **118**, 2865–2868; *Angew. Chem. Int. Ed.*, (2006), **45**, 2799–2802.

84. Himmel, H.-J., Hübner, O., Klopper, W., and Manceron, L. (2006) *Phys. Chem. Chem. Phys.*, **8**, 2000.
85. Manceron, L., Hübner, O., and Himmel, H.-J. (2009) *Eur. J. Inorg. Chem.*, **2009**, 595–598.
86. Duchateau, R., Gambarotta, S., Beydoun, N., and Bensimon, C. (1991) *J. Am. Chem. Soc.*, **113**, 8986–8988.
87. See, for example: (a) Yildirim, T., Íñiguez, J., and Ciraci, S. (2005) *Phys. Rev. B: Condens. Matter*, **72**, 153403–153406; (b) Sun, Q., Wang, Q., Jena, P., and Kawazoe, Y. (2005) *J. Am. Chem. Soc.*, **127**, 14582–14583; (c) Zhao, Y., Kim, Y.- H., Dillon, A.C., Heben, M.J., and Zhang, S.B. (2005) *Phys. Rev. Lett.*, **94**, 155504-1–155504-4.
88. Navaratnarajah, K., Green, J.C., and Himmel, H.-J. (2006) *New J. Chem.*, **30**, 1253–1262.
89. Muñoz, F., Cardenas, C., Rogan, J., Valdivia, J.A., Fuentealba, P., and Kiwi, M. (2013) *J. Phys. Chem. C*, **117**, 4287–4291.
90. Souvi, S.M., Berkaïne, N., Alikhani, M.E., and Manceron, L. (2009) *Phys. Chem. Chem. Phys.*, **11**, 9831–9839.
91. (a) Gutsev, G.L., Bauschlicher, C.W., and Andrews, L. (2003) *J. Chem. Phys.*, **119**, 3681–3690; (b) Tremblay, B. and Manceron, L. (2008) *Inorg. Chem.*, **47**, 4531–4535.
92. Tremblay, B., Manceron, L., Gutsev, G.L., Andrews, L., and Partridge, H. III, (2002) *J. Chem. Phys.*, **117**, 8479–8485.
93. For theoretical study on the carbonyl complexes of 3d metal dimers, see: Jiang, L. and Xu, Q. (2008) *J. Chem. Phys.*, **128**, 124317-1–124317-9.
94. Addicoat, M.A., Lim, K.F., and Metha, G.F. (2012) *J. Chem. Phys.*, **137**, 034301-1–034301-9.
95. (a) Pullumbi, P., Mijoule, C., Manceron, L., and Bouteiller, Y. (1994) *Chem. Phys.*, **185**, 13–24; (b) Knight, L.B. Jr., Banisaukas, J.J. III,, Babb, R., and Davidson, E.R. (1996) *J. Chem. Phys.*, **105**, 6607–6615.
96. Köhn, A., Himmel, H.-J., and Gaertner, B. (2003) *Chem. Eur. J.*, **9**, 3909–3919.
97. Moc, J. (2004) *Chem. Phys. Lett.*, **395**, 38–43.
98. Harrison, J.F. (2000) *Chem. Rev.*, **100**, 679–716.
99. Kretzschmar, I., Schröder, D., Schwarz, H., and Armentrout, P.B. (2001) *Advances in Metal and Semiconductor Clusters*, vol. 5, Elsevier Science B. V., New York, pp. 347–395.
100. (a) Beattie, I.R., Ogden, J.S., and Price, D.D. (1987) *Inorg. Chem.*, **17**, 3296; (b) Herwig, C. and Limberg, C. (2008) *Inorg. Chem.*, **47**, 2937–2939.
101. Gong, Y., Zhou, M., Kaupp, M., and Riedel, S. (2009) *Angew. Chem.*, **121**, 8019–8023; *Angew. Chem. Int. Ed.*, (2009), **48**, 7879–7883.
102. Danset, D., Alikhani, M.E., and Manceron, L. (2005) *J. Phys. Chem. A*, **109**, 97–104.
103. Danset, D., Alikhani, M.E., and Manceron, L. (2005) *J. Phys. Chem. A*, **109**, 105–114.
104. Marzouk, A., Danset, D., Zhou, M.F., Gong, Y., Alikhani, M.E., and Manceron, L. (2011) *J. Phys. Chem. A*, **115**, 9014–9021.
105. Gong, Y., Zhou, M., Andrews, L., Schlöder, T., and Riedel, S. (2011) *Theor. Chem. Acc.*, **129**, 667–676.
106. Zhao, Y., Su, J., Gong, Y., Li, J., and Zhou, M. (2008) *J. Phys. Chem. A*, **112**, 8606–8611.
107. Gong, Y., Wang, G., and Zhou, M. (2009) *J. Phys. Chem. A*, **113**, 5355–5359.
108. Stößer, G. and Schnöckel, H. (2005) *Angew. Chem.*, **117**, 4334–4336; *Angew. Chem. Int. Ed.*, (2005), **44**, 4261–4264.
109. (a) Citra, A., Chertihin, G.V., Andrews, L., and Neurock, M. (1997) *J. Phys. Chem. A*, **101**, 3109–3118; (b) Danset, D., Manceron, L., and Andrews, L. (2001) *J. Phys. Chem. A*, **105**, 7205–7210; (c) Allouti, F., Manceron, L., and Alikhani, M.E. (2006) *Phys. Chem. Chem. Phys.*, **8**, 448–455.
110. Köhn, A., Gaertner, B., and Himmel, H.-J. (2005) *Chem. Eur. J.*, **11**, 5575–5588.
111. Andrews, L., Kushto, G.P., Yustein, J.T., Archibong, E., Sullivan, R., and Leszczynski, J. (1997) *J. Phys. Chem. A*, **101**, 9077–9084.

112. (a) Pradhan, K. and Jena, P. (2011) *J. Chem. Phys.*, **135**, 144305-1–144305-6; (b)The also the discussion of superhalogens in: Scheller, M.K., Compton, R.N., and Cederbaum, L.S. (1995) *Science*, **270**, 1160–1166.
113. Gaertner, B., Köhn, A., and Himmel, H.-J. (2006) *Eur. J. Inorg. Chem.*, **2006**, 1496–1504.
114. For a theoretical work on isomers of GaO_4, see: Archibong, E.F. and Ramasami, P. (2011) *Comp. Theor. Chem.*, **964**, 324–328.
115. Schnöckel, H. (1978) *Angew. Chem.*, **90**, 638–639; *Angew. Chem. Int. Ed.*, (1978), **17**, 616 – 617.
116. Schnöckel, H. (1980) *Z. Anorg. Allg. Chem.*, **460**, 37–50.
117. Tremblay, B., Roy, P., Manceron, L., Alikhani, M.E., and Roy, D. (1996) *J. Chem. Phys.*, **104**, 2773–2781.
118. Schnöckel, H., Mehner, T., Plitt, H.S., and Schunck, S. (1989) *J. Am. Chem. Soc.*, **111**, 4578–4582.
119. aFor some previous studies, see: (a) Anderson, J.S. and Ogden, J.S. (1969) *J. Chem. Phys.*, **51**, 4189–4196; (b) Hastie, J.W., Hauge, R.H., and Margrave, J.L. (1969) *Inorg. Chim. Acta*, **3**, 601–606.
120. Balteanu, I., Balaj, P., Beyer, M.K., and Bondybey, V.E. (2004) *Phys. Chem. Chem. Phys.*, **6**, 2910–2913.
121. Danset, D. and Manceron, L. (2005) *Phys. Chem. Chem. Phys.*, **7**, 583–591.
122. Staemmler, V., Reinhardt, P., Allouti, F., and Alikhani, M.E. (2008) *Chem. Phys.*, **349**, 83–90.
123. Hübner, O. and Himmel, H.-J. (2009) *Chem. Phys. Phys. Chem.*, **11**, 2241–2248.
124. (a) Asmis, K.R., Brümmer, M., Kaposta, C., Santambrogio, G., von Helden, G., Mejer, G., Rademann, K., and Wöste, L. (2002) *Phys. Chem. Chem. Phys.*, **4**, 1101–1104; (b) Feyel, S., Döbler, J., Schröder, D., Sauer, J., and Schwarz, H. (2006) *Angew. Chem.*, **118**, 4797–4801; *Angew. Chem. Int. Ed.*, (2006), **45**, 4681 – 4685.
125. Schröder, D., Souvi, S.O., and Alikhani, E. (2009) *Chem. Phys. Lett.*, **470**, 162–165.
126. Li, R.-Z., Xu, H.-G., Cao, G.-J., Zhao, Y.-C., and Zheng, W.-J. (2011) *J. Chem. Phys.*, **135**, 134307-1–134307-5.
127. Schröder, D., Schwarz, H., Clemmer, D.E., Chen, Y., Armentrout, P.B., Baranov, V.I., and Böhme, D.K. (1997) *Int. J. Mass Spectrom.*, **161**, 175–191.
128. Rollason, R.J. and Plane, M.C. (2000) *Phys. Chem. Chem. Phys.*, **2**, 2335–2343.
129. Zhou, M.F., Wang, C.W., Li, Z.H., Zhuang, J., Zhao, Y.Y., Zheng, X.M., and Fan, K.N. (2010) *Angew. Chem.*, **122**, 7923–7927; *Angew. Chem. Int. Ed.*, (2010), **49**, 7757 – 7761.
130. Zhou, M., Wang, C., Zhuang, J., Zhao, Y., and Zheng, X. (2011) *J. Phys. Chem. A*, **115**, 39–46.
131. Gong, Y., Andrews, L., Jackson, V.E., and Dixon, D.A. (2012) *Inorg. Chem.*, **51**, 11055–11060.
132. Rozanska, X. and Sauer, J. (2009) *J. Phys. Chem. A*, **113**, 11586–11594.
133. Koszinowski, K., Schlangen, M., Schröder, D., and Schwarz, H. (2005) *Eur. J. Inorg. Chem.*, **2005**, 2464–2469.
134. Hübner, O. and Himmel, H.-J. (2011) *Phys. Chem. Chem. Phys.*, **13**, 2963–2971.
135. Hübner, O. and Himmel, H.-J. (2013) *J. Phys. Chem. A.*, **117**, 12635–12641.

3
Toward Single-Molecule Catalysis

Arina Rybina , Marcel Wirtz, Dominik Brox, Roland Krämer, Gregor Jung, and Dirk-Peter Herten

3.1
Introduction

More than two decades ago, the possibility for optically detecting a single molecule made possible the direct observation of the stochasticity of molecular processes and led to a variety of different tools based on single-molecule fluorescence spectroscopy (SMFS) [1–3]. In the beginning, single-molecule studies at room temperature were mainly focused on biochemical and biophysical processes involving proteins and nucleic acids [4–6]. One important reason for this was that general approaches for labeling proteins and nucleic acids with fluorescent dyes were already well established. But even before these techniques had been established, already in 1961, Rotman [7] developed a way to measure the activity of single enzymes and successfully addressed the question of molecular heterogeneity. His assay was based on time-lapse fluorescence microscopy of microdroplets that were prepared from a very dilute sample of β-D-galactosidase from *Escherichia coli.* According to Poisson distribution, most of the droplets contained no enzyme molecule, a few of them only one, and very few more than one. This expectation was confirmed with the finding that the activities present in individual microdroplets were a multiple of the slowest activity measured. These measurements were possible because the microdroplets contained an excess of the nonfluorescent substrate 6-hydroxyfluoran-13-D-galactopyranoside, which is converted by the galactosidase to the fluorescent 6-hydroxyfluoran that accumulated in the droplets. The activity of individual enzymes is indicated as a steady increase of fluorescence with time. An important finding was that the slope of the fluorescence increase was varying with time, thereby indicating dynamic heterogeneities probably due to conformational fluctuations of the enzyme which had an influence on its activity. Rotman's approach was then used to study enzymatic activity at the level of single cells [8] and to quantify membrane-bound immunoglobulins on single lymphocytes [9]. In 1976, Hirschfeld [10] constructed an improved fluorescence microscope that allowed him to observe single antibodies that were labeled with 80–100 fluorescein isocyanate molecules each. Later, a microscope-fluorometer was invented consisting of a glass capillary with

Molecular Catalysts: Structure and Functional Design, First Edition. Edited by Lutz H. Gade and Peter Hofmann.

picoliter volume, allowing the detection of 0.3 attomol of fluorescein (~200 000 molecules) [11]. Finally, this concept led to the development of enzyme-amplified diagnostics using (auto-)catalytic conversion of fluorescent substrates [12], which was later extended to fluorescence-based enzyme assays for high-throughput screening [13].

3.1.1
Single-Molecule Enzymology

Only with the advent of SMFS in the early 1990s have the studies of molecular heterogeneities returned into the focus of research. In one of the early examples again, the formation of fluorescent product has been used to study the catalytic activity of single enzyme molecules and also of single metal ions in microcavities [14]. The first real-time observation of enzymatic turnover has been demonstrated in the early studies of the cholesterol oxidase. Here, the state of the enzyme was, for the first time, directly observed as changes in fluorescence emission of the cofactor flavin adenine dinucleotide (FAD) upon its conversion to the nonfluorescent $FADH_2$. Temporal cross-correlation of individual single-molecule transients revealed the static and dynamic disorder of the cholesterol reaction rates [15]. The use of the fluorogenic substrate is limiting because only the final step of a whole series of elementary reaction steps can be observed. A consequent extension is a fluorescent substrate capable of reporting multiple states along the reaction trajectory. This was realized in live-cell experiments by a BODIPY (boron-dipyrromethene) derivative that has a double bond in conjugation with the chromophoric center such that oxidation of the double bond shifts the emission wavelength from red through yellow to green [16]. A similar and more versatile readout is provided by the Förster resonance energy transfer (FRET) between a donor and an acceptor dye molecule provided the structural changes to be observed occur on the same length scale of 1–10 nm. This is the case for ribozymes where single-pair Förster resonance energy transfer (spFRET) was used to correlate the structural dynamics with their heterogeneous cleavage dynamics [17]. Later, the concept of fluorogenic substrates was extended to the observation of individual turnover events at immobilized β-galactosidase and lipase enzymes [18–20]. Both studies revealed a memory effect in catalytic turnover, that is, temporal fluctuations of the conversion rates, which could be explained by slow conformational fluctuations of the enzymes that were observed in single-molecule experiments using FRET [21]. In a similar manner, fluctuations in the protein interaction of single metallochaperones could be related to the presence of metal ions [22]. These studies of weak protein–protein interaction by spFRET were made possible by protein trapping in nano vesicles allowing higher local concentrations of the labeled protein to increase the probability for binding without increasing background fluorescence [23]. The same problem was tackled by making use of the fact that most proteins are charged. Here, an anti-Brownian electrokinetic (ABEL) trap was used to apply an oscillating field in which the charged protein was kept in a confined area so small that the cooperativity of adenosine triphosphate (ATP) hydrolysis in multi-subunit enzymes could be studied by probing the FRET between the enzyme and the substrate [24]. These studies point to the problem that isolation of molecules is frequently achieved by diluting the sample, which, at the same time, dramatically influences the probability for molecular interaction. Thus, novel approaches for isolating molecules, for example, droplets, traps, cavities, or microchannels, will be the key for future studies in catalysis. Most single-molecule studies discussed so far use fluorescence intensity (as a function of the excitation or emission wavelength) to discriminate between different molecular states. In another study, also the excited-state lifetime of the fluorescent label was used to sense the state of the nitrite reductase during catalytic turnover using confocal fluorescence lifetime microscopy [25]. Together, these examples show that single-molecule experiments are well established in the field of biochemical and biomolecular research and give a deep insight into molecular mechanisms and molecular heterogeneities.

3.1.2 Single-Molecule Studies in Chemistry

In contrast to single-molecule studies in enzymology, approaches to study reactions in pure chemistry and especially in catalysis were rare for different reasons except for a few examples [14, 26, 27]. Only recently, several applications of SMFS in heterogeneous catalysis and few in homogeneous chemical reactions could demonstrate the potential of single-molecule applications in chemistry [28–35]. One difficulty when studying the dynamics of single molecules is their isolation, which is usually achieved by confining the observation volume, for example, by use of confocal or total internal reflection fluorescence (TIRF) microscopy. This works well for time-resolved studies of molecules immobilized in a matrix or at a surface. For molecules in solution, however, the observation time is limited by their diffusion, which is usually in the order of a few milliseconds when using a confocal fluorescence microscope. The observation of homogeneous reactions therefore requires some kind of mechanical confinement, such as in droplets [7], vesicles [23], microcavities [14], or electrokinetic traps [24]. Another alternative is the electrochemical creation of two opposing concentration gradients by two electrodes forming a narrow reaction zone, like in the following example. Here, chemiluminescence invoked by recombination of radical ions generated from 9,10-diphenylanthracence by potential pulses with microelectrodes was detected with single-photon sensitivity [26]. The dimensions of the electrochemical cells and the pulse duration were tuned to create a very narrow zone with a volume of about 20 fl (1 fl = 10^{-15} l) in which the electrogenerated radical ions recombine. Individual photons were detected on the picosecond time scale to measure the recombination kinetics. Despite this exceptional example, most other single-molecule experiments are based on SMFS and, so far, most of them make use of fluorogenic substrates. In one of the early examples already mentioned above, the reaction rate of the Ce(IV) and As(III) catalyzed by Os(VIII) was measured as buildup of the luminescent Ce(III)-species in microcavities using a highly diluted sample of OsO_4 [14]. Then, only in 2001 the oxidation of terrylene molecules in *p*-terphenyl host crystals was directly observed as discrete jumps in the fluorescence emission of individual terrylene molecules by time-resolved SMFS. Time-resolved studies of the emission spectra revealed the formation of different adducts with oxygen, supported by quantum chemical calculations [27]. These time-resolved studies were possible because the reacting molecules were immobilized in a matrix. For similar reasons, polymers doubling of properties were studied early on by SMFS, yielding insights into the molecular and electronic properties of polymer networks [36–40]. The component of reaction dynamics was added by tracking individual monomers during the radical polymerization, which allowed visualizing heterogeneities arising from the growing polymer network [29, 41]. In a similar approach, olefin polymerization using the Grubbs II catalyst was followed using fluorescently labeled monomers which could be visualized when incorporated in growing polymer particles because of their reduced mobility [42]. The major finding was that growing polymer particles were never found to form at crystal interfaces of the Grubbs II catalyst as observed by white-light microscopy. The authors therefore assumed that metathesis was catalyzed only by the homogeneously dissolved catalyst. In these experiments, tracking of individual fluorescent monomers by wide-field microscopy was possible because they become part of a growing polymer particle that slows down their diffusion such that sufficient photons can be collected from one observed spot within a few milliseconds. In that sense, heterogeneous catalysts are an ideal target for SMFS, as the catalytic center is immobile by the nature of the catalyst and can be observed over long time periods. Still, the sample of interest needs to be optically transparent and should not show autofluorescence unless it is connected to the state of a catalytic site.

3.1.2.1 Single-Molecule Studies in Heterogeneous Catalysis

So far, most SMFS studies of heterogeneous catalysts have used fluorescent or fluorogenic substrates. Layered double hydroxide (LDH) is frequently used as a component of washing powders because it can catalyze hydrolysis of lipids. Aiming at a better understanding of specific binding and transport mechanisms, the sorption kinetics of an anionic fluorophore has been studied on single LDH crystals using time-lapse fluorescence microscopy [43]. Individual binding and unbinding events could be observed and related to the properties of the crystallites. This study was further extended by mapping the spatial distribution of catalytically active centers at the surface of LDH crystals. The signal was caused by cleavage of ester groups from the nonfluorescent leukobase of a 5-carboxyfluorescein derivative, yielding a fluorescent product that occurred specifically at catalytic sites of the crystals. Product formation was observed as the temporary appearance of a bright spot in individual frames acquire by time-lapse fluorescence microscopy. Spatially resolved analysis revealed static heterogeneities where the rate of product formation varied with a factor 10 at different crystal faces [44]. In a similar manner, individual zeolite particles used in fluid catalytic cracking were selectively stained with different fluorescent compounds and have been studied with confocal microscopy to identify the catalytically active sites [45]. Zeolites find even more versatile applications in heterogeneous catalysis because they can be doped with a variety of metal species. In this context, a series of interesting studies on the diffusion and the structures of mesoporous sieves have been carried out using SMFS. For example, the translational diffusion of single terrylene diimide molecules was studied in a mesostructured molecular sieve [46]. By application of fluorescence correlation spectroscopy (FCS), different modes of motion were revealed that varied in differently prepared mesoporous materials [47]. Such changes in molecular transport might be connected with structural features of the material as was shown by single-molecule tracking experiments where details on accessibility and connectivity of the mesoporous channel systems were revealed in differently prepared silica materials [48]. Ideally, one would like to connect knowledge about the accessibility with information about the precise location of the catalytic site in the material. A first step in that direction had been taken by measuring the buildup of the fluorescent product at zeolite crystals with wide-field fluorescence microscopy, revealing the spatial heterogeneity of acid catalysis at different crystal surfaces [49]. The microscopic resolution of such studies was then significantly improved by combining the principles of super-resolution microscopy with the turnover of fluorogenic substrates [50]. Here, the acid-catalyzed condensation of the nonfluorescent furfuryl alcohol yields fluorescent poly(furfuryl) chains at the catalytic sites of zeolite crystals, which can be observed for a short time as diffraction-limited spot until photobleaching occurs. Individual spots are then localized by mathematical modeling with nanometer precision as in other localization-based super-resolution microscopy methods, such as photo-activated localization microscopy (PALM), stochastic optical reconstruction microscopy (STORM), direct stochastical optical reconstruction microscopy (dSTORM), and so on [51–53]. The data from time-lapse microscopy contains multiple turnover events that are localized and used for reconstructing a map of the catalytically active sites with nanometer precision, which can then be compared with other microscopy techniques with higher resolution, such as electron microscopy. This method, which is based on the formation of a fluorescent product, allows not only measuring catalytic turnover but also mapping zones of varying catalytic activity at a resolution of ~10 nm [50]. In a similar experiment, the diffusion limitation of catalytic epoxidation of a fluorescent substrate was studied in mesoporous titanosilicates by high-resolution mapping of single turnover events. Turnover was signaled by a change in the emission spectrum of a fluorescent substrate induced by the epoxidation of a double bond in conjugation to the chromophoric center [54]. The concept has also been picked up to probe the activity of photocatalytic sites on single titanosilicate zeolites using a redox-sensitive fluorescein derivative [55]. Taken together, these studies have shown that the kinetic of catalytic conversions in heterogeneous catalysts can be assessed by SMFS. Moreover, the precise localization of the catalytic centers along with information about the transport processes involved opens new perspectives for the rational design of catalysts made of mesoporous materials.

Nanoparticles are among the materials with the highest surface-to-volume ratio and have therefore also become an important field of research in heterogeneous catalysis. The finding that colloidal spherical Au nanoparticles can catalyze the reduction of the nonfluorescent resazurin to the fluorescent resorufin allowed the study of their catalytic turnover by single-molecule-sensitive TIRF microscopy [56]. It was observed that the measured kinetics follows the Langmuir–Hinshelwood mechanism but also that single Au nanoparticles exhibit activity fluctuations which can be explained by dynamic surface restructuring. Additionally, concentration variation of the substrate revealed heterogeneities among different nanoparticles [57], which were found to be dependent on the particle size [58]. Further analysis of the data implies that the dynamic fluctuations at different active sites on one particle are correlated [59]. The previously discussed oxidation of terrylene by Basché *et al.* [27] was used to gain further insights into the mechanism of photo-induced oxidation by TiO_2 nanoparticles [60]. The formed products characterized at the single-molecule level gave indication that the oxidation proceeded by photo-induced singlet-oxygen formation. Likewise, photocatalytic oxidation at TiO_2 nanoparticles was studied using a redox-sensitive BODIPY derivative [61]. The same dye was used for high-resolution mapping of the catalytic site on titanium-based nanoparticles [62]. Another application of redox-sensitive dyes at the single-molecule level was found in the electrocatalytic conversion of resazurin by single-walled carbon nanotubes [63, 64].

Aside from fluorescence microscopy, a few other methods also are used to study the properties of heterogeneous catalysts even at the atomic scale. Scanning tunneling microscopy (STM) is used for determining the interfacial properties of surfaces. Such examples include defined crystalline surfaces, such as that of RuO_2 [65, 66]. Recently, also surface plasmon spectroscopy has been used to follow the oxidation of ascorbate on single gold nanoparticles by directly measuring the rates of electron transfer and reaction with O_2 [67]. More recently, soft X-ray microscopy has been used to study the distribution of aluminum coordination environments in zeolites with nanometer resolution [68]. If the probe consists of very bright and highly photostable fluorophores embedded in a matrix, even pump–probe techniques can be used to observe ultrafast processes, such as exciton coupling of single terrylene molecules [69]. As has already been stated, the integration of multiple techniques with SMFS will open new perspectives for observing chemical conversions at individual catalytic sites [70].

3.1.2.2 Single-Molecule Chemistry in Homogeneous Catalysis

Difficulties arise when turning to homogeneous reactions. Neither substrates nor catalysts sit still but diffuse freely in solution, making it impossible with current microscopy techniques to follow their fate for more than few milliseconds. Moreover, fast diffusion of the molecules significantly reduces the fluorescence intensity that can be detected in one pixel of a charge-coupled device (CCD) camera, yielding blurred and noisy signals. Thus, to study homogeneous reactions the molecular systems need to be confined in their movement. Currently, immobilization of the catalyst or the substrate is the by far simplest and most frequently used approach in SMFS studies. It is suited for confocal microscopy, allowing time-resolved sampling of the fluorescence emission of a single entity at the microseconds time scale, and also for TIRF or wide-field microscopy, enabling the simultaneous observation of many individual entities at a glass/liquid interface at the milliseconds time scale. However, the potential interaction of the different compounds with the surface might lead to artifacts especially when the molecules show unspecific adsorption at the interface. If the catalyst is immobilized, this approach might also be limited by the buildup of fluorescent products formed by other catalysts nearby but not in the field of view, leading to a steady increase of background fluorescence. This is not so much of problem when using a highly diluted sample of catalyst in microcavities, small droplets, or vesicles [7, 14, 23]. Here, the buildup of fluorescence emission directly reports the enzymatic conversion rate. Still, the catalyst could interact with the surface, but this can be kept at a minimum by tuning the experimental conditions, such as solvent, buffer, or surface passivation. A very elegant way for providing the conditions of homogeneous solutions is the ABEL trap discussed previously, where electrokinetic forces are used to trap the molecule of interest [24]. This approach requires charged molecules and has so far been applied only to the trapping of enzymes. The trapping of smaller molecular entities in solution has so far not been reported. The difficulty in isolating individual catalyst is probably one reason why reports on single-molecule experiments with homogeneous (catalytic) reactions are rare.

Another difficulty is that the state of interest of catalyst or substrate has to be connected to specific signature in fluorescence emission that can be detected by SMFS. The binding of fluorescently labeled ligands to immobilized Pd and Pt centers is reported by the occurrence of fluorescent spots at the glass interface [71–73]. This approach is useful for recording the binding and, in the case of a reversible reaction also the unbinding, of ligands. In order to gain deeper insights into the mechanism of catalytic reactions, more complex probes have to be developed. An early example that will be discussed in more detail in the following was the reversible complex formation with Cu(II) reported as reversible quenching of the fluorophore-labeled ligand immobilized on glass cover slides [74]. Another study found the redox state of a perylene diimide dye to be connected with its emission, and showed its potential use for studying reversible redox reactions at the single-molecule level [75]. In a very recent example, the formation of fluorescent nanoparticles was observed at the single-molecule level by controlled mixing in microchannels [76]. These examples show that homogeneous catalytic reactions can, in principle, be addressed by the methods of SMFS, although probe and experimental design might be more demanding as in heterogeneous catalysis. Consequently, we identified probes that can be used in single-molecule experiments as an important factor to advance in single-molecule chemistry.

3.2 Probes for Single-Molecule Chemistry

Besides the appropriate experimental setup for particular chemical problems, even more care has to be taken in selecting the right combination of substrate and chemical reaction. If one refuses to reinvent the wheel but prefers to rely on available knowledge, it appears useful to check, for example, chemical databases. Figure 3.1 shows a recent compilation of 12 well-known dye classes, their name-dropping in single-molecule-related publications, and how many references are devoted to synthesis. Although the chart should be interpreted with care for several reasons, some useful conclusions can be drawn: most synthetic papers exist for blue fluorescing dyes, but distinctly fewer publications describe synthesis or transformation of orange/red fluorescent dyes. Blue fluorescent dyes are, however, not very suitable for single-molecule investigations because of higher background contributions from Raman scattering and impurities. As a rule of thumb, the stability of those dyes is worse as well. Terrylene dyes, on the other side, exhibit exceptional photostabilities under oxygen-free conditions but offer limited synthetic accessibility [27]. The trade-off between photostability, accessibility, and known photophysical behavior therefore favors fluorescent dyes

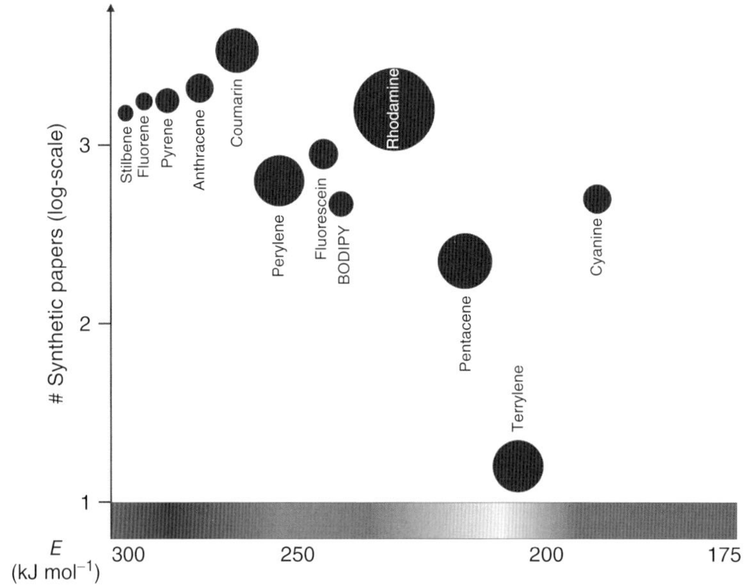

Figure 3.1 Compilation of different chromophore families with regard to their utility for single-molecule chemistry, based on Web-of-Science database search. Fluorophores with excitation in the green to red region of the visible electromagnetic spectrum are superior to blue and violet fluorescing dyes because of the lower amount of deposited energy. Among these, especially fluorescein, BODIPY, and rhodamine derivatives appear suitable because of their occurrence in synthetic publications. Relative citation frequency within single-molecule work is reflected by the area.

in the green to yellow region for single-molecule chemistry. It is therefore not surprising that most single-molecule studies on chemical topics were carried out with xanthene-(like) and BODIPY dyes [30, 35, 44, 71].

3.2.1 Fluorescence Properties: Overall Considerations

For accessing chemically relevant information from single-molecule methods, several fluorescence properties are at disposal (Figure 3.2). One has to distinguish whether reversible reactions, such as binding and complexation, or irreversible transformations should be studied. In the first case, the absolute starting point of measurements is irrelevant and there is no hurry to start the experiment. Therefore, point detection as in confocal microscopy can be used for single-molecule mapping. The use of photon-counting devices is preferred, as it allows the determination of the macroscopic time scale and the microscopic photon arrival time after pulsed excitation. In principle, changes of fluorescence lifetime and fluorescence on–off dynamics due to intersystem crossing or cis/trans isomerization are useful to the analysis of reversible reactions. Irreversible reactions can also be followed if the excitation light induces the reaction. However, confocal setups completely fail at following irreversible courses when their location cannot be predicted. Imaging setups are superior because of the highly parallelized detection, but rare and sudden events will not be seen in the field of view. However, only different fluorescence colors, polarizations, or the generation of fluorescence can be simultaneously detected. Pitfalls due to spontaneous events, typical for single-molecule observations, have to be taken into account. We therefore conclude that only color changes

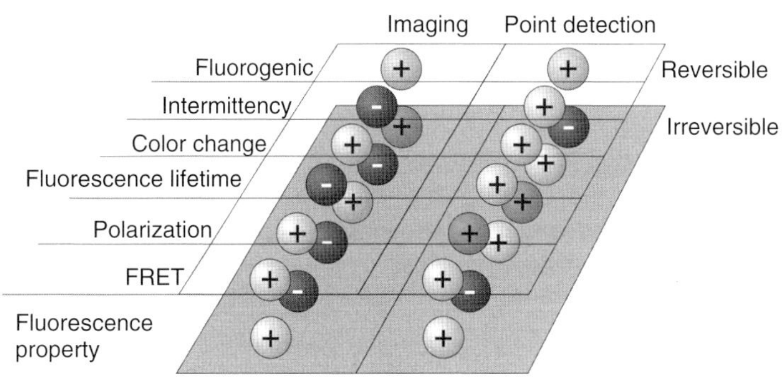

Figure 3.2 Fluorescence properties for different methods of single-molecule fluorescence spectroscopy. The principal utility of the particular feature for studying reversible (upper level) or irreversible reactions (lower level) is signed. Only fluorogenic substrates or molecules exhibiting a distinct color change apply for studying irreversible reactions when the reaction spot cannot be foreseen. Experiments by other researchers are highlighted by dark gray.

or fluorogenic substrates are promising for investigating irreversibility in chemistry.

3.2.2
Fluorogenic Substrates

The established approaches so far mostly rely on dark substrates which are converted to fluorescent products. Both colorless and strongly absorbing substances were used for visualizing turnover in heterogeneous catalysis. This concept was exemplified in condensation, trans-esterification, and hydrolysis as well as the redox reactions on active nanoparticles [44, 56]. Differing activities of various crystal surfaces and the kinetics of substrate adsorption and product desorption are the outcome of those studies. Moreover, the discovery of catalytic hot spots provides unprecedented insight into the nonuniformity of heterogeneous catalysis, and nanoscopy allows precise localization. Current limitations concern the solid systems that can be studied. Conducting particles and strongly colored transition-metal compounds likely quench fluorescence close to the surface. Consequently, only already desorbed products can be detected in these cases, blurring the position of the active site. Also, all events preceding the last reaction step in which the fluorophore is generated remain invisible and deserve careful interpretation (Figure 3.3a). The overall yield of the fluorogenic reaction cannot be determined, and one can imagine that under particular conditions only side products of the reaction are detected. A fluorogenic reaction, not strictly in its meaning, involves the immobilization of freely diffusing dyes by catalysts [72]. Molecules are not detected before a flexible linker is copolymerized.

3.2.3
Substrates for Reversible Reactions

Probes for reversible binding are exploited in bioanalysis for quantifying ion concentrations. Dual emissive probes, that is, molecules that change their emission color upon binding, are known for detecting protons [78]. Mechanistic investigations with other ions can rely on other photophysical parameters. Both fluorescence quenching and generation can occur upon binding, depending on the respective system, but also changes of, for example, the fluorescence lifetime might be susceptible to those reversible reactions. Owing to the continuous cycling of back and forth transitions, the undergoing changes of the fluorescence properties can be separated from occasional fluctuations such as blinking and spectral and orientational jumps. Such probes can be used in single-molecule studies once the immobilization issue is solved, and provide the kinetics for association and dissociation [74]. One could imagine that such chromophoric systems could be exploited as ligands in homogeneous catalysis. Another remarkable manifestation in single-molecule chemistry was found in isomerization by tautomerism [79]. Other reversible changes by chemical reactions also might be found.

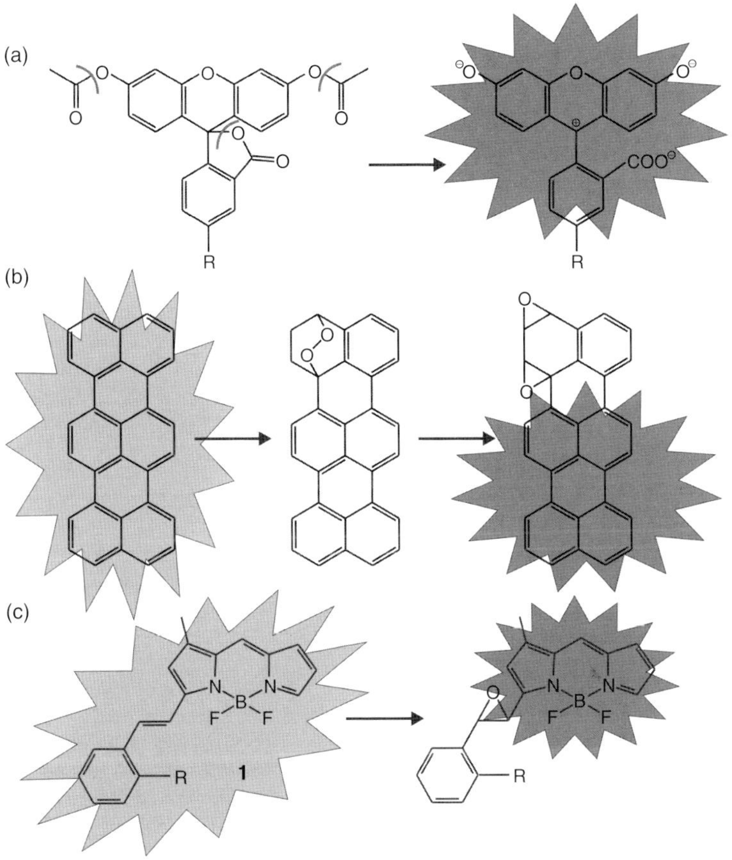

Figure 3.3 Three examples of single-molecule reaction studies. (a) Fluorogenic fluorescein diacetate allowing the visualization of active sites of heterogeneous catalysis [44]. Several elementary reaction steps, as indicated by black lines, have to happen before fluorescence can be observed. (b) The photochemical oxygenation of terrylene monitored by orange to green fluorescence conversion with luminescence intermission in between [27]. (c) The recently established versatile reactive BODIPY dye **1** exhibiting a fluorescence color change upon epoxidation [77]. The substituent R was exploited for immobilization.

3.2.4 Substrates for Irreversible Reactions

Irreversibility is not meant in a purely thermodynamic sense but more from a kinetic understanding: we denote reactions as irreversible when the backreaction is unlikely to occur within the time frame of observation. Those transformations impose several obstacles to their implementation in single-molecule research. Under practical considerations, only reactions in which their onset is triggered by the excitation light are susceptible to continuous single-molecule observation by the widely used photon-counting point detectors. All other reactions demand imaging setups and, concomitantly, distinct photophysical changes of the probe. Besides the lighting up of fluorogenic substrates, only strong spectral shifts during the reaction are unambiguous for a subsequent interpretation. Some systems can be found in literature where substrate and product are both fluorescent, for example see Ref. [80]. Among them is also the pair resazurin/resorufin [81], which was used in a fluorogenic assay in the gold-catalyzed reduction by hydroxylamine [56]. A compilation of fluorescent convertible dyes is lacking, to our knowledge.

Another challenge is the conversion number within the observation volume of ~1 fl. Assuming an elementary bimolecular reaction where both molecules are fluorescing, their concentration should not exceed 1 nM. Even if diffusion-controlled conditions are assumed, that is, $k_{\text{reaction}} > 10^9\ \text{M}^{-1}\ \text{s}^{-1}$, only one reactive collision will be detected within $10^8 - 10^9$ s. Alternatively, 10^6 molecule pairs must be studied at the same time, and one reaction event will be detected every 100 s or longer. It is obvious from this rough estimate that one substrate must be added in large excess for providing useful kinetics. This compound cannot be fluorescent but colorless or fluorogenic. Accordingly, it is not surprising that irreversible reactions are rarely investigated by SMFS.

The very first experiment was the photochemical oxygenation of terrylene (Figure 3.3b) [27]. Bis-epoxidation via an endoperoxide was concluded to result in a blue shift, which was based on theoretical computations. However, a thorough investigation of the product or the selectivity was not carried out. We are not aware of other examples (see Section 3.3).

3.3
Approaching Single-Molecule Studies in Homogeneous Catalysis

Experiments with single chemical objects reveal features that are hidden in the ensemble. With respect to chemical reactions other than enzymatic turnover, single-molecule observation can provide unprecedented insights into chemical reaction dynamics. Branching, parallel reactions to the same product (bifurcation), the existence of intermediates that are faster decaying than being formed, or correlated product formation is more directly detected by single-molecule observations. Other examples could be imagined where the interpretation is facilitated by this kind of investigations. In combination with temperature-dependent experiments, rough sketches of the energy landscape can be drawn and compared with the outcome of theoretical chemistry calculations. Many textbook examples of synthetic pathways lack in a detailed picture of their complex energetics, and this is even more true for catalytic reactions.

3.3.1
Fluorophore-Labeled Cu(II) Chelators and Substrates

A main idea of this project was the observation of immobilized metal complexes and their catalytic activity at the single-molecule level with the aid of fluorophores that are attached to the metal complex. In principle, two strategies may be used to build such metal complex–fluorophore conjugates: (i) coupling the fluorophore to a robust, exchange-inert metal complex, or (ii) coupling the fluorophore to a chelating ligand and subsequent *in situ* preparation of the (exchange labile) complex. Since multiple functionalizations (fluorophore labeling and anchor for immobilization) of metal complexes appeared rather challenging, we focused on the latter approach – taking into account that dissociation/association of labile complexes may complicate the observation and interpretation of catalysis-related processes in SMFS experiments. Concerning the fluorophores, it was important to select bright and photostable compounds that match available excitation and emission wavelengths of the microscope settings.

Copper(II) ions have a relatively high affinity to chelating ligands and readily form complexes *in situ*. In addition, the activity of Cu(II) complexes as group transfer or oxidation catalysts in aqueous solution, a preferred medium for our SMFS studies, is well established. While the d^9 Cu(II) ion typically quenches the fluorescence of proximal fluorophores by electron or energy transfer, the d^{10} Cu(I) ion does not, so that reporter groups may signal a change of the metal oxidation state during redox catalysis.

In the beginning, we have focused on copper complexes that catalyze the hydrolysis of carboxylic esters, and at a later stage of the project, oxidation reactions by O_2 or H_2O_2. Figure 3.4 summarizes the newly synthesized fluorophore-labeled chelators, and their properties are briefly discussed below. A significant advantage of the DNA-linked chelators is that they can be readily combined with both a reporter fluorophore and a functionality for surface immobilization by hybridization with a complementary DNA strand that is 3′ terminally functionalized with, for example, tetramethylrhodamine (TMR) and 5′-functionalized with biotin for interaction with surface-bound streptavidin.

Another approach to the observation of catalytic events by SMFS is the use of fluorophore-labeled substrates that change fluorescence intensity upon conversion to products. This type of substrates has previously been applied to the monitoring of enzymatic catalysis. In the context of Cu(II)-triggered catalysis, we have developed the fluorophore-labeled substrates shown in Figure 3.5.

Compound **2a** is a DNA conjugate of the chelator 2,2′-bipyridine. When hybridized to a complementary DNA strand terminally labeled with TMR, the reporter fluorophore signals the binding of Cu(II) ions to the bipyridine ligand by fluorescence quenching. The surface-immobilized DNA hybrid allowed the first observation of a metal complex dissociation/association equilibrium on the single-molecule level [74, 82]. Lability of the complex and the need of excess free Cu(II) ions to promote complex formation interfere with the observation of catalytic processes. While the cleavage of the fluorogenic ester substrate **10** by 2,2-bipyridine–Cu(II) is readily monitored in ensemble experiments by the fluorescence of released fluorescein [83], the activity individual immobilized **2a**–Cu(II) complexes could not be analyzed because of strong interference with background ester cleavage by free Cu(II) ions. The related terpyridine DNA

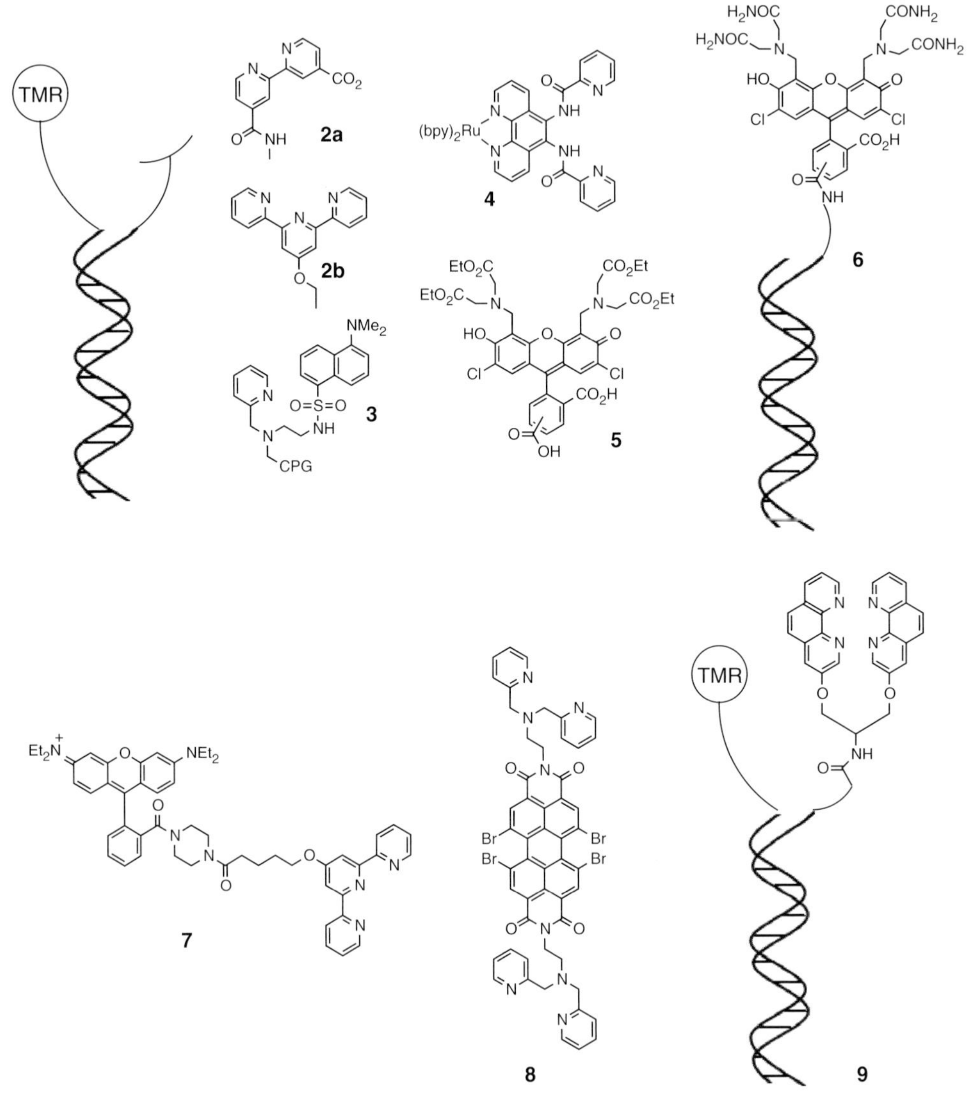

Figure 3.4 Fluorophore-labeled Cu(II) chelators **2–9**.

conjugate **2b** has a higher affinity to copper ions but is a rather poor catalyst and was not further considered for SMFS studies [84].

A catalytic reaction involves several molecular interactions between the catalyst and the reactant. In order to study these complex interactions, we first need precise knowledge of the dynamics in the ligand sphere of transition metals. We used model compound **2a** that translates the complexation dynamics of a Cu(II)–bipyridine complex into changes of the fluorescence emitted by

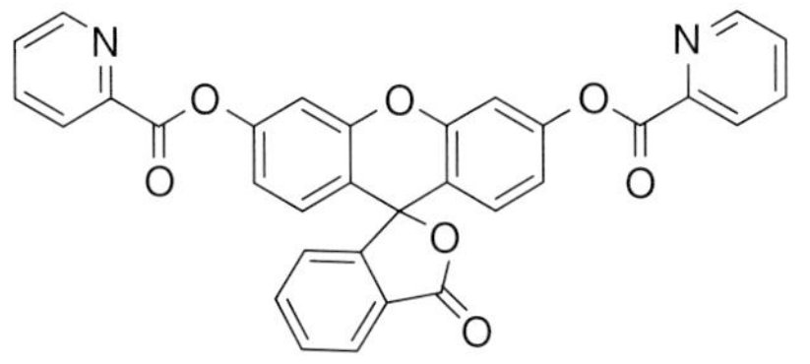

10

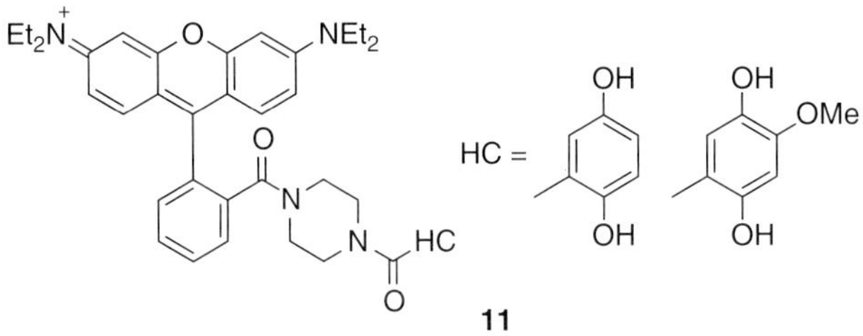

11

Figure 3.5 Fluorophore-labeled substrates **10–11** for Cu(II) complex catalysis.

the attached TMR dye. The fluorescence of TMR is strongly quenched upon formation of the Cu(II)–bipyridine complex in the near vicinity.

For single-molecule experiments, **2a** was immobilized on a glass surface (Figure 3.6a). The location of single probe molecules was determined from an overview image (Figure 3.6b), and distinct molecules were then brought in the focus of a single-molecule-sensitive confocal microscope to record the fluorescence transients. Compound **2a** shows a constant emission in the absence of Cu(II) (Figure 3.6c). In contrast to that, the probe cycles between a bright "on" and a dark "off" state in the presence of Cu(II). With increasing concentration of Cu(II), the length of the "on" states decreases (Figure 3.6d–f). We were able to calculate the association and dissociation rates by statistical analysis of the intensity fluctuations and thereby measure the complexation constant in thermal equilibrium. The determined value from single-molecule experiments of $K = (2.7 \pm 0.9) \times 10^6\ M^{-1}$ corresponds well with the value $K = (4.2 \pm 0.3) \times 10^6\ M^{-1}$ determined from ensemble fluorescence measurements [74]. From molecular dynamics simulations, we identified the minimum energy geometries where the Cu(II)–bipy complex comes into close contact with the TMR dye. In principle, this renders either FRET or photo-induced electron transfer as possible mechanisms for fluorescence quenching. We could rule out energy transfer because of the low absorption of the complex, leaving electron transfer as the most probable quenching mechanism [85].

Besides the observation of single metal complexation dynamics, **2a** has proven to be a much more universal tool. The stochastic blinking of single probe molecules has been exploited for an alternative approach in localization-based super-resolution microscopy (CHIRON, CHemically Improved Resolution for Optical Nanoscopy), which, in contrast to other methods, does not depend on

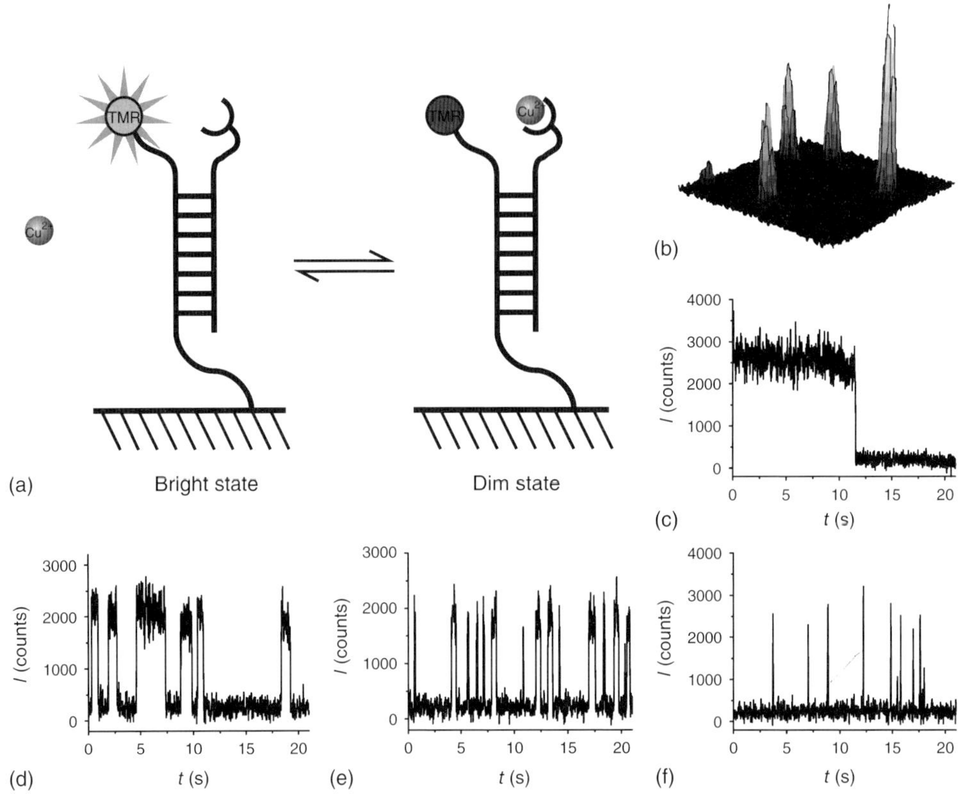

Figure 3.6 (a) Schematic illustration of **2a**: reversible transition between the metal-free bright state (left) and the complexed dim state (right). (b) Fluorescence image of single **2a** molecules immobilized on a glass substrate. (c–f) Representative fluorescence intensity transients of single probe molecules in the absence of copper (c) and in the presence of 200 nM (d), 1 μM (e), and 5 μM (f) $CuSO_4$.

the excitation intensity. We could show that single emitters could be localized with a full-width at half-maximum (FWHM) of ∼30 nm and that this technique can in principle be applied to image microtubules in fixed cells with an increased resolution (Figure 3.7) [82].

Also, tri- and tetradentate chelators **3** [86], **4** [87], and **7**, in which a dansyl, Ru(II)–bipyridine, or rhodamine fluorophore is directly coupled to the chelating moiety, strongly bind Cu(II) with fluorescence quenching, but display little or no hydrolytic activity against carboxylate esters. Immobilization of **3** on glass surfaces (CPG, controlled porous glass) is straightforward. The Cu(II) complex of **7** slowly catalyzes the hydrolysis of the methyl ester of pyridine-2-carboxylic acid. In ensemble studies, subtle changes in fluorescence intensity of the weakly fluorescent Cu(II) coordinated state are observed on coordination of this ester substrate and carboxylate product to the complex: that is, the fluorescent reporter signals two key intermediates of the catalytic cycle.

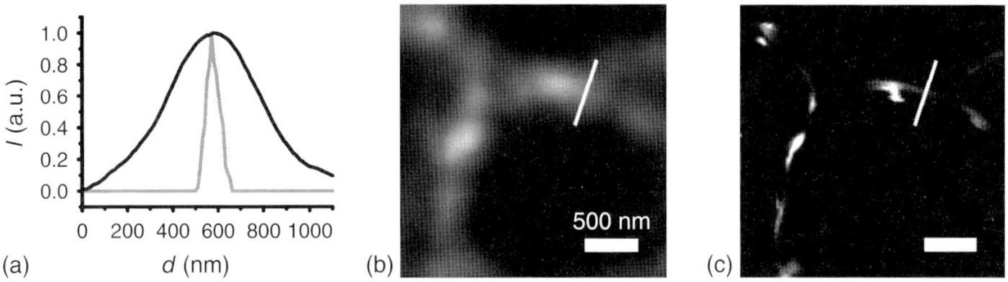

Figure 3.7 Application of the probe for super-resolution microscopy: microtubules in fixed cells were labeled with **2a** and imaged in presence of Cu(II). The diffraction-limited (a) and the reconstructed (b) images demonstrate the increased resolution of CHIRON, which becomes even clearer when looking at the intensity profiles (c) along the white lines in (a,b).

The fluorescein derivative **5** combines a fluorescent reporter, a chelating moiety, and an ester substrate in the same molecule [88]. The poor Cu(II) chelator **5** binds just a small fraction of the Cu(II) ions present in solution, and strong fluorescence of free **5** is observed. Cu(II)-promoted ester cleavage turns **5** into a tetradentate chelator, which results in strong, possibly irreversible, binding of Cu(II), and fluorescence quenching. Attempts to couple this ester substrate to DNA for surface immobilization failed but was successful with an amide analog of **5**, leading to **6**. The latter was developed further into a FRET-based fluorescent reporter with tunable emission wavelength by hybridization with complementary DNA labeled with a second fluorophore [89]. While the fluorescein derivatives enable the monitoring of Cu binding and Cu-promoted reactions in ensemble experiments, initial attempts of single-molecule experiments indicated insufficient photostability, so the fluorescein labels have not been further considered for continuous observation of immobilized metal complexes.

Perylene diimides (PDIs), in contrast, are very photostable fluorophores, and their amino derivatives have been applied to the monitoring of proton-transfer catalysis by SMFS in organic solvents [75]. The strong aggregation tendency via pi-stacking of PDIs in polar solvents [90] was reduced by introducing bulky substituents at the bay positions. In **8** and related compounds, the PDI fluorophore is coupled to tridentate chelators. Fluorescence is partially quenched on complexation by Cu(II). The Cu(II) complex of **8** catalyzes the oxidation of hydroquinone by H_2O_2. In contrast to the related naphthalene diimide chelators [91], **8** in both metal-free or complexed (Cu(II), Cu(I), Zn(II)) state has a surprisingly low fluorescence, thus limiting their performance in SMFS studies.

Compound **9** is a DNA conjugate of the tetradentate chelator "Clip-Phen" that was selected to overcome lability toward metal exchange as observed for the bipyridine conjugate **2a** [92]. Compound **9** in its hybrid with a complementary TMR labeled DNA strand allows monitoring the binding of Cu(II) to Clip-Phen by fluorescence quenching of the TMR reporter. The complex **9**–Cu(II) has a formation constant of about 10^{12} M^{-1}, and dissociation of the complex takes minutes rather than seconds [92], that is, the complexes are much more stable compared

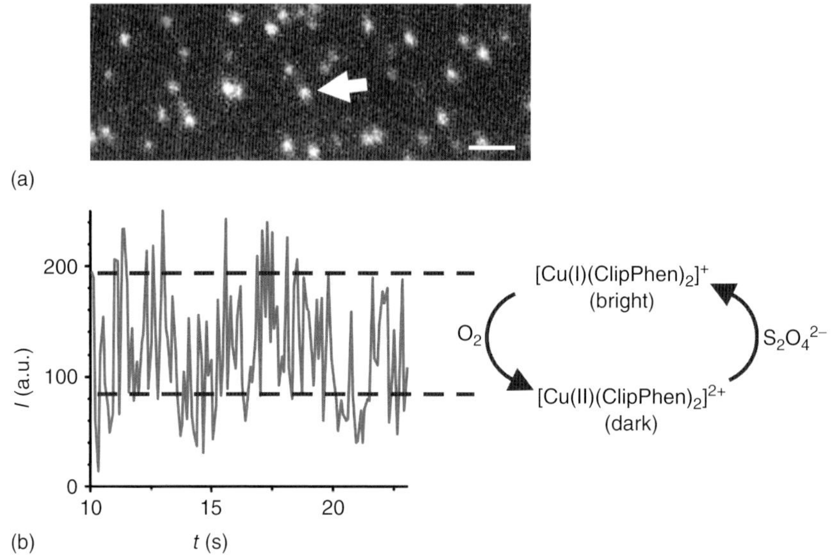

Figure 3.8 (a) **9**–TMR–DNA hybrids immobilized on a glass cover slide and imaged by TIRF microscopy after complexation by Cu(II), leading to quenching and addition of $Na_2S_2O_4$ which recovers the fluorescence emission. (b) Under these conditions, the fluorescence intensity of the molecules show rapid fluctuations as exemplified for the molecule signified by an arrow in (a), which can be attributed to a rapid turnover between Cu(II) and Cu(I) due to reduction by thionite and oxidation by oxygen present in the aqueous solution.

to the 2,2′-bipyridine conjugate **2a**–Cu. Clip-phen–Cu(II) catalyzes air oxidation of thiols and dithionite, and the oxidation of hydroquinone to benzoquinone by H_2O_2 with a turnover rate of $10\,min^{-1}$ in aqueous solution at pH 7. The *in situ* prepared orange Cu(I) complex of Clip-Phen in acetonitrile is very oxygen sensitive and readily oxidized to its Cu(II) form. Immobilized **9**–TMR–DNA hybrids were observed by single-molecule techniques (Figure 3.8). Compared to **2a**–Cu, much lower Cu(II) concentrations are needed for *in situ* metal binding and effective fluorescence quenching. On addition of the reductant dithionite, fluorescence of individual molecules is restored and rapid blinking is observed (Figure 3.8b). This might be attributed to successive redox cycling between the unquenched Cu(I) state due to reduction by dithionite and the quenched Cu(II) state due to reoxidation by O_2. Unfortunately, the complicated solution chemistry of Clip-phen–Cu complexes under reducing conditions makes validation of this hypothesis by ensemble studies difficult.

Rhodamine-labeled hydroquinone substrates **11** were also prepared with the aim of monitoring redox catalysis. H_2O_2 oxidation to the corresponding benzoquinones by Cu ions or Cu(II) phenanthroline complexes results in a strong reduction of rhodamine fluorescence [93]. Compound **11** is not suitable as a substrate for SMFS studies with immobilized individual Cu complexes because affinity for the catalyst is low and a large excess of **11** required for effective conversion, resulting in an unacceptable fluorescence background by the freely diffusing substrate molecules. However, a derivative of **11** was surface-immobilized, and H_2O_2 oxidation of individual molecules by excess Cu(II) ions in solution could be observed in preliminary SMFS studies.

3.3.2
BODIPY Substrates for Probing Reactions of Double Bonds

In one of our projects, we have chosen to start with substrates based on the framework of BODIPY dyes [94]. In comparison to other well-known dye classes in the blue-green region, fewer publications dealing with their synthesis are available (Figure 3.1). A closer look to the references, however, reveals that the majority of those on xanthene dyes (rhodamine, fluorescein) deal with their application as labels. Other alternative dye classes such as cyanine dyes exhibit detrimental photophysical properties like low fluorescence quantum yield, pronounced cis-trans isomerization, and so on [95]. Further advantages of the BODIPY dyes arise from a chemical point of view. They are easy to synthesize within a few reaction steps from cheap starting materials. Their neutral charge is also crucial for purification in large amounts. Owing to their compact size, even minor chemical changes result in detectable changes of their fluorescence properties, and knowledge about it is described in several reviews [96, 97]. Their rigid tricyclic chromophore structure with some features of cyanine dyes, and also their photophysical properties, such as high fluorescence quantum yields, long fluorescence lifetimes, and good extinction coefficients, make them promising candidates for application in single-molecule chemistry. In comparison to fluorescein dyes, which possess similar spectroscopic properties, we found higher photostability for BODIPY dyes [98].

BODIPY dyes were subjected to modifications in their meso position as the previous fluorescein dyes. Aromatic moieties at this position are perpendicular to the molecular plane but, nevertheless, affect the fluorescence properties of the core [99]. Subtle manipulation of the redox potential can be also read out by changes of the fluorescence lifetime as exemplified in a substrate for phosphoester cleavage [100]. Depending on of the redox properties of the substituents, fluorescence can be turned on and off [101]. When the electrochemical potential of the substituent is altered during a reaction, a fluorogenic probe for this particular reaction is tailored. This was exploited for studying interfacial electron transfer [61]. More applications appear feasible.

According to Figure 3.2, irreversible reactions can be investigated by two convenient ways. Fluorogenic substrates are useful for localizing catalytically active sites; their use can thus be called *activity driven*. "Reaction driven" assays, where the mechanism is subject of investigation, however, require an approach where the fluorescence color changes during the reaction. This can be exemplified by comparison of fluorescein-diacetate with our styryl-BODIPY **1** (Figure 3.3c). Fluorescence emergence requires at least three different reaction steps, that is, twice cleavage of the ester moiety and once the opening of the lactone, but their sequence after adsorption remains hidden. With a color change, at least the duration of such a complete sequence as well as desorption and turnover kinetics can be quantified. These parameters are essential for a Michaelis–Menten-like description [62].

A plethora of synthetic instructions exist where fluorescent BODIPY dyes are turned into fluorophores with shifted fluorescence. Even if these do not run already under soft conditions, they provide at least access to substrates for irreversible transformations. Hence, the side groups that are introduced by these reactions are not mandatory for fluorescence but might act as targets for chemical agents. Based on a ratiometric substrate for reactive oxygen species [16], which was also used as substrate for heterogeneous catalysis [54], we developed a styryl-expanded BODIPY core **1** as substrate for oxidation reactions [94]. From a conceptual point of view, the conjugation via the exocyclic double bond provides a unique opportunity for examining the rich chemistry of double bonds by SMFS.

As a first approach, we studied the epoxidation of the exocyclic double bond of **1** by *meta*-chloroperbenzoic acid (mCPBA) in dichloromethane. IR spectra provided evidence for a fluorescent epoxide [94]. Later on, surface-catalyzed epoxidation was ascribed to the formation of further fluorescent BODIPY derivatives [54]. The bimolecular rate constant had been determined beforehand, and pointed to the concentration range of mCPBA for maintaining useful pseudo-first-order kinetics. Immobilization turned out to be a major issue because, while the oxidizing species should have free access to the double bond, the translational mobility of the substrate must be widely suppressed. Introduction of a further terminal double bond permitted subsequent immobilization on polymeric silicone [102–104].

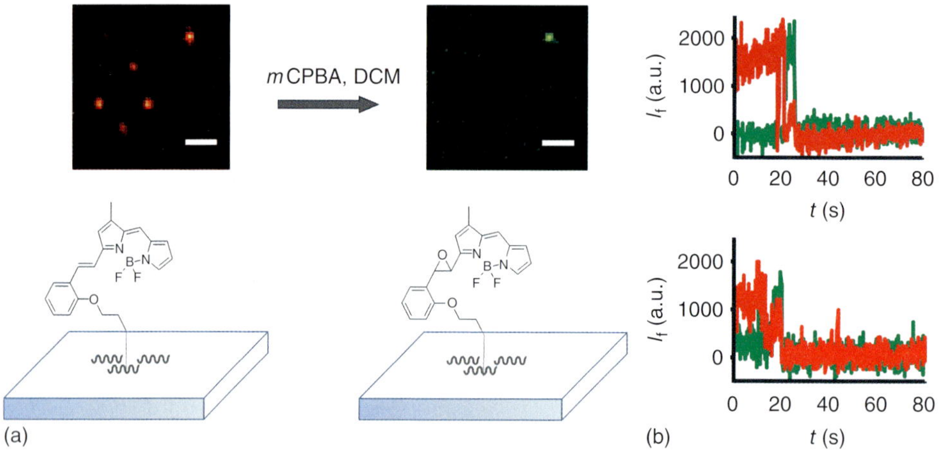

Figure 3.9 Photostability of BODIPY dyes in a single-molecule experiment. (a) The images show single, immobilized BODIPY fluorophores at different times points which were recorded by a TIRF microscope upon continuous illumination (scale bar: 1 μm). (b) Histogram representing the normalized probability density distribution of the survival time *t* of 192 single fluorophores until photobleaching.

Most surprisingly, the BODIPY dyes tolerated the harsh conditions (especially high temperature) of immobilization and turned out to be stable enough for long-lasting observation (Figure 3.9). Although the glass surface, was covered with the polymeric material, its meshes are loosely distributed and reaching into the solution thus probably maintaining a surrounding comparable to free solution.

The epoxidation reaction was studied by SMFS by pursuing more than 1000 time traces [77]. We found that roughly 7% of all molecules underwent the expected color change from orange to green (Figure 3.10), whereas all other molecules faded. Considering the cyanine-like structure of BODIPY dyes with alternating double bonds, this low yield could be interpreted as a negligible preference of mCPBA for attacking the exocyclic double bond. Nevertheless, the benefits of single-molecule studies become evident as reactive trajectories can be selected among a multitude of others. Therefore, selectivity is not a big obstacle to interpretation in the present case. Analysis of the recorded single-molecule transients revealed two different reactive pathways to the final product, as

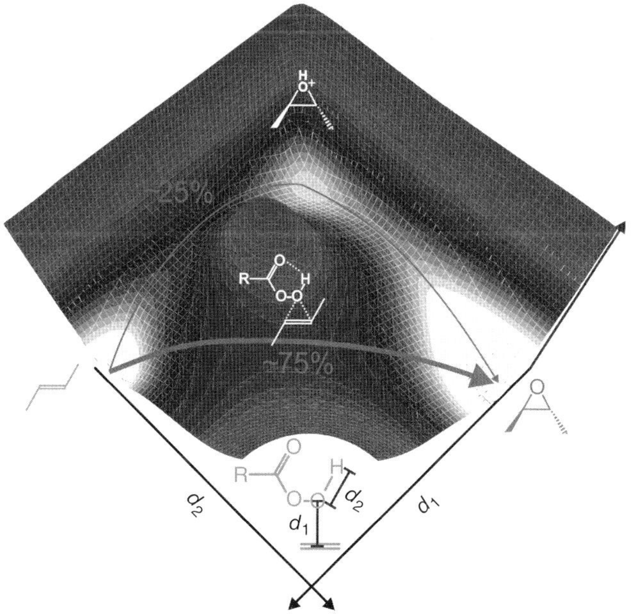

Figure 3.10 Single-molecule observation of the epoxidation reaction of single immobilized BODIPY substrate molecules. (a) TIRFM images taken from the same region at two different time points and two distinct spectral signatures of the substrate (left: 570–615 nm) and the product (right: 500–525 nm). The product emission was recorded after reaction with 0.2 mM mCPBA in dichloromethane (DCM). (b) Two exemplary single-molecule transients showing the conversion from substrate to product emission. The upper transient shows instantaneous conversion on the time scale of our measurement, which is in good agreement with the concerted epoxidation mechanism, while a short dim state occurs in between indicative for an alternative reaction pathway [108].

indicated by two distinctly different time constants for the fluorescence color change. The direct conversion via a single transition state proceeds most probably according to text-book description [105]. Its efficiency amounts to 75%. With a probability of 25%, however, an alternative pathway to the same product is taken. Following the results of theoretical chemistry computations [106, 107], the intermediate with a lifetime of 1–2 s is assigned to a protonated oxirane structure (Figure 3.11).

Being the cornerstone of forthcoming oxidation studies, the experiments demand for further advances in the probe design. At the moment, microscopic reversibility cannot be unambiguously distinguished from ubiquitous blinking. Also, the selectivity of the reaction might be improved by additional modifications of the BODIPY core, from which we expect higher yields of successful trajectories. Our ultimate goal is, however, to design more such substrates with defined reactivity, which will enable us to study mechanisms in homogeneous catalysis, like metathesis.

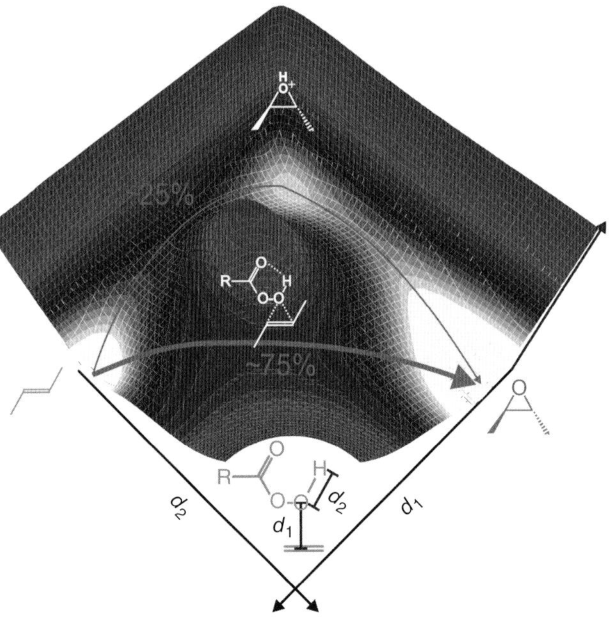

Figure 3.11 Simplified sketch of the potential energy surfaces (PESs) according to the results of Ref. [77]. The PES is spanned by two relevant distances, that is, distance d_1 of the peroxy O atom to the target double bond and O–H distance of the acid d_2. The preponderant concerted mechanism is evidenced by the simultaneous disappearance of orange and immediate emergence of green luminescence. In one-fourth of all reactive trajectories, continuous fluorescence is interrupted by population of the intermediate state. The protonated oxirane had been previously postulated [106, 107].

3.4
Discussion and Perspectives

Application of SMFS in molecular catalysis relies on the visualization of the chemical reaction or a reaction step by linking changes in the reaction center to spectroscopic changes of a fluorophore. Therefore, either the reaction must involve a fluorescent dye, or the fluorophore has to be (covalently) attached to the reaction center. Both cases might be demanding from the view point of synthesis. The involvement of a fluorophore is, however, not sufficient to make a reaction observable by fluorescence microscopy. Moreover, the emission properties must be specifically linked to the chemical transformation of interest: that is, the emission properties have to change immediately when the reaction under study occurs. This does not only require a proper synthesis route but at the same time demands well-defined photophysical and/or photochemical processes to be implemented into the molecular structure. Therefore, molecular structures suitable for specific signaling of a chemical conversion are still rare. Other important problems are related to the experimental realization of how to observe chemical conversions of single reactant molecules. This turns out to be technically demanding, especially for homogeneous reactions and it is currently limited by the time resolution of the SMFS techniques, which is in the order of $10^{-3}-10^{-6}$ s depending on the type of microscope used. Today, problems in heterogeneous catalysis can be more easily addressed with the current toolbox of SMFS. As most of these experiments make use of fluorogenic substrates, they are at present mostly limited to the measurement of turnover and the localization of the active centers. Although this is already used in the rational design of heterogeneous catalysts, more sophisticated probes reporting specific elementary reaction steps are highly demanded for single-molecule studies of both heterogeneous and homogeneous catalyses. Finally, it should be noted that, because of the fact that only specific states can be observed, implementation of other analytical techniques is highly demanded to enable interpretation within the chemical context. Despite the technical barriers and difficulties in designing suitable probes, first promising steps have been taken toward single-molecule studies in homogeneous catalysis. The probes developed so far exemplify different possible elementary reaction steps that can be followed by SMFS, starting with the formation of transition-metal complexes up to the successful conversion of a substrate in an epoxidation reaction as well as in a reversible redox reaction. With our studies, we could show that data can be acquired that is complementary to any other analytic technique. The most recognized examples so far are the kinetic measurements of a complexation reaction in thermal equilibrium and the discovery of an alternative reaction pathway in the epoxidation reaction of a double bond. With the development of further probes, such as differently labeled ligands, we envision that more complex systems can be approached with the tools of SMFS, allowing, for example, the discrimination between different successive steps occurring at catalytically active transition-metal centers.

Acknowledgments

We thank our colleagues Oliver Trapp, Peter Comba, and Stephen Hashmi for the very productive collaborations and the fruitful discussions. We gratefully acknowledge the Deutsche Forschungsgemeinschaft (DFG, Grants SFB623 and EXC81) for their generous financial support.

References

1. Moerner, W.E. and Kador, L. (1989) *Phys. Rev. Lett.*, **62**, 2535.
2. Orrit, M. and Bernard, J. (1990) *Phys. Rev. Lett.*, **65**, 2716.
3. Shera, B.E., Seitzinger, N.K., Davis, L.M., Keller, R.A., and Soper, S.A. (1990) *Chem. Phys. Lett.*, **174**, 553.
4. Xie, X.S. (2002) *J. Chem. Phys.*, **117**, 11024.
5. Smiley, R.D. and Hammes, G.G. (2006) *Chem. Rev.*, **106**, 3080.
6. Chen, P., Andoy, N.M., Benítez, J.J., Keller, A.M., Panda, D., and Gao, F. (2010) *Nat. Prod. Rep.*, **27**, 757.
7. Rotman, B. (1961) *Proc. Natl. Acad. Sci. U.S.A.*, **47**, 1981.
8. Wudl, L. and Paigen, K. (1975) *Science*, **184**, 992.
9. Hösli, P., Avrameas, S., Ullmann, A., Vogt, E., and Rodrlgot, M. (1978) *Clin. Chem.*, **24**, 1325.
10. Hirschfeld, T. (1976) *Appl. Opt.*, **15**, 2965.
11. Mroz, E.A. and Lechene, C. (1980) *Anal. Biochem.*, **102**, 90.
12. Bates, D.L. (1987) *TIBTECH*, **5**, 204.
13. Goddard, J. and Reymond, J. (2004) *Curr. Opin. Biotechnol.*, **15**, 314.
14. Tan, W. and Yeung, E.S. (1997) *Anal. Chem.*, **69**, 4242.
15. Lu, H.P., Xun, L., and Xie, X.S. (1998) *Science*, **282**, 1877.
16. Pap, E.H.W., Drummen, G.P.C., Winter, V.J., Kooji, T.W.A., Rijken, P., Wirtz, K.W.A., Op den Kamp, J.A.F., Hage, W.J., and Post, J.A. (1999) *FEBS Lett.*, **453**, 278.
17. Zhuang, X., Kim, H., Pereira, M.J.B., Babcock, H.P.W.N.G., and Chu, S. (2002) *Science*, **296**, 1473.
18. English, B.P., Min, W., van Oijen, A.M., Lee, K.T., Luo, G., Sun, H., Cherayil, B.J., Kou, S.C., and Xie, X.S. (2005) *Nat. Chem. Biol.*, **2**, 87.
19. Flomenbom, O., Velonia, K., Loos, D., Masuo, S., Cotlet, M., Engelborghs, Y., Hofkens, J., Rowan, A.E., Nolte, R.J.M., Van der Auweraer, M., De Schryver, F.C., and Klafter, J. (2005) *Proc. Natl. Acad. Sci. U.S.A.*, **102**, 2368.
20. Velonia, K., Flomenbom, O., Loos, D., Masuo, S., Cotlet, M., Engelborghs, Y., Hofkens, J., Rowan, A.E., Klafter, J., Nolte, R.J.M., and De Schryver, F.C. (2005) *Angew. Chem. Int. Ed.*, **44**, 560.
21. Min, W., English, B.P., Luo, G., Cherayil, B.J., Kou, S.C., and Xie, X.S. (2005) *Acc. Chem. Res.*, **38**, 923.
22. Benítez, J.J., Keller, A.M., Huffman, D.L., Yatsunyk, L.A., Rosenzweig, A.C., and Chen, P. (2010) *Faraday Discuss.*, **148**, 71.
23. Benítez, J.J., Keller, A.M., and Chen, P. (2010) *Methods Enzymol.*, **472**, 41.
24. Jiang, Y., Douglas, N.R., Conley, N.R., Miller, E.J., Frydman, J., and Moerner, W.E. (2011) *Proc. Natl. Acad. Sci. U.S.A.*, **108**, 16962.
25. Tabares, L.C., Kostrz, D., Elmalk, A., Andreoni, A., Dennison, C., Aartsma, T.J., and Canters, G.W. (2011) *Chem. Eur. J.*, **17**, 12015.
26. Collinson, M.M. and Wightman, R.M. (1995) *Science*, **268**, 1883.
27. Christ, T.K.F., Bordat, P., and Basché, T. (2001) *Angew. Chem. Int. Ed.*, **40**, 4192.
28. De Vos, D.E., Dams, M., Sels, B.F., and Jacobs, P.A. (2002) *Chem. Rev.*, **102**, 3615.
29. Wöll, D., Braeken, E., Deres, A., De Schryver, F.C., Uji-i, H., and Hofkens, J. (2009) *Chem. Soc. Rev.*, **38**, 313.
30. Xu, W., Shen, H., Liu, G., and Chen, P. (2009) *Nano Res.*, **2**, 911.

31. Chen, P., Xu, W., Zhou, X., Panda, D., and Kalininskiy, A. (2009) *Chem. Phys. Lett.*, **470**, 151.
32. Chen, P., Zhou, X., Shen, H., Andoy, N.M., Choudhary, E., Han, K.S., Liu, G., and Meng, W. (2010) *Chem. Soc. Rev.*, **39**, 4560.
33. De Cremer, G., Sels, B.F., De Vos, D.E., Hofkens, J., and Roeffaers, M.B.J. (2010) *Chem. Soc. Rev.*, **39**, 4703.
34. Buurmans, I.L.C. and Weckhuysen, B.M. (2012) *Nat. Chem.*, **4**, 873.
35. Tachikawa, T. and Majima, T. (2012) *Langmuir*, **28**, 8933.
36. Zumbusch, A., Fleury, L., Brown, R., Bernard, J., and Orrit, M. (1993) *Phys. Rev. Lett.*, **70**, 3584.
37. Macklin, J.J., Trautman, J.K., Harris, T.D., and Brus, L.E. (1996) *Science*, **272**, 255.
38. Hu, D., Yu, J., and Barbara, P.F. (1999) *J. Am. Chem. Soc.*, **121**, 6936.
39. Huser, T., Yan, M., and Rothberg, L.J. (2000) *Proc. Natl. Acad. Sci. U.S.A.*, **97**, 11187.
40. Uji-i, H., Melnikov, S.M., Deres, A., Bergamini, G., De Schryver, F.C., Herrmann, A., Müllen, K., Enderlein, J., and Hofkens, J. (2006) *Polymer*, **47**, 2511.
41. Wöll, D., Uji-i, H., Schnitzler, T., Hotta, J.-I., Dedecker, P., Herrmann, A., De Schryver, F.C., Müllen, K., and Hofkens, J. (2008) *Angew. Chem. Int. Ed.*, **47**, 783.
42. Esfandiari, N.M. and Blum, S.A. (2011) *J. Am. Chem. Soc.*, **133**, 18145.
43. Roeffaers, M.B.J., Sels, B.F., Loos, D., Kohl, C., Müllen, K., Jacobs, P.A., Hofkens, J., and De Vos, D.E. (2005) *ChemPhysChem*, **6**, 2295.
44. Roeffaers, M.B.J., Sels, B.F., Uji-i, H., De Schryver, F.C., Jacobs, P.A., De Vos, D.E., and Hofkens, J. (2006) *Nature*, **439**, 572.
45. Buurmans, I.L.C., Ruiz-Martínez, J., Knowles, W.V., van der Beek, D., Bergwerff, J.A., Vogt, E.T.C., and Weckhuysen, B.M. (2011) *Nat. Chem.*, **3**, 862.
46. Seebacher, C., Hellriegel, C., Deeg, F.-W., Bräuchle, C., Altmaier, S., Behrens, P., and Müllen, K. (2002) *J. Phys. Chem. B*, **106**, 5591.
47. Fu, Y., Ye, F., Sanders, W.G., Collinson, M.M., and Higgins, D.A. (2006) *J. Phys. Chem. B*, **110**, 9164.
48. Kirstein, J., Platschek, B., Jung, C., Brown, R., Bein, T., and Bräuchle, C. (2007) *Nat. Mater.*, **6**, 303.
49. Roeffaers, M.B.J., Sels, B.F., Uji-i, H., Blanpain, B., L'hoëst, P., Jacobs, P.A., De Schryver, F.C., Hofkens, J., and De Vos, D.E. (2007) *Angew. Chem. Int. Ed.*, **46**, 1706.
50. Roeffaers, M.B.J., De Cremer, G., Libeert, J., Ameloot, R., Dedecker, P., Bons, A.-J., Bückins, M., Martens, J.A., Sels, B.F., De Vos, D.E., and Hofkens, J. (2009) *Angew. Chem. Int. Ed.*, **48**, 9285.
51. Betzig, E., Patterson, G.H., Sougrat, R., Lindwasser, O.W., Olenych, S., Bonifacino, J.S., Davidson, M.W., Lippincott-Schwartz, J., and Hess, H.F. (2006) *Science*, **313**, 1642.
52. Rust, M.J., Bates, M., and Zhuang, X. (2006) *Nat. Methods*, **3**, 793.
53. Heilemann, M., van de Linde, S., Schüttpelz, M., Kasper, R., Seefeldt, B., Mukherjee, A., Tinnefeld, P., and Sauer, M. (2008) *Angew. Chem. Int. Ed.*, **47**, 6172.
54. De Cremer, G., Roeffaers, M.B.J., Bartholomeeusen, E., Lin, K., Dedecker, P., Pescarmona, P.P., Jacobs, P.A., De Vos, D.E., Hofkens, J., and Sels, B.F. (2010) *Angew. Chem. Int. Ed.*, **49**, 908.
55. Tachikawa, T., Yamashita, S., and Majima, T. (2010) *Angew. Chem. Int. Ed.*, **49**, 432.
56. Xu, W., Kong, J.S., Yeh, Y.-T.E., and Chen, P. (2008) *Nat. Mater.*, **7**, 992.
57. Xu, W., Kong, J.S., and Chen, P. (2009) *Phys. Chem. Chem. Phys.*, **11**, 2767.
58. Zhou, X., Xu, W., Liu, G., Panda, D., and Chen, P. (2010) *J. Am. Chem. Soc.*, **132**, 138.
59. Ochoa, M.A., Zhou, X., Chen, P., and Loring, R.F. (2011) *J. Chem. Phys.*, **135**, 174509.
60. Tachikawa, T., Fujitsuka, M., and Majima, T. (2007) *J. Phys. Chem. C*, **111**, 5259.
61. Tachikawa, T., Wang, N., Yamashita, S., Cui, S.-C., and Majima, T. (2010) *Angew. Chem. Int. Ed.*, **49**, 8593.

62. Tachikawa, T., Yonezawa, T., and Majima, T. (2013) *ACS Nano*, **7**, 263.
63. Xu, W., Shen, H., Kim, Y.J., Zhou, X., Liu, G., Park, J., and Chen, P. (2009) *Nano Lett.*, **9**, 3968.
64. Shen, H., Xu, W., and Chen, P. (2010) *Phys. Chem. Chem. Phys.*, **12**, 6555.
65. Over, H., Kim, Y.D., Seitsonen, A.P., Wendt, S., Lundgren, E., Schmid, M., Varga, P., Morgante, A., and Ertl, G. (2000) *Science*, **287**, 1474.
66. Lauritsen, J.V., Vang, R.T., and Besenbacher, F. (2006) *Catal. Today*, **111**, 34.
67. Novo, C., Funston, A.M., and Mulvaney, P. (2008) *Nat. Nanotechnol.*, **3**, 598.
68. Aramburo, L.R., Liu, Y., Tyliszczak, T., De Groot, F.M.F., Andrews, J.C., and Weckhuysen, B.M. (2013) *ChemPhysChem*, **14**, 496.
69. van Dijk, E., Hernando, J., García-López, J.-J., Crego-Calama, M., Reinhoudt, D., Kuipers, L., García-Parajó, M., and van Hulst, N. (2005) *Phys. Rev. Lett.*, **94**, 78302.
70. Roeffaers, M.B.J., De Cremer, G., Uji-i, H., Muls, B., Sels, B.F., Jacobs, P.A., De Schryver, F.C., De Vos, D.E., and Hofkens, J. (2007) *Proc. Natl. Acad. Sci. U.S.A.*, **104**, 12603.
71. Canham, S.M., Bass, J.Y., Navarro, O., Lim, S.-G., Das, N., and Blum, S.A. (2008) *Organometallics*, **27**, 2172.
72. Esfandiari, N.M., Wang, Y., Bass, J.Y., Cornell, T.P., Otte, D.A.L., Cheng, M.H., Hemminger, J.C., McIntire, T.M., Mandelshtam, V.A., and Blum, S.A. (2010) *J. Am. Chem. Soc.*, **132**, 15167.
73. Esfandiari, N.M., Wang, Y., McIntire, T.M., and Blum, S.A. (2011) *Organometallics*, **30**, 2901.
74. Kiel, A., Kovacs, J., Mokhir, A., Krämer, R., and Herten, D.-P. (2007) *Angew. Chem. Int. Ed.*, **46**, 3363.
75. Ameloot, R., Roeffaers, M., Baruah, M., De Cremer, G., Sels, B., De Vos, D.E., and Hofkens, J. (2009) *Photochem. Photobiol. Sci.*, **8**, 453.
76. Schütze, F., Stempfle, B., Jüngst, C., Wöll, D., Zumbusch, A., and Mecking, S. (2012) *Chem. Commun.*, **48**, 2104.
77. Rybina, A., Lang, C., Wirtz, M., Grußmayer, K., Kurz, A., Maier, F., Schmitt, A., Trapp, O., Jung, G., and Herten, D.-P. (2013) *Angew. Chem. Int. Ed.*, **52**, 6322–6325.
78. Johnson, I. and Spence, M. (2010) *The Molecular Probes Handbook*, Life Technologies Corporation, Paisley.
79. Piwoński, H., Stupperich, C., Hartschuh, A., Sepioł, J., Meixner, A., and Waluk, J. (2005) *J. Am. Chem. Soc.*, **127**, 5302.
80. Montano, L.A. and Ingle, J.D. Jr., (1979) *Anal. Chem.*, **51**, 919.
81. Bueno, C., Villegas, M.L., Bertolotti, S.G., Previtali, C.M., Neumann, M.G., and Encinas, M.V. (2002) *Photochem. Photobiol.*, **76**, 385.
82. Schwering, M., Kiel, A., Kurz, A., Lymperopoulos, K., Sprödefeld, A., Krämer, R., and Herten, D.-P. (2011) *Angew. Chem. Int. Ed.*, **50**, 2940.
83. Kierat, R. and Krämer, R. (2005) *Bioorg. Med. Chem. Lett.*, **15**, 4824.
84. Graf, N., Goeritz, M., and Krämer, R. (2006) *Angew. Chem. Int. Ed.*, **45**, 4013.
85. Brox, D., Kiel, A., Wörner, S.J., Pernpointner, M., Comba, P., Martin, B., and Herten, D.-P. (2013) *PLoS One*, **8**, e58049.
86. Kovbasyuk, L. and Krämer, R. (2006) *Inorg. Chem. Commun.*, **9**, 586.
87. Comba, P., Krämer, R., Mokhir, A., Naing, K., and Schatz, E. (2006) *Eur. J. Inorg. Chem.*, **21**, 4442.
88. Mokhir, A. and Krämer, R. (2005) *Chem. Commun.*, 2244.
89. Mokhir, A., Kiel, A., Herten, D.-P., and Krämer, R. (2005) *Inorg. Chem.*, **44**, 5661–5666.
90. Szelke, H., Schübel, S., Harenberg, J., and Krämer, R. (2010) *Chem. Commun.*, **46**, 1667.
91. Kim, S.Y. and Hong, J.-I. (2009) *Tetrahedron Lett.*, **50**, 2822.
92. Sprödefeld, A., Kiel, A., Herten, D.-P., and Krämer, R. (2013) *Z. Anorg. Allg. Chem.*, **639**, 1636–1639.
93. Thaler, B., Kierat, R., and Krämer, R. (2010) *Bioorg. Med. Chem. Lett.*, **20**, 1457.
94. Schmitt, A., Hinkeldey, B., Hötzer, B., and Jung, G. (2009) *J. Phys. Org. Chem.*, **22**, 1233.
95. Widengren, J. and Schwille, P. (2000) *J. Phys. Chem. A*, **104**, 6416.

96. Loudet, A. and Burgess, K. (2007) *Chem. Rev.*, **107**, 4891.
97. Ulrich, G., Ziessel, R., and Harriman, A. (2008) *Angew. Chem. Int. Ed.*, **47**, 1184.
98. Hinkeldey, B., Schmitt, A., and Jung, G. (2008) *ChemPhysChem*, **9**, 2019.
99. Boens, N., Leen, V., and Dehaen, W. (2012) *Chem. Soc. Rev.*, **41**, 1130.
100. Jacob, M., Schmitt, A., and Jung, G. (2008) *J. Fluoresc.*, **18**, 639.
101. Gabe, Y., Urano, Y., Kikuchi, K., Kojima, H., and Nagano, T. (2004) *J. Am. Chem. Soc.*, **126**, 3357.
102. Lang, C., Gärtner, U., and Trapp, O. (2010) *Chem. Commun.*, **47**, 391.
103. Spallek, M.J., Storch, G., and Trapp, O. (2012) *Eur. J. Org. Chem.*, **2012**, 3929.
104. Cousin, H., Trapp, O., Peulon-Agasse, V., Pannecoucke, X., Banspach, L., Trapp, G., Jiang, Z., Combret, J.C., and Schurig, V. (2003) *Eur. J. Org. Chem.*, **2003**, 3273.
105. Bartlett, P.D. (1950) *Rec. Chem. Progr.*, **11**, 47.
106. Shi, H., Zhang, Z., and Wang, Y. (2005) *J. Mol. Catal. A: Chem.*, **238**, 13.
107. Bach, R.D., Canepa, C., Winter, J.E., and Blanchette, P.E. (1997) *J. Org. Chem.*, **62**, 5159.
108. Callan, J.F., de Silva, A.P., and Magri, D.C. (2005) *Tetrahedron*, **61**, 8551.

4
Intermediates and Elementary Reactions in Gold Catalysis

A. Stephen K. Hashmi

4.1
Introduction

The success of homogeneous gold catalysis [1] was accompanied by intensive efforts to understand the mechanisms of the new conversions [2]. None of the many new gold-catalyzed reactions proceeds through only one transition state, we always deal with multistep mechanisms. Many of the reactions have the first steps in common, and then diverge at a later stage of the sequence of elementary steps of the catalytic cycles, which ultimately leads to the manifold of different products that can be obtained.

These initial steps involve the π-coordination of the unsaturated substrate which activates the substrate for the attack of an intra- or intermolecular nucleophile. At the other end of the catalytic cycle, typically the last step is a proto-deauration, setting free the LAu^+ fragment for the next turnover of the catalytic cycle.

4.2
The Initial Step: π-Coordination of the Substrate

In most gold-catalyzed reactions, the substrates are unsaturated organic molecules. Then, π-coordination to gold for alkynes, allenes, and olefins as substrates activates the molecules for the subsequent reaction steps [3]. These organometallic π-complexes of unsaturated organic substrates have been known for a long time; because of the dominance of gold(I) catalysts in gold catalysis, most examples have been prepared with LAu^+ fragments typical for catalysis reactions [4]. Previous reports on gold–alkyne complexes lack an unambiguous proof of their structure [5]. A few of the catalytic cycles are assumed to involve gold(III) [6], and for these strongly oxidizing gold fragments, X-ray crystal structure analyses are known [7] and thorough NMR studies have been published [8].

By a detailed computational investigation, including relativistic treatment, insights into the mechanism of activation by π-complexation to gold could be obtained. For both gold(I) and gold(III), the π-coordination to propyne has been

Molecular Catalysts: Structure and Functional Design, First Edition. Edited by Lutz H. Gade and Peter Hofmann.

calculated [9]. This study also allowed explaining the experimentally observed large rate differences [10] between gold(III) and the isoelectronic platinum(II) species. For both gold(III)–alkyne or gold(I)–alkyne complexes, the lowest unoccupied molecular orbitals (LUMO) energies, which are the crucial frontier orbitals for the typical nucleophilic attack, are significantly lower than those of the corresponding platinum(II)–alkyne complexes [9a], which also lowers the barrier for the subsequent attack of a nucleophile to this frontier orbital [9].

4.3 The Nucleophilic Addition: Vinylgold and Alkylgold Intermediates

After activation of the alkyne or allene by the interaction with the gold catalyst (complex **2**) [9], a backside attack of the nucleophile (**Nu**), either intramolecular or intermolecular, delivers a vinylgold species [9, 11]. Such vinylgold(I) compounds were assumed to be intermediates in the catalytic cycles of gold-catalyzed reactions of alkynes and allenes. The experimental observation of these central vinylgold intermediates **3** was not possible, and the subsequent conversion with the electrophile **E** seemed to be fast and no detectable concentration of **3** could be accumulated (Scheme 4.1) [12]. On the other hand, the stoichiometric preparation of vinylgold compounds by other metalorganic routes, for example, from vinyllithium compounds, vinylmagnesium compounds, or vinylboronic acids, and gold chloro complexes, was successful [13]. This indicated that this class of organogold compounds should be isolable from catalytic cycles, too.

Only in 2009/2010, a proof for their participation in catalytic cycles could be obtained by isolation of some of these species. This, in the end, opened up a preparative access to these species, which for the first time enabled chemists to expand this aspect of organogold chemistry in new directions.

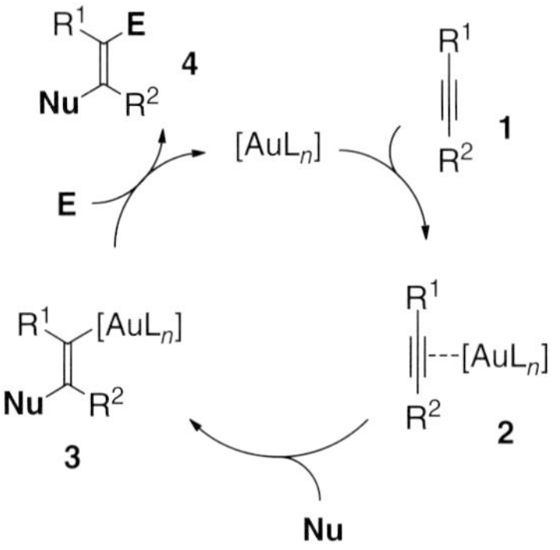

Scheme 4.1 Not detected *in situ*: vinylgold intermediates as key intermediates in reactions of alkynes **1**.

The first and important evidence came from the group of Hammond [14]. They used allenic esters **5**. Stoichiometric experiments with the cationic gold(I) species

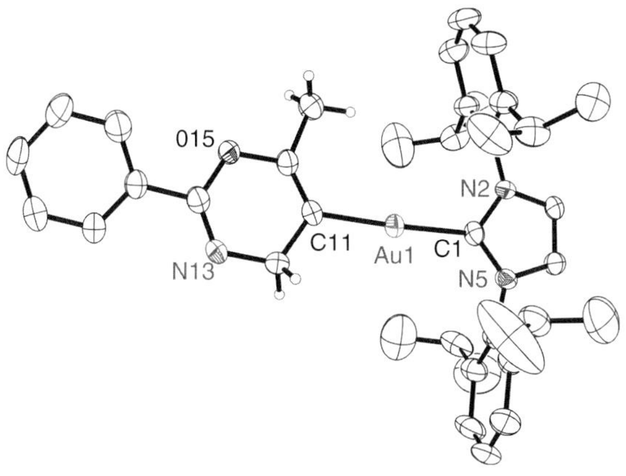

Figure 4.1 Molecular structure of a vinylgold(I) species **12** in the solid state.

provided good yields of the stable vinylgold(I) complexes **6** as shown in Figure 4.2. With the phosphane ligands used in the investigation, the compounds **6** are thermally stable and, probably because of the acceptor group on the alkene, need 80 °C and 2 equiv of a strong Brønsted acid for protonation. Thus, the compounds **6** are definitively not part of an efficient catalytic cycle (as the stability of the gold(I) catalyst at this temperature is limited and usually efficient homogeneous gold catalysis reactions often even proceed readily below room temperature). They are stoichiometrically prepared organogold compounds. Indeed, allenic esters **6**, very different from the corresponding allenyl ketones [15], cannot be converted in a catalytic manner. Nevertheless, the elemental reactions of the formation of **6**, the coordination of the allene at the more electron-rich double bond in the intermediate **A**, and the subsequent attack of the nucleophilic carbonyl oxygen atom (shown in bold) to deliver intermediate **B** parallel the single steps for the catalytic reactions of alkynes or allenes shown in Figure 4.1. Only the electrophilic attack of the proton (shown in bold) does not occur readily; this is probably an effect of the reduced electron density in the Michael acceptor substructure, especially at the carbon atom bearing the gold in the C–C double bond of **6**. This allows the easy isolation of **6** without any precautions in high preparative yields. A number of representatives of **6** could even be characterized by crystal structure analyses (Scheme 4.2).

The group of Gagné [16], soon after Hammond's publication, was able to isolate a closely related vinylgold(I) species **8**, again with phosphane ligand on gold and from an intramolecular reactions of an allene 7, this time a hydroarylation reaction (Scheme 4.3). One significant and important difference is the fact that **8** is a true intermediate of a catalytic reaction, which is due to the lack of an acceptor on the double bond, while Hammond's reactions are stoichiometric in gold. On the other hand, this did not represent a systematic synthetic access to these compounds; in this single case, the vinylgold compound crystallized spontaneously, and thus was "intercepted" by the low solubility. The single-crystal structure analysis unambiguously confirmed the structural assignment. The value of this investigation

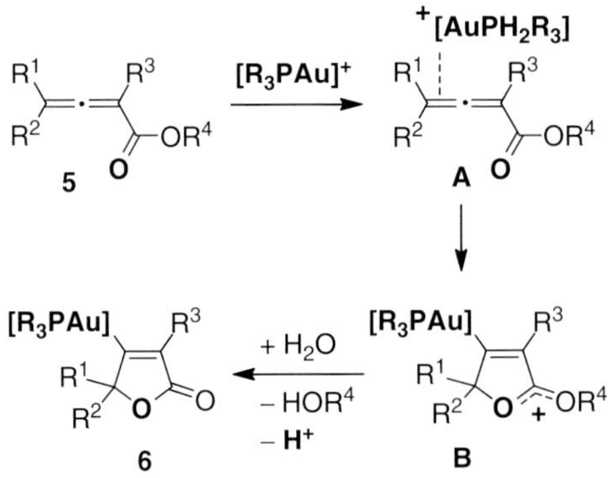

Scheme 4.2 Stoichiometry from allenes with phosphane ligands: Hammond's stable acceptor substituted vinylgold compounds **6**.

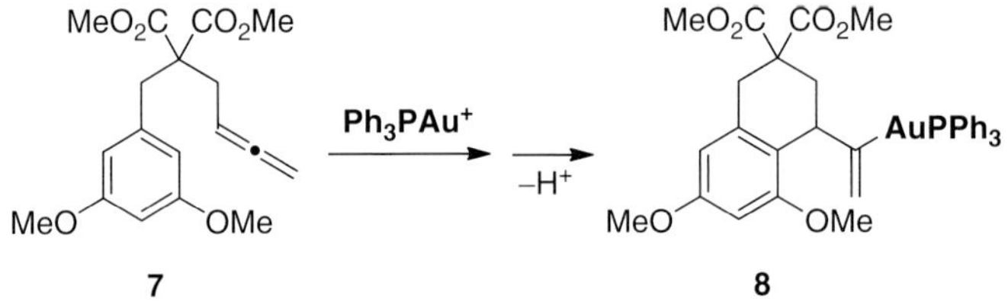

Scheme 4.3 From allenes with phosphane ligands: Gagné's vinylgold complexes without acceptor substituent.

lies mainly in the first characterization of an allene-derived vinylgold intermediate without a stabilizing acceptor substituent and, additionally, the experimental evidence for intermediates with two gold atoms as potential (previously unrecognized) resting states.

Our group turned to alkynes as substrates, the much more frequently used class of substrates than the allenes. Scheme 4.4 shows the cycloisomerization of *N*-propargylcarboxamides **9**, which in the case of terminal alkynes ($R^2 = H$) form oxazoles **10** with gold(III) catalysts [17] and alkylidene oxazolines **11** with gold(I) catalysts [18]. With internal alkynes ($R^2 =$ alkyl), the final product is not a five-membered ring (as in **10** and **11**). Instead of a 5-*exo-trig* cyclization, a 6-*endo-dig* cyclization leads to a six-membered ring in the product **13**. This not only represented a remarkable switch in the chemoselectivity of the catalytic conversion; when we used an N-heterocyclic carbene (NHC) ligand in the presence of a simple base like triethylamine to lower the proton activity (protons are set free from intermediate **C**), we were able isolate the vinylgold complexes **12** in very good yields on preparative scale in experiments stoichiometric in gold [19]. Compound **12** really is an isolable intermediate of the catalytic cycle because, as one would demand for such a species, both **12** in non-absolute solvent and **12** in the presence of acid still catalyze the transformation of **9**. The simple trick of using a tertiary amine

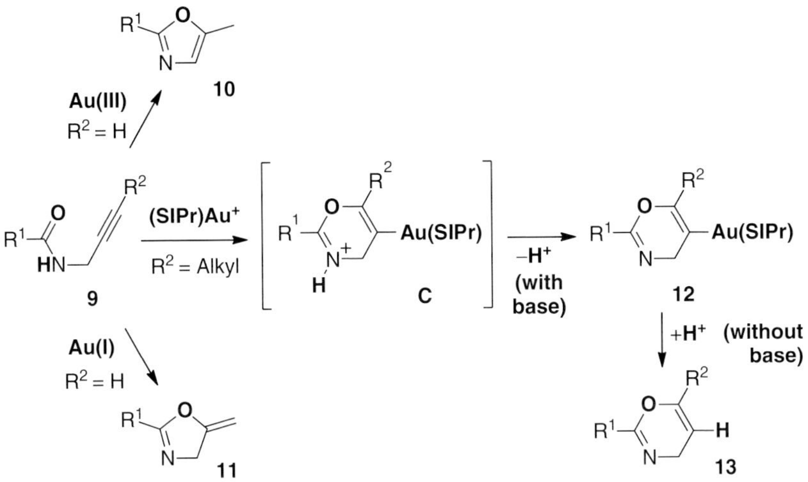

Scheme 4.4 Stable intermediates and stoichiometric access: vinylgold complexes of N-heterocyclic ligands obtained in the presence of a base.

as a base allowed the isolation of related intermediates of gold-catalyzed conversions of alkynes, for example, hydroaminations and hydroalkoxylation reactions [20]. And in cooperation with Nolan's group [21], the combination gold catalyst/triethylamine could also be replaced by the corresponding gold(I) hydroxides.

The structure of one derivative **12** ($R^1 = Ph$, $R^2 = Me$) was proven by an X-ray crystal structure analysis. A picture is shown in Figure 4.1. In this structure, it is interesting to note that both gold–carbon distances, that is, the one to the vinyl carbon and the one to the carbon atom of the NHC, are almost identical.

With a phosphane ligand instead of the NHC ligand, and in the presence of triethylamine, the vinylgold intermediates could be detected by *in situ* NMR spectroscopy, but all attempts for the isolation failed [19].

Subsequent investigations also tried to address the corresponding alkylgold intermediates. These can be formed in two ways: either by reactions with alkenes, or as allylgold intermediates by reactions at allenes in which the nucleophile ends up at the central position of the allene.

By using our protocol with triethylamine as the auxiliary base, the group of Toste [22] succeeded in isolating and investigating such an alkylgold(I) intermediate (**15**) of the intramolecular hydroamination in substrate **14**, using a different transfer reagent for the Ph_3PAu^+ fragment (Scheme 4.5).

We investigated the cyclization of allenic amides **16** in detail (Scheme 4.6). Unlike the related propargylic amides, which for terminal alkynes deliver oxazoles [17, 18] and for internal alkynes deliver oxazines [19], the allenic substrates always gave oxazines as products [23]. If one considers the mechanism, it immediately becomes obvious that an allylgold(I) intermediate must be involved. Depending on the ligand on gold, the product **17** was obtained either exclusively or together with **18**. But which of the two conceivable σ-allyl intermediates, gold

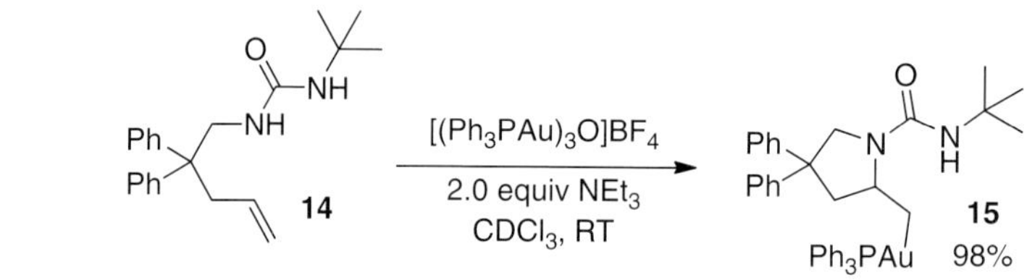

Scheme 4.5 Isolable alkylgold(I) compound **15** from alkene **14**.

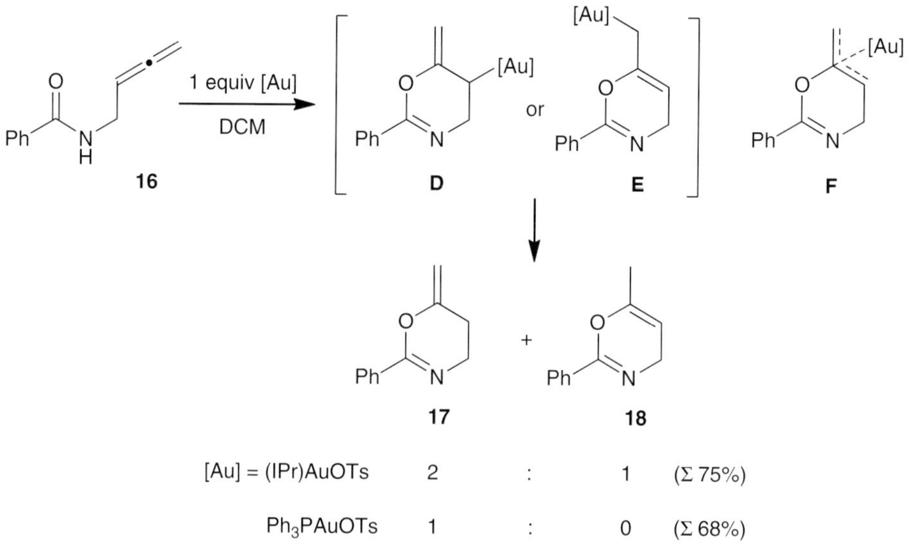

Scheme 4.6 Which allylgold(I) species is involved?

either on the exocyclic (**D**) or the endocyclic (**E**) position, is involved and what about a π-allyl species **F**? We first conducted *in situ* ^{31}P NMR spectroscopy on the catalysis mixture (for the case of the triphenylphosphane ligand). These experiments indicated that there was only one allyl-type intermediate in the reaction mixture, and low-temperature NMR did not show any dynamic effects. Quenching with D^+ instead of H^+ showed that the electrophile exclusively attacked at the endocyclic terminus of the allyl system. Finally, computational studies (Figure 4.2; with Me_3P instead of Ph_3P, previous benchmark studies had shown that this simplification is allowed) [24] showed that an exocyclic position of the gold was preferred by about 4 kcal mol^{-1} and that the π-allyl gold(I) species was a transition state of a 1,3-metallotropic shift in the allyl system. Overall, obviously the allylic proto-deauration process proceeds by an S_E' reaction, which involves a replacement of the LAu^+ species by the proton with inversion of the allylic system.

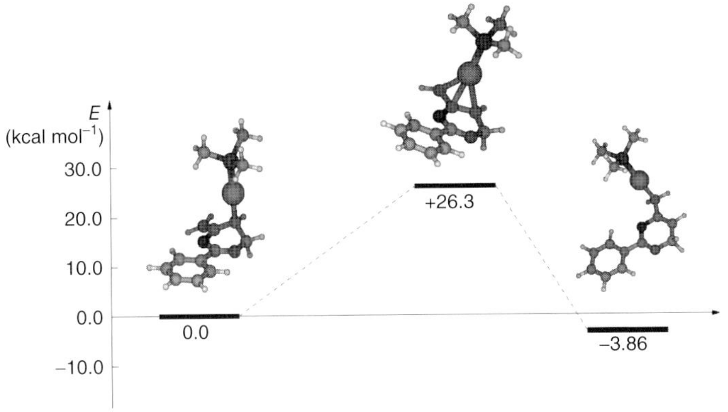

Figure 4.2 The π-allylgold(I) species is the transition state of the 1,3-metallotropic shift.

4.4 The Reaction of the Organogold Intermediates with Electrophiles

The vinylgold intermediate **3** (Scheme 4.1) then typically reacts with an electrophile (**E**); in most of the literature this is a proton. Only a few cases of the use of halogen electrophiles have been reported [25]. This use of halogen electrophiles is problematic because, similar to halolactonization reactions, the electrophilic halide can possibly induce the cyclization without the gold.

Subsequently, palladium(II) was investigated as a potential electrophile in transmetallation reactions [26].

The Hammond-type organogold compounds **19**, even though not highly reactive toward electrophiles, can efficiently be used in cross-coupling chemistry (Scheme 4.7). This adds a new dimension to the synthetic potential and efficiency of the organic chemistry of gold [26]. Also, other organogold complexes could be used. Scheme 4.7 shows the catalytic cycle for these reactions, which form new C–C bonds. Initially, palladium(0) is formed by stoichiometric reduction of a stable palladium(II) catalyst precursor. The subsequent reaction is catalytic

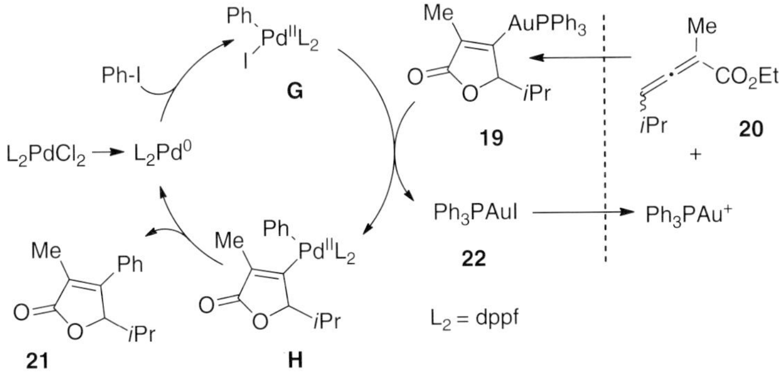

Scheme 4.7 Transmetallation from gold(I) to palladium(II) as a new key step.

in palladium, forming the intermediate **G** by oxidative addition of an aryl halide such as Ph–I to a palladium(0) complex bearing a bidentate ligand (dppf, 1,1′-bis(diphenylphosphino)ferrocene). **G** now forms the diorganopalladium(II) species **H** by a transmetalation reaction with the vinylgold compound **19**. The choice of the ligand is crucial; a ligand exchange between the two different transition metal centers can be avoided if palladium bears a bidentate ligand which, due to the chelate effect, has a high binding constant and gold(I) bears a monodentate ligand. Compound **H** then, by reductive elimination, sets free the cross-coupling product **21**. This last step then closes the catalytic cycle by re-forming palladium(0). The scope of the method is impressive; a number of different aryl groups R were successfully converted in excellent yields. Most unfortunately, these reactions are stoichiometric in gold. During the transmetallation step, Ph_3PAuI (**22**) is also formed. It can be re-isolated in almost quantitative yield and, after removal of the halide by Ag(I), be reused. Instead of such a batchwise conversion, a process catalytic in both metals would be desirable. Still, the current results are of high interest, as in synthesis they extend the scope from a simple cycloisomerization to an extension of the carbon framework by additional C–C bond formation.

Possible variations for a conversion using catalytic amounts of both metals were also investigated. Since the halide of the organic substrate undergoing oxidative addition at palladium ultimately poisons the gold catalyst, the first obvious idea was the use of the corresponding triflates. This never worked in our hands, but another group published successful conversions using this method [27]. When reinvestigating this, we could never repeat such experiments, but from the supporting material we learned that the other group had prepared the organogold compounds *in situ*, which meant that there was 1 equiv of halide ions in the solution (which certainly once again would poison the gold "catalyst"). Thus we investigated this transmetallation step in detail. Subsequently, and with regard to the transition state, Espinet [28] published a paper. But with regard to the question of the use of triflate or similar leaving groups, the overall thermodynamics was more important. In a detailed computational study of the whole catalytic cycle, we could show that a crucial contribution for the overall thermodynamics originates from the high bonding energy between the LAu^+ and the X^- fragments in the case of the halides. With triflates, this energy is much lower, and the overall thermodynamics does not allow the conversion any more. Additional and specifically designed experiments confirmed these computational results. The interaction of gold(I) with hard oxygen anions in general is weak; for example, gold(I) phenolate complexes are a class of compounds that are quite difficult to prepare [29].

Another possibility would be the use of allylic esters, which after a gold-catalyzed cycloisomerization with the carbonyl oxygen atom as the nucleophile deliver activated allylic intermediates which at the same time contain a vinylgold substructure. After transfer of an allyl cation to palladium(0), an oxidative addition to palladium, the vinylgold intermediate could transfer the organic moiety to palladium(II). A final reductive elimination would close the catalytic cycle. At the same time, no halide that potentially could deactivate the cationic gold(I) catalyst would be present. Indeed, Blum *et al.* [30] presented such systems. But

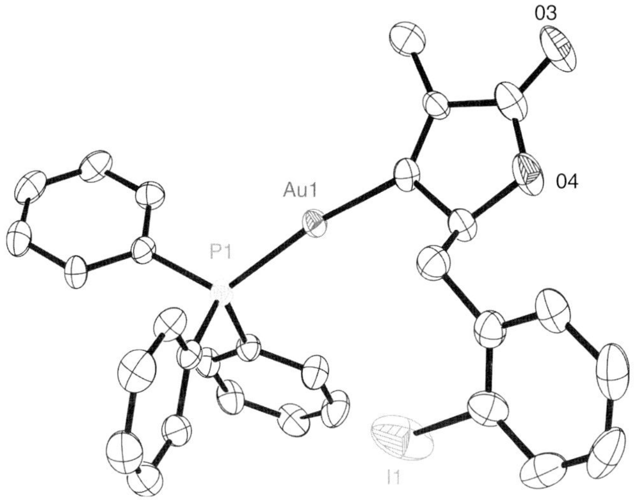

Figure 4.3 A stable iodoarylgold(I) compound that does not undergo intramolecular oxidative addition via a six-membered transition state.

subsequent control experiments showed that there are three scenarios [26c]. First of all, in the case of less electron-rich Hammond-type vinylgold intermediates the reaction does not need gold(I), and it can be conducted with palladium only. Second, in the case of more nucleophilic, electron-rich, enolester-type vinylgold intermediates, palladium is not needed, and gold(I) alone can catalyze the whole reaction. Third, only in the case of the unsubstituted allyl groups and similar systems the assistance of the palladium is needed in the case of the enolester-type intermediates.

In the context of that investigation, a vinylgold(I) compound bearing an aryl iodide in the side chain could be isolated as a perfectly stable compound (Figure 4.3). This nicely demonstrated the synthetic orthogonality of the d^8 systems gold(I) and palladium(0). A di-coordinated palladium(0) species that could undergo an intramolecular oxidative addition via a six-membered transition state of an aryl iodide would not be stable; but the corresponding gold(I) compound is.

4.5 "Vinylidene" Gold(I) Intermediates

4.5.1 Setting the Stage

Five years ago, Houk, Toste, and coworkers [31] suggested a fascinating new reaction mechanism. The cycloisomerization of 1,5-allenynes **23** by the $[(Ph_3PAu)_3O]BF_4$ catalyst delivers cross-conjugated trienes **24** (Scheme 4.8).

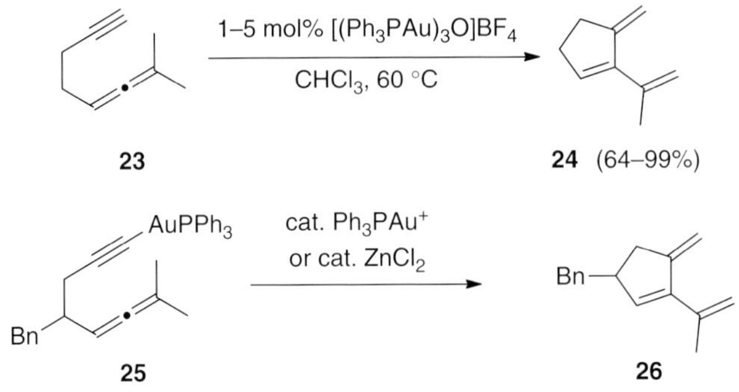

Scheme 4.8 Cyclization of allenynes **23** and **25**.

An important piece of evidence was that the reaction was limited to terminal alkynes. Furthermore, a deuterium labeling experiment showed that the hydrogen transfer during the reaction was diastereoselective. A crossover experiment indicated an intramolecular hydrogen transfer. The observation that the hydrogen atom of the terminal alkyne was exchanged for deuterium if the reaction was run in MeOD was crucial. The primary kinetic isotope effect of this formal 1,5-H-shift was 1.8 – 1.9, which is significantly smaller than the concerted pericyclic 1,5-sigmatropic shifts. A more detailed investigation of different mechanistic possibilities by computational chemistry was accompanied by additional experiments, one of them being the preparation of the gold(I) acetylide **25**. Compound **25** is stable, but when catalytic amounts of cationic gold(I) catalysts or Lewis acids such as zinc(II) are added, the product **26** is formed.

As the final conclusion, the authors assumed that a mechanism "involving nucleophilic addition of an allene double bond to a phosphinegold-complexed phosphinegold acetylide" to be more likely than an "oxidative cyclization or simple nucleophilic addition to phosphinegold-complexed substrate." The corresponding catalytic cycle for the more probable pathway is shown in Scheme 4.9.

Only a year later, Gagosz and coworkers [32] published a synthesis of 10-membered medium-sized cycloalkynes **28** from 1,10-diynes **27**. Like in the earlier investigation of Houk and Toste, the ligand was a bulky phosphane ligand. Two important mechanistic proposals were included in this publication: the one shown in Scheme 4.10a which is based on the proposal of Houk and Toste [31], and also the alternative pathway shown in Scheme 4.10b. Both pathways are in accordance with the deuterium isotope labeling experiments conducted. The Houk/Toste catalyst transfer step was not considered by Gagosz. On the other hand, Gagosz already addressed a possible role of the counterion as a proton acceptor, a motive later discussed by Corma [33] and Widenhoefer [34] for the formation of species related to **I** and **Q**.

The supporting information of Gagosz contains a detailed discussion of several different mechanistic possibilities, including two pathways not based on dual activation, but those two pathways could be excluded by the results of the labeling studies.

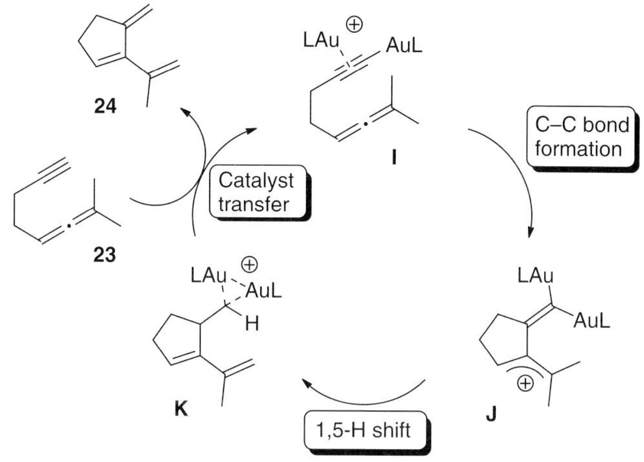

Scheme 4.9 Mechanism proposed for the conversion of **23** and **25**.

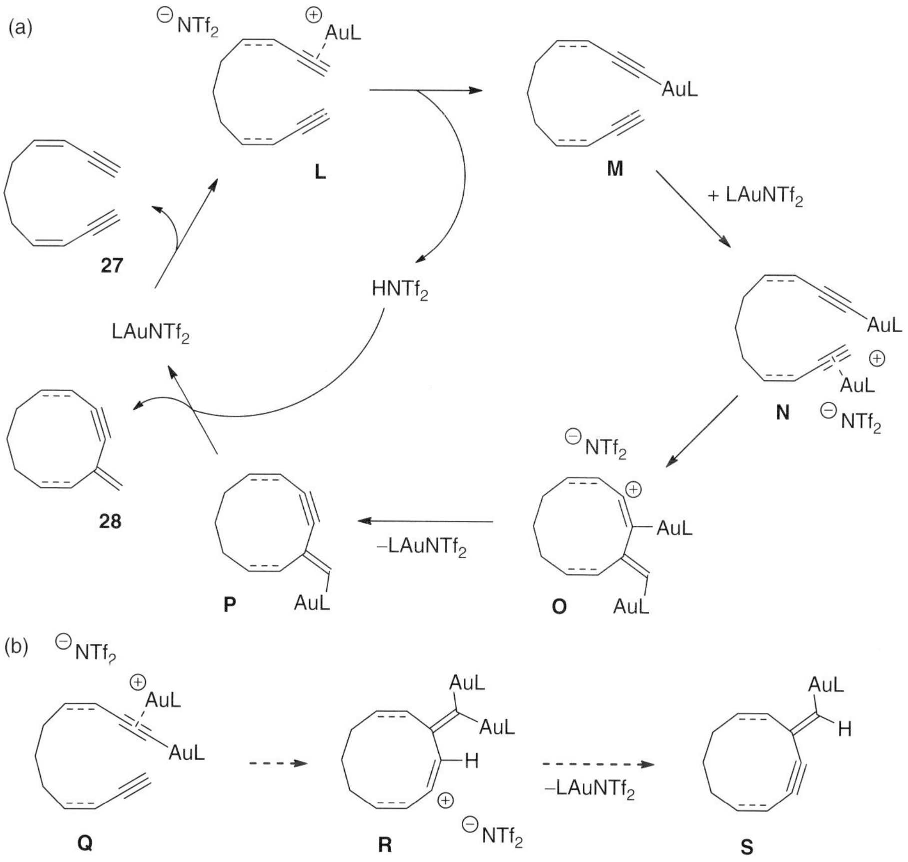

Scheme 4.10 (a,b) Two mechanistic proposals for the macrocyclization to **28**.

In 2012, Zhang and coworkers published an intermolecular version of Gagosz's alkyne-homodimerization. This paper did not provide new mechanistic insights but focused on the synthetic aspects of that chemistry [35].

Looking at this family of transformations, one open question was the initial formation of the gold acetylide. Laguna and coworkers [36] had published a paper on the photochemical properties of σ,π-digold acetylides bearing two phosphane ligands at the π-coordinated gold which induced a quite unusual, unsymmetrical coordination of the alkyne going along with a gold–gold interaction. In 2011, the groups of Corma [33] and Widenhoefer [34] also addressed the gold acetylides. The latter group could show that the π-complex, generated at low temperatures from $NHCAu^+$ and phenylacetylene, at temperatures above −20 °C is converted to σ,π-digold acetylides. Similar σ,π-digold acetylides bearing bulky phosphane ligands were prepared by Corma *et al.* They could also show that these complexes possessed excellent catalytic activity in the intermolecular alkene/alkyne [2 + 2] cycloaddition, a reaction discovered by Echavarren [37]. But more recent work of Fensterbank, Gimbert, and coworkers [38] suggested that these species were not involved in the cycloisomerization reaction, but for intramolecular enyne cycloisomerizations such dinuclear complexes could be detected by electrospray ionization mass spectrometry. They also confirmed that the gold acetylide shows a high affinity for the σ-coordination of a LAu^+ fragment.

4.5.2 An Unexpected Regioselectivity Raises Questions

In 2012, our group investigated diynes of type **29** [39]. We were quite surprised that not only α-**30** but also β-**30** was isolated. The compound β-**30** was neither formed by a rearrangement of α-**30** nor by an initial hydroarylation of one of the two alkynes and a subsequent enyne cycloisomerization. Control experiments clearly excluded that possibility. Obviously, the reaction pathway leading to β-**30** was unprecedented (Scheme 4.11).

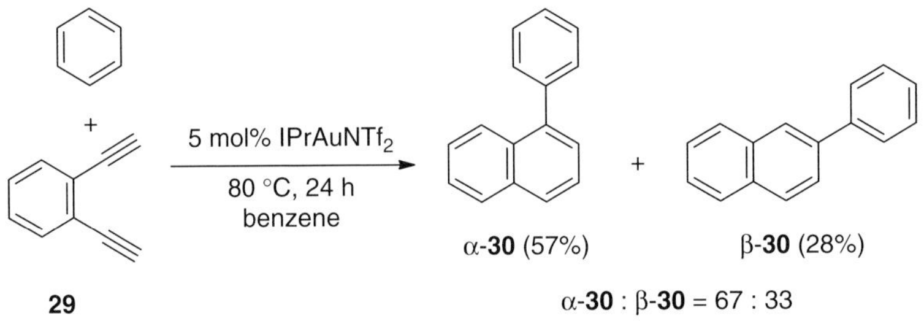

Scheme 4.11 The formation of β-**30** was unexpected.

We conducted many experiments, and finally three key findings allowed us to understand the reaction mechanism. A first crucial evidence for the mechanism of formation of β-**30** originated from a kinetic analysis. In the initial phase of the

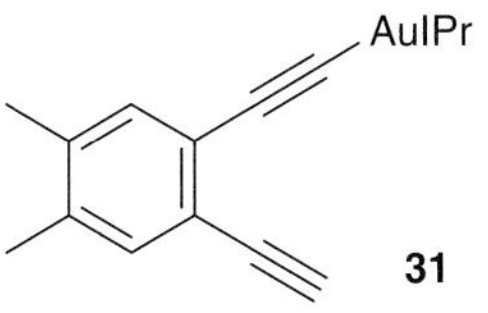

31

Figure 4.4 Gold acetylide **31** as a stable compound.

reaction, the formation of α-**30** is fast; after about 40–50% conversion, suddenly the generation of α-**30** slows down significantly. Then the product β-**30** begins to appear with a rate even slightly higher than the formation of α-**30**. This indicated that the active species had changed and a different second catalytic cycle was dominating in the second phase of the conversion.

The presence of the terminal alkyne in the substrate as one possibility suggested a slow generation of the gold(I)-acetylide **31** (Figure 4.4). We prepared **31**, but it was a stable compound, easily surviving heating to 80 °C, which is the temperature of formation of β-**30**.

Conducting additional experiments with **31**, we found that a catalytic amount of gold(I) catalysts with a stoichiometric amount of **31** exclusively gave β-**30**.

These two key findings clearly indicated that two different catalytic cycles compete. In the initial phase of the reaction, a classical electrophilic activation of the substrate **29** by only one gold complex forms α-**30** in high selectivity. Then, and seemingly much more slowly than in Fensterbank's [38] experiments using phenylacetylene as substrate, some of the cationic gold(I) complex is transferred into the gold acetylide, which then in good selectivity provides β-**30**. As one would expect, the presence of simple bases such as triethylamine or even Al_2O_3 accelerated this formation of **31**, thus under these conditions mainly β-**30** (ratio β-**30**/α-**30** = 98 : 2) was isolated.

At this point, we only had understood the initial formation of the gold acetylide; the other end of the catalytic cycle, namely the "catalyst transfer," was crucial, too. The arylgold(I) intermediate **32** should be the last species before closing the catalytic cycle. Then, for actually closing the catalytic cycle, a direct proton transfer from the alkyne proton of the next alkyne substrate molecule **29**–**32**, going along with the formation of the gold acetylide **31**, should be essential to stay β-selective. If **31** would not, as described above, be regenerated directly by using the next substrate molecule as the proton source, but by any other proton source, this would generate free LAu^+ and the α-selectivity of LAu^+ would not allow it to stay β-selective. A corresponding control experiment demonstrated that indeed the direct and highly selective protonation of a mimic of intermediate **32**, namely LAu-Ph (**33**), with **29** readily proceeded under the normal reaction conditions for the formation of β-**30** (Scheme 4.12). The pK_a value of the terminal alkyne is only about +25, but still the formation of the thermodynamically favored gold acetylide **31** provides the thermodynamic driving force for that step.

But there still is one open question: why does **31** with LAu^+ provide β-**30**? The answer to this crucial third evidence came from the fact that reaction of substrates of type **34** with a *tert*-butyl group did not lead to a naphthalene, an anellation of two six-membered rings, but to benzofulvene **35** with an anellation of one six-

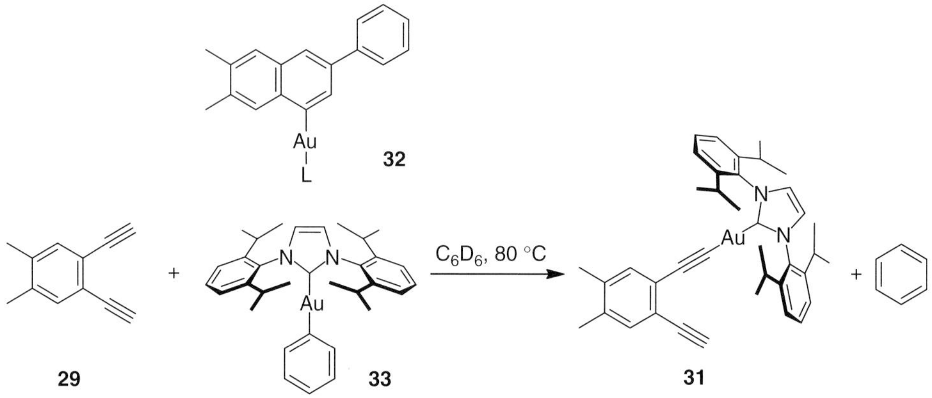

Scheme 4.12 Control experiment for the "catalyst transfer" step.

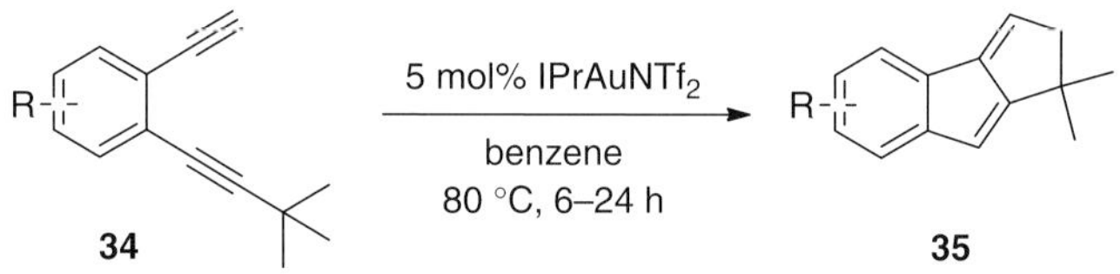

Scheme 4.13 Formation of five-membered rings in **35**.

and two five-membered rings (Scheme 4.13) [40]. This suggested the formation of the first five-membered ring by the a cyclization, followed by the formation of a second five-membered ring by a C,H-activation of one of the primary non-activated C–H bonds at a *tert*-butyl group. In parallel work, Zhang's group [41] found similar reactions and conducted a detailed computational analysis addressing the mechanism. They found a highly interesting bifurcation situation with two possible pathways: either the formation of a vinylidene species, or the formation of aurated aryl cations as reactive species. In our investigation, we used NHC ligands, and Zhang's group had very good success with the BrettPhos ligand in the presence of N-oxides.

Another piece of evidence for the initial formation of a five-membered ring rather than a six-membered ring was provided by the reaction of the substrates **36** bearing an aryl group instead of the *tert*-butyl group. Dibenzopentalenes **37** were isolated (Scheme 4.14) and, once more, two five-membered rings were found in the product [42].

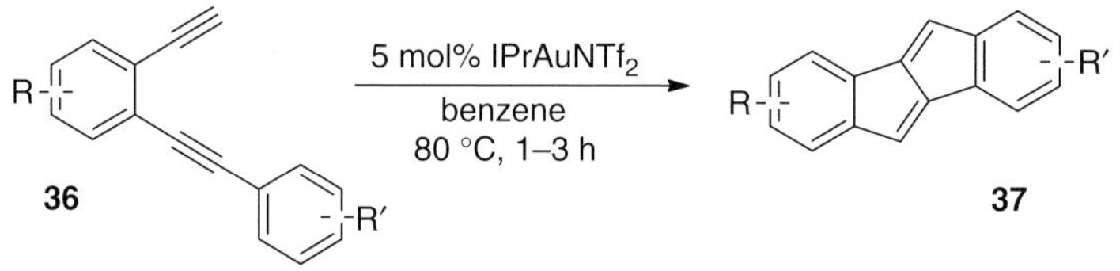

Scheme 4.14 Formation of five-membered rings in **37**.

4.5.3
The Mechanistic Hypothesis

These three key findings, along with a series of deuterium isotope labeling experiments and the important additional computational results of Zhang's group, led to the following new mechanistic cycle for the formation of **35** (Scheme 4.15).

The catalytic cycle starts with the formation of the gold(I) acetylide **T**. The "dual activation" is then initiated by the π-coordination of a LAu^+ unit to the other triple bond (**U**). This activates the system for an electrophilic attack in the β-position of the gold acetylide, providing the gold(I) vinylidene-like species **V**. Vinylidene-gold(I) complexes have not been described in the literature, so we tried to prepare the corresponding allenylidenes. But these compounds showed no Au–C double bond character (see Section 4.5.5) [43]. As a result, these intermediates are highly electrophilic carbenoids. Next follows the insertion into the C–H bond of the *tert*-butyl group. Considering related reactions, for example, the concerted reaction with olefins (see below), the C,H-insertion should also proceed in a concerted way. The stepwise alternative involving a hydride abstraction and a subsequent electrophilic attack [44] of the carbenium ion at the vinylgold would generate a primary carbenium ion **W**, which should show competing Wagner–Meerwein rearrangements and thus deliver other products also. But this was not observed.

Then the vinylgold(I) species **Y** would be formed by elimination of LAu^+ from species **X**; a reversible addition of LAu^+ to the vinylgold(I) unit at the ipso-position of gold is the basis for the side equilibrium with the gem-diaurated side product **38**. Such digold species could be isolated and characterized by crystal structure analyses in this and many other related reactions. The restart of the catalytic conversion when adding some of the digold species to the substrate proved the reversibility of the side equilibrium with the diaurated species. As discussed above, the protodeauration by the alkyne of the next substrate molecule **34** closes the catalytic cycle (as investigated in Scheme 4.12).

Since in these reactions two new five-membered rings were formed, a ring expansion to a six-membered ring, that is, a naphthalene core, was not possible. Such a ring expansion of the central five-membered ring at the same time would demand a ring contraction of the outer five-membered ring to deliver a four-membered ring, which is overall thermodynamically not feasible (Scheme 4.16).

If the reactions involving C,H activation were conducted in an intermolecular manner, a ring expansion was possible, as only one new five-membered ring was formed [45].

Since in the intermolecular case by the C,H activation and in the case of the synthesis of the β-substituted naphthalene **30** no second five-membered ring is generated, naphthalenes are accessible by the ring expansion of the five-membered ring that is formed in the initial cyclization step. The catalytic cycle for the naphthalene synthesis is shown in Scheme 4.17. Very important is the proof of the intramolecular H-shift in the step from **AB** to **AC** by deuterium labeling studies. The ring expansion would proceed from the carbenoid species **AC**.

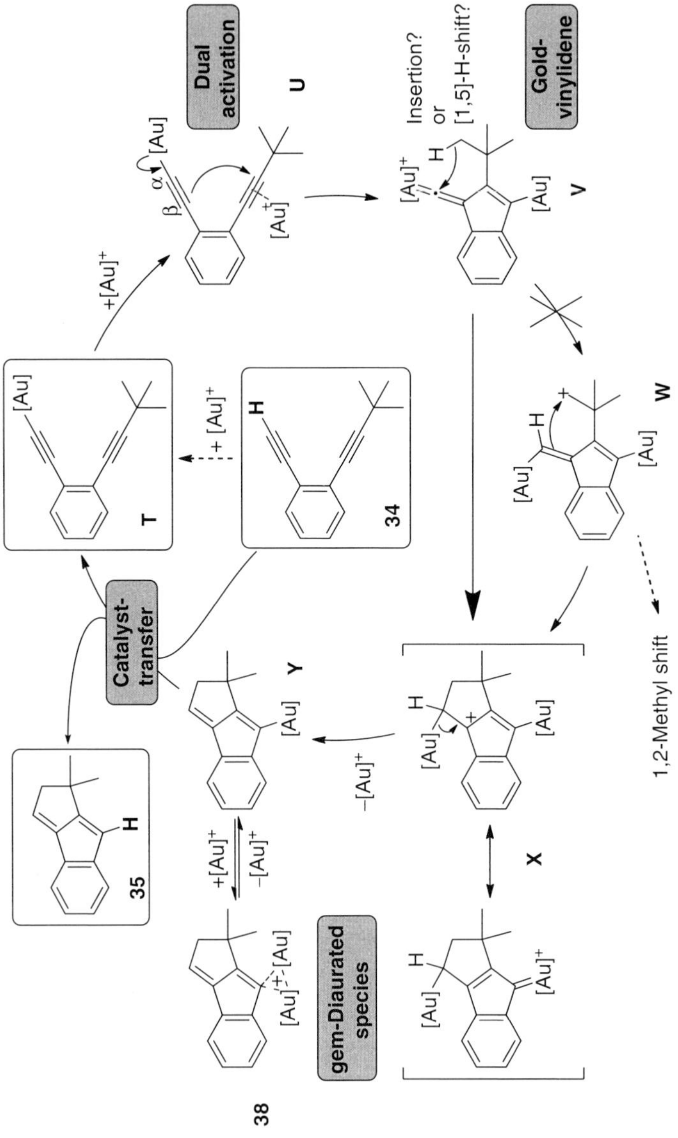

Scheme 4.15 Catalytic cycle for the formation of benzofulvenes.

Scheme 4.16 There is no thermodynamic driving force for a ring expansion/ring contraction in the anellation of two five-membered rings.

Alkenes also could be intermolecular reaction partners of the diynes **29**, and thus benzocyclobutenes were formed [46]. Two important observations were made in this context:

1) The reaction is diastereoselective; a (*Z*)-olefin like **40** delivers the cis product **41**, and the (*E*)-olefin **42** delivers the trans product **43** (Scheme 4.18). A concerted reaction pathway can be assumed. This is the reason why for the C,H-insertion also a direct concerted insertion rather than a stepwise process initiated by a [1,5]-H shift was assumed in Scheme 4.15 (species **V** directly to **X** and not **V** via **W** to **X**).
2) Bergman-like cyclization pathways of the ene-diynes substructures in the substrates, which would involve radical intermediates, can be excluded as the typical interceptor for such radicals, the hydrogen atom donor 1,4-cyclohexadiene (**44**), also provided benzocyclobutene **45** and not products of a hydrogen transfer.

4.5.4 The Other Pathway

But there was also an alternative pathway for the synthesis of the six-membered ring. In fact, the calculations of Zhang and coworkers [41] showed that an initial cyclization to a six-membered ring should be preferred, but the five-membered ring was observed in the reaction, forming benzofulvenes. In order to explore five-membered hetarenes with a larger distance between the two reactive alkyne units, we changed to a thiophene instead of the benzene ring as the backbone (**46**). Now, indeed, the six-membered ring in **47** was formed (Scheme 4.19) [46]. Once more, the reaction went along with a C,H-activation. More important was the fact that the calculations showed that in a bifurcation pathway the six-membered ring formation (**AG** and not the vinylidene **AK** with an anellation of two five-membered rings) was preferred and the intermediates **AH**/**AI** possessed carbene character for the C,H-activation. For substrates bearing the alkyl group on the alkyne at 2-position of the thiophene, the yields were much higher, which suggested an additional interaction of the gold catalyst with the thiophene sulfur atom.

For the 1,2-[Au]-shift converting species **AG** to **AH**/**AI**, a gold(I) aryne species could be located as the transition state [46]. Recent findings of Aue, Zhang, and coworkers [47] for en-diyne substrates fully confirm this.

As a consequence, the possibility that species **AH**/**AI** is not being formed via the vinylidenegold(I) complex offered an alternative explanation for the formation of six-membered rings in the other reactions (like β-**30**, the intermolecular

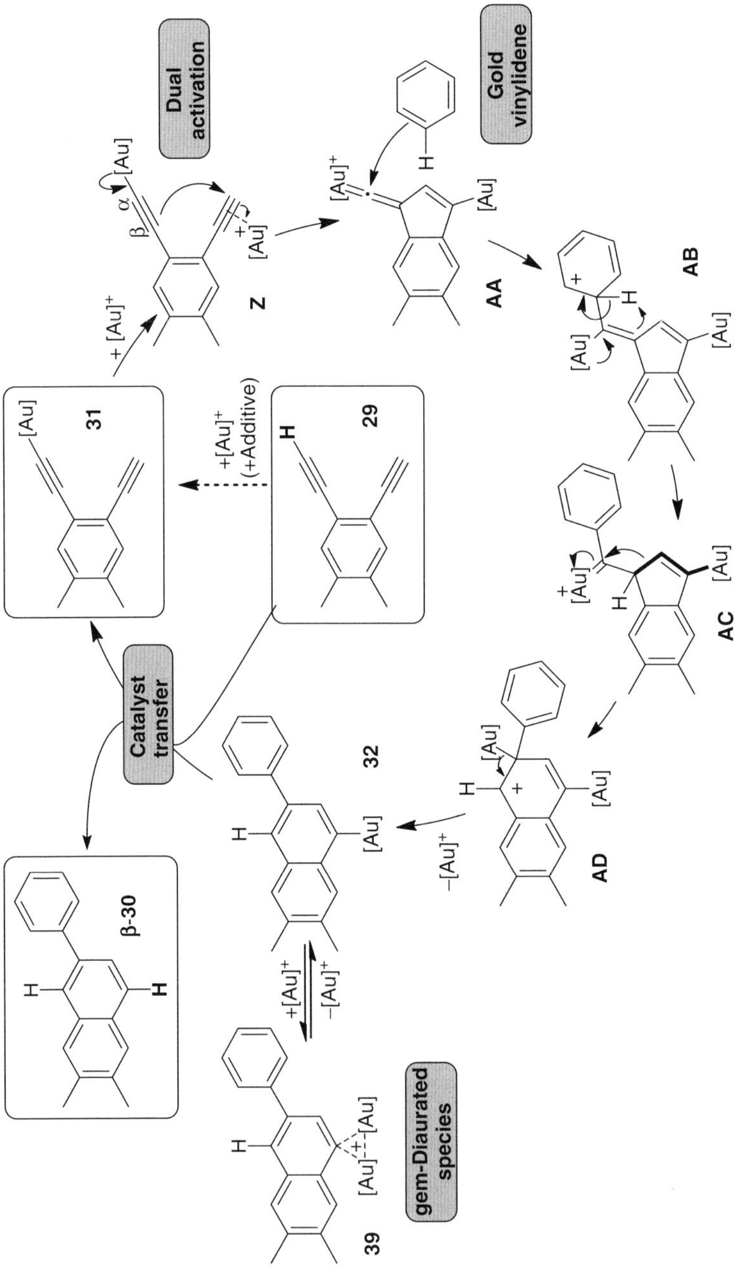

Scheme 4.17 Catalytic cycle for the formation of β-substituted naphthalenes.

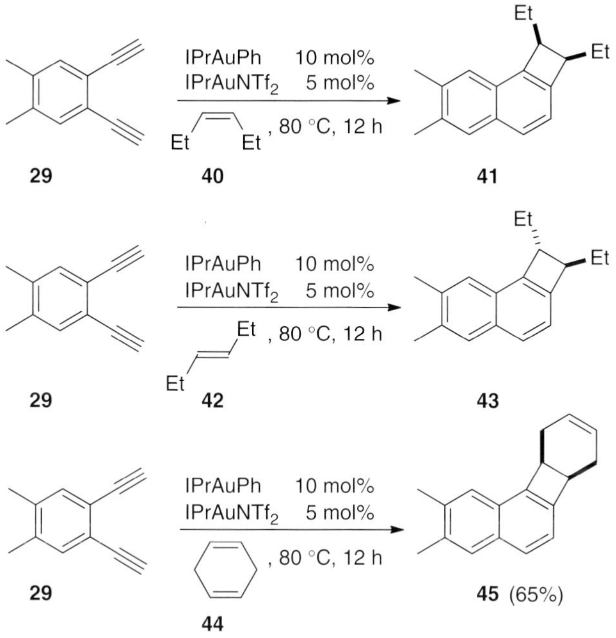

Scheme 4.18 With alkenes as intermolecular partners benzocyclobutene substructures are formed.

reactions providing **41**, **43**, or **45**). The initially assumed ring expansion is not the only possibility (e.g., Scheme 4.17, species **AC**).

4.5.5 Gold Allenylidenes as Analogs of Gold Vinylidenes?

Based on the analysis of the electronic structure of vinylidene intermediates, we wanted to test whether allenylidene complexes of gold(I) could be prepared. First, the gold(I) acetylides **49** were prepared; then an alkylation delivered **50** (Scheme 4.20). A detailed structure investigation, based on experimental and computational methods, followed [43]. The strong donors, which allow **50** to be isolated, stabilize the carbenium ion with the lone pair of the heteroatom (O or N). Thus no significant Au–C double bond character could be detected.

4.5.6 Dual Activation Catalysts

The optimal catalyst for this new family of "dual activation" reactions was an open problem. For these reactions, the ideal ratio of gold acetylide to LAu^+ is 1 : 1. On the typical small scale of methodology development or mechanistic studies, this was difficult to adjust by mixing different precatalysts or a precatalyst and a

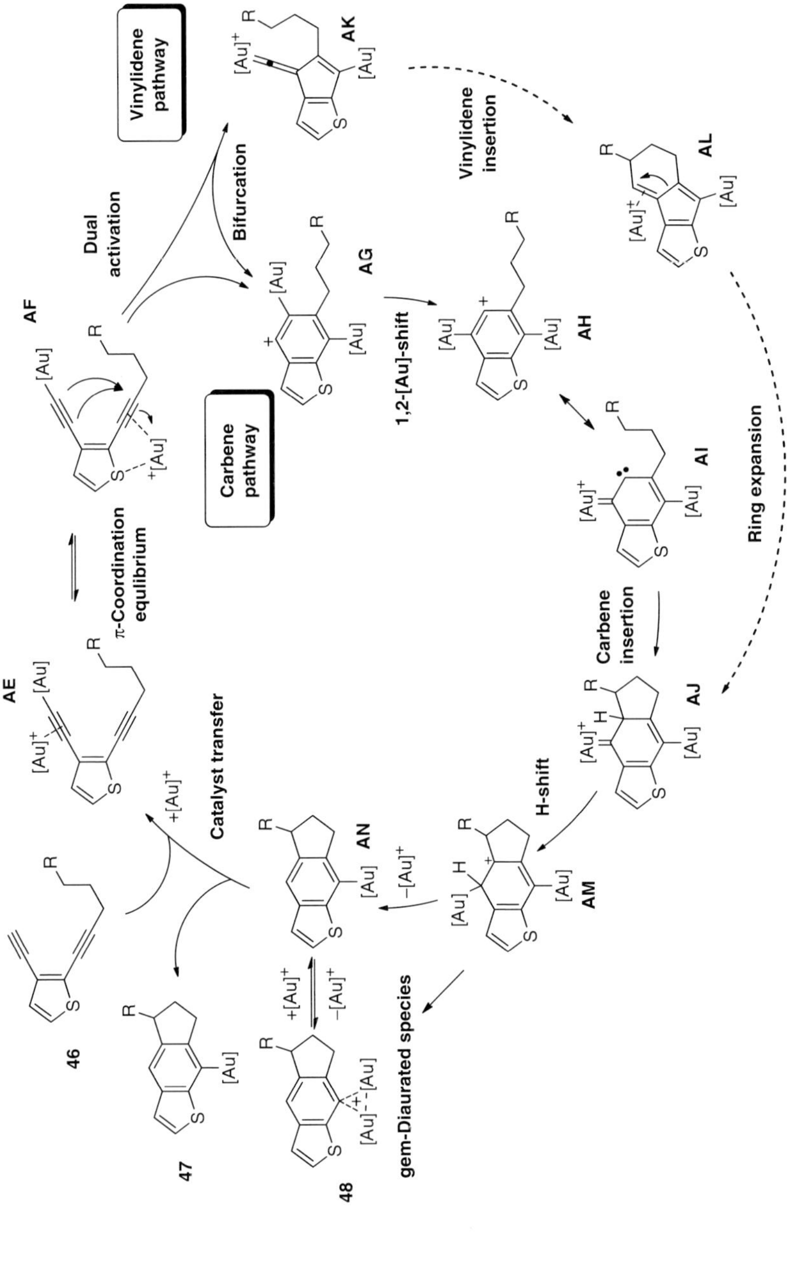

Scheme 4.19 Mechanistic cycle for the formation anellated benzothiophenes.

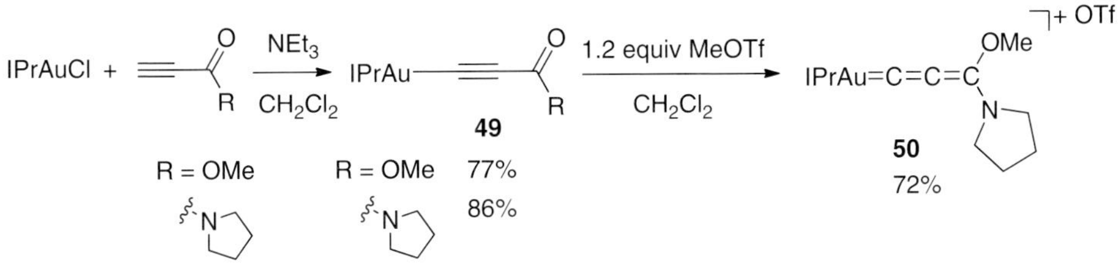

Scheme 4.20 Synthesis of "gold(I) allenylidenes."

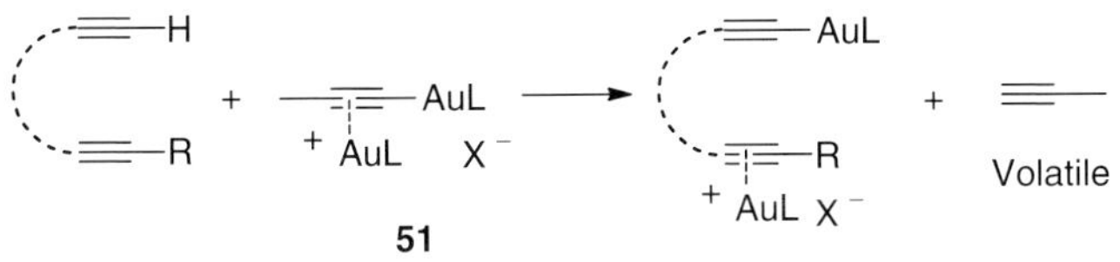

Scheme 4.21 Dual activation catalysts (DACs).

base. If the amount of base is too small, an excess of LAu^+ would remain, which in the naphthalene synthesis and similar reactions would ruin the β-selectivity (Figure 4.1). An excess of base would convert all the LAu^+ to the gold acetylide, and then no catalysis would be observed. The use diaurated species like **38** or **39** would be one option, but these will set free 1 equiv (with regard to the catalyst) of a product-like side product (unless a specific precatalyst derived from the substrate would be used, which would demand the preparation of a new precatalyst for each reaction). Very good success was obtained with a family of different catalysts **51** (dual activation catalysts DACs, Scheme 4.21). These compounds were easily accessible, could be stored, and proved to be active in a number of reactions following the dual activation mechanism [48].

4.6
Protons and Hydride in Gold Catalysis

At the stage of alkylgold intermediates, the question of β-H elimination becomes relevant. In the field of organotransition metal compounds, this pathway is considered to be a classical prototype of reactivity. In the field of gold catalysis, especially for gold(I) catalysis, a number of reports specifically mentioned that, while other metals gave β-H elimination products, gold did not show this behavior [49]. Initially it was assumed that the gold hydrides would not be thermodynamically stable (combination of a strong oxidant with a strong reductant), but in the meantime gold hydrides had been established as stable compounds [50].

At that point, we had a closer look at the possibility of β-H eliminations at gold(I) alkyl compounds [51]. We chose the IPr NHC ligand, as this was the ligand that allowed the isolation [52] of the first stable gold(I) hydrides.

As the most important finding of the computational study, it was revealed that there is a pathway for the β-H elimination, but the overall barrier is so high that this reaction should proceed only at temperatures of about 220 °C, which means that this step indeed should be irrelevant for all homogeneous gold-catalyzed reactions published so far.

The subsequent experimental verification of these results indeed proved that IPrAu-alkyl compounds start to decompose only at temperature of about 180 °C. In analogy with organomercury compounds, this decomposition could be based on a homolytic metal–carbon bond cleavage.

At the stage of vinylgold or arylgold intermediates, most catalytic cycles show the proto-deauration as the final step. In many of these reactions, rather than a direct proton transfer from the nucleophile which was added, a long-distance proton transfer is necessary. A detailed investigation revealed that gold catalysis not only tolerates water but also utilizes clusters of four or more water molecules as proton shuttles [53].

4.7
Future Perspectives

Fourteen years ago, when the papers initiating the current 'gold rush' appeared, [54] there existed barely any insight into the mechanisms of homogeneous gold catalysis. The numerous contributions in the last years have changed that significantly. Now, even new reaction principles, which were previously unknown in homogeneous catalysis, have been developed.

The continuously growing family of gold-catalyzed reactions, which still is allowing the development of new reaction possibilities for the different types of intermediates, will add new elementary reactions and new modes of activation of the organic substrates to the toolbox of the synthetic chemist. In the years to come, we will see more exciting new results of this type.

References

1. (a) Hashmi, A.S.K. and Hutchings, G.J. (2006) *Angew. Chem.*, **118**, 8064–8105; *Angew. Chem. Int. Ed.*, (2006), **45**, 7896–7936; (b) Rudolph, M. and Hashmi, A.S.K. (2012) *Chem. Soc. Rev.*, **41**, 2448–2462.
2. Hashmi, A.S.K. (2010) *Angew. Chem.*, **122**, 5360–5369; *Angew. Chem. Int. Ed.*, (2010), **49**, 5232–5241.
3. Cinellu, M.-A. (2012) in *Modern Gold Catalzed Synthesis* (eds A.S.K. Hashmi and F.D. Toste), Wiley-VCH Verlag GmbH, Weinheim, p. 153–175.
4. For X-ray crystal structure analyses, see:(a) Schulte, P. and Behrens, U. (1998) *Chem. Commun.*, 1633–1634; (b) Shapiro, N.D. and Toste, F.D. (2008) *Proc. Natl. Acad. Sci. U.S.A.*, **105**, 2779–2782; (c) Wu, J., Kroll, P., and Dias, H.V.R. (2009) *Inorg. Chem.*, **48**, 423–425; (d) Dias, H.V.R., Fianchini, M., Cundari, T.R., and Campana, C.F. (2008) *Angew. Chem.*, **120**, 566–569; *Angew. Chem. Int. Ed.*, (2008), **47**, 556–559; (e) Flügge, S., Anoop, A., Goddard, R., Thiel, W., and Fürstner, A. (2009) *Chem. Eur. J.*, **15**, 8558–8565; (f) Dias,

H.V.R. and Wu, J. (2007) *Angew. Chem.*, **119**, 7960–7962; *Angew. Chem. Int. Ed.*, (2007), **46**, 7814–7816.

5. (a) Chalk, A.J. (1986) *J. Am. Chem. Soc.*, **108**, 4733–4734. (b) Hüttel, R. and Forkl, H. (1972) *Chem. Ber.*, **105**, 1664–1673; (c) Wittig, G. and Fischer, S. (1972) *Chem. Ber.*, **105**, 3542–3552; assigned by NMR data only: d) Reetz, M.T. and Sommer, K. (2003) *Eur. J. Org. Chem.*, **2003**, 3485–3496; (e) Casado, R., Contel, M., Laguna, M., Romero, P., and Sanz, S. (2003) *J. Am. Chem. Soc.*, **125**, 11925–11935; (f) Lavallo, V., Frey, G.D., Kousar, S., Donnadieu, B., and Bertrand, G. (2007) *Proc. Natl. Acad. Sci. U.S.A.*, **104**, 13569–13573; (g) Lavallo, V., Frey, G.D., Donnadieu, B., Soleilhavoup, M., and Bertrand, G. (2008) *Angew. Chem.*, **120**, 5302–5306; *Angew. Chem. Int. Ed.*, (2008), **47**, 5224–5228; (h) Garca-Mota, M., Cabello, N., Maseras, F., Echavarren, A.M., Perez-Ramirez, J., and Lopez, N. (2009) *ChemPhysChem*, **9**, 1624–1629.
6. (a) Hashmi, A.S.K., Blanco, M.C., Fischer, D., and Bats, J.W. (2006) *Eur. J. Org. Chem.*, **2006**, 1387–1389; (b) Sahoo, A.K., Nakamura, Y., Aratani, N., Kim, K.S., Noh, S.B., Shinokubo, H., Kim, D., and Osuka, A. (2006) *Org. Lett.*, **8**, 4141–4144; (c) Zhang, G., Peng, Y., Cui, L., and Zhang, L. (2009) *Angew. Chem.*, **121**, 3158–3161; *Angew. Chem. Int. Ed.*, (2009), **48**, 3112–3115.
7. Cinellu, M.A., Minghetti, G., Stoccoro, S., Zuccaa, A., and Manassero, M. (2004) *Chem. Commun.*, 1618–1619.
8. Zuccaccia, D., Belpassi, L., Tarantelli, F., and Macchioni, A. (2009) *J. Am. Chem. Soc.*, **131**, 3170–3171.
9. (a) Pernpointner, M., Hashmi, A.S.K., and Chem, J. (2009) *Theor. Comput.*, **5**, 2717–2725; (b) Hashmi, A.S.K., Pernpointner, M., and Hansmann, M.M. (2011) *Faraday Discuss.*, **152**, 179–184.
10. (a) Hashmi, A.S.K., Frost, T.M., and Bats, J.W. (2001) *Org. Lett.*, **3**, 3769–3771; (b) Hashmi, A.S.K. (2003) *Gold Bull.*, **36**, 3–9.
11. Lein, M., Rudolph, M., Hashmi, A.S.K., and Schwerdtfeger, P. (2010) *Organometallics*, **29**, 2206–2210.
12. Hashmi, A.S.K. (2009) *Gold Bull.*, **42**, 275–279.
13. (a) Hashmi, A.S.K., Ramamurthi, T.D., and Rominger, F. (2009) *J. Organomet. Chem.*, **694**, 592–597; (b) Hashmi, A.S.K., Ramamurthi, T.D., Todd, M.H., Tsang, A.S.-K., and Graf, K. (2010) *Aust. J. Chem.*, **63**, 1619–1626.
14. Liu, L.-P., Xu, B., Mashuta, M.S., and Hammond, G.B. (2008) *J. Am. Chem. Soc.*, **130**, 17642–17643.
15. Hashmi, A.S.K., Schwarz, L., Choi, J.-H., and Frost, T.M. (2000) *Angew. Chem.*, **112**, 2382–2385; *Angew. Chem. Int. Ed.*, (2000), **39**, 2285–2288.
16. Weber, D., Tarselli, M.A., and Gagné, M.R. (2009) *Angew. Chem.*, **121**, 5843; *Angew. Chem. Int. Ed.*, (2009), **48**, 5733.
17. Hashmi, A.S.K., Weyrauch, J.P., Frey, W., and Bats, J.W. (2004) *Org. Lett.*, **6**, 4391.
18. (a) Weyrauch, J.W., Hashmi, A.S.K., Schuster, A., Hengst, T., Schetter, S., Littmann, A., Rudolph, M., Hamzic, M., Visus, J., Rominger, F., Frey, W., and Bats, J.W. (2010) *Chem. Eur. J.*, **16**, 956–963; (b) Hashmi, A.S.K., Rudolph, M., Schymura, S., Visus, J., and Frey, W. (2006) *Eur. J. Org. Chem.*, **2006**, 4905–4909.
19. Hashmi, A.S.K., Schuster, A., and Rominger, F. (2009) *Angew. Chem.*, **121**, 8396; *Angew. Chem. Int. Ed.*, (2009), **48**, 8247.
20. Hashmi, A.S.K., Ramamurthi, T.D., and Rominger, F. (2010) *Adv. Synth. Catal.*, **352**, 971–975.
21. Hashmi, A.S.K., Schuster, A.M., Gaillard, S., Cavallo, L., Poater, A., and Nolan, S.P. (2011) *Organometallics*, **30**, 6328–6337.
22. LaLonde, R.L., Brenzovich, W.E. Jr., Benitez, D., Tkatchouk, E., Kelley, K., Goddard, W.A. III, and Toste, F.D. (2010) *Chem. Sci.*, **1**, 226–233.
23. Hashmi, A.S.K., Schuster, A.M., Litters, S., Rominger, F., and Pernpointner, M. (2011) *Chem. Eur. J.*, **17**, 5661–5667.
24. Vilhelmsen, M.H., Hashmi, A.S.K. (2014) *Chem. Eur. J.*, **20**, 1901–1908.
25. (a) Buzas, A. and Gagosz, F. (2006) *Org. Lett.*, **8**, 515–518; (b) Buzas, A. and Gagosz, F. (2006) *Synlett*, 2727–2730; (c) Buzas, A., Istrate, F., and Gagosz, F. (2006) *Org. Lett.*, **8**, 1958–2006; (d) Kirsch, S.F. (2007) *Angew. Chem. Int. Ed.*, **46**, 2310–2313; (e) Zhang, L. (2007) *Org. Lett.*, **9**, 2147–2150; (f) Crone, B.

and Kirsch, S.F. (2007) *J. Org. Chem.*, **72**, 5435–5438; (g) Bhargava, S.K., Mohr, F., Bennett, M.A., Welling, L.L., and Willis, A.C. (2000) *Organometallics*, **19**, 5628–5635; (h) Shi, Z. and He, C. (2004) *J. Am. Chem. Soc.*, **126**, 3596–13597.

26. (a) Hashmi, A.S.K., Lothschütz, C., Döpp, R., Rudolph, M., Ramamurthi, T.D., and Rominger, F. (2009) *Angew. Chem.*, **121**, 8392–8395; *Angew. Chem. Int. Ed.*, (2009), **48**, 8243–8246; (b) Hashmi, A.S.K., Döpp, R., Lothschütz, C., Rudolph, M., Riedel, D., and Rominger, F. (2010) *Adv. Synth. Catal.*, **352**, 1307–1314; (c) Hashmi, A.S.K., Lothschütz, C., Döpp, R., Ackermann, M., De Buck, J., Rudolph, M., Scholz, C., and Rominger, F. (2012) *Adv. Synth. Catal.*, **354**, 133–147.
27. Pena-López, M., Ayán-Varela, M., Sarandeses, L.A., and Pérez Sestelo, J. (2010) *Chem. Eur. J.*, **16**, 9905–9909.
28. (a) Pérez-Temprano, M.H., Casares, J.A., de Lera, Á.R., Álvarez, R., and Espinet, P. (2012) *Angew. Chem. Int. Ed.*, **51**, 4917–4920; (b) delPozo, J., Carrasco, D., Pérez-Temprano, M.H., García-Melchor, M., Álvarez, R., Casares, J.A., and Espinet, P. (2013) *Angew. Chem. Int. Ed.*, **52**, 2189–2193.
29. Ibrahim, N., Vilhelmsen, M.H., Pernpointner, M., Rominger, F., and Hashmi, A.S.K. (2013) *Organometallics*, **32**, 2576–2583.
30. Shi, Y., Roth, K.E., Ramgren, S.D., and Blum, S.A. (2009) *J. Am. Chem. Soc.*, **131**, 18022–18023.
31. Cheong, P.H.-Y., Morganelli, P., Luzung, M.R., Houk, K.N., and Toste, F.D. (2008) *J. Am. Chem. Soc.*, **130**, 4517–4526.
32. Odabachian, A., Le Goff, X.F., and Gagosz, F. (2009) *Chem. Eur. J.*, **15**, 8966–8970.
33. Grirrane, A., Garcia, H., Corma, A., and Álvarez, E. (2011) *ACS Catal.*, **1**, 1647–1653.
34. Brown, T.J. and Widenhoefer, R.A. (2011) *Organometallics*, **30**, 6003–6009.
35. Sun, S., Kroll, J., Luo, Y., and Zhang, L. (2012) *Synlett*, 54–56.
36. Blanco, M.C., Cámara, J., Gimeno, M.C., Jones, P.G., Laguna, A., López-de-Luzuriaga, J.M., Olmos, M.E., and Villacampa, M.D. (2012) *Organometallics*, **31**, 2597–2605.
37. Lopez-Carrillo, V. and Echavarren, A.M. (2010) *J. Am. Chem. Soc.*, **132**, 9292–9294.
38. Simonneau, A., Jaroschik, F., Lesage, D., Karanik, M., Guillot, R., Malacria, M., Tabet, J.-C., Goddard, J.-P., Fensterbank, L., Gandon, V., and Gimbert, Y. (2011) *Chem. Sci.*, **2**, 2417–2422.
39. Hashmi, A.S.K., Braun, I., Rudolph, M., and Rominger, F. (2012) *Organometallics*, **31**, 644–661.
40. Hashmi, A.S.K., Braun, I., Nösel, P., Schädlich, J., Wieteck, M., Rudolph, M., and Rominger, F. (2012) *Angew. Chem.*, **124**, 4532–4536; *Angew. Chem. Int. Ed.*, (2012), **51**, 4456–4460.
41. Ye, L., Wang, Y., Aue, D.H., and Zhang, L. (2012) *J. Am. Chem. Soc.*, **134**, 31–34.
42. Hashmi, A.S.K., Wieteck, M., Braun, I., Nösel, P., Jongbloed, L., Rudolph, M., and Rominger, F. (2012) *Adv. Synth. Catal.*, **354**, 555–562.
43. Hansmann, M.M., Rominger, F., and Hashmi, A.S.K. (2013) *Chem. Sci.*, **4**, 1552–1559.
44. Hashmi, A.S.K., Häffner, T., Yang, W., Pankajakshan, S., Schäfer, S., Schultes, L., Rominger, F., and Frey, W. (2012) *Chem. Eur. J.*, **18**, 10480–10486.
45. Hashmi, A.S.K., Wieteck, M., Braun, I., Rudolph, M., and Rominger, F. (2012) *Angew. Chem.*, **124**, 10785–10789; *Angew. Chem. Int. Ed.*, (2012), **51**, 10633–10637.
46. Hansmann, M.M., Rudolph, M., Rominger, F., and Hashmi, A.S.K. (2013) *Angew. Chem.*, **125**, 2653–2659; *Angew. Chem. Int. Ed.*, (2013), **52**, 2593–2598.
47. Wang, Y., Yepremyan, A., Ghorai, S., Todd, R., Aue, D.H., and Zhang, L. (2013) *Angew. Chem. Int. Ed.*, **125**, 7949–7953.
48. Hashmi, A.S.K., Lauterbach, T., Nösel, P., Vilhelmsen, M.H., Rudolph, M., and Rominger, F. (2013) *Chem. Eur. J.*, **19**, 1058–1065.
49. Baker, R.T., Nguyen, P., Marder, T.B., and Westcott, S.A. (1995) *Angew. Chem.*, **107**, 1451–1452; *Angew. Chem., Int. Ed. Engl.*, (1995), **34**, 1336–1338.

50. Hashmi, A.S.K. (2012) *Angew. Chem.*, **124**, 13109–13110; *Angew. Chem. Int. Ed.*, (2012), **51**, 12935–12936.
51. Klatt, G., Xu, R., Pernpointner, M., Molinari, L., Hung, T.Q., Rominger, F., Hashmi, A.S.K., and Köppel, H. (2013) *Chem. Eur. J.*, **19**, 3954–3961.
52. Tsui, E.Y., Müller, P., and Sadighi, J.P. (2008) *Angew. Chem.*, **120**, 9069–9072; *Angew. Chem. Int. Ed.*, (2008), **47**, 8937–8940.
53. Krauter, C.M., Hashmi, A.S.K., and Pernpointner, M. (2010) *ChemCatChem*, **2**, 1226–1230.
54. (a) Hashmi, A.S.K., Frost, T.M., Bats, J.W. (2000) *J. Am. Chem. Soc.*, **122**, 11553–11554; (b) Hashmi, A.S.K., Schwarz, L., Choi, J.-H., Frost, T.M. (2000) *Angew. Chem.*, **112**, 2382–2385; (2000) *Angew. Chem. Int. Ed. Engl.*, **39**, 2285–2288.

5
Diastereoselectivity in Alkene Metathesis

Bernd F. Straub and Achim Häußermann

5.1
Introduction

Alkene metathesis leads to the formal exchange of carbene fragments of alkenes. This reaction is of high relevance in organic synthesis, be it for organic synthesis with homogeneous catalysts [1] or in the large-scale conversion of alkene feedstock in the SHOP (shell higher olefin process) [2]. Three chemists – Chauvin, Grubbs, and Schrock – were awarded the Nobel Prize in 2005 for their contributions in this field [3]. The general mechanism of alkene metathesis – for both heterogeneous and homogeneous catalyses – has been proposed by Hérisson and Chauvin in 1971 [4] (Scheme 5.1).

Ruthenium – carbene complexes, mainly developed by Grubbs's research group, feature a high tolerance of "hard" functional groups such as esters, ethers, amides, amines, and phenols [5]. Tungsten– and molybdenum – carbene complexes, mainly investigated by Schrock's research group [6], are more stable toward "soft" functional groups such as thioethers and phosphanes. Catalyst stability and activity have been optimized by tuning the ligands and their substituents in an "evolutionary" process, where successive small structural changes in the catalyst system have led to vast improvements in catalytic properties (Figure 5.1).

This chapter is based in part on the results of Häußermann *et al.* [12], on Häußermann's PhD thesis [13], and on unpublished density functional theory (DFT) calculations by Straub.

5.2
Stereoselective Alkene Metathesis Catalysts

In recent years, catalysts and protocols for the Z-selective formation of alkenes via alkene metathesis have been reported. Most progress has been made in the successful optimization of molecular molybdenum – alkylidene complexes. The substituents at their imido ligand fragment and at the electron-poor alcoholate/

Molecular Catalysts: Structure and Functional Design, First Edition. Edited by Lutz H. Gade and Peter Hofmann.

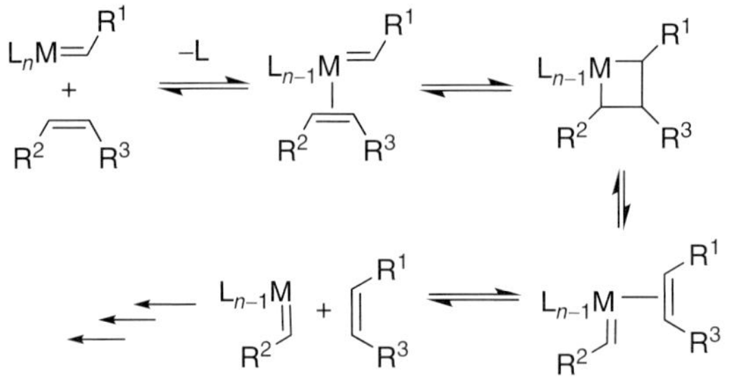

Scheme 5.1 Chauvin mechanism of alkene metathesis [3a].

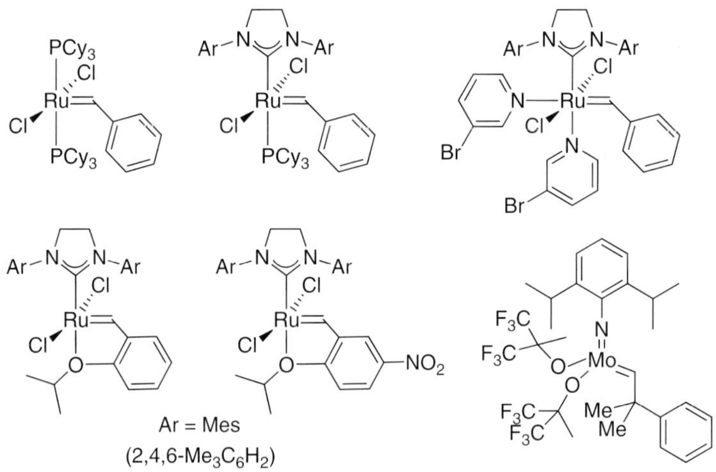

Figure 5.1 Alkene metathesis catalysts: Grubbs-type first generation, second generation, and third generation. (From left side, top row to right side, second row) Grubbs–Blechert–Hoveyda catalyst, Grela's nitro derivative, and Schrock-type catalyst [7–11].

phenolate/pyyrolide ligands proved to be perfectly suited for tuning the stereoselectivity of alkene metathesis.

The steric shielding of biaryloxide ligands induces a higher stability of *cis*-substituted molybdacyclobutane intermediates and thus leads selectively to *Z*-alkenes as kinetic products (Scheme 5.2) [14].

The steric repulsion of a large aryloxide ligand in combination with a comparatively small substituent at the imido ligand favors an *all-cis* metallacyclobutane (Scheme 5.3). The cycloaddition/cycloreversion sequence inverts the absolute configuration at the stereogenic metal center.

The origin of the diastereoselectivity of Schrock's olefin metathesis catalysts can be traced back to the preferential syn and anti conformations of the alkylidene fragment (Scheme 5.4). The steric repulsion of the alkylidene substituents with phenoxide ligands (top row) or with imido substituents (bottom row) determines the diastereoselectivity of the alkene metathesis.

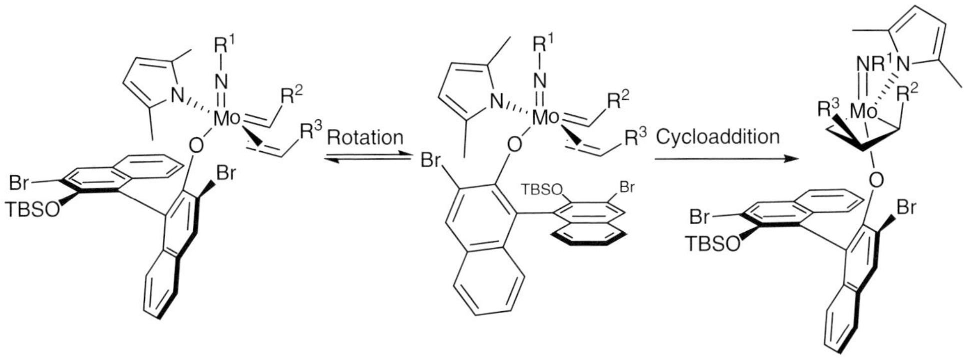

Scheme 5.2 Schematic structure of a *Z*-selective molybdenum catalyst; $R^1 = 2,6$-Xylyl or $2,6\text{-}^i\text{Pr2C}_6\text{H}_3$ [14].

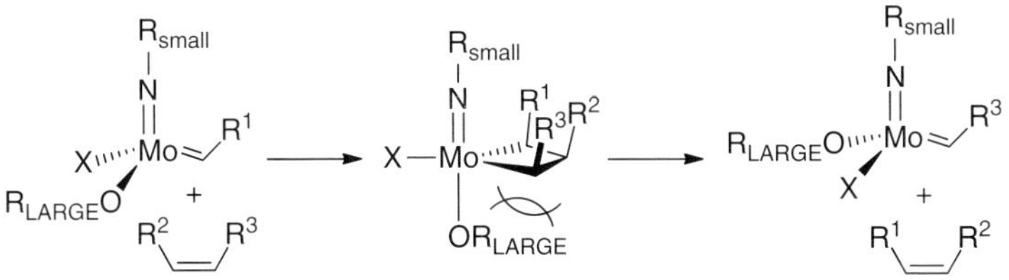

Scheme 5.3 General pathway for *Z*-selective olefin metathesis with molybdenum alkylidene catalysts.

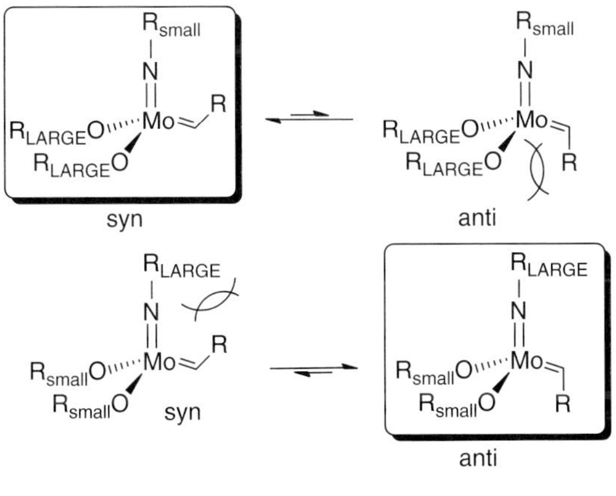

Scheme 5.4 General pathway for *Z*-selective olefin metathesis with molybdenum alkylidene catalysts.

Besides catalysts based on biaryloxide ligands, also the use of immensely sterically demanding di-ortho-substituted phenolate ligands results in kinetically controlled *Z*-selective alkene formation (Scheme 5.5) [15].

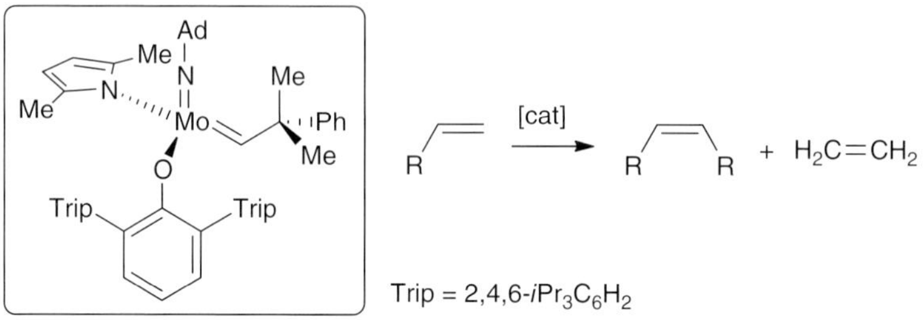

Scheme 5.5 Molybdenum pyrrolide catalyst in the *Z*-selective cross metathesis of terminal alkenes [15, 16].

Blechert *et al.* successfully introduced a desymmetrization approach. An asymmetric ring-opening cross metathesis of norbornenedicarboxylic anhydride was rendered possible by a Hoveyda–Blechert type ruthenium–carbene complex with a chiral N-heterocyclic carbene (NHC) ligand (Scheme 5.6). The chiral information of the NHC backbone was translated into ruthenium's active coordination sphere by the steric repulsion of an isopropyl group with a skewed *ortho*-biphenyl substituent [17].

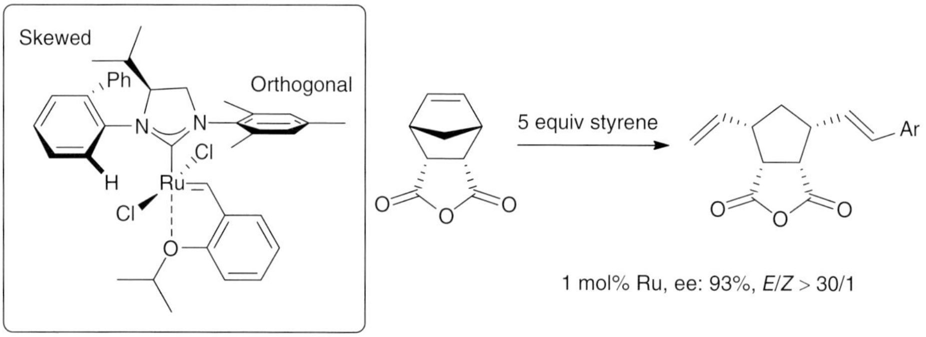

Scheme 5.6 Alkene desymmetrization by a chiral ruthenium catalyst of the Blechert–Hoveyda type [17].

A ruthenium catalyst with a chelating carbanionic NHC ligand and a pivaloate ligand was synthesized by Grubbs's research group [18a] (Scheme 5.7). The comparatively low catalytic activity meets a mediocre catalyst stability and a reasonable stereoconservation of *Z*-alkene substrate. Ligand exchange from pivaloate to nitrate led to a highly active class of catalysts that exhibits much higher activity and Z-selectivity.

A complete Chauvin catalyst cycle of olefin metathesis at ruthenium–carbene complexes with different anionic ligands X leads to an inversion of configuration at the stereogenic metal center (Scheme 5.8).

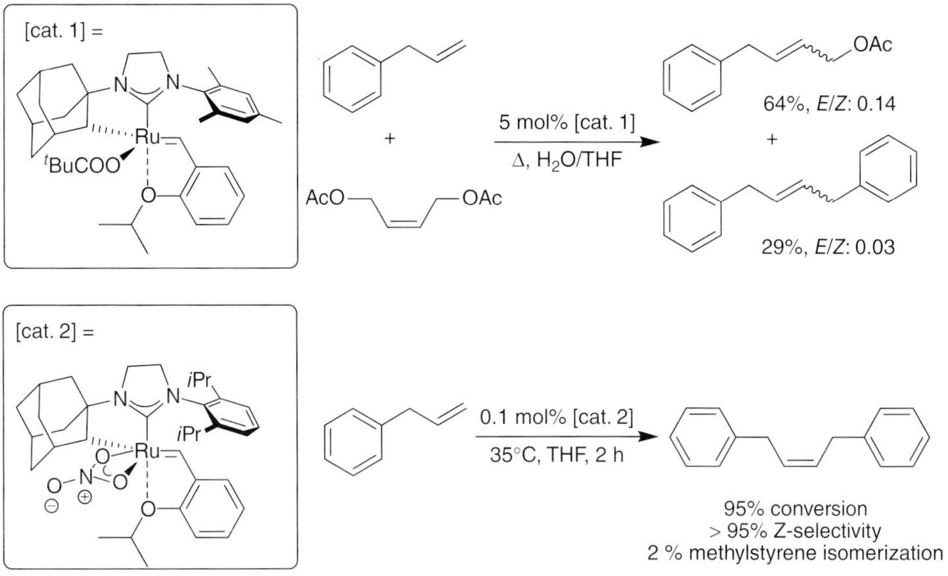

Scheme 5.7 C–H activated adamantyl-substituted NHC-containing catalyst with high *Z*-selectivity [18].

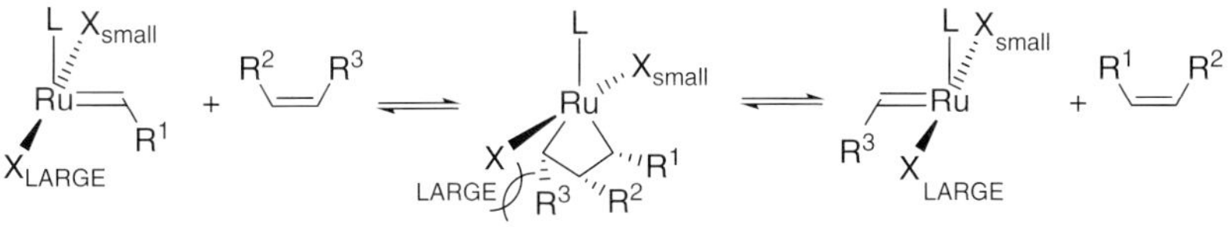

Scheme 5.8 Pathway for *Z*-selective olefin metathesis with ruthenium carbene catalysts.

Thus, the pathways of Schrock and Grubbs catalysts are analogous in the number of coordinating ligand atoms at the metal and the inversion of the absolute configuration. The active catalyst species differ, however, in their formal metal oxidation states (II vs IV) and in the coordination environment (pseudo-tetrahedral vs pseudo-octahedral without two ligands).

5.3 Combining Catalytic Activity and Stereoselectivity in Ruthenium Carbenes: an Antagonism?

For the catalytic activity of ruthenium–carbene complexes, the relative stability of the carbene conformations plays a pivotal role. After the dissociation of a victim ligand, followed by alkene coordination, the carbene has to adopt an active conformation. Only then can the new C–C bond in the ruthena(IV)cyclobutane evolve (green traffic light row in Figure 5.2) [19, 20].

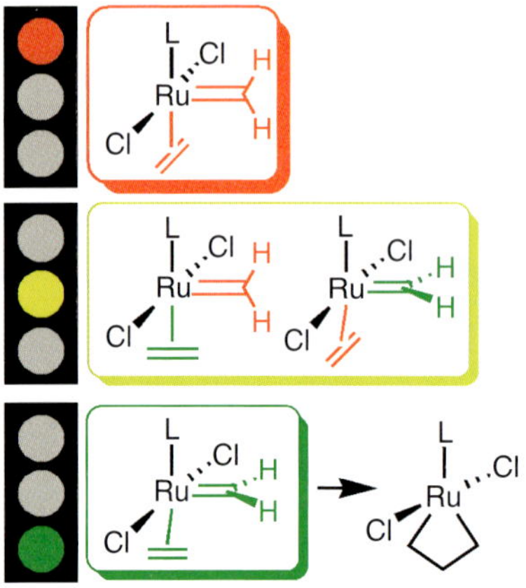

Figure 5.2 Role of the carbene conformation in the C–C bond formation [19, 20].

Strong σ-donor ligands (e.g., NHC) in trans position to the alkene ligand lead to a more stable “active” carbene conformation, while weak σ-donor ligands (e.g., triaryl phosphines) in trans position to the alkene ligand lead to a more stable “inactive” carbene conformation. Weak σ-donor ligands (e.g., chloride, sulfonate, carboxylate) in cis position to the alkene ligand lead to a more stable “active” carbene conformation, while strong σ-donor ligands (e.g., iodide, alkoxide, alkyl) in cis position to the alkene ligand lead to a more stable “inactive” carbene conformation [19, 21].

E-alkenes are usually only slightly more stable than the diastereomeric *Z*-alkenes. Reactions near the thermodynamic equilibrium thus lead to mixtures of alkene diastereomers. The formation of pure *E*- or *Z*-alkene product by kinetic control offers a strategy to avoid *E*/*Z* equilibria. Thus, stereoconservative alkene metathesis catalysis is characterized by a rapid conversion of the *E*-alkene substrate into *E* product, and of the *Z*-alkene substrate into *Z* product. Additionally, the interconversion of *E*-alkene into *Z*-alkenes (and vice versa, according to the principle of microscopic reversibility) has to proceed as slowly as possible, since it would lead to an *E*/*Z* equilibrium mixture (Scheme 5.9).

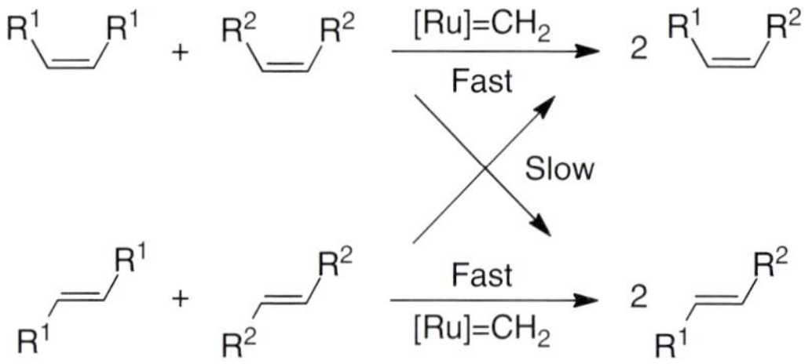

Scheme 5.9 Concept of stereoconservative alkene metathesis.

For a combination of high catalyst activity (i.e., low energy barriers for ruthenacyclobutane formation) and an effective discrimination of *E*/*Z* interconversion,

Table 5.1 Coordination mode of neutral and anionic ligands at uncharged and cationic ruthenium–carbene complexes [8, 22].

	Uncharged	Cationic
trans-L or *trans*-X	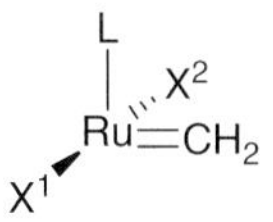 R. H. Grubbs, W. A. Herrmann	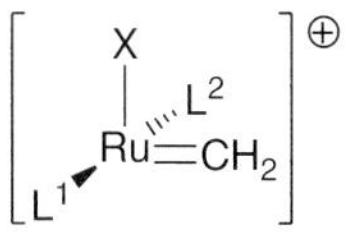 (Unknown)
cis-L or *cis*-X	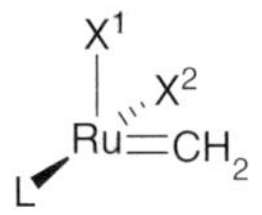 (Unknown)	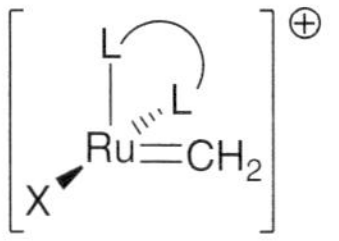 P. Hofmann

a ruthenium(II)–carbene complex with a distinct left/right differentiation of the ligand's steric demand and a favorable electronic situation is essential (Table 5.1).

A catalyst complex with beneficial steric and electronic influence of the ancillary ligand backbone has not yet been successfully put forward. Ruthenium–carbene catalysts with a chelating *ortho*-phosphino phenolate ligand feature low activity in olefin metathesis (Scheme 5.10). The catalysts showed little or no reactivity in the ring-opening metathesis polymerization (ROMP) of cyclooctene, even when dissolved in neat cyclooctene for up to 7 h at 60 °C for the most extreme cases [23a]. Presumably, the carbene ligand is electronically locked in an inactive conformation because of the weak phosphine σ-donor ligand and the strong phenolate σ-donor. The highly reactive norbornene, however, is converted efficiently [23b].

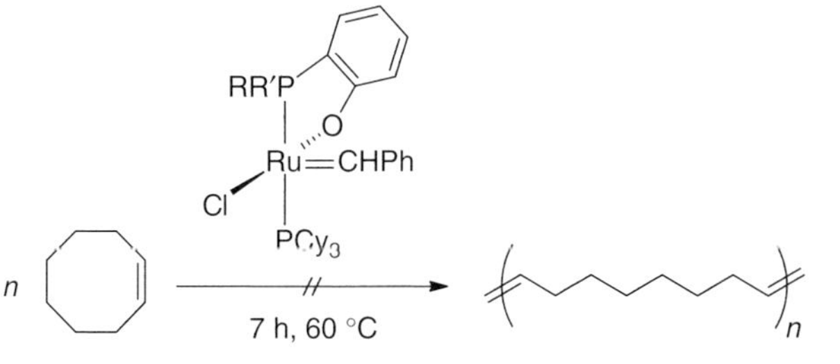

Scheme 5.10 Ruthenium(II)–carbene complex with low catalytic activity; R, R′ = Ph, Cy, ^{t}Bu [23].

A large number of second-generation Grubbs catalysts with unsymmetrically substituted NHC ligands have been synthesized [24]. However, the main challenge for diastereoselective olefin metathesis remains the incorporation of two different anionic ligands, namely pseudohalides [25].

5.4
Stereoselectivity in Ring-Opening Metathesis Polymerization (ROMP)

The advantage of modern alkene metathesis catalysts of the Schrock type becomes particularly apparent in polymerization reactions. With simple alkene substrates, the different tolerance of Grubbs and Schrock catalyst systems toward functional groups becomes almost insignificant. With a sterically demanding sulfonate instead of the chloride ligand, an alternating ROMP of cyclooctene and norbornene yields up to 51% *Z*-configuration of the polymer's alkene moieties (Scheme 5.11).

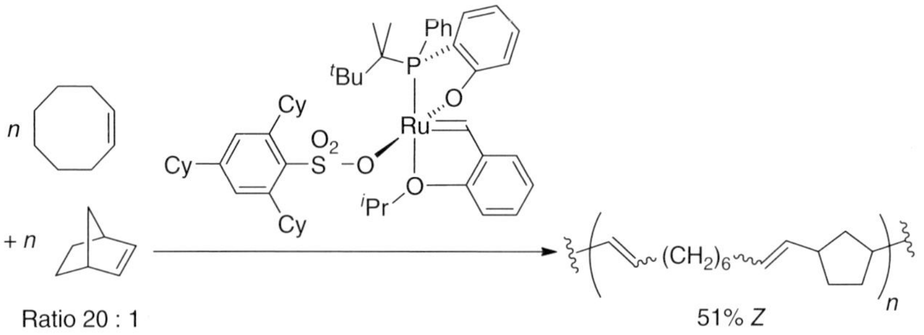

Scheme 5.11 Alternating copolymerization of cyclooctene and norbornene with an unsymmetric ruthenium catalyst.

In the ROMP of norbornene, four types of stereoregular polynorbornene products are theoretically possible. The alkene moiety can be either in the *E* or *Z* (trans or cis) orientation. Furthermore, the relative orientation of the tertiary stereocenters can be either isotactic or syndiotactic (Figure 5.3) [26]. With

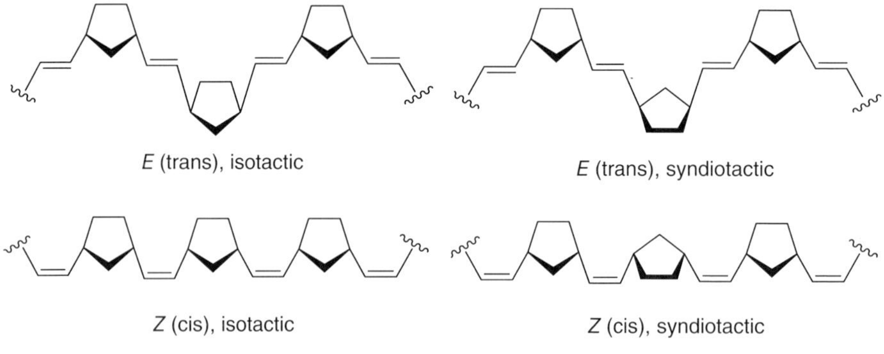

Figure 5.3 Stereoisomers of ROMP products of norbornene [26].

Molybdenum alkylidene catalysts, three of the four polymers have indeed been successfully synthesized.

As outlined in Scheme 5.4, bulky aryl groups at the imido ligand enforce an anti orientation of the alkylidene substituent with respect to the imido spectator ligand. The resulting molybdenacyclobutanes cyclorevers to E (trans) ROMP products (Scheme 5.12).

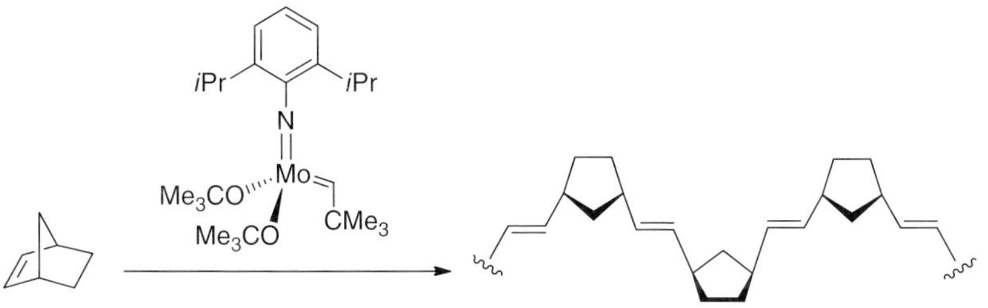

Scheme 5.12 Chain-end control in norbornene ring-opening metathesis polymerization.

With small aryl groups at the imido ligand, a syn orientation of the alkylidene substituent with respect to the imido spectator ligand is favored. This leads to *all-cis* molybdenacyclobutanes, and cycloreversion to Z (cis) ROMP products (Scheme 5.13).

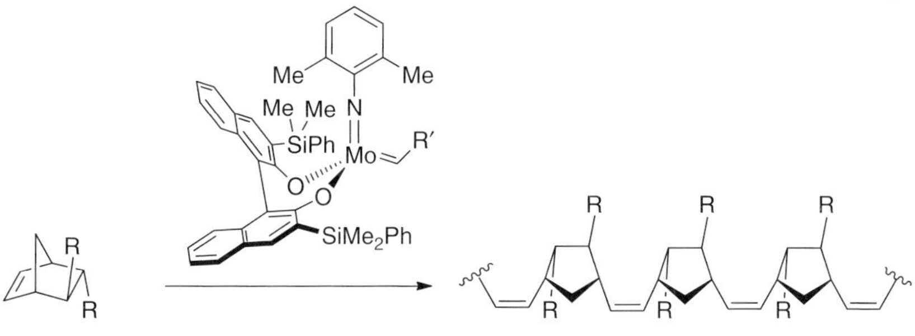

R = CO_2Me, CH_2OMe, Me

Scheme 5.13 Enantiomorphic site control in norbornene ring-opening metathesis polymerization.

Based on the inversion of the stereogenic metal center's absolute configuration after each metathesis cycle, monoalkoxide pyrrolide (MAP) complexes undergo Z-selective ROMP to syndiotactic polynorbornene (Scheme 5.14).

With molybdenum imido alkylidene complexes, three of four possible stereoregular ROMP polymers can be synthesized [26].

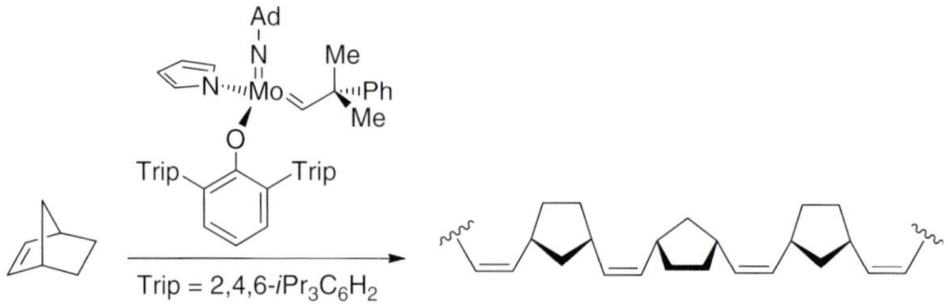

Scheme 5.14 Stereogenic metal control in norbornene ring-opening metathesis polymerization.

5.5
Outlook

Since halides and pseudohalides are mostly sterically undemanding, steric repulsion is difficult to induce in ruthenium–carbene complexes of the type $LRuX_2(=CH_2)$. With a chelating pincer-type backbone, a sterically and electronically tailored ancillary ligand can be designed (Scheme 5.15). A strong σ-donor indolide ligand trans to the alkene substrate and two weak σ-donor ligands (carboxylate and uncharged amine or heterocycle) cis to the alkene substrate's

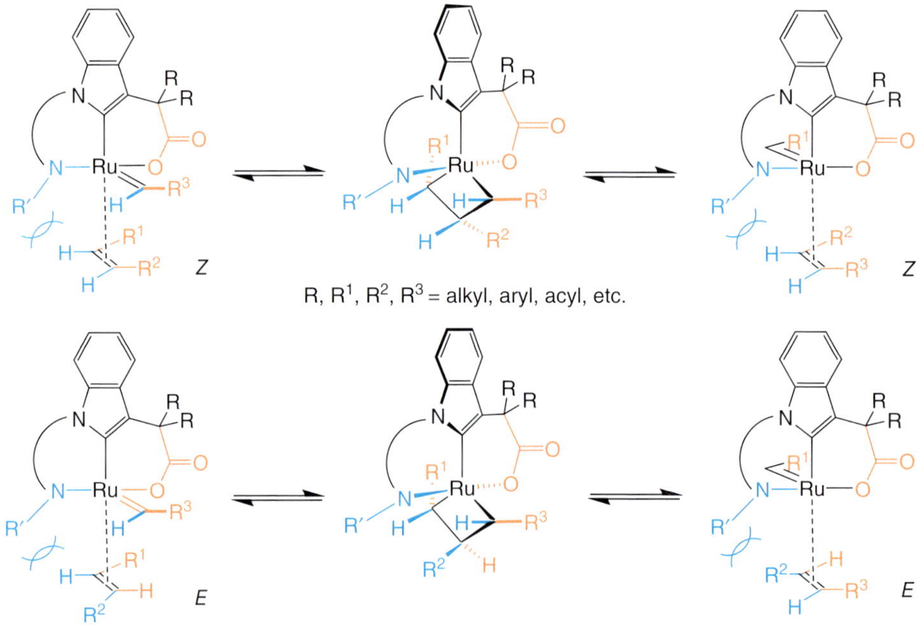

Scheme 5.15 Mechanistic scenario for a proposed sterically induced diastereoconservation in ruthenium(II)–carbene indolide complexes. Color code: light blue (nitrogen side) and orange (oxygen side).

coordination site fulfill the requirements for a favorable energy hypersurface of carbene rotation (Figure 5.2). The nitrogen atom ligand side (left molecule side, blue color code) is sterically demanding, forcing carbene substituents toward the sterically undemanding carboxylate side (right molecule side, orange color code). With such an approach, the combination of diastereoselectivity and catalytic activity appears to be feasible. Catalyst stability, however, will remain a major challenge for ruthenium–carbene catalysts [20, 27].

5.6 Summary

The development of Schrock-type metathesis catalysts has made significant progress in recent years. Both stereoselectivity and tolerance toward "hard" functional groups have improved by optimization of ligands. Molecular molybdenum–carbene complexes have evolved into valuable catalysts for natural product syntheses – a field that has been dominated by ruthenium-based metathesis catalysts. For the latter, the improvement of activity, stability, chemoselectivity, and stereoselectivity remains a formidable goal [27]. The question is whether successful design, preparation, and testing of superior ruthenium catalysts require new or novel ancillary ligand backbones.

References

1. (a) Vougioukalakis, G.C. and Grubbs, R.H. (2010) *Chem. Rev.*, **110**, 1746; for a review on stereocontrolled alkene metathesis; see (b) Kress, S. and Blechert, S. (2011) *Chem. Soc. Rev.*, **41**, 4389; for a german language review on Z-selective alkene metathesis; see (c) Wegner, H.A. (2012) *Nachr. Chem.*, **60**, 738.
2. Reuben, B. and Wittcoff, H. (1988) *J. Chem. Ed.*, **65**, 605.
3. (a) Chauvin, Y. (2006) *Angew. Chem.*, **118**, 3825; *Angew. Chem. Int. Ed.*, 2006, **45**, 3740; (b) Schrock, R.R. (2006) *Angew. Chem.*, **118**, 3832; *Angew. Chem. Int. Ed.*, (2006), **45**, 3748; (c) Grubbs, R.H. (2006) *Angew. Chem. Int. Ed.*, **45**, 3760.
4. Hérisson, J.-L. and Chauvin, Y. (1971) *Makromol. Chem.*, **141**, 161.
5. Toste, F.D., Chatterjee, A.K., and Grubbs, R.H. (2002) *Pure Appl. Chem.*, **74**, 7.
6. Schrock, R.R. and Hoveyda, A.H. (2003) *Angew. Chem. Int. Ed.*, **42**, 4592.
7. (a) Nguyen, S.T., Johnson, L.K., and Grubbs, R.H. (1992) *J. Am. Chem. Soc.*, **114**, 3974; (b) Wu, Z., Benedicto, A.D., and Grubbs, R.H. (1993) *Macromolecules*, **26**, 4975; (c) Nguyen, S.T., Ziller, J.W., and Grubbs, R.H. (1993) *J. Am. Chem. Soc.*, **115**, 9858.
8. (a) Huang, J., Stevens, E.D., Nolan, S.P., and Petersen, J.L. (1999) *J. Am. Chem. Soc.*, **121**, 2674; (b) Weskamp, T., Kohl, F.J., Hieringer, W., Gleich, D., and Herrmann, W.A. (1999) *Angew. Chem. Int. Ed.*, **38**, 2416; (c) Scholl, M., Trnka, T.M., Morgan, J.P., and Grubbs, R.H. (1999) *Tetrahedron Lett.*, **40**, 2247.
9. Love, J., Morgan, J., Trnka, R., and Grubbs, R.H. (2002) *Angew. Chem. Int. Ed.*, **41**, 4035.
10. (a) Garber, S.B., Kingsbury, J.S., Gray, B.L., and Hoveyda, A.H. (2000) *J. Am. Chem. Soc.*, **122**, 8168; (b) Gessler, S., Randl, S., and Blechert, S. (2000) *Tetrahedron Lett.*, **41**, 9973.

11. Michrowska, A., Bujok, R., Harutyunyan, S., Sashuk, V., Dolgonos, G., and Grela, K. (2004) *J. Am. Chem. Soc.*, **126**, 9318.
12. Häußermann, A., Rominger, F., and Straub, B.F. (2012) *Chem. Eur. J.*, **18**, 14174.
13. Häußermann, A. (2013) Synthese, Decarboxylierung, Umlagerung und Koordination von tridentaten Arylliganden. PhD thesis. Universität Heidelberg.
14. Meek, S.J., O'Brien, R.V., Llaveria, J., Schrock, R.R., and Hoveyda, A.H. (2011) *Nature*, **471**, 461.
15. Jiang, A.J., Zhao, Y., Schrock, R.R., and Hoveyda, A.H. (2009) *J. Am. Chem. Soc.*, **131**, 16630.
16. Marinescu, S.C., Schrock, R.R., Müller, P., Takase, M.K., and Hoveyda, A.H. (2011) *Organometallics*, **30**, 1780.
17. Tiede, S., Berger, A., Schlesiger, D., Rost, D., Lühl, A., and Blechert, S. (2010) *Angew. Chem.*, **122**, 4064; *Angew. Chem. Int. Ed.*, (2010), **49**, 3972.
18. (a) Endo, K. and Grubbs, R.H. (2011) *J. Am. Chem. Soc.*, **133**, 8525; (b) Rosebrugh, L.M., Herbert, M.B., Marx, V.M., Keitz, B.K., Grubbs, R.H. (2013) *J. Am. Chem. Soc.*, 1**35**, 1276.
19. Straub, B.F. (2005) *Angew. Chem.*, **117**, 6129; *Angew. Chem. Int. Ed.*, (2005), **44**, 5974.
20. Straub, B.F. (2007) *Adv. Synth. Catal.*, **349**, 204.
21. (a) Sanford, M.S., Ulman, M., and Grubbs, R.H. (2001) *J. Am. Chem. Soc.*, **123**, 749; (b) Sanford, M.S., Love, J.A., and Grubbs, R.H. (2001) *J. Am. Chem. Soc.*, **123**, 6543.
22. (a) Volland, M.A.O., Adlhart, C., Kiener, C.A., Chen, P., and Hofmann, P. (2001) *Chem. Eur. J.*, **7**, 4621; (b) Hansen, S.M., Volland, M.A.O., Rominger, F., Eisenträger, F., and Hofmann, P. (1999) *Angew. Chem.*, **111**, 1360; *Angew. Chem., Int. Ed. Engl.*, (1999), **38**, 1273; (c) Hansen, S.M., Rominger, F., Metz, M., and Hofmann, P. (1999) *Chem. Eur. J.*, **5**, 557.
23. (a) Bornand, M., Torker, S., and Chen, P. (2007) *Organometallics*, **26**, 3585; (b) Bornard, M. and Chen, P. (2005) *Angew. Chem.*, **117**, 8123; *Angew. Chem. Int. Ed.*, (2005), **44**, 7909; (c) Torker, S., Müller, A., and Chen, P. (2010) *Angew. Chem.*, **122**, 3850; *Angew. Chem. Int. Ed.*, (2010), **49**, 3762.
24. Hamad, F.B., Sun, T., Xiao, S., and Verpoort, F. (2013) *Coord. Chem. Rev.*, **257**, 2274.
25. Anderson, E.B. and Buchmeiser, M.R. (2012) *Synlett*, 185.
26. Schrock, R.R. (2011) *Dalton Trans.*, **40**, 7484.
27. (a) Hong, S.H., Wenzel, A.G., Salguero, T.T., Day, M.W., and Grubbs, R.H. (2007) *J. Am. Chem. Soc.*, **129**, 7961; (b) Vehlow, K., Gessler, S., and Blechert, S. (2007) *Angew. Chem. Int. Ed.*, **46**, 8082; (c) Dinger, M.B. and Mol, J.C. (2003) *Organometallics*, **22**, 1089.

Part II
New Catalysts – New and Old Reactions

Peter Hofmann

If one takes a look at the present situation of academic and industrial state-of-the-art catalysis research dealing with structurally well-defined molecular catalysts in homogeneous phase or in their immobilized form (the general theme of the Heidelberg SFB "Molecular Catalysts: Structure and Functional Design"), the discovery and development of useful, new, unprecedented catalytic reactions and the search for novel, highly active, and selective catalyst systems represent the main challenge and focus. We are certainly justified in saying that the large number of newly established catalytic reactivity patterns, especially (but not exclusively) the ones involving transition-metal catalysis, has revolutionized the field of organic synthesis over the past three decades. This became possible because more and more mechanistic insight into and understanding of the complete catalytic cycles and the molecular species involved have become available through increasingly sophisticated experimentation (instrumentation) and theory (e.g., quantum chemistry) over the years. This progress has begun to enable the rational or at least semi-rational design of, *inter alia*, tailor-made ligand systems capable of efficiently shaping a broad variety of transition metals for their desired catalytic task. In this context, it is quite remarkable that, since the turn of the century, the Nobel Prize in Chemistry has been awarded to nine researchers working in the field of molecular transition-metal catalysis. Their discoveries, which originally evolved from mostly curiosity-driven basic academic research of the awardees themselves and of their numerous scientific predecessors, have rather rapidly found their way into industrial applications, spanning a range from fine-chemicals and drug synthesis to larger scale production. Indeed, catalysis must be viewed as one of chemistry's most important technologies for the twenty-first century, and successful catalysis research is a prerequisite for dealing with Nature and our environment in a responsible, sustainable, and energy- and resource-saving manner.

Despite the obvious impetus and motivation to devise and to optimize new, innovative molecular catalyst systems for synthetic transformations identified and recognized as relevant and thermodynamically feasible, many of them could not yet be brought to realization and remain as a challenge. The selective activation and functionalization of C–H bonds, selective oxidation chemistry of

Molecular Catalysts: Structure and Functional Design, First Edition. Edited by Lutz H. Gade and Peter Hofmann.

hydrocarbons, utilization of CO_2 as a cheap C1 building block, and many more so-called "dream-reactions" can be mentioned here.

There is another important field of focus for catalysis research both in academia and industrial laboratories. Although a number of industrial production units for obvious reasons are preferentially operated employing heterogeneous catalysts, there are also large-scale and medium-scale catalytic reactions that deliver high production volumes of commodity chemicals or of high-value fine chemicals, respectively, and which are run on the basis of homogeneous catalysis. Typical examples of large-scale reactions are the Co- and Rh-catalyzed hydroformylation of olefins (the oxo-reaction), which today represents the biggest homogeneous metal-catalyzed organic synthesis process operated in industry, the Ni-catalyzed hydrocyanation of 1,3-butadiene to adiponitrile using HCN (the Du Pont process), the hydrosilylation of alkenes, the Monsanto and Cativa processes for the carbonylation of methanol, the Wacker–Hoechst oxidation of ethylene to acetaldehyde, the Shell higher olefin process (SHOP), the $FeCl_3$/S-heterocycle cocatalyzed *p*-selective chlorination of toluene, and several homogeneous transition-metal-catalyzed polymer or copolymer production processes with homogeneous, single-site catalysts (ring-opening metathesis polymerization (ROMP), metallocene catalysis, and late-transition-metal-catalyzed CO/olefin copolymerization). Smaller scale, but equally important industrial organic synthesis strategies have been developed and are operated commercially for enantioselective C=C and C=X hydrogenation and for stereoselective isomerization reactions, for Pd-catalyzed C–C and C–heteroatom bond formation, or for enantioselective cyclopropanation, to mention just a few. These and related catalytic reactions often lead to high-value products for the chemical or pharmaceutical industry. The essential point here is that, for both large and smaller scale successful implementation of an industrial production plant and the required investment, highly demanding conditions have to be met, which inevitably contrast the typical scenarios and the boundary conditions of lab-scale homogeneous catalysis experiments in typical academic research environments, as they are usually reported and of course remain acceptable during the discovery and development phase of new catalytic reactions. The settings mandatory for an industrial application of molecular catalysis in order to deliver products competitive on a global market, however, are much more difficult to achieve. Most often, a catalyst concentration in the 5–10 mol% range is far from acceptable, and only parts per million concentrations are tolerated. Consequently, the turnover frequency (TOF) and turnover number (TON), that is, catalyst activities and catalyst lifetimes, have to reach very high limits, and of course this is also true for catalyst selectivity leading to the desired products. Catalyst decomposition, often inevitable during extremely long running times of large-scale continuous industrial production units, must not lead to a loss of precious metals, so recycling strategies and processes have to be optimized. To give just one example, in the standard Rh/PPh_3-catalyzed hydroformylation of olefins with reactor volumes of 200–300 m^3, Rh is present in the 200–300 ppm range (multi-kilogram scale, worth several million Euros), the PPh_3 ligand is present at the multi-ton scale, and TOF values way above 1000 and TON around 400 000 are achieved. Quite obviously, most sophisticated chemical engineering aspects are at least as important here as the chemistry involved.

Even for hydroformylation and similar industrial-scale catalytic organic syntheses of large-volume bulk products, although established and operating for quite some time, there is still a continuous need for basic research directed toward improvement of known catalysts and for research aiming at new and more efficient catalysis. In the context of the following chapter, research projects will be described that deal with various aspects of new types of ligands and their transition-metal chemistry for molecular catalysis in "old and new reactions."

6
Oxidation Catalysis with High-Valent Nonheme Iron Complexes

Peter Comba

6.1
Introduction

Iron proteins are ubiquitous, and it appears that the bioavailability of iron together with the accessibility of various oxidation and spin states and the fine-tuning of their relative stabilities by subtle modification of the coordination geometries are the main reasons for the prevalence of iron sites in Nature. The important family with mononuclear nonheme iron(II) centers combines dioxygen activation with a broad variety of substrate oxidation processes [1]. Consequently, the search for economically proficient and environmentally benign methods for chemical synthesis has resulted in an increasing interest for iron-catalyzed reactions, in particular but not exclusively for oxidation processes [2]. Much of the extensive knowledge of nonheme iron-oxo sites in enzymes [1a,b] and low molecular weight model compounds [1c, 3] has been assembled in the last dozen years. The milestones include the first report of the spectroscopic [4] and the first structural characterization of intermediate-spin ferryl complexes [5], the detailed mechanistic study of the α-keto acid-dependent TauD enzyme with a high-spin ferryl active site [6], the report of high-spin low molecular weight ferryl complexes [7], the characterization of oxo–iron(V) complexes [8], the proposal of catalytically active intermediate spin dihydroxo–iron(IV) complexes [7b], the catalytic halogenation of unactivated substrates with ferryl model systems [9], the stabilization and reactivity of ferryl complexes in water [10], water oxidation by mononuclear nonheme iron–oxo complexes [11], and oxygenation processes with molecular dioxygen [12]. A decade of extensive experiments, including structural work, the trapping and spectroscopic characterization of metastable intermediates, and thorough mechanistic studies [1] supported by detailed quantum chemical work [13], has led to a detailed understanding of low molecular weight and enzyme active sites. Part of this exciting development is reviewed in this chapter.[1)]

1) Owing to the special character of this book as an account of work done in the 12 years of the Heidelberg network of molecular catalysis (SFB 623), I will primarily concentrate on published and ongoing work based on high-valent iron complexes of bispidine ligands and compare it, where appropriate, with that of other ligand platforms.

Molecular Catalysts: Structure and Functional Design, First Edition. Edited by Lutz H. Gade and Peter Hofmann.

6.2 Bispidine Ligands

The salient features of the bispidine platform (3,7-diazabicyclo[3.3.1]nonane, see Scheme 6.1 for ligands discussed in this chapter and Figure 6.1 for the structural properties of the metal-free ligands and their transition metal complexes) are a high degree of preorganization for cis-octahedral coordination geometry, the rigidity of the adamantane-derived backbone, and the modular synthesis with a large variety of tetra-, penta- and hexadentate ligands available in good yields from simple two-step procedures (apart from the ligands discussed here, bispidine-based macrocyclic ligands, chiral bispidines, and second-generation bispidines enforcing trigonal geometries have been described and used in transition-metal coordination chemistry) [14]. Bispidines were first prepared by Mannich and Mohs [15], and the first metal complexes were obtained by Stetter

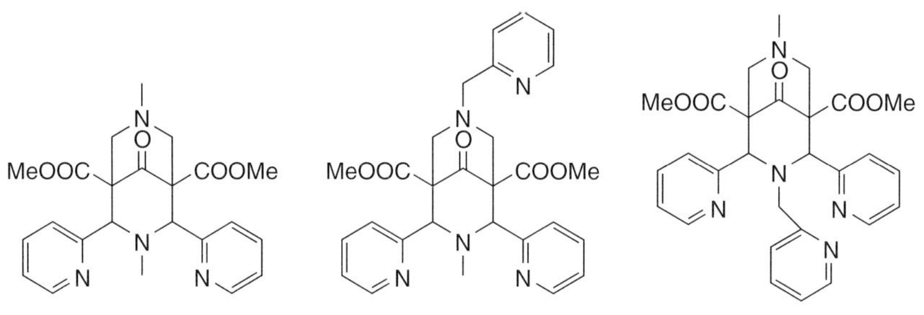

Scheme 6.1 Structures of the bispidine ligands discussed in this chapter.

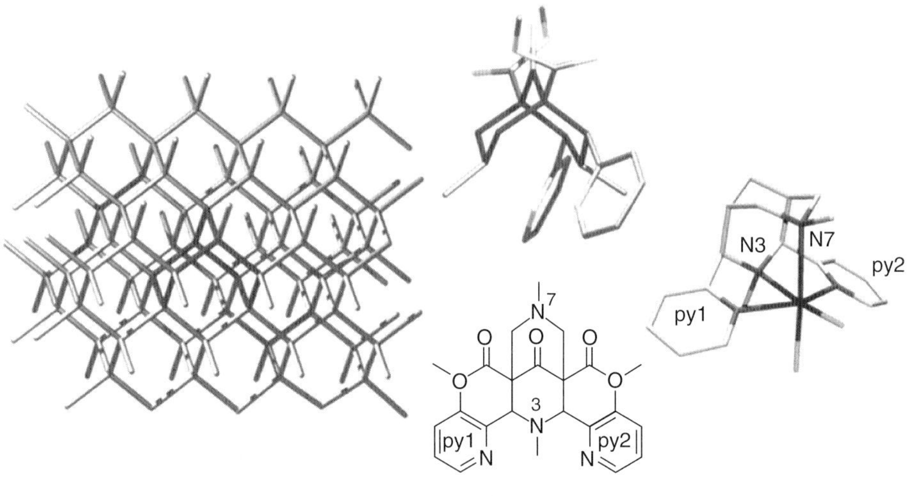

Figure 6.1 The rigidity of the bispidine scaffold: the structures of diamond, the tetradentate bispidine ligand L^1, and a corresponding metal complex.

and Haller [16], but it was not until about 15 years ago that the potential of the bispidines for coordination chemistry was discovered [17]; the development of the iron–bispidine chemistry [18] and that of high-valent iron bispidines followed soon after [19]. An important feature of the bispidine scaffold is that bonds from the metal center to the two tertiary amines are geometrically different (see Figure 6.1) – that to N3 is short and very rigid (part of two five-membered chelate rings, involving a pyridine group each) and that to N7 generally is longer and more flexible (part of more flexible six-membered chelate rings). Consequently, coligands trans to N3 generally are more tightly bound than those trans to N7. This leads to interesting and predictable differences between complexes of the isomeric pentadentate ligands L^2 and L^3. The redox potentials, electron transfer kinetics, and oxidation/oxygenation reactivities of the corresponding high-valent iron complexes as well as the Fe^{II}-H_2O_2 chemistry have been studied in detail (K. Benzing, P. Comba, A.-M. Rensland, and S. Wunderlich, unpublished data) [19, 20]. Importantly, differences in the metal–ligand bonding to the two tertiary amine donors have been shown to be of importance for the tuning of the spin state of the ferryl complexes (see Section 6.4) [13g] and the redox potentials and catalytic activities of $Cu^{II/I}$ couples [21]; the bispidine-derived geometry is particularly well suited for the Jahn–Teller active Cu^{II} ion and has led to a rich Cu^{II} coordination chemistry [22] with interesting applications in bioinorganic model chemistry (hemocyanin [23], catecholase[24]), catalytic aziridination [21a, 25], molecular magnetism [26], and medicinal chemistry [27].

Bispidine ligands and, in particular, their iron complexes have attracted some attention in industrial processes. It appears that the high-valent iron-bispidine compounds are among the most efficient bleaching catalysts (note that in bleaching processes the catalyst stability is of minor importance (see Section 6.6 for problems related to the stability of the ferryl complexes)) [28] and a number of patents have been filed in this area [29] (the corresponding manganese complexes have also been patented as bleaching agents [30, 31]). With Ni^{II}–bispidine complexes, interesting applications in homogeneous catalysis have been described [32]. So far, the iron–bispidine bleaching catalysts have not been produced commercially but they are used as efficient catalysts for liquid hardening (processing of coatings) [33].

6.3
Oxidation of the Ferrous Precursors

The iron(II) precatalysts are generally oxidized by H_2O_2 or oxygen transfer agents such as iodosylbenzene (PhIO, see Scheme 6.2), but alkylhydroperoxides, peracids, oxone, N_2O, XO^- (X = Cl, Br), ozone, dioxygen, and Ce^{IV} in aqueous solution have also been used [12c, 34]. The H_2O_2 oxidation of the iron(II) precursors with the pentadentate ligands L^2 and L^3 in methanol at low temperature leads to metastable Fe^{III} intermediates which were spectroscopically characterized; the L^2-based system with the hydroperoxo group trans to N7 is,

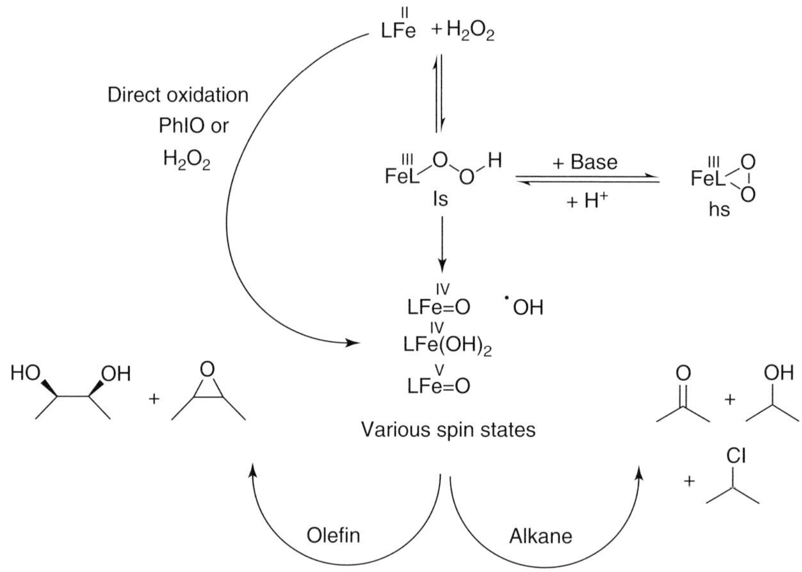

Scheme 6.2 Oxidation of the ferrous precatalyst to the catalytically active species for the oxidation of alkanes and olefins.

as expected, less stable than its isomer with L^3. With the L^3-based Fe^{III} – peroxo complex it was possible to interconvert the low-spin hydroperoxo and high-spin peroxo complexes reversibly by titration with base or acid, respectively, resulting in $4.5 < pK < 5.0$ (MeOH, −40 °C) [19]. In acetonitrile at ambient temperature, the oxidation of the L^2- and L^3-based Fe^{II} precursors produce intermediate-spin ($S = 1$) Fe^{IV}=O complexes, which were spectroscopically characterized [10, 20a, 35] and shown to be very efficient oxidation catalysts (see Section 6.6) [20a–c, 35, 36]. Trapping of Fe^{III} and Fe^{IV} oxygen adducts with the tetradentate bispidine L^1 has been much more difficult, although the $Fe^{II}/L^1/H_2O_2$ or PhIO systems were shown to be catalytically very active with a variety of substrates [7b, 9, 20c,d]. The first intermediates that were trapped and spectroscopically studied with tetradentate bispidine ligands were obtained by oxidation of the Fe^{II} precursor with *t*BuOOH, and the corresponding alkylperoxo ferric complex was shown to be a spin-crossover compound with interesting spectroscopic properties [37]. This indicated that with the tetradentate ligand system it might be possible to obtain high-spin ($S = 2$) ferryl complexes, and this was confirmed with a density functional theory (DFT) based theoretical analysis (see Section 6.4) [13g].

The first nonheme ferryl complex in pure water was observed with the pentadentate bispidine L^3 at neutral to acidic pH at 0 °C, and this had some implications for catalytic water oxidation with ferryl compounds (Section 6.6.5) [10]. Interestingly, the kinetic analysis of the formation of $[(L^3)Fe^{IV}{=}O]^{2+}$ in water indicates that this is a clean monophasic reaction (see Figure 6.2); that is, there

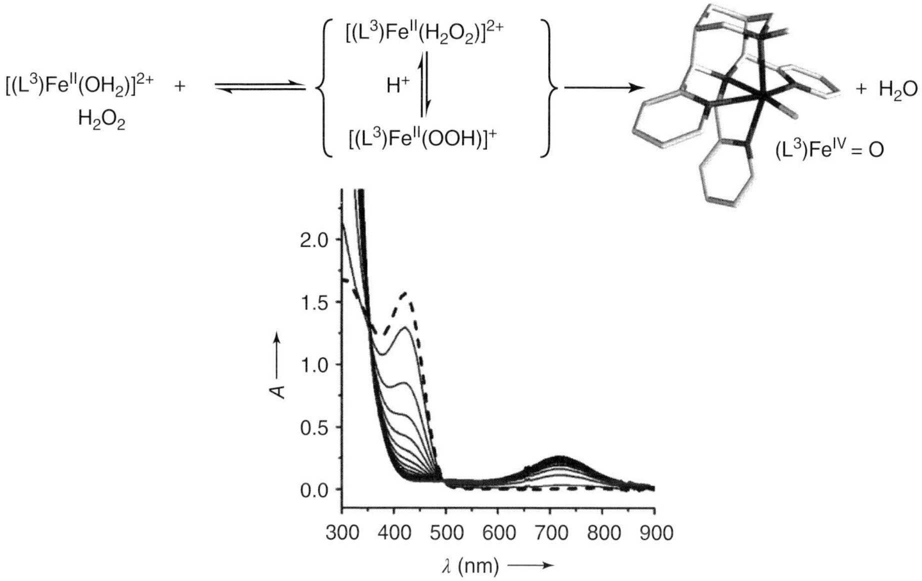

Figure 6.2 Oxidation of the ferrous complex of L^3 with H_2O_2 in water (0 °C, citrate buffer, pH 5.4). The dotted line is the electronic spectrum of the ferric complex, and the dark line corresponds to the ferryl complex, developed within 80 s) [10].

is no Fe^{III}–hydroperoxo intermediate and no formation of OH radicals. This has important implications for Fenton-type chemistry, where OH radical-based (Haber–Weiss) [38] and Fe^{IV}=O-based (Bray and Gorin) [39] processes are discussed. A DFT-based analysis indicates that the mechanism of the Fe^{II}-catalyzed hydrogen peroxide decomposition in general, that is, radical versus non-radical pathways, strongly depends on the ligand-enforced perturbation of the electronic structure of the iron center [13e].[2)]

A thorough DFT-based theoretical analysis of the formation of $[(L^1)Fe^{IV}{=}O(OH)]^+$ indicates that the most favorable process in this system is a direct transformation of the initially formed Fe^{II}/H_2O_2 adduct to the novel $[(L^1)Fe^{IV}(OH)_2]^{2+}$ product [13f]. This may explain why in the L^1-based iron chemistry it is difficult to trap Fe^{III} intermediates. However, reaction pathways and stabilities of intermediates depend on the solvent; this also emerges from the spin equilibrium of the $[(L^1)Fe^{III}(OOBu\textit{t})]^{2+}$ system described above [37] and from the DFT analysis of the L^1-based Fe^{II}/H_2O_2 system which indicates that a number of steps are assisted by H_2O molecules in the second coordination sphere [13f].

2) Note that in an excess of H_2O_2 the L^3-based ferryl complex is stable for hours even at ambient temperature. Note also, however, that the kinetics depicted in Figure 6.2 show a clean isosbestic point but this has only been studied over a time period of 1980s, that is, until the ferryl complex was fully formed. Side reactions may occur at a later stage; for example, it is known that bispidine-ferryl complex can oxidize water, see Section 6.6.5.

6.4 Spin States of the Ferryl Catalysts

The current paradigm is that high-spin ($S = 2$) ferryl complexes are significantly more reactive than intermediate-spin ($S = 1$) systems (see Figure 6.3 below; the low-spin ($S = 0$) form is much higher in energy and has not yet been reported). Part of this perception emerges from the fact that all ferryl enzyme sites are of high spin while, for a long time, all low molecular weight models were of intermediate spin. It is widely believed that the $S = 1$ species gain much of their reactivity by a crossover to the high-spin surface. The concept that describes this pathway is widely known as *two-state reactivity* (TSR, or multistate reactivity, blended spin states) [40], and this is supported by a number of experimental studies and a large body of computational work [13d, 41]. A possible critical point in some of the theory-based studies is that only rarely has spin–orbit coupling been included in the computation of the crossover energetics [42]. Therefore, for some time there was a run for "the first high-spin ferryl species," and a number of well-characterized high-spin $Fe^{IV}{=}O$ complexes are now available.[3] A thorough spectroscopic and computational analysis of the crystallographically characterized and extensively investigated (in terms of the reactivity) trigonal bipyramidal high-spin complex indicates that the two available reaction channels ($\sigma(d_{z^2})$ and $\pi(d_{xz,yz})$ with obviously differing orientation dependencies) [43] have similar reactivities because both develop a high level of oxyl character in the transition state [44].

It has been argued that for the reactivity the driving force involving the ferryl oxidant, that is, its redox potential, might be of more importance than its spin state [45]. However, information on the redox potentials of ferryl complexes only started to appear recently, and there is still some dispute on the experimental techniques to measure these potentials (see Section 6.5). Moreover, so far there is no experimental data that allows the unambiguous comparison of the oxidation

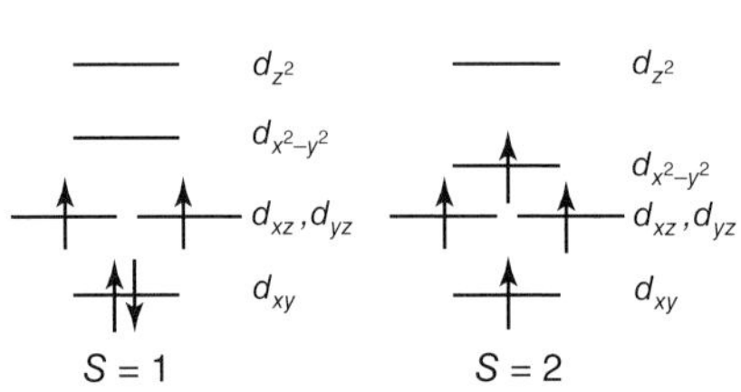

Figure 6.3 (a,b) Splitting and electron occupation of the d orbitals for intermediate-spin and high-spin Fe^{IV} centers.

3) Apart from the ferryl aqua ion which has been characterized by Mössbauer spectroscopy [7a], the L^1-based ferryl complex may be considered the first reported high-spin nonheme $Fe^{IV}{=}O$ model system [7b], although a full spectroscopic characterization is still missing, and a crystal structure and full spectroscopic characterization of other $S = 2$ ferryl complexes have been published since [7c–f, 44]. Note also that the spin state of $[(L^1)Fe^{IV}{=}O(X)]^{n+}$ depends on the coligand X, and this has not yet been fully analyzed (see also Sections 6.5 and 6.6).

power as a function of the spin state of the ferryl compound – in fact, so far there is no report of an experimentally determined redox potential of a high-spin ferryl species. An interesting recent observation is that the first low-spin ($S = 0$) $Ru^{IV}{=}O$ complex was shown to be an efficient catalyst for the oxidation of various organic substrates and in fact similar in efficiency and selectivity as a very similar $Ru^{IV}{=}O$ species with $S = 1$ electronic configuration [46]. The interpretation that the spin state of the metal–oxo oxidant (at least in this case) is not decisive for the reactivity led to some dispute [47].

To enforce a high-spin state in octahedral ferryl complexes, the in-plane ligand field needs to be reduced, that is, the $d_{x^2-y^2}$ orbital needs to be stabilized, see Figure 6.3 (note that, in general, the enzyme active sites are five-coordinate (distorted trigonal bipyramidal), and this also leads to a stabilization of the high-spin state; this concept has been used to create one of the recently published high-spin ferryl species [7c]). With the tetradentate bispidine ligand L^1, the ferryl complex has two possible configurations, that is, with the oxo group trans to N3 or trans to N7 (see e.g., Figure 6.1 for a visualization of this isomerism). In the high-spin electronic configuration, the axially distorted octahedral species is pseudo Jahn–Teller active, and a corresponding distortion leads to the required stabilization of the $d_{x^2-y^2}$ orbital and potentially to a preference for the high-spin electronic configuration (note that with the z-axis along the Fe=O bond, the elongation must occur in direction of the x- or y-axis). Owing to the rigid bispidine scaffold, this elongation is possible only along Fe–N7 and therefore, the isomer with the Fe=O bond trans to N3 leads to the more stable high-spin structure (see Table 6.1); this, indeed, is the most stable species of all, that is, $[(L^1)Fe^{IV}{=}O(OH)]^+$ is predicted to have high-spin electronic configuration [7b, 13g].

Note that the energy difference between the two spin states of $[(L^1)Fe^{IV}{=}O(OH)]^+$ of less than $10\,kJ\,mol^{-1}$ is within the error limit of this type of DFT calculations; that is, the two electronic configurations are basically degenerate with a possible preference for the high-spin one. Importantly, therefore, support for the assignment of the high-spin ground state arises from the

Table 6.1 Selected bond distances (see Figure 6.1 for atom numbering) and relative energies emerging from a DFT-based analysis of the configurational isomers and relevant spin states of the tetradentate bispidine L^1-based ferryl complex (note that the high-spin configurations are Jahn–Teller active and the elongation is expected to occur perpendicular to the Fe–O axis) [13g; reprinted with permission from Inorganic Chemistry, Copyright 2001 by American Chemical Society].

$(L^1)Fe^{IV}{=}O$	S	Bond length (Å) (DFT)					E_{rel} ($kJ\,mol^{-1}$)
		Fe–O	Fe–N7	Fe–N3	Fe–Npy	Fe–OH	
Fe=O	1	1.617	2.099	2.100	1.987	2.069	+7.1
trans-N3	2	1.615	2.281	2.082	2.030	2.335	0.0
Fe=O	1	1.612	2.270	1.970	2.000	2.084	+22.3
trans-N7	2	1.612	2.230	2.167	2.090	2.177	+31.8

fact that this oxidant is very efficient (see Section 6.6) and there is no experimental spectroscopic evidence for an intermediate-spin ferryl complex [7b].[4] Note that the corresponding catalyst in acetonitrile $[(L^1)Fe^{IV}{=}O(NCMe)]^{2+}$ has an intermediate-spin electronic configuration (NMR spectroscopy (Evans method) [20e]; electronic spectroscopy, dd transition at 760 nm; MeCN is a stronger ligand than OH_2 and OH^-). More recent data indicate that the addition of Cl^- or F^- to $[(L^1)Fe^{IV}{=}O(NCMe)]^{2+}$ leads to a complete loss of the dd transition at 760 nm with retained oxidation power, indicating that the corresponding chloro and fluoro ferryl complexes indeed have high-spin electronic configuration (K. Benzing, P. Comba, A. Waleska, and S. Wunderlich, unpublished results). An unambiguous spectroscopic analysis (e.g., Mössbauer and magnetic circular dichroism (MCD) spectroscopies) in combination with quantum chemical calculations of these species has so far been prevented by the high reactivity of these efficient oxidants (these ferryl complexes are extremely unstable, even at very low temperature, and for various and obvious reasons only very low concentrations are available for spectroscopy (relatively slow formation of the ferryl complex at low temperature, combined with a relatively fast formation of oxo-bridged diiron(III) species in an excess of Fe^{II} complex)). Interesting open questions that need further experimental and computational work are (i) how the redox potentials of the high-spin chloro, fluoro, and hydroxo complexes compare with that of the intermediate-spin acetonitrile complex and (ii) how the reactivities of the two electronic forms compare with each other.

6.5 Redox Properties of the Ferryl Oxidants

The first redox potentials of ferryl complexes were measured by cyclic voltammetry in dry acetonitrile [48] but the instability of the reduced form "$[(L)Fe^{III}-O]^+$" leads to observed data that are not unambiguous. Two other types of experiments have been described more recently to obtain valuable information on the oxidation power of ferryl complexes: (i) the spectropotentiometric oxidation of the $Fe^{III}-OH$ complex in acetonitrile with added water [49] and (ii) the titration of the ferryl complex with ferrocene derivatives (Fc) in dry acetonitrile to determine the $Fc + Fe^{IV}{=}O/Fc^+ + Fe^{III}-O$ electron transfer equilibrium constant and, together with the known redox potential of the Fc derivative used, the $Fe^{IV}{=}O/Fe^{III}-O$ potential [48b]. Note that the two potentials (i) and (ii) describe two entirely different processes, both of importance for ferryl-catalyzed oxidation reactions, that is, (i) a H^+-coupled electron transfer and (ii) a pure electron transfer. That is,

4) An interesting very recent observation is that oxidation of $[Fe(L^1)(OH_2)_2]^{2+}$ in water at 10 °C with H_2O_2, PhIO, or cer(IV) ammonium nitrate (see also Section 6.6.4) produces a transient (max. concentration after ~0.2 s) with an absorption maximum at 804 nm; at present there is no unambiguous characterization of this species (P. Comba, A. Waleska, unpublished results) [48].

Table 6.2 Redox potentials and reorganization energies for the $Fe^{IV/III}$ couples of bispidine–iron complexes (P. Comba and S. Wunderlich, unpublished results) [20a,e, 50].

	H^+-coupled e-transfer *V* versus SCE	Pure e-transfer *V* versus SCE	Reorganization energy (eV)
L^1	—	0.73	2.05
L^2	1.61	0.42	2.06
L^3	1.32	0.40	2.29
L^{2a} a)	—	0.50	2.07

a) L^{2a} is an L^2 derivative with a methyl substituent at the methylene bridge of the N7-appended pyridine group [50; reprinted with permission from J. Am. Chem. Soc., Copyright 2001 by American Chemical Society].

different potential values are expected and indeed observed (see Table 6.2 for the potentials available for some of our bispidine-ferryl complexes) [20a,e].

Among all the published redox potentials of iron(IV)–oxo compounds so far, those of the bispidine complexes are the highest. An important and not often appreciated point is that the very high potentials indicate that the ferryl complexes are unstable. Obviously, this not only involves the reduction to Fe^{III} but also the stability of the bispidine–iron–oxo complexes; that is, it emerges that these very efficient oxidation catalysts have a high propensity to decay (e.g., by decomplexation) and this could be one of the reasons why only a limited number of turnovers is observed in various of the catalytic reactions (see Section 6.6).

The difference of over 1 V between the two types of potentials (see Table 6.2, H^+-coupled vs pure electron transfer), as well as the significantly lower potentials for the pure electron transfer processes, is not unexpected because of the fact that the putative Fe^{III}–O species are very unstable. The potentials for the H^+-coupled electron transfer of the L^2- and L^3-, and by extrapolation also of the L^1-based systems, are such that water oxidation should formally be feasible (see Section 6.6.5), and oxidation of the very unreactive C–H bonds in alkanes has been demonstrated experimentally (see Section 6.6.2). An interesting observation is that methyl (or phenyl) substitution of L^2 at the methylene bridge of the N7-appended pyridine group (ligand L^{2a} in Table 6.2) leads to an increase of the oxidation power of nearly 100 mV. We interpret this as a result of increased ligand rigidity: the dangling pyridine group is the only part of ligand L^2 with any significant flexibility and the substitution of the methylene bridge of the five-membered chelate ring locks the conformational flexibility [50].

There are interesting differences between the two isomeric ferryl complexes based on L^2 and L^3: the L^2-based iron system is known to have less stable Fe^{III} intermediates (see above) [19] and more reactive ferryl complexes (see Section 6.6) [20a,b, 35]. It therefore is not unexpected that the H^+-coupled electron transfer potentials differ by about 300 mV (Table 6.2) [20a]. More surprising is the fact that the pure electron transfer potentials are nearly identical but that the related reorganization energies differ by about 0.3 eV.

6.6
Reactivity of the Ferryl Compounds

The high-valent iron-oxo sites of nonheme iron enzymes catalyze a variety of reactions (halogenation and hydroxylation of alkanes, desaturation and cyclization, electrophilic aromatic substitution, and cis-dihydroxylation of olefins) [1b]. Most of these (and other) reactions have also been achieved and studied with model systems [1c, 2a–c]. With the bispidine complexes, we have primarily concentrated on olefin epoxidation and dihydroxylation, alkane hydroxylation and halogenation, and sulfoxidation and demethylation processes. The focus in these studies so far has been on a thorough analysis of the reaction mechanisms rather than the substrate scope and catalyst optimization.

6.6.1
Olefine Oxidation

A standard reaction to study olefin epoxidation and dihydroxylation is the oxidation of cyclooctene as substrate and H_2O_2 as oxidant, in MeCN at ambient temperature (standard procedure as also used by other groups: 1000 equiv substrate, 10 equiv oxidant (Turnover number maximum (TON^{max}) = 10), reaction time 0.5 h, standard work-up procedures, GC/GC–MS analysis) [7b, 35, 51]. Separate experiments were also done with ^{18}O-labeled H_2O, H_2O_2, or O_2, and the determined product ratios (epoxide:diol (*cis:trans* diol)) and the amount of ^{18}O label transferred to the product were measured as a function of various parameters. A selection of relevant data is assembled in Table 6.3 (the data of the isomeric system with L^3, not shown, is very similar but the rate of product formation is significantly slower). In contrast to most reports with other catalysts, there is a striking difference of the product ratios in the presence or absence of dioxygen.[5)]

For the pentadentate ligand systems, it emerges that only epoxide is formed under dioxygen (5 TON of a maximum of 10) but about 80% of this is lost under inert atmosphere with roughly an identical amount of diol (50/50 ratio of cis and trans product) also formed under argon [35]; note that diol formation is unusual for pentadentate-ligand-derived ferryl catalysts. The product ratios and their dependence on the presence of dioxygen are supported by the labeling experiments, that is, the extra 85% of epoxide in $^{18}O_2$ is fully labeled. Moreover, the total scrambling of the diol product (cis/trans) also emerges from scrambling

5) From most published studies, it does not emerge whether or not the observed dependence from dioxygen is unique for the bispidine platform since often it is not reported whether or not aerobic or strictly anaerobic conditions apply, and data in both aerobic and anaerobic conditions are only rarely available.

Table 6.3 Product ratios and ^{18}O labeling data of the catalytic olefin oxygenation with bispidine–ferryl complexes (cyclooctene, MeCN, H_2O_2, $TON^{max} = 10$) [7b, 35].

	L^2 under O_2	L^2 under Ar	L^1 under O_2	L^1 under Ar
Diol; epoxide	0.0; 5.0	1.0; 1.0	1.6; 2.2	1.5; 1.0
Epoxide % ^{18}O [$H_2{}^{18}O_2/H_2{}^{18}O/{}^{18}O_2$]	15/≈0/85	82/18/–	45/0/55	100/–/–
cis-Diol % ^{18}O from $H_2{}^{18}O_2$[noO/10/20] $H_2{}^{18}O$[noO/10/20]	No diol	0/23/77	0/5/95	0/4/96
		0/30/0	92/8/0	93/7/0
trans-Diol % ^{18}O from $H_2{}^{18}O_2$[noO/10/20] $H_2{}^{18}O$[noO/10/20]	—	0/56/44	Selectivity for *cis*-diol	
		0/50/0		

in the labeling data, and this indicates that a carbon-based radical intermediate might be involved. Specifically with octene, it is known that radical-based intermediates are fully quenched by molecular oxygen and selectively yield the epoxide product [52]. The mechanism derived from these data involves an asymmetric attack of the ferryl species to yield the substrate radical intermediate complex (L)Fe^{III}–O–C–C, which has three possible decay pathways (see Scheme 6.3). This mechanism is fully confirmed by an extensive DFT-based mechanistic study [13h].

The product ratio and labeling characteristics with the tetradentate ligand L^1-based catalyst are strikingly different: again there is over 50% reduction of epoxide produced under argon but there is a quite constant yield of diol product which is selectively cis-configurated. Importantly, the labeling data suggest that both oxygen atoms from the H_2O_2 oxidant are transferred to the diol product (see Table 6.3) [7b]. It emerges that the epoxide formation follows the mechanism proposed in Scheme 6.3 but that another pathway is responsible for the diol formation, and this is not radical-based (i.e., diol formation is independent of dioxygen and this also arises from experiments in $^{18}O_2$). From the above-mentioned direct formation of $[(L^1)Fe^{IV}(OH)_2]^{2+}$ emerges a possible candidate for the observed reactivity (see Scheme 6.4) [7b, 13f]. This is a novel and interesting pathway, specifically also in view of a possible optimization of this reaction for the chemically important selective *cis*-dihydroxylation of olefins [2g, 53]. The mechanism proposed in Scheme 6.4 is fully supported by DFT calculations [7b]. Support for the mechanism and the novel intermediate-spin dihydroxo complex also comes from the fact that oxidation of the Fe^{II} precursor with the oxygen transfer agent PhIO selectively produces epoxide (see Scheme 6.4) (P. Comba and M. Menzel, unpublished results).

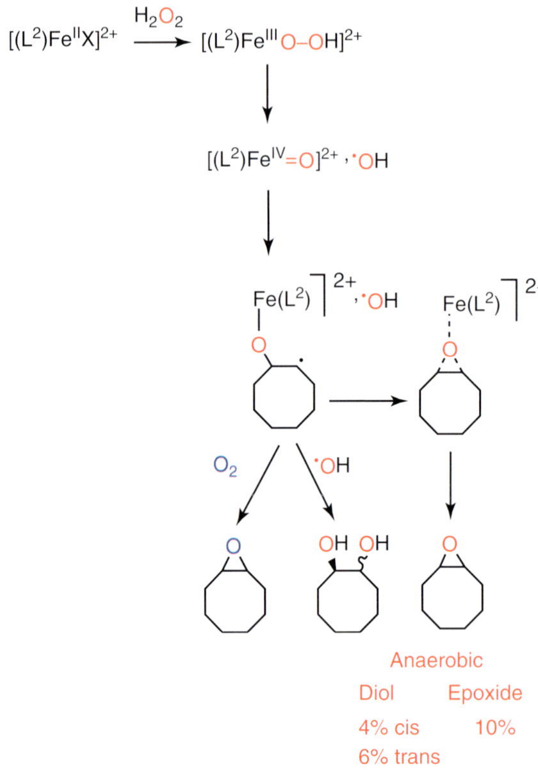

Scheme 6.3 Pathways for the oxidation of cyclooctene with H_2O_2, catalyzed by iron complexes of the pentadentate bispidine ligands L^2 and L^3 [35].

6.6.2
Alkane Oxidation

The very high $Fe^{IV/III}$ redox potentials suggest that the bispidine-based ferryl complexes should be able to oxidize the strong C–H bonds of alkanes. With bispidines and other ligand systems for nonheme ferryl complexes [54], various possible species were proposed to be responsible for substrate activation in general, that is, $Fe^{V}=O$ (heterolytic O–O cleavage in the Fe^{III}–OOH intermediate), $Fe^{IV}=O$ together with OH radicals (or *t*BuOO radicals, depending on the oxidant used; homolytic O–O cleavage in the Fe^{III} precursor), or the products of direct oxidation of the Fe^{II} precursor to the catalytically active Fe^{IV} oxidant [9, 20b, 51, 55]. In the catalytic oxidation of alkanes (typically cyclohexane[6)]), it is believed that metal-based oxidation (either $Fe^{IV}=O$ or $Fe^{V}=O$) predominantly yields the

6) In our experiments, we used a similar general experimental setup as for the olefin oxidation reactions and as generally used in literature, that is, catalyst/oxidant/substrate ratio of 1 : 100 : 1000 (TON^{max} = 100), MeCN, ambient temperature, anaerobic conditions, 30 min reaction time, GC or GC/MS analysis [9, 20b, 20d].

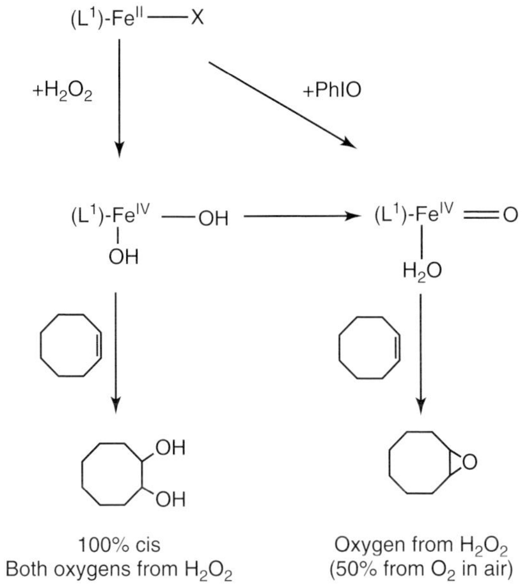

Scheme 6.4 Mechanism of the $[Fe^{II}(L^1)(X)]^{n+}$-catalyzed oxidation of cyclooctene [7b].

alcohol products, that is, hydrogen abstraction by the ferryl oxidant leads to carbon-based radicals which are rebound to the Fe–OH species to produce the alcohol product [58]; oxidation of the substrate by OH or OR radicals, in contrast, is suggested to lead to carbon-based substrate radicals that are trapped by O_2 to generate equimolar amounts of alcohol and ketone (see Scheme 6.5)

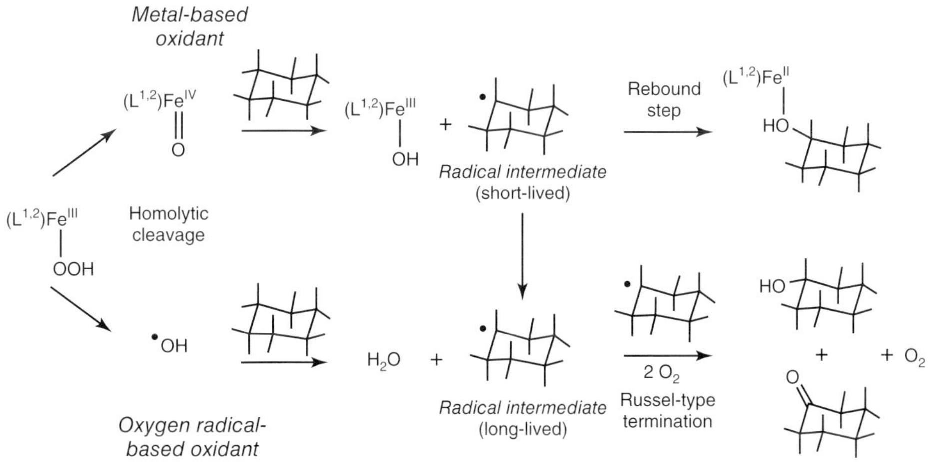

Scheme 6.5 Mechanism of the iron–bispidine-catalyzed oxidation of cyclohexane (H-atoms represented as small sticks) [20b; reproduced by permission of The Royal Society of Chemistry].

[56]. Therefore, in a series of experiments, involving the tetradentate bispidine L^1 as well as the two isomeric pentadentate bispidine L^2- and L^3-based systems (various oxidants (PhIO, H_2O_2, *t*BuOOH; stoichiometric as well as catalytic reactions), aerobic or anaerobic conditions and also involving ^{18}O labeling studies, the determination of kinetic isotope effects (KIEs), and experiments with adamantane as substrate to determine the preference of tertiary over secondary alcohol formation (stabilization of the tertiary radical intermediate)), various possible pathways were probed, and the following general conclusions emerge [9, 20b,d]: (i) bispidine–iron(II) complexes are efficient precatalysts for the oxidation of unactivated alkanes; (ii) carbon-based radical intermediates are involved and, by the usual rebound mechanism [58], these produce alcohol products; (iii) with H_2O_2 as an oxidant, $Fe^V{=}O$ is accessible, and this leads to differences in product ratios; with alkylperoxides and H_2O_2 there are also oxygen radical-based pathways with a loss of specificity.

The alkane C–H activation reactivities of the ferryl complexes with the two isomeric pentadentate bispidines L^2 and L^3 were compared with other extensively studied pentadentate amine/pyridine ligands [20a]. The interesting observation is that, in contrast to the sulfoxidation reaction (see Section 6.6.3), there is no correlation of the $Fe^{IV/III}$ redox potentials (H^+-coupled electron transfer) with the reactivities. So far, there is no unambiguous interpretation of this result[7] but it emerges that, more often than appreciated, there are various possible pathways and a generalization therefore is very often not warranted (see also Section 6.7).

Biological iron-catalyzed halogenation reactions occur with heme haloperoxidases or α-ketoglutarate-dependent nonheme halogenases [59]. Chloroperoxidases generate a metal-bound hypochlorite ion which is capable of halogenating electron-rich organic substrates but the αKG-dependent nonheme halogenases can also halogenate less reactive electron-poor alkanes, and spectroscopic studies have revealed the existence of high-spin ferryl centers as reactive intermediates [59]. These are believed to abstract a hydrogen atom from the substrate, and the rebound step then transfers a halogen atom from the Fe^{III} center to the substrate radical (Scheme 6.6). An analogous reaction sequence produces hydroxylated products in the corresponding oxygenases (e.g., TauD). Given the similarity of the active sites and relevant intermediates, it is conceivable that the radical intermediate competes in the rebound step with the $Fe^{III}OH$ and $Fe^{III}Cl$ sites. However, until recently all attempts to detect alcohol products in halogenases failed [51, 55]. There was one earlier report of a stoichiometric halogenation process with a low molecular weight ferryl model complex [58], and the ferryl complex of the tetradentate ligand L^1 is the first example in which the halogenation of alkanes is achieved catalytically; the rebound-based mechanism (see Scheme 6.6) was confirmed by DFT calculations [9].

7) An interesting observation is that the major experimentally detected difference between the L^2- and L^3-based systems in terms of the electron transfer properties is the reorganization energy (see Table 6.2), and the corresponding data are not available for the other pentadentate ligand systems.

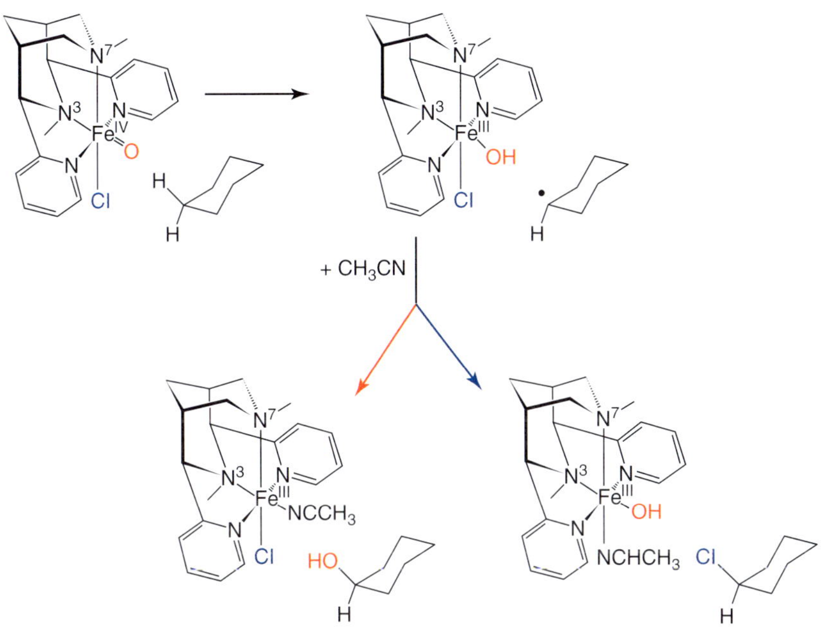

Scheme 6.6 Mechanism of the iron–bispidine-catalyzed halogenation of cyclohexane [9].

An interesting question is whether low molecular weight halogenase model systems are selective. From our results, it appears that $[(Cl)(L^1)Fe^{IV}{=}O]^+$ (or $[(Cl)(L^1)Fe^{V}{=}O]^{2+}$) show a high halogenation selectivity (up to nearly 100%), and it emerges that decreasing selectivities are generally due to radical-based side reactions [9]. The total selectivity required in biological reactions seems to be accomplished by enzyme-based substrate positioning [57]. A problem in terms of our model system is the extremely low TON, and this seems to be related to the re-formation of the $(L^1)Fe^{II}Cl$ precursor, that is, to the ligand exchange – a problem that has some scope for optimization. An interesting question is related to the spin state and its relevance for the reactivity and selectivity (see Section 6.4; note that the reactions described here generally were done in acetonitrile, where the intermediate-spin configuration is stabilized but the corresponding chloro complex required for the halogenation is believed to have high-spin electronic configuration).

6.6.3 Sulfoxidation

The selective oxidation of sulfides to sulfoxides is an important reaction in biological systems. It is a well-studied model reaction for oxo atom transfer and it is of importance for applications in pharmaceutical and preparative organic chemistry [1, 2, 60]. Iron- and ruthenium-based sulfoxidation catalysts have also been explored in the area of asymmetric synthesis and enantioselective catalysis

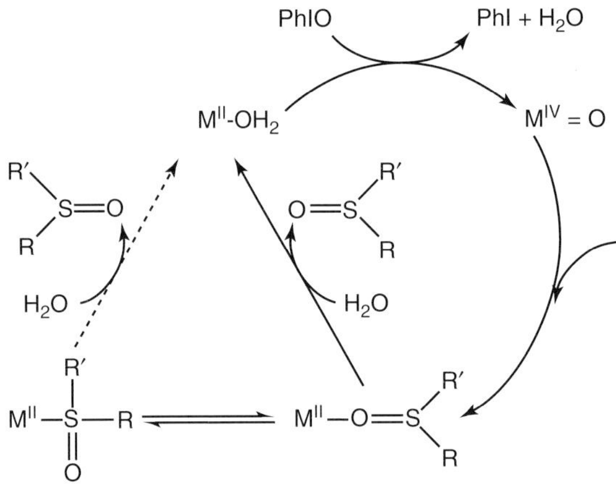

Figure 6.4 Sulfoxidation of a thiaether substrate with M^{IV}=O (M = Fe, Ru), showing linkage isomerization of the coordinated oxygenated substrate, leading to product inhibition with the softer Ru^{II} but not with the Fe^{II} center [21b].

[61], and mechanistic studies in biomimetic heme- and nonheme iron-mediated sulfoxidation catalysis have been supported by DFT-based theoretical work [62]. An interesting recent observation is that iron-catalyzed sulfoxidation might be an important initial step for the abiotic methanogenesis from biomolecules [63].

Both the tetradentate (L^1) and pentadentate (L^2) bispidine-based systems have been found to be efficient in iron-catalyzed thioanisole oxidation reactions [20a,c]. Interestingly, there is a good correlation between the $Fe^{IV/III}$ redox potentials (H^+-coupled electron transfer) and the sulfoxidation reactivity of five nonheme iron systems with pentadentate amine–pyridine-based ligands [20a], suggesting that these reactivities primarily depend on the driving force (see Section 6.6.2 for the corresponding C–H activation studies). An interesting observation is that the corresponding L^2-based ruthenium complex shows subtle differences in its sulfoxidation reactivity (the epoxidation reactivity of the ruthenium and iron systems of L^2 has also been studied and analyzed as a function of the metal center [64]). An important feature is O-to-S linkage isomerism of the sulfoxide product complex, and this may lead to product inhibition in the catalytic cycle (see Figure 6.4) [65]. The increased efficiency of bispidine in comparison to other ligand systems is probably due to the partial suppression of this inhibition pathway [65]; for the iron-based chemistry, linkage isomerization does not occur, as expected [20c].

6.6.4 Water Oxidation

The conversion of solar energy into chemical energy (artificial photosynthesis) may include light harvesting, charge separation, water oxidation, and water reduction [66]. High-valent nonheme iron systems have been proposed as promising candidates for one of the most important and difficult parts of this cycle, namely the catalytic water oxidation [11, 67]. Indeed, the redox potentials of many iron-based nonheme model systems are high enough to allow the thermal oxidation of water (see e.g., Table 6.2), and systems based on cer(IV) ammonium nitrate (CAN) as sacrificial oxidant in acidic aqueous solution as well as light-driven $[Ru(bpy)_3]^{2+}$/persulfate systems (bpy (2,2′-bipyridine), $S_2O_8{}^{4-}$ as sacrificial oxidant) have been described [11, 67b]. Based on the properties and performance of the bispidine-based ferryl complexes, specifically their redox properties (see Section 6.5), there was some promise in using these compounds as catalysts for artificial photosynthesis. However, as already appears from previous sections, the stability of the bispidine-based catalyst systems is a major problem (see also Section 6.7). It is therefore no surprise that (i) there is an efficient bispidine–iron-catalyzed oxidation of water to dioxygen but (ii) the reaction is severely limited by catalyst degradation.[8] In general, it emerges that in acidic solution (as generally occurs with CAN) the catalytic turnover is limited by degradation of the catalyst (decomplexation and subsequent oxidation of the free ligand). In neutral to basic solution, where light-driven $[Ru(bpy)_3]^{2+}$-based systems have been extensively studied, it appears that the degradation of the ferryl catalysts leads to the formation of Fe_2O_3 nanoparticles which are the actual catalyst for water oxidation [67b]. With this preliminary information, there is currently not much of an urge for the optimization of mononuclear iron-based water oxidation catalysts.

6.6.5
Dioxygen as Oxidant

Nature's oxidant in nonheme iron-catalyzed systems is dioxygen, and there are only very few low molecular weight iron complexes where this has been modeled [12, 36, 68]. Interestingly, the bispidine–Fe^{II} precursors were shown some time ago to be able to oxidize *cis*-1,2-dimethylcyclohexane with dioxygen alone [69]. In the light of recent work with the ferryl complex of the classical tetramethylcyclam ligand systems [12c, 70], where the formation of very reactive Fe^{III}–superoxo complexes was proposed [12e], we were interested in testing whether the observed oxidation processes with our systems are due to putative iron–superoxo intermediates. The results of this investigation are not yet published and will not be discussed here in detail [71]. The important general result, based on a detailed kinetic analysis with the Fe^{II} complexes of the tetra- and pentadentate bispidine ligand systems, including experiments with radical scavengers and -starters, in combination with mass spectrometry, UV–vis–NIR, and electron paramagnetic resonance (EPR) spectroscopy, is that the substrate oxidation is not based on dioxygen chemistry but is initiated by carbon-based

8) The results of this study [71a] include the spectroscopic characterization of the ferryl complexes (ligands L^1, L^2, L^3), a kinetic analysis of formation and decay of the ferryl complexes, ^{18}O labeling studies, and product analysis (TON and yield of O_2 vs CO_2) as a function of the three ligands and various oxidants.

radicals. Interestingly, the tetramethylcyclam–iron(II) complex reported to produce active Fe^{III}–superoxo complexes has also been described more recently to be involved in carbon-radical-based oxidation reactions [72].

6.7 Conclusion

Much has been accomplished in the last decade in the field of high-valent nonheme iron chemistry and, in addition to the milestones noted in Section 6.1, the most relevant iron–bispidine-based achievements are that (i) with the bispidine ligands the highest $Fe^{IV/III}$ potentials have been reported and that these potentials have been shown to be tunable (i.e., higher potentials than those reported so far may still be accessible) [14e] and (ii) that the spin state of the bispidine-based $Fe^{IV}{=}O$ centers is, within limits, tunable. An interesting, still open, and demanding task therefore is to get experimental information and theory-based support related to the question of how the redox potential of the ferryl site varies with its spin state, and how this translates to reactivities. Experiments to approach these problems should now be possible with bispidine-based systems, and the corresponding results may lead to new arguments in the ongoing discussion on the importance of the spin of ground and transition states, the driving force, and correlations between oxidation power and reactivities. Related to this is one of the remaining challenges, that is, to produce an $S = 0$ ferryl species – my bold guess is that a bispidine-based ligand might get us there.

Two important questions related to this high-valent iron chapter remain to be addressed:

1) How relevant are the bispidine–ferryl-based data presented here for high-valent nonheme iron chemistry in general? A number of observations and corresponding interpretations in this chapter demonstrate that, as usual, one needs to be extremely careful with extrapolations, and this is not always appreciated enough. One of the examples is the clean oxidation of $[(L^3)Fe^{II}(OH_2)]^{2+}$ with H_2O_2 in water to $[(L^3)Fe^{IV}{=}O]^{2+}$ (see Figure 6.2): it is important to stress again that this observation does not indicate that Fenton chemistry does not involve any OH radicals but it has important general implications for the Fenton process. Another example is the observed carbon-radical-based pathways in the bispidine–ferryl-catalyzed olefin epoxidation (see Scheme 6.3): this mechanism may not be general but it is a pathway that needs to be considered for any ferryl–olefin system. The third relevant example is the bispidine–iron(II)-based oxidation with molecular oxygen: it does not, of course, indicate that this type of reaction generally is due to autoxidation rather than iron-based oxygen activation, but initiation by carbon-based radicals is a general problem that can be difficult to exclude experimentally.

2) Are there possible applications for iron–bispidine-based catalytic processes emerging from the material presented here? For industrial catalytic processes, where a large number of turnovers are of importance, the answer clearly is no. An important and not generally appreciated reason for this is the fundamental relation between complex stability and redox potential [73]. There are possibilities to improve the lifetime of a catalyst without significantly reducing the oxidation power, but very strong enzymatic oxidation catalysts such as the cytochromes are also not optimized for a long lifetime. However, there are possible industrial applications for high-valent iron–bispidine systems, and these are not necessarily related to "real catalysis"; bispidine complexes may be and are already used as actuators, where the "catalyst" only needs to be active for a short time and then may decay such as in various types of bleaching processes [28] and in paint processing. In these applications, the decay products (i.e., the metal-free ligands) are preferentially "biodegradable," and the inherent instability of the Mannich-type ligands clearly is of advantage here.

Acknowledgment

Financial support by the German Science Foundation (DFG, SFB 623) and the hard, creative, and intelligent work by the coworkers who have coauthored the publications cited in this review are gratefully acknowledged.

References

1. (a) Abu-Omar, M.M., Loaiza, A., and Hontzeas, N. (2005) *Chem. Rev.*, **105**, 2227–2252; (b) Krebs, C., Galonici-Fujimori, D.G., Walsh, C.T., and Bollinger, J.M. Jr., (2007) *Acc. Chem. Res.*, **40**, 484–492; (c) McDonald, A.R. and Que, L. Jr., (2013) *Coord. Chem. Rev.*, **257**, 414–428.
2. (a) Meunier, B. (2000) *Metal-Oxo and Metal-Peroxo Species in Catalytic Oxidations*, vol. 97, Springer, Berlin. (b) Chen, A.S. and White, M.C. (2000) *Science*, **318**, 783–787; (c) Bigi, M.A., Reed, S.A., and White, M.C. (2011) *Nat. Chem.*, **3**, 216–222; (d) Plietker, B. (2008) *Iron Catalysis in Organic Chemistry*, Wiley-VCH Verlag GmbH, Weinheim. (e) Schröder, K., Junge, K., Bitterlich, B., and Beller, M. (2011) in *Iron Catalysis: Fundamentals and Applications*, vol. 33 (ed. B. Plietker), Springer, Berlin, pp. 83–109; (f) Bolm, C., Legros, J., Le Paih, J., and Zani, L. (2004) *Chem. Rev.*, **104**, 6217–6254; (g) Prat, I., Font, D., Company, A., Junge, K., Ribas, X., Beller, M., and Costas, M. (2013) *Adv. Synth. Catal.*, **355**, 947–956; (h) Gomez, L., Canta, M., Font, D., Prat, I., Ribas, X., and Costas, M. (2013) *J. Org. Chem.*, **78**, 1428–1433; (i) Sun, C.-L., Li, B.-J., and Shi, Z.-J. (2011) *Chem. Rev.*, **111**, 1293–1314.
3. Hohenberger, J., Ray, K., and Meyer, K. (2012) *Nat. Commun.*, **3**, 720.
4. Grapperhaus, C.A., Mienert, B., Bill, E., Weyhermüller, T., and Wieghardt, K. (2000) *Inorg. Chem.*, **39**, 5306–5317.
5. Rohde, J.-U., In, J.-H., Lim, M.H., Brennessel, W.W., Bukowski, M.R., Stubna, A., Münck, E., Nam, W., and Que, L. Jr., (2003) *Science*, **299**, 1037–1039.

6. (a) Proshlyakov, D.A., Henshaw, T.F., Monterosso, G.R., Ryle, M.J., and Hausinger, R.P. (2004) *J. Am. Chem. Soc.*, **126**, 1022–1023; (b) Riggs-Gelasco, P.J., Price, J.C., Guyer, R.B., Brehm, J.H., Barr, E.W., Bollinger, J.M. Jr., and Krebs, C. (2004) *J. Am. Chem. Soc.*, **126**, 8108–8109; (c) Price, J.C., Barr, E.W., Tirupati, B., Bollinger, J.M. Jr., and Krebs, C. (2003) *Biochemistry*, **42**, 7497–7508; (d) Price, J.C., Barr, E.W., Glass, T.E., Krebs, C., and Bollinger, J.M. Jr., (2003) *J. Am. Chem. Soc.*, **125**, 13008–13009.
7. (a) Pestovsky, O., Stoian, S., Bominaar, E.L., Shan, X., Münck, E., Que, L. Jr., and Bakac, A. (2005) *Angew. Chem. Int. Ed.*, **44**, 6871–6874; (b) Bautz, J., Comba, P., Lopez de Laorden, C., Menzel, M., and Rajaraman, G. (2007) *Angew. Chem. Int. Ed.*, **46**, 8067–8070; (c) England, J., Martinho, M., Farquhar, E.R., Frisch, J.R., Bominaar, E.L., Münck, E., and Que, L. Jr., (2009) *Angew. Chem. Int. Ed.*, **48**, 3622–3626; (d) Lacy, D.C., Gupta, R., Stone, K.L., Greaves, J., Ziller, J.W., Hendrich, M.P., and Borovik, A.S. (2010) *J. Am. Chem. Soc.*, **132**, 12188–12190; (e) England, J., Guo, Y., Heuvelen, K.M.V., Cranswick, M.A., Rohde, G.T., Bominaar, E.L., Münck, E., and Que, L. Jr., (2011) *J. Am. Chem. Soc.*, **133**, 11880–11883; (f) Bigi, J.P., Harman, W.H., Lassalle-Kaiser, B., Robles, D.M., Stich, T.A., Yano, J., Britt, R.D., and Chang, C.J. (2012) *J. Am. Chem. Soc.*, **134**, 1536–1542.
8. (a) Prat, I., Mathieson, J.S., Guell, M., Ribas, X., Luis, J.M., Cronin, L., and Costas, M. (2011) *Nat. Chem.*, **3**, 788–793; (b) Lyakin, O.Y., Bryliakov, K.P., and Talsi, E.P. (2011) *Inorg. Chem.*, **50**, 5526–5538; (c) Lyakin, O.Y., Bryliakov, K.P., Britovsek, G.J.P., and Talsi, E.P. (2009) *J. Am. Chem. Soc.*, **131**, 10798–10799; (d) Lyakin, O.Y., Prat, I., Bryliakov, K.P., Costas, M., and Talsi, E.P. (2012) *Catal. Commun.*, **29**, 105–108; (e) Tiago de Oliveira, F., Chanda, A., Banerjee, D., Shan, X., Mondal, S., Que, L. Jr., Bominaar, E.L., Münck, E., and Collins, T.J. (2007) *Science*, **315**, 835–838.
9. Comba, P. and Wunderlich, S. (2010) *Chem. Eur. J.*, **16**, 7293–7299.
10. Bautz, J., Bukowski, M., Kerscher, M., Stubna, A., Comba, P., Lienke, A., Münck, E., and Que, L. Jr., (2006) *Angew. Chem. Int. Ed.*, **45**, 5681.
11. Fillol, J.L., Codola, Z., Garcia-Bosch, I., Gomez, L., Pla, J.J., and Costas, M. (2011) *Nat. Chem.*, **3**, 807–813.
12. (a) Schröder, K., Join, B., Amali, A.J., Junge, K., Ribas, X., Costas, M., and Beller, M. (2011) *Angew. Chem. Int. Ed.*, **50**, 1425–1429; (b) Thibon, A., England, J., Martinho, M., Young, V.G. Jr., Frisch, J.R., Guillot, R., Girerd, J.-J., Münck, E., Que, L. Jr., and Banse, F. (2008) *Angew. Chem. Int. Ed.*, **47**, 7064–7067; (c) Lee, Y.-M., Hong, S., Morimoto, Y., Shin, W., Fukuzumi, S., and Nam, W. (2010) *J. Am. Chem. Soc.*, **132**, 10668–10670; (d) Mukherjee, A., Cranswick, M.A., Chakrabarti, M., Paine, R.K., Fujisawa, K., Münck, E., and Que, L. Jr., (2010) *Inorg. Chem.*, **49**, 3618–3628; (e) Chen, H., Cho, K.-B., Lai, W., Nam, W., and Shaik, S. (2012) *J. Chem. Theory Comput.*, **8**, 915–926.
13. (a) Neese, F. (2006) *J. Inorg. Biochem.*, **100**, 716–726; (b) Decker, A., Rohde, J.-U., Klinker, E.J., Wong, S.D., Que, L. Jr., and Solomon, E.I. (2007) *J. Am. Chem. Soc.*, **129**, 15983–15996; (c) Lundberg, M. and Borowski, T. (2013) *Coord. Chem. Rev.*, **257**, 277–289; (d) Usharani, D., Janardanan, D., Li, C., and Shaik, S. (2012) *Acc. Chem. Res.*, **46**, 471–482; (e) Anastasi, A., Lienke, A., Comba, P., Rohwer, H., and McGrady, J.E. (2007) *Eur. J. Inorg. Chem.*, 65; (f) Comba, P., Rajaraman, G., and Rohwer, H. (2007) *Inorg. Chem.*, **46**, 3826; (g) Anastasi, A., Comba, P., McGrady, J., Lienke, A., and Rohwer, H. (2007) *Inorg. Chem.*, **46**, 6420; (h) Comba, P. and Rajaraman, G. (2008) *Inorg. Chem.*, **47**, 78.
14. (a) Comba, P., Kerscher, M., and Schiek, W. (2007) *Prog. Inorg. Chem.*, **55**, 613–704; (b) Comba, P., Kerscher, M., Merz, M., Müller, V., Pritzkow, H., Remenyi, R., Schiek, W., and Xiong, Y. (2002) *Chem. Eur. J.*, **8**, 5750–5760; (c) Comba, P., Haaf, C., and Wadepohl, H. (2009) *Inorg. Chem.*, **48**, 6604–6614. (d) Comba, P., Morgen, M., and

Wadepohl, H. (2013) *Polyhedron*(100 years Werner Nobel Prize), **52**, 1239–1245; (e) Comba, P., Morgen, M., and Wadepohl, H. (2013) *Inorg. Chem.*, **52**, 6481–6501.

15. Mannich, C. and Mohs, P. (1930) *Chem. Ber.*, **B63**, 608–612.
16. (a) Stetter, H. and Merten, R. (1957) *Chem. Ber.*, **90**, 868–875; (b) Haller, R. (1969) *Arch. Pharm.*, **302**, 113–118.
17. Comba, P., Nuber, B., and Ramlow, A. (1997) *J. Chem. Soc., Dalton Trans.*, 347–352.
18. Börzel, H., Comba, P., Hagen, K.S., Merz, M., Lampeka, Y.D., Lienke, A., Linti, G., Pritzkow, H., and Tsymbal, L.V. (2002) *Inorg. Chim. Acta* (Karl Wieghardt Special Issue), **337**, 407–419.
19. Bukowski, M.R., Comba, P., Limberg, C., Merz, M., Que, L. Jr., and Wistuba, T. (2004) *Angew. Chem. Int. Ed.*, **43**, 1283–1287.
20. (a) Wang, D., Ray, K., Collins, M.J., Farquhar, E.R., Frisch, J.R., Gomez, L., Jackson, T.A., Kerscher, M., Waleska, A., Comba, P., Costas, M., Münck, E., and Que, L. Jr., (2013) *Chem. Sci.*, **4**, 282–291; (b) Comba, P., Maurer, M., and Vadivelu, P. (2009) *Inorg. Chem.*, **48**, 10389–10396; (c) Jaccob, M., Comba, P., Maurer, M., Vadivelu, P., and Venuvanalingham, V. (2011) *Dalton Trans.*, **40**, 11276–11281; (d) Comba, P., Maurer, M., and Vadivelu, P. (2008) *J. Phys. Chem. A*, **112**, 13028; (e) Comba, P., Fukuzumi, S., Kotani, H., and Wunderlich, S. (2010) *Angew. Chem. Int. Ed.*, **49**, 2622–2625.
21. (a) Comba, P., Merz, M., and Pritzkow, H. (2003) *Eur. J. Inorg. Chem.*, **1711**, 1718; (b) Comba, P. and Kerscher, M. (2004) *Cryst. Eng.*, **6**, 197–211.
22. (a) Comba, P. and Lienke, A. (2001) *Inorg. Chem.*, **40**, 5206–5209; (b) Comba, P., Hauser, A., Kerscher, M., and Pritzkow, H. (2003) *Angew. Chem. Int. Ed.*, **42**, 4536–4540; (c) Comba, P., Lopez de Laorden, C., and Pritzkow, H. (2005) *Helv. Chim. Acta*, **88**, 647–664; (d) Bentz, A., Comba, P., Deeth, R.J., Kerscher, M., Pritzkow, H., Seibold, B., and Wadepohl, H. (2008) *Inorg. Chem.*, **47**, 9518.
23. (a) Börzel, H., Comba, P., Katsichtis, C., Kiefer, W., Lienke, A., Nagel, V., and Pritzkow, H. (1999) *Chem. Eur. J.*, **5**, 1716–1721; (b) Börzel, H., Comba, P., Hagen, K.S., Katsichtis, C., and Pritzkow, H. (2000) *Chem. Eur. J.*, **6**, 914–919; (c) Comba, P., Haaf, C., Helmle, S., Karlin, K.D., Pandian, S., and Waleska, A. (2012) *Inorg. Chem.*, **51**, 2841–2851.
24. (a) Börzel, H., Comba, P., and Pritzkow, H. (2001) *J. Chem. Soc., Chem. Commun.*, 97–98. (b) Born, K., Comba, P., Daubinet, A., Fuchs, A., and Wadepohl, H. (2007) *J. Biol. Inorg. Chem.*, **12**, 36; (c) Comba, P., Martin, B., Muruganantham, A., and Straub, J. (2012) *Inorg. Chem.*, **51**, 9214–9225.
25. (a) Comba, P., Lang, C., Lopez de Laorden, C., Muruganantham, A., Rajaraman, G., Wadepohl, H., and Zajaczkowski, M. (2008) *Chem. Eur. J.*, **14**, 5313–5328; (b) Comba, P., Haaf, C., Lienke, A., Muruganantham, A., and Wadepohl, H. (2009) *Chem. Eur. J.*, **15**, 10880–10887.
26. (a) Atanasov, M., Busche, C., Comba, P., El Hallak, F., Martin, B., Rajaraman, G., van Slageren, J., and Wadepohl, H. (2008) *Inorg. Chem.*, **47**, 8112; (b) Atanasov, M., Comba, P., Hanson, G.R., Hausberg, S., Helmle, S., and Wadepohl, H. (2011) *Inorg. Chem.*, **50**, 6890–6901; (c) Atanasov, M., Comba, P., and Helmle, S. (2012) *Inorg. Chem.*, **51**, 9357–9368.
27. (a) Juran, S., Walther, M., Stephan, H., Bergmann, R., Steinbach, J., Kraus, W., Emmerling, F., and Comba, P. (2009) *Bioconj. Chem.*, **20**, 347–359. (b) Comba, P., Hunoldt, S., Morgen, M., Pietsch, J., Stephan, H., and Wadepohl, H. (2013) *Inorg. Chem.*, **52**, 8131–8143; (c) Comba, P., Kerscher, M., Pietzsch, H.-J., Spies, H., Stephan, H., and Juran, S. (2006) Radioaktive metallkomplexe auf der basis von bispidin und dessen derivate. Deutsches Patent WO 2006072336 A2.
28. Hage, R. and Lienke, A. (2006) *Angew. Chem. Int. Ed.*, **45**, 206–222.
29. (a) Börzel, H., Comba, P., Hage, R., Kerscher, M., Lienke, A., and Merz, M. (2002) Germany; Patent WO 0248301 A1 (2002-06-20); Patent US

2002/014900 (2002-10-17), 'Ligand and complex for catalytically bleaching a substrate' (b) Comba, P., Koek, J.H., Lienke, A., Merz, M., Tsymbal, L. (2003) EP-Patent PCT/EP03/04906; 'Ligand and complex for catalytically bleaching a substrate.' Patent WO 2003104234 A1, 2003 *Chem. Abstr.*, **140**, 28783 (Germany); (c) Comba, P., Haaf, C., Kerscher, M., and Lienke, A. (2009) Bispidinliganden und ihre Metallkomplexe. PCT WO 2009/010129; d(d) Caron, A. and Dücker, B. (2012) "Verfahren zur Herstellung von 3,7-diazabicyclo[3.3.1]nonan-Metallkomplexen" Patent WO 2012048815 A1.
30. Comba, P., Kanellakopulos, B., Katsichtis, C., Lienke, A., Pritzkow, H., and Rominger, F. (1998) *J. Chem. Soc., Dalton Trans.*, 3997–4002.
31. Perkins, M.C. and Mark, C. (2000) Patent WO 2000060045 A1 Transition metal bleaching agents.
32. Haberberger, M., Someya, C.I., Company, A., Irran, E., and Enthaler, S. (2012) *Catal. Lett.*, **142**, 557–565.
33. Hage, R. and Wesenhag, V.P. (2008) "Liquid hardening; see also Jennifer Hostetler, OMGroup, Borchi^R OXY coat, a novel oxidative catalyst for alkyd coatings", WO 2008003652 A1.
34. (a) Que, L. Jr., (2007) *Acc. Chem. Res.*, **40**, 493–500; (b) Lee, Y.-M., Dhuri, S.N., Sawant, S.C., Cho, J., Kubo, M., Ogura, T., Fukuzumi, S., and Nam, W. (2009) *Angew. Chem. Int. Ed.*, **48**, 1803–1806.
35. Bukowski, M.R., Comba, P., Lienke, A., Limberg, C., Lopez de Laorden, C., Mas-Balleste, R., Merz, M., and Que, L. Jr., (2006) *Angew. Chem. Int. Ed.*, **45**, 3446.
36. Comba, P., Wadepohl, H., and Wunderlich, S. (2011) *Eur. J. Inorg. Chem.*, **2011**, 5242–5249.
37. Bautz, J., Comba, P., and Que, L. Jr., (2006) *Inorg. Chem.*, **45**, 7077.
38. Haber, F. and Weiss, J. (1934) *Proc. R. Soc. (London)*, **147**, 332.
39. Bray, W.C. and Gorin, M.H. (1932) *J. Am. Chem. Soc.*, **554**, 2124.
40. (a) Shaik, S., Danovich, D., Fiedler, A., Schröder, D., and Schwarz, H. (1995) *Helv. Chim. Acta*, **78**, 1393–1407; (b) Schröder, D., Shaik, S., and Schwarz, H. (2000) *Acc. Chem. Res.*, **33**, 139–145.
41. (a) Poli, R. and Harvey, J.N. (2003) *Chem. Soc. Rev.*, **32**, 1–8; (b) Shaik, S., Hirao, H., and Kumar, D. (2007) *Acc. Chem. Res.*, **40**, 532–542; (c) Hirao, H., Que, L. Jr., Nam, W., and Shaik, S. (2008) *Chem. Eur. J.*, **14**, 1740–1756.
42. (a) Pau, M.Y.M., Lipscomb, J.D., and Solomon, E.I. (2007) *Proc. Natl. Acad. Sci. U.S.A.*, **104**, 18361–18362; (b) Shiota, Y. and Yoshizawa, K. (2003) *J. Chem. Phys.*, **118**, 5872–5879.
43. Geng, C., Ye, S., and Neese, F. (2010) *Angew. Chem. Int. Ed.*, **49**, 5717–5720.
44. Srnec, M., Wong, S.D., England, J., Que, L. Jr., and Solomon, E.I. (2012) *Proc. Natl. Acad. Sci. U.S.A.*, **109**, 14326–14331.
45. Mayer, J.M. (2011) *Acc. Chem. Res.*, **44**, 36–46.
46. Kojima, T., Hirai, Y., Ishizuka, T., Shiota, Y., Yoshizawa, K., Ikemura, K., and Ogura, T. (2010) *Angew. Chem. Int. Ed.*, **49**, 8449–8453.
47. (a) Schröder, D. and Shaik, S. (2011) *Angew. Chem. Int. Ed.*, **50**, 3850–3851; (b) Kojima, T. and Fukuzumi, S. (2011) *Angew. Chem. Int. Ed.*, **50**, 3852–3853.
48. (a) Sastri, C.V., Oh, K., Lee, Y.J., Seo, M.S., Shin, W., and Nam, W. (2006) *Angew. Chem. Int. Ed.*, **45**, 3992–3995; (b) Lee, Y.-M., Kotani, H., Suenobu, T., Nam, W., and Fukuzumi, S. (2008) *J. Am. Chem. Soc.*, **130**, 434–435.
49. Collins, M.J., Ray, K., and Que, L. Jr., (2006) *Inorg. Chem.*, **45**, 8009–8011.
50. Comba, P., Wadepohl, H., and Wiesner, S. (2011) *Eur. J. Inorg. Chem.*, **2011**, 2610–2615.
51. Chen, K., Costas, M., and Que, L. Jr., (2002) *J. Chem. Soc., Dalton Trans.*, 672.
52. van Sickle, D.E., Mayo, F.R., and Arluck, R.M. (1965) *J. Am. Chem. Soc.*, **87**, 4824.
53. Bataille, C.J.R. and Donohoe, T.J. (2011) *Chem. Soc. Rev.*, **40**, 114–128.
54. (a) Roelfes, G., Lubben, M., Hage, R., Que, L. Jr., and Feringa, B.L. (2000) *Chem. Eur. J.*, **6**, 2152–2159; (b) Kim, C., Chen, K., Kim, J., and Que, L. Jr., (1997) *J. Am. Chem. Soc.*, **119**, 5964; (c) Chen, K. and Que, J.L. (1999) *J. Chem. Soc., Chem. Commun.*, 1375–1376.

55. (a) Jackson, T.A. and Que, L. Jr., (2006) in *Concepts and Models in Bioinorganic Chemistry* (eds H.-B. Kraatz and N. Metzler-Nolte), Wiley-VCH Verlag GmbH, Weinheim, p. 259.
56. (a) Burger, R.M. (1998) *Chem. Rev.*, **98**, 1153; (b) Stubbe, J., Kozarich, J.W., Wu, W., and Vanderwall, D.E. (1996) *Acc. Chem. Res.*, **29**, 322–330.
57. Blasiak, L.C., Vaillancourt, F.H., Walsh, C.T., and Drewman, C.L. (2006) *Nature*, **440**, 368.
58. Groves, J.T. and McClusky, G.A. (1976) *J. Am. Chem. Soc.*, **98**, 859–861.
59. Fujimori, D.G., Barr, E.W., Matthews, M.L., Koch, G.M., Yonce, J.R., Walsh, C.T., Bollinger, J.M. Jr., Krebs, C., and Riggs-Gelasco, P.J. (2007) *J. Am. Chem. Soc.*, **129**, 13408–13409.
60. Patai, S. and Rappoport, Z. (1994) *Synthesis of Sulfones, Sulfoxides, and Cyclic Sulfides*, John Wiley & Sons, Ltd, Chichester.
61. Zhou, X.T., Ji, H.B., Cheng, Z., Xu, J.C., Pei, L.X., and Wang, L.F. (2007) *Bioorg. Med. Chem. Lett.*, **17**, 4650.
62. (a) Kumar, D., de Visser, S.P., Sharma, P.K., Hirao, H., and Shaik, S. (2005) *Biochemistry*, **44**, 8148; (b) Li, C., Zhang, L., Zhang, C., Wu, W., Hirao, H., and Shaik, S. (2008) *Angew. Chem. Int. Ed.*, **47**, 8148; (c) Shaik, S., Wang, Y., Chen, H., Song, J., and Meir, R. (2010) *Faraday Discuss.*, **145**, 49.
63. Althoff, F., Benzing, K., Comba, P., McRoberts, C., Boyd, D.R., Greiner, S., and Keppler, F. *Commun.*, submitted.
64. Benet-Buchholz, J., Comba, P., Llobet, A., Roeser, S., Vadivelu, P., Wadepohl, H., and Wiesner, S. (2009) *Dalton Trans.*, 5910–5923.
65. Benet-Buchholz, J., Comba, P., Llobet, A., Roeser, S., Vadivelu, P., and Wiesner, S. (2010) *Dalton Trans.*, **39**, 3315.
66. (a) Nocera, D.G. (2012) *Acc. Chem. Res.*, **45**, 767; (b) Fukuzumi, S. (2008) *Phys. Chem. Chem. Phys.*, **10**, 2283.
67. (a) Kundu, S., Matito, E., Walleck, S., Pfaff, F.F., Heims, F., Rabay, B., Luis, J.M., Company, A., Braun, B., Glaser, T., and Ray, K. (2012) *Chem. Eur. J.*, **18**, 2787; (b) Chen, G., Chen, L., Ng, S.-M., Man, W.-L., and Lau, T.-C. (2013) *Angew. Chem. Int. Ed.*, **52**, 1789–1791.
68. Mandon, D., Jaafar, H., and Thibon, A. (2011) *New J. Chem.*, **35**, 1986.
69. Merz, M. (2002) Dissertation thesis, "Synthese neuer substituierter 3,7-Diazabicyclononanone und die katalytischen Eigenschaften ihrer Kupfer- und Eisenkomplexe" Universität Heidelberg, Heidelberg.
70. Hong, S., Lee, Y.-M., Shin, W., Fukuzumi, S., and Nam, W. (2009) *J. Am. Chem. Soc.*, **131**, 13910.
71. (a) Waleska, A. (2013), "Untersuchungen zur Wasseroxidationskatalyse und zu Sauerstoffaktivierungs- und Oxidationsprozessen mit Bispidin-Eisenkomplexen", Universität Heidelberg, Heidelberg; (b) Comba, P., Lee, Y.-M., Nam, W., and Waleska, A. (2014), **50**, 412–414.
72. Morimoto, Y., Lee, Y.-M., Nam, W., and Fukuzumi, S. (2013) *Chem. Commun.*, **49**, 2500–2502.
73. Comba, P. and Kerscher, M. (2009) *Coord. Chem. Rev.*, **253**, 564–574.

Lewis acid is feasible. A number of other external donor groups have been tested, but the breakthrough in terms of catalyst activity combined with high molecular weight PE was achieved with neutral donors covalently attached to the Cp ring.

7.4.1 Cp Systems with Covalently Bound Additional Donor Functions

Jolly and coworkers [16] reported highly active chromium complexes such as compound **9** in 1997. The corresponding dimethyl complex shows even higher activity, and ultrahigh molecular weight PE ($M_n = 2 \times 10^6$ g mol^{-1}) is formed within only 4 min of reaction time [17]. From these data, a turnover frequency of 300 s^{-1} can be calculated with 37% of the Cr centers to be active. In addition to the N-donor systems, P, O, and S donors have also been tested but with much lower activities and lower molecular weights [18].

Compounds **9**–**10** differ in the length of the spacer between the Cp ring and the N donor atom. Only the complexes with a C_2 spacer and a neutral N donor give both high activities and high molecular weights. Complexes of the type **11** were activated with $AlMe_3$ or $AlEt_3$. The active catalyst from complex **12** suffers from low stability.

We developed rigid donor-functionalized Cp ligands in 1996 [19] and were able to coordinate them in to Cr(III) complexes of the type **13** in 2001 [20]. These complexes give excellent polymerization catalysts when activated with MAO.

7.5 Polymerization Behavior of Donor-Functionalized Cp Chromium Complexes Developed in Heidelberg

Three representative examples of complexes discussed in this chapter are shown in Figure 7.3 (**13**–**15**). They have several properties in common, which are summarized in the following paragraphs.

7.5.1 Structural Features

The ligands give rise to complexes with very similar coordination geometries because of their rigid and predefined geometry, which is a result of three common attributes:

- A neutral N donor, preferably incorporated into an aromatic ring;
- A linker unit consisting of two atoms between the N donor atom and the Cp ring;
- Alkyl, silyl, or aryl substituents at the Cp ring.[2)]

2) We have also synthesized several complexes with an indenyl instead of cyclopentadienyl ring and some of them showed exciting polymerization behavior. However, in solution the catalysts suffered from insufficient stability over a longer polymerization time, especially at higher temperatures.

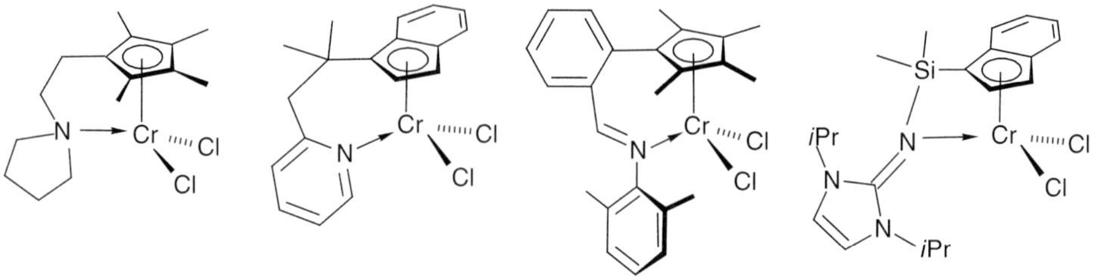

9 (Jolly[17]) 3590/900 000 | **10** (Huang[21]) 2000/25 000 | **11** (Mu[22]) 3120/61 000 | **12** (Tamm[23]) 2125/19 000

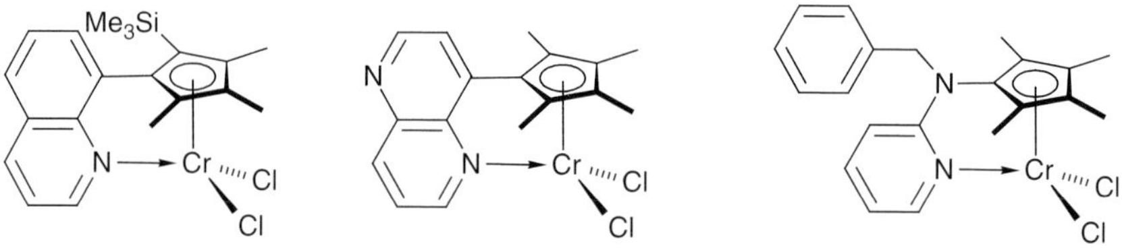

13[20, 24, 25] 3240/170 000 | **14**[26] 5200/430.000 | **15**[27] 4550/1 700 000

Figure 7.3 Precatalysts based on cyclopentadienide ligands with covalently bound neutral donor functions.

These structural features lead to a constrained "piano stool" geometry where one leg (N donor) is slightly bent toward the Cp ring. In the quinoline or naphthyridine systems, the N lone pair is perfectly oriented for interaction with the metal center. This enhances the binding strength and suppresses the formation of coordination polymers. Consequently, metals in higher oxidation states are strongly bound by these ligands without the occurrence of hemilabile behavior. We never observed decoordination from a chromium(III) center in a precatalyst or an activated catalyst. Steric bulk introduced at the Cp ring has no negative effect on polymerization behavior as will be discussed later. However, small substituents at the N donor moiety already lower the accessibility of the olefin monomer and lead to a decrease in catalyst activity.

7.5.2
Catalyst Activation and Catalytic Activities

The prerequisite for an olefin polymerization catalyst is a metal–alkyl bond. Hence the first step in catalyst activation is an alkylation reaction. However, with trialkylaluminum compounds alone, no catalytic polymerization activity is observed. Only cocatalysts that are able to abstract anionic ligands to generate the corresponding cationic complexes give good catalysts (e.g., MAO or metal alkyls in combination with $[HNMe_2Ph]^+[(C_6F_5)_4B]^-$). The activation is accompanied by a change of color from blue-green to pink which can be monitored by UV–vis spectroscopy. A minimum of 50 equiv Al per Cr atom is necessary for a complete

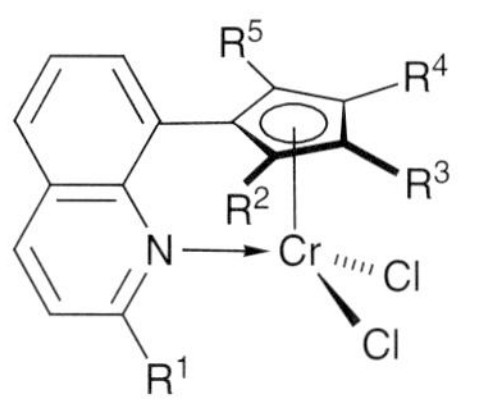

Figure 7.4 Precatalyst derivatives for comparison of catalyst activities.

activation, and 100 equiv already give high polymerization activities. Increasing the MAO amount leads to a slight increase in catalyst activity with an optimum around 1000 equiv. The active catalyst is stable at room temperature for several hours with only slight decrease in activity. Donor solvents suppress the catalyst activity. The tolerance to donor impurities is low, with 4 equiv of tetrahydrofuran (THF) leading to complete catalyst deactivation.

A comparison of catalyst activities for a series of quinolone-substituted Cp chromium complexes, presented in Figure 7.4, indicates the following trends:

- Substitution at the quinoline ortho position leads to lower catalytic activity ($R^1 = CH_3 < R^1 = H$).
- Alkyl, aryl, or silyl groups attached to the Cp ring increase the activities. Such substituents are also responsible for the higher solubility of the precatalysts in nonpolar solvents such as toluene. The best activity is obtained for complexes with $R^2 - R^5 = CH_3$, $R^2 - R^4 = CH_3$, $R^5 = SiMe_3$, $R^2 - R^5 = C_6H_5$ (T., Gehrmann and M., Enders, unpublished). Similar trends have been observed for the naphthyridine derivatives such as complex **14**.

The activity profiles of polymerization catalysts provide important insight into catalyst activation and deactivation. In a simplified model, the addition of the cocatalyst turns all precatalyst molecules into the active form (100% of active centers). This system should be stable during the polymerization process and the rate of insertion of the monomer into the chromium–alkyl bond should be independent of the length of the chromium–alkyl chain. If there is no chain termination (e.g., in a living system or at the beginning of a typical chain growth polymerization), this will lead to a constant monomer uptake until the first polymer chains with the catalyst attached to it precipitate out of the solution. Indeed, most homogeneous polymerization catalysts show the maximum activity at the beginning of the reaction. Steep decreases of monomer uptake within the first minutes are quite common and are usually a consequence of catalyst deactivation. An ideal catalyst system for technical processes should show constant activity with a smooth buildup at the start of the reaction.

We evaluated the activity profiles of some of our catalyst systems, and a representative example is shown in Figure 7.5a. In this experiment, the precatalyst **14** had been activated with MAO 10 min before the polymerization run. The solvent (toluene) was presaturated with ethylene at room temperature and the activated catalyst solution was introduced. The ethylene absorption maximum was

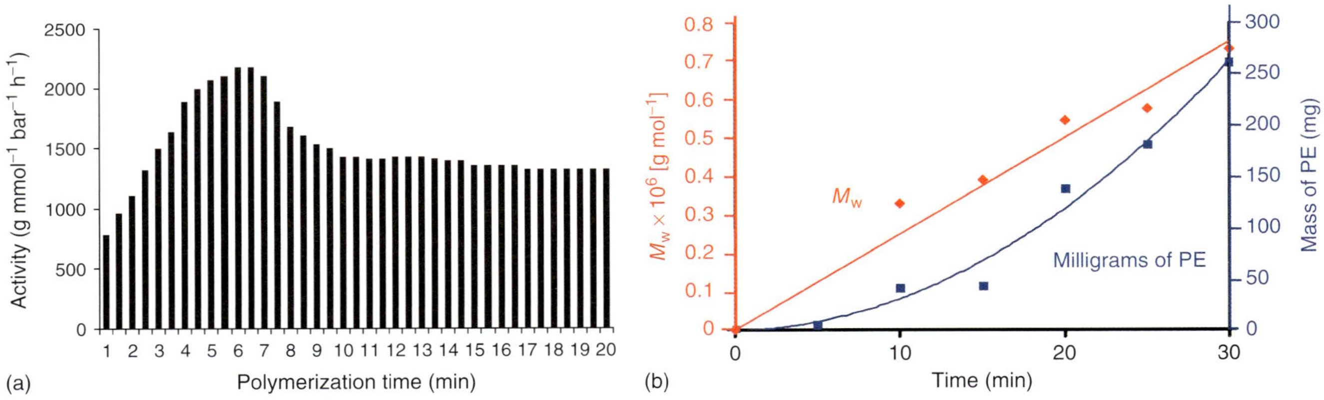

Figure 7.5 (a) Activity profile obtained in solution polymerization with precatalyst **14**/MAO at room temperature. (b) Increase in polymer chain length and polymer mass obtained with **14**/MAO at −30 °C [26].

found not at the beginning but after reaction time of 5–7 min.[3] Between 10 and 20 min, the absorption rate was almost constant. The smooth buildup at the start of the reaction was verified with a polymerization run at −30 °C (Figure 7.5b). At this temperature, the ethylene insertion was sufficiently slow, which allowed us to monitor the quasi-living behavior at the start of typical chain-growth polymerizations. The chain length increased linearly with time, but the increase in the amount of polymer was nonlinear. Assuming quasi-living behavior, the calculation of the number of active centers revealed that only 7% of the chromium centers were active after 10 min. This number increased to 14% after 30 min at −30 °C. This means that the number of active catalyst centers was small at the beginning of the polymerization reaction but increased during the run which readily explains the activity profile shown in Figure 7.5. These experimental observations could be explained by a substantial difference in the ethylene insertion barrier into the Cr–methyl bond compared to Cr–propyl (or Cr–polymer) bonds. Once the ethylene has been inserted into the first Cr–Me complex, the subsequent insertions are facilitated so that the Cr–Me complexes serve as a reservoir for more active catalyst species. This interpretation was supported by comprehensive density functional theory (DFT) calculations carried out in cooperation with Köppel and coworkers [28].

7.5.3 Chain Termination and Molecular Weights

An important chain-termination process in olefin polymerization is β-hydrogen elimination and the formation of a metal hydride species or β-hydrogen transfer to an incoming monomer which would give a metal-ethyl moiety in the case of ethylene. Both processes lead to unsaturated end groups of the polymer chain. However, β-elimination in chromium alkyl complexes with our ligands has never been observed. We synthesized dialkyl derivatives by replacing both chlorido substituents in complex **15** by CH_3 or C_2H_5, respectively [27]. The diethyl complex has a higher stability than the corresponding dimethyl derivative, showing the inertness of such alkyl chromium complexes toward β-hydrogen elimination. For MAO-activated complexes, no experimental evidence points to β-hydrogen elimination or transfer. However, chain transfer to aluminum has been observed quite often with our complexes as well as in comparable systems by other authors [17]. Consequently, the molecular weight of the polyethylene depends strongly on the nature of the cocatalyst. Conventional MAO contains small amounts of $AlMe_3$ which is able to act as chain-transfer reagent. With polymethyl aluminoxane (PMAO), which is a non-hydrolytically prepared variant of MAO [29], the chain transfer is suppressed, leading to slightly higher molecular weights. In solution polymerizations, the efficiency of chain transfer to aluminum, and hence

3) Ethylene diffusion effects may distort the picture to some extent but can be minimized by lowering the catalyst concentration.

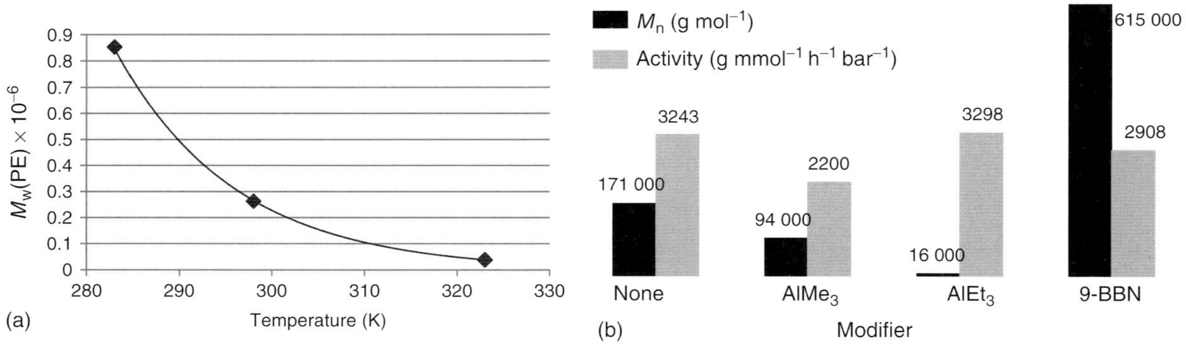

Figure 7.6 (a) Temperature dependence of molecular weight of PE with **13**/MAO and (b) polymerization data in the presence of modifiers.

the molecular weight of the polymer, is strongly temperature dependent as can be seen in Figure 7.6.

We have also evaluated the chain-transfer efficiency of different chain-transfer reagents. $AlMe_3$ displayed relatively low efficiency, whereas $AlEt_3$ dramatically decreased the molecular weight. With longer α-olefins, the transfer to aluminum dominated over olefin insertion so that catalytic carbalumination was possible with such catalysts [30]. The effect of typical chain-transfer reagents as modifiers is presented in Figure 7.6b. The catalyst activity is not affected seriously by the modifier. However, 9-borabicyclononane (9-BBN), which is a chain-transfer reagent for metallocene-type catalysts [31], leads to a dramatic increase in molecular weight up to $M_w = 3.14 \times 10^6$ g mol^{-1} ($M_n = 615\,000$ g mol^{-1}). This effect can be explained by a preferential coordination of 9-BBN to the active catalyst, thus preventing the attack of Al-alkyls. Chain transfer of the polymer chain from chromium to boron and formation of a chromium hydride species is energetically disfavored as we showed by DFT modeling [24].

7.6
En Route to Tunable Catalysts

Catalyst properties can be altered and optimized by small changes in the ligand environment. An important step in this procedure is the laborious introduction of different substituents in the ligand backbone before complex formation. Changing catalyst properties after complex formation is an attractive complementary approach for tuning of catalyst properties. The prerequisite for a realization of this concept is an efficient bond-forming process which does not interfere with vital catalyst functionalities. We have chosen Lewis acid–base reactions in order to accomplish that task. A distal Lewis basic center at the ligand backbone cannot interact with the catalytic center but it can accommodate Lewis acids. This concept is sketched in Figure 7.7, together with a system that provided the proof of principle.

The 1,5-naphthyridine complex **14** was reacted with Lewis acids shown in Figure 7.7. The reaction enthalpies could be estimated by DFT calculations, which were consistent with our experimental results. The reaction with triphenylborane, for example, was calculated to be endothermic, and indeed we were not able to identify the addition product with this reagent[4]. The formation of 1 : 1 adducts was derived from a combination of analytical data. IR spectroscopy confirmed the coordination of the $Rh(CO)_2Cl$ fragment to the naphthyridine moiety with two CO ligands in a cis arrangement. The paramagnetic ^1H NMR spectra showed slight shifts of the resonances upon coordination of the different Lewis acids. The assignment of the signals was based on DFT calculation of the

4) Calculation of the reaction enthalpies for the formation of the compounds shown in Figure 7.7 gave negative values whereas for the Lewis-Acid $B(C_6H_5)_3$ it is with +2.5 kcal mol^{-1} positive (DFT; B3LYP/6-311G(d) (2009) Gaussian09, Revision A.02, Gaussian Inc., Wallingford, CT.

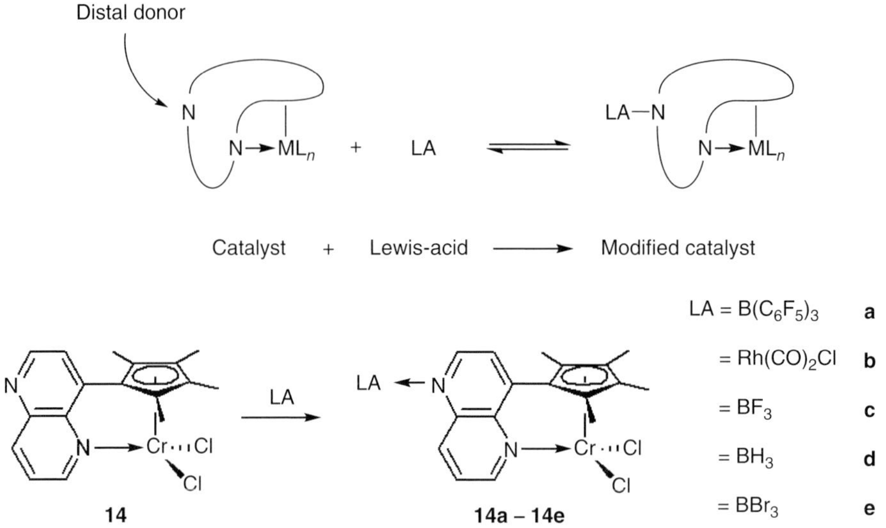

Figure 7.7 Concept and realization of tunable complexes.

Fermi-contact shifts and was further verified by comparison of NMR spectra of similar derivatives [25]. A solid-state molecular structure of **14a** verified the interaction of the Lewis acid $B(C_6F_5)_3$ with the distal nitrogen atom.

The complexes were evaluated in olefin polymerization. The unmodified complex **14** (without addition of Lewis acid) displayed an activity for ethylene of 5170 g $mmol^{-1}$ h^{-1} bar^{-1} and gave rise to a molecular weight which was in the ultrahigh range ($M_w = 1.24 \times 10^6$ g mol^{-1}, polydispersity index (PDI = 2.9)). Compared to the analogous quinoline chromium complex (similar to **13**, see Figure 7.4), an increase in molecular weight was observed upon the introduction of the second nitrogen atom. The Lewis acid-modified complexes showed catalyst activities between 1000 and 5800 g $mmol^{-1}$ h^{-1} bar^{-1}, and the molecular weight distribution was moved to higher or lower molecular weight, respectively. The PDIs between 2 and 3 indicated one active species, which excluded a partial substitution or decoordination of the Lewis acids during the activation process. These results demonstrate that catalyst tuning by introduction of a distal donor as docking position for modifiers is possible.

7.7 Conclusion

Many single-site chromium catalysts exist that are able to polymerize ethylene at a high rate. However, only a few produce long polymer chains at the same time. Chromium(III) complexes with cyclopentadienyl ligands have been identified as catalysts more than 20 years ago, and with the introduction of chelate N-donor-Cp ligands such complexes became competitive for industrial application. The stability of the complexes and hence the catalyst lifetime and productivity could be increased by using rigid and well-defined linker and N-donor moieties. This also allowed a more detailed investigation of the mechanisms of catalyst activation, olefin insertion, and chain termination processes, which account for the overall polymerization behavior of these and related complexes.

References

1. (a) Hogan, J.P. and Banks, R.L. (1958) Solid polymers of olefins and production of such polymers. (Phillips Petroleum Co.) US Patent 2825,721, (priority Jan. 27, 1953); (b) Ziegler, K., Breil, H., Holzkamp, E., and Martin, H. (1960) Verfahren zur Herstellung von hochmolekularem Polyäthylenen. German Patent 973626, 1960 (priority Nov. 17, 1953).
2. (a) Karapinka, G.L. (1973) Polymerization of Ethylene using supported Bis-(cyclopentadienyl) Chromium(II) Catalysts. (Union Carbide Co.) US Patent 3,709,853; (b) Karol, F.J., Karapinka, G.L., Wu, C., Dow, A.W., Johnson, R.N., and Garrick, W.L. (1972) *J. Polym. Sci. Part., A: Polym. Chem.*, **10**, 2621–2637.
3. (a) Cossee, P. (1964) *J. Catal.*, **3**, 80; (b) Arlman, E. (1964) *J. Catal.*, **3**, 89; (c) Arlman, E. and Cossee, P. (1964) *J. Catal.*, **3**, 99.
4. (a) McGuinness, D.S. (2011) *Chem. Rev.*, **111**, 2321–2341; (b) Agapie, T. (2011) *Coord. Chem. Rev.*, **255**, 861–880.
5. (a) Theopold, K.H. (1998) *Eur. J. Inorg. Chem.*, **1998**, 15–24. (b) Britovesk, G.J.P., Gibson, V.V., and Wass, D.F. (1999) *Angew. Chem. Int. Ed.*, **38**, 428–447; (c) Gibson, V.C. and Spitzmesser, S.K. (2003) *Chem. Rev.*, **103**, 283–316; (d) Smith, K.M. (2006) *Curr. Org. Chem.*, **10**, 955–963; (e) McGuinness, D.S. (2011) in *Olefin Upgrading Catalysis by Nitrogen-based Metal Complexes I*, Catalysis by Metal Complexes, vol. **35** (eds G. Giambastiani and J. Cámpora), Springer p. 1–26.
6. Esteruelas, M.A., López, A.M., Méndez, L., Oliván, M., and Oñate, E. (2003) *Organometallics*, **22**, 395–406.
7. Köhn, R.D., Haufe, M., Mihan, S., and Lilge, D. (2000) *Chem. Commun.*, 1927–1928.
8. Hurtado, J., Mac-Leod Carey, D., Muñoz-Castro, A., Arratia-Pérez, R., Quijada, R., Wu, G., Rojas, R., and Valderrama, M. (2009) *J. Organomet. Chem.*, **694**, 2636–2641.
9. (a) Mihan, S., Enders, M., and Fritz, O. (2004) Transition-Metal complexes with tridentate, Nitrogen-containing Ligands. WO 2004/104052; (b) Fritz, O. (2005) Dissertation Thesis, (Dr. rer. nat.) University of Heidelberg Metallkomplexe des Tris(imidazol-2-yl)-phosphins – Darstellung, Reaktivität und katalytische Anwendung.
10. Jones, D.J., Gibson, V.C., Green, S.M., Maddox, P.J., White, A.J.P., and Williams, D.J. (2005) *J. Am. Chem. Soc.*, **127**, 11037–11046.
11. T. Xu, H. An, W. Gao, Y. Mu, *Eur. J. Inorg. Chem.* 2010, 2010 3360–3364.
12. Hanaoka, H., Imamoto, Y., Hino, T., Kohno, T., Yanagi, K., and Oda, Y. (2007) *J. Polym. Sci. Part A: Polym. Chem.*, **45**, 3668–3676.
13. McGowan, K.P. and Veige, A.S. (2012) *J. Organomet. Chem.*, **711**, 10–14.
14. Thomas, B.J., Kyun Noh, S., Schulte, G.K., Sendlinger, S.C., and Theopold, K.H. (1991) *J. Am. Chem. Soc.*, **113**, 893–902.
15. (a) Xu, T., Mu, Y., Gao, W., Ni, J., Ye, L., and Tao, Y. (2007) *J. Am. Chem. Soc.*, **129**, 2236–2237; (b) Huang, Y.-B., Yu, W.-B., and Jin, G.-X. (2009) *Organometallics*, **28**, 4170–4174.
16. Emrich, R., Heinemann, O., Jolly, P.W., Krüger, C., and Verhovnik, G.P.J. (1997) *Organometallics*, **16**, 1511–1513.

17. Döhring, A., Göhre, J., Jolly, P.W., Kryger, B., Rust, J., and Verhovnik, G.P.J. (2000) *Organometallics*, **19**, 388–402.
18. (a) Döhring, A., Jensen, V.R., Jolly, P.W., Thiel, W., and Weber, J.C. (2001) *Organometallics*, **20**, 2234–2245; (b) Schuhen, K., Sieb, D., Wadepohl, H., and Enders, M. (2009) *Z. Anorg. Allg. Chem.*, **635**, 1560–1567.
19. Enders, M., Rudolph, R., and Pritzkow, H. (1996) *Chem. Ber.*, **129**, 459–463.
20. (a) Enders, M., Fernández, P., Ludwig, G., and Pritzkow, H. (2001) *Organometallics*, **20**, 5005–5007. (b) Mihan, S., Lilge, D., De Lange, P., Schweier, G., Schneider, M., Rief, U., Handrich, U., Hack, J., Enders, M., Ludwig, G., and Rudolph, R. (2001) Monocyclopentadienyl complexes of Chromium, Molybdenum or Tungsten with a Donor Bridge. WO 01/12641 A1.
21. Zhang, H., Ma, J., Qian, Y., and Huang, J. (2004) *Organometallics*, **23**, 5681–5688.
22. Zhang, L., Gao, W., Tao, X., Wu, Q., Mu, Y., and Ye, L. (2011) *Organometallics*, **30**, 433–440.
23. Randoll, S., Jones, P.G., and Tamm, M. (2008) *Organometallics*, **27**, 3232–3239.
24. (a) Mark, S., Kurek, A., Mülhaupt, R., Xu, R., Klatt, G., Köppel, H., and Enders, M. (2010) *Angew. Chem.*, **122**, 8933–8936; *Angew. Chem. Int. Ed.*, (2010), **49**, 8751–8754. (b) Kristen, M.O., Mark, S., and Enders, M. (2010) Catalyst system for the polymerization of Olefins. WO 2010112165 (A1).
25. Fernández, P., Pritzkow, H., Carbo, J.J., Hofmann, P., and Enders, M. (2007) *Organometallics*, **26**, 4402–4412.
26. Sieb, D., Baker, R.W., Wadepohl, H., and Enders, M. (2012) *Organometallics*, **31**, 7368–7374.
27. Sieb, D. (2012) Naphthyridin- und Aminopyridin-funktionalisierte Cyclopentadienyl-Chrom(III)-Komplexe und deren katalytische Anwendung in der Olefinpolymerisation. Dissertation Thesis, (Dr. rer. nat.) University of Heidelberg.
28. Xu, R., Klatt, G., Enders, M., and Köppel, H. (2012) *J. Phys. Chem.*, **116**, 1077–1085.
29. Smith, G.M., Palmaka, S.W., Rogers, J.S., and Malpass, D.B. (1998) Polyalkylaluminoxane compositions formed by non-hydrolytic means. US Patent 5.831.109.
30. Mark, S., Gaidzik, N., Doye, S., and Enders, M. (2009) *Dalton Trans.*, 4875–4877.
31. Amin, S.B. and Marks, T. (2008) *Angew. Chem Int. Ed.*, **47**, 2006–2025.

8
Ligand Design and Mechanistic Studies for Ni-Catalyzed Hydrocyanation and 2-Methyl-3-Butenenitrile Isomerization Based upon Rh-Hydroformylation Research

Peter Hofmann and Michael E. Tauchert

8.1
Introduction

Seventy-five years ago, Roelen (1897–1993), then working at the chemical company Ruhr-Chemie in Germany, based upon his experience and expertise gathered as a Ph.D. student at the Kaiser-Wilhelm Institut für Kohlenforschung in Mülheim/Ruhr (renamed as Max-Planck Institut für Kohlenforschung after World War II) in his collaboration with Franz Fischer and Hans Tropsch on converting synthesis gas (CO/H_2) to liquid fuels [1], found and developed a cocatalyzed reaction [2], which in its variants over the years has become and still is the largest volume production process operated using homogeneous transition-metal catalysis (with predominantly Co or Rh as metals) in the chemical industry worldwide [3].

At the ACS National Meeting in the spring of 2013 in New Orleans, the 75th anniversary of this fundamental catalytic reaction [4], named "hydroformylation" (or "oxo-reaction," Scheme 8.1), was celebrated through a special symposium.

The annual industrial production volume of aldehydes and their downstream products (alcohols, carboxylic acids, esters, plasticizers, detergents, surfactants, lubricants, fuel additives, solvents, fine chemical intermediates, etc.) is rather impressive. Between 9 and 10 million tons of aldehydes are produced worldwide. Over 5 million tons just of C_4 oxo products per year come from chemical industry's hydroformylation plants. In the light of today's discussion about sustainability issues of chemical synthesis, the atom economy of the oxo process as a (formally) waste-free intermolecular addition reaction of three molecular components is well worth mentioning.

The simplified general picture displayed in Scheme 8.1 employing simple 1-alkenes as feedstock reflects the main issues and mandatory prerequisites of successful hydroformylation catalysis: chemoselectivity (avoiding alkene isomerization, alkene hydrogenation, aldehyde hydrogenation, aldehyde condensation, ligand degradation, etc.), regioselectivity (yielding a defined ratio of linear to branched [$l/b = n/i$] aldehydes as needed, usually with the linear product preferred), and enantioselectivity (if *i*-aldehydes are the desired products). In addition, in contrast to laboratory-scale hydroformylation reactions (for example,

Molecular Catalysts: Structure and Functional Design, First Edition. Edited by Lutz H. Gade and Peter Hofmann.

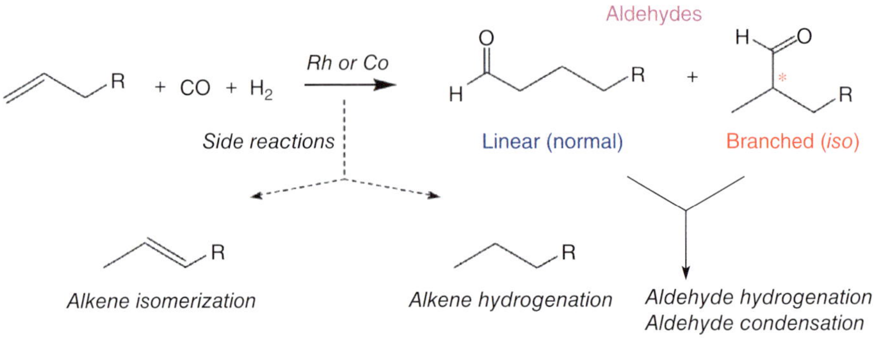

Scheme 8.1 A (simplified) educt/product representation of the hydroformylation reaction.

in standard research labs or in commercial fine chemicals synthesis where this reaction surprisingly is much less in use than it could be), typical large-scale industrial applications are operated with up to 250 000 tpa (tons per annum) aldehyde capacities and with reactor volumes of 200–300 m^3. In Rh-catalyzed hydroformylation plants, the metal concentration is normally limited to the 200–300 ppm range, amounting, however, to kilogram quantities of Rh worth millions of dollars, while ligand quantities, as for instance of the broadly used triphenylphosphine ligand PPh_3, may well reach the level of tons. It should be also realized in this context that running large-scale industrial hydroformylation reactions economically successful makes it mandatory to reach turnover frequency (TOF) values way above 1000 mol mol^{-1} h^{-1} and turnover number (TON) values of around 4 000 000 mol product/mole Rh or higher. This sets a scene that is far from comparable to most normal and accepted homogeneously catalyzed, standard, small-scale reactions in synthetic laboratories.

Starting out from ligand-free cobalt catalysis in high-pressure reactions initiated by the formation of $HCo(CO)_4$ and from early basic mechanistic understanding of the catalytic cycle (which – most remarkably – was developed by Breslow and Heck [5], the latter one of course is known much more for his discovery of the famous Pd-catalyzed "Heck reaction," which earned him (together with Suzuki and Negishi) the Nobel Prize of 2010), Rh catalysis for hydroformylation using appropriate ligand systems appeared on the scene in the 1960s, and today it is the predominant hydroformylation process in industry. This became possible because mechanistic understanding of the catalytic cycle increased steadily and the important role of ligands in this reaction led to an immense effort to develop tailor-made ligand families for the desired catalyst performance. Still today, PPh_3, the prototype of a monodentate ligand, for various reasons is a major player in industrial hydroformylation plants, and has played a privileged role in mechanistic studies. Somewhat later, chelating ligands were introduced into Rh-based oxo chemistry. Given the significance and elegant efficiency of hydroformylation catalysis, it is understandable that the amount of patent literature as well as the number of original research publications, reviews, and books dealing with this reaction and with ligand synthesis has always been and still is enormous and thus sometimes not easy to follow.

In this chapter, we do not intend to give a survey covering the present status of contemporary Rh-catalyzed hydroformylation and related ligand research. We will, however, focus on a specific group of chelate ligand systems designed and developed in our group within the research project SFB 623, which was carried out with the intention to boost the performance of Rh hydroformylation with respect to its *n*-selectivity and catalyst activity. These efforts turned out to be quite successful [6]. Moreover, and of primary relevance for this chapter, the ligand class that originally was tailor-made for Rh-hydroformylation catalysis meanwhile has found a number of other interesting applications.

One of them has to do with research directed toward the mechanism of Ni-catalyzed hydrocyanation of butadiene to adiponitrile (the DuPont or ADN process) and the hydrocyanation of olefins, as well as research related to 2-methyl-3-butene nitrile (2M3BN) isomerization, an important step of catalytic adiponitrile (ADN) production (Scheme 8.2, Equation II). Our findings in the field of ligand design for Rh hydroformylation catalysis soon turned our attention to the chemistry of this also homogeneously catalyzed, large-scale, C–C bond-forming reaction, which has been established as an industrial high volume production process leading to Nylon-6,6. The work we will describe in the following has in fact evolved as a direct consequence of our efforts in Rh-hydroformylation ligand design.

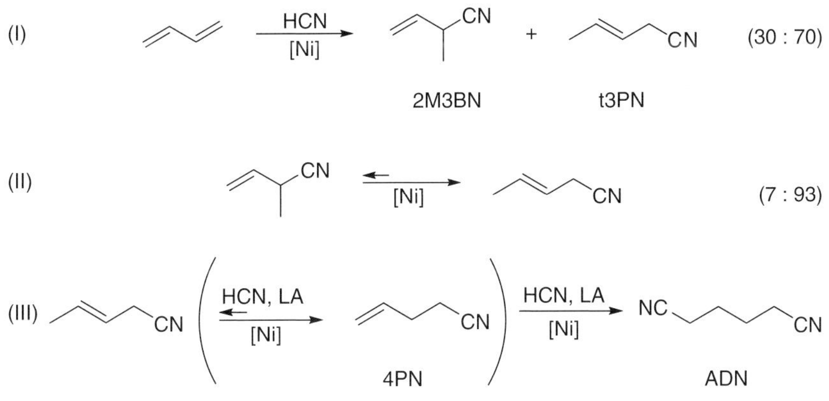

Scheme 8.2 The DuPont adiponitrile process.

So we will limit this review to certain aspects of catalytic carbon–carbon bond formation brought about by addition reactions of the C_1-building block HCN to unsaturated hydrocarbon substrates, for example, to alkenes and to dienes, and to the mechanistic details thereof.

The famous Ni-catalyzed hydrocyanation of 1,3-butadiene to adiponitrile (the DuPont or ADN process) opens the pathway to the homogeneously catalyzed, large-scale production of Nylon-6,6 via adipic acid and 1,6-diaminohexane, both downstream products of adiponitrile. This reaction today, of course, represents textbook knowledge [7]. Overall, it is a more complicated process than hydroformylation because it is actually performed as a three-step reaction sequence [8], as shown in Scheme 8.2.

After the first Ni-catalyzed addition step of HCN to 1,3-butadiene leading to a 30 : 70 nonequilibrium mixture of the 1,2- and the 1,4-monoadducts 2M3BN and t3PN (trans-3-pentenenitrile), isomerization to the 7 : 93 equilibrium mixture of these two unsaturated mononitriles, again Ni-catalyzed, is necessary before in a final step the second addition of HCN, after Ni-catalyzed, Lewis acid-promoted isomerization of t3PN to 4PN, results in adiponitrile.

The Ni catalysts utilized in the industrial DuPont ADN synthesis are tetrakis phosphite complexes of Ni(0) as, for example, tetrakis(tri-2-tolylphosphite)nickel (0), the phosphite ligand of which is easily accessible (i.e., cheap) even in large quantities. Despite the availability of this established ligand class for the large-scale production of ADN, there is of course still quite some research interest in academia and industry directed toward the theory- and experiment-based design of new ligand systems. This simply has to do with the fact that the overall reaction sequence of Scheme 8.2 describes a much less efficient and much more complicated case of large-scale homogeneous transition-metal catalysis than established in other cases as, for example, in hydroformylation. Probably and supposedly, there is still room for improvement, with respect to catalyst activity, selectivity, and stability if, on the basis of a detailed mechanistic understanding, rational or at least semi-rational design for creating novel ligand families is used. An ultimate goal, of course, would be to find appropriate ligand structures for Ni(0) precatalyst complexes that are able to replace the three-step procedure of Scheme 8.2 by a one-pot reaction, or which will allow performing at least step I and step II in one reaction.

The following sections describe some of our research activities aiming toward ligand design, ligand and catalyst synthesis including structure determination, and mechanistic understanding in the field of Ni-catalyzed hydrocyanation chemistry.

8.2
Recent Advances in Ni-Catalyzed Hydrocyanation and Isomerization Reactions

The mechanism and the scope of the hydrocyanation and 2-methyl-3-butenenitrile isomerization reaction has been studied at DuPont in great detail for nickel catalysts featuring monodentate phosphite ligands, such as $P(Otolyl)_3$. The results of these studies have been published in a review by Tolman and coworkers [9]. Later advances in the field have been summarized in recent reviews [10, 11], and book chapters [12, 13]. The aim of this chapter is to give an account of those developments that have not been covered in these reviews yet or are directly relevant to our own SFB-based research conducted in the field.

8.2.1 Hydrocyanation of Vinylarenes

Hydrocyanation of vinylarenes provides an interesting access to anti-inflammatory and pain killer drugs such as naproxen and dexibuprofen (the pharmacologically active enantiomer of ibuprofen, Scheme 8.3). However, the enantioselective hydrocyanation of these substrate types remains a challenge. In 1994, RajanBabu and Casalnuovo [14] reported the application of glucophosphinite ligands **1** for the hydrocyanation of β-vinylnaphathalene **3**. Upon optimization, nitrile **4** was obtained in 84% ee and good yield (>99% ee after crystallization). Application of the reaction conditions to other substrates, such as dexibuprofen precursor **5**, resulted only in low or moderate enantioselectivities.

Our own group recently presented a new class of chelating phosphonite ligands featuring chemically and thermally inert cycloalkane spacers (7) [15]. This class of ligands was originally developed for Rh-catalyzed hydroformylation [16], but also displayed good activity and excellent selectivity in the (racemic) hydrocyanation of styrene.

Another promising approach to tackle the challenge of vinylarene hydrocyanation was presented by de Greef and Breit in 2009 [17]. They employed a library of self-assembled bidentate phosphine ligands for the (racemic) hydrocyanation of styrene. Remarkably, they observed increased yields for electron-rich phosphine ligands, which is in contrast to earlier reports with Xantphos phosphine derivatives [18].

Very recently, Schmalz and coworkers [19] demonstrated the successful utilization of mixed Taddol-derived phosphine-phosphite ligands **2**. Employing a mixture of trimethylsilyl cyanide (TMSCN) and MeOH for *in situ* generation of HCN, they obtained for the first time in high yielding conversions excellent enantiomeric excesses for substituted styrenes. The best results were reported for dexibuprofen precursor **6**, with an enantiomeric excess of 92% and 100% isolated yield (Scheme 8.3).

8.2.2 Hydrocyanation of 1,3-Dienes

In addition to hydrocyanation of styrene, our phosphonite ligands **7** also showed excellent performance in the technically relevant hydrocyanation of 1,3-butadiene (Scheme 8.2, Equation I) featuring t3PN and 2M3BN. Utilization of ligand **7b** resulted in an observed TOF of >400 h^{-1}, with a total TON >600, using as little as 0.1 mol% catalyst [15].

Similarly, RajanBabu's glucophosphinite ligands **1** show high activity and moderate to good enantioselectivity in the hydrocyanation of substituted 1,3-dienes (Scheme 8.4) [20].

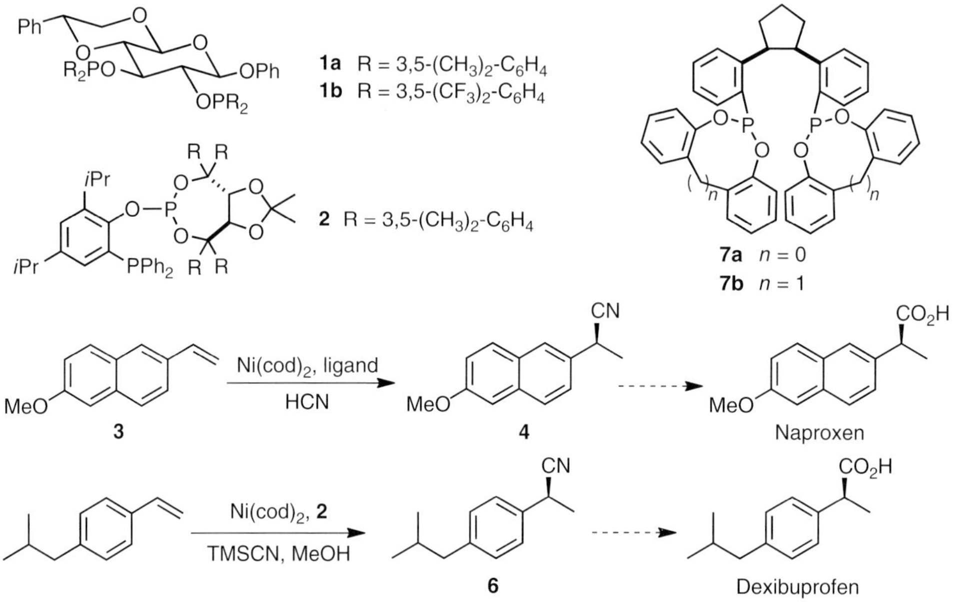

Scheme 8.3 State-of-the-art enantioselective hydrocyanation of vinylarenes [14, 15, 19].

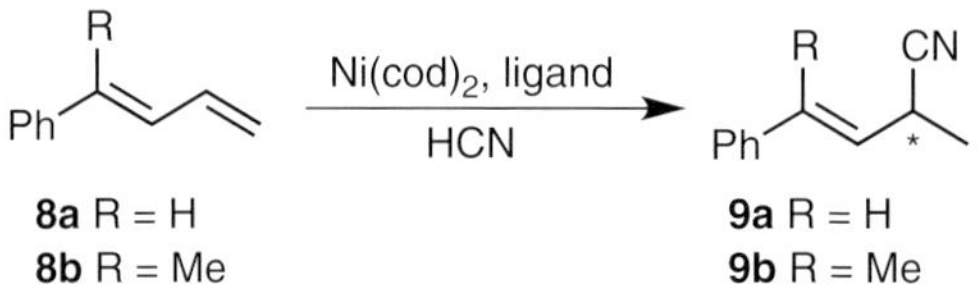

R	Ligand	Yield (%)	ee (%)
H	**1a**	87	78
Me	**1b**	56	68

Scheme 8.4 Examples for enantioselective hydrocyanation of substituted 1,3-dienes [20].

8.2.3
Hydrocyanation of *trans*-3-Pentenenitrile

The technically relevant hydrocyanation of t3PN (Scheme 8.2, Equation III) represents the most difficult step of the ADN process, and reports in the literature tackling this problem are very scarce. The requirements for the catalyst are quite challenging, as high activity in pentenenitrile isomerization and high substrate selectivity and *n*-selectivity for the hydrocyanation are essential for the production of the desired adiponitrile (Scheme 8.5).

Pringle and coworkers introduced cage phosphinites such as **10** [21] and a cage fluorophosphine **11** [22] as new ligand classes for this reaction (Scheme 8.5). Using acetone cyanohydrin as a source for HCN and various Lewis acids as cocatalysts, they achieved up to 80% yield for dinitriles with up to 85% selectivity for the desired adiponitrile. Fluorophosphine **11** has also been used successfully in Rh-catalyzed low-pressure hydroformylation of 1-heptene, showing similar activity to PPh_3 with slightly increased *n*-selectivity [22].

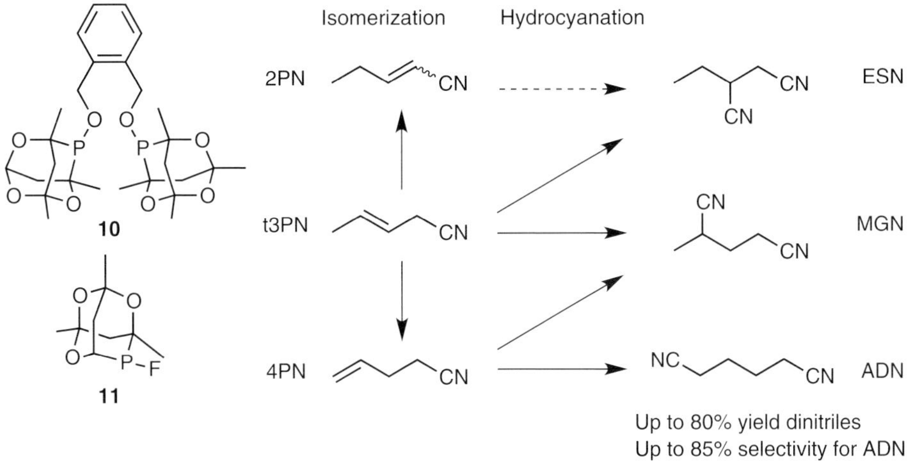

Scheme 8.5 Possible reactions involved under t3PN hydrocyanation conditions and Pringle ligands **10** and **11** [21, 22]. ESN, ethylsuccinonitrile; MGN, 2-methylglutaronitrile.

8.2.4
Isomerization of 2-Methyl-3-Butenenitrile

In the past few years, application of electron-rich phosphine [23–27] and carbene [28] ligands has attracted some attention for the Ni-catalyzed 2M3BN isomerization reaction. These catalyst systems commonly yield 2-pentenenitriles (2PN) and branched 2-methyl-2-butenenitriles (2M2BN). Although the formation of α,β-unsaturated nitriles 2PN and 2M2BN is of no technical relevance, it has been studied in quite some detail because the transformation is an interesting model reaction to study Ni-catalyzed C–H bond activation (Scheme 8.6).

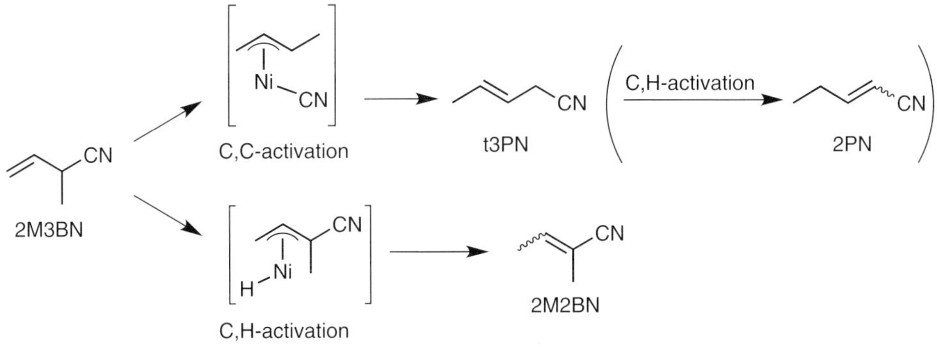

Scheme 8.6 Mechanisms involved in Ni-catalyzed 2M3BN isomerization [26].

The C–H bond activation step has been particularly well studied by Jones and coworkers [24–26] using $[Ni(dippe)H]_2$ (dippe, 1,2-bis(diisopropylphosphino) ethane) as a catalyst. A remarkable solvent dependence of the mechanism was observed. In nonpolar solvents, such as *n*-decane, the major product of 2M3BN isomerization is t3PN (82.5%), while in polar solvents, such as MeCN or dimethylformamide (DMF), the branched nitriles 2M2BN (97.9% and 89.4%, respectively) predominate [26].

8.3 Recent Advances in Ni-Catalyzed Hydrocyanation and Isomerization Reactions Employing the TTP-Ligand Family

8.3.1 Genesis of the TTP-Ligand Family

One of the most important ligands for Rh-catalyzed low-pressure hydroformylation still is PPh_3, which features a high TOF of 8600 h^{-1} with a moderate *n*/*i* ratio of 3 in the hydroformylation of *n*-octene (Scheme 8.7). Chelating diphosphine ligands often produce better *n*/*i* ratios, but are less active [34]. This discovery led to the development of Xantphos by van Leeuwen [35], which till today is a sort of a benchmark ligand in terms of activity and *n*-selectivity in Rh-catalyzed hydroformylation of terminal alkenes (Scheme 8.7).

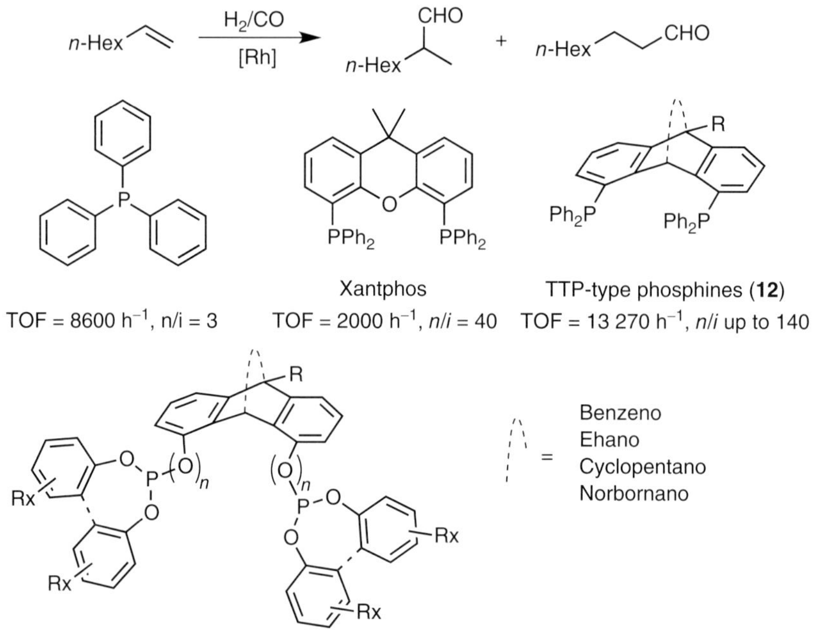

$n = 0$: TTP-type phosphonites (**13**), TOF = 2900 h^{-1}, *n*/*i* up to 140
$n = 1$: TTP-type phosphites (**14**), TOF = 9040 h^{-1}, *n*/*i* up to 177

Scheme 8.7 Phosphorous ligands for Rh-catalyzed hydroformylation of *n*-octene [29–33].

The excellent selectivity was believed to derive from the rigid xanthene backbone and from a favorable bite angle, with a sufficiently large flexibility range allowing the catalyst to adapt to all the different complex geometries of the catalytic cycles, without high energy demand [35]. Subsequent density functional theory (DFT) calculations showed that regioselectivity was governed by nonbonding interactions between the diphenylphosphino groups and the substrate [36]. The xanthene backbone adopts a roof-shaped conformation in metal complexes with a low inversion barrier (Scheme 8.8) [36, 37]. The phenyl rings at the phosphorus atoms are not hampered by the xanthene backbone and thus lead to numerous ligand conformers and various coordination geometries, as shown for the trigonal-bipyramidal model complex (Xantphos)Rh(CO)(PPh_3)H (Scheme 8.8).

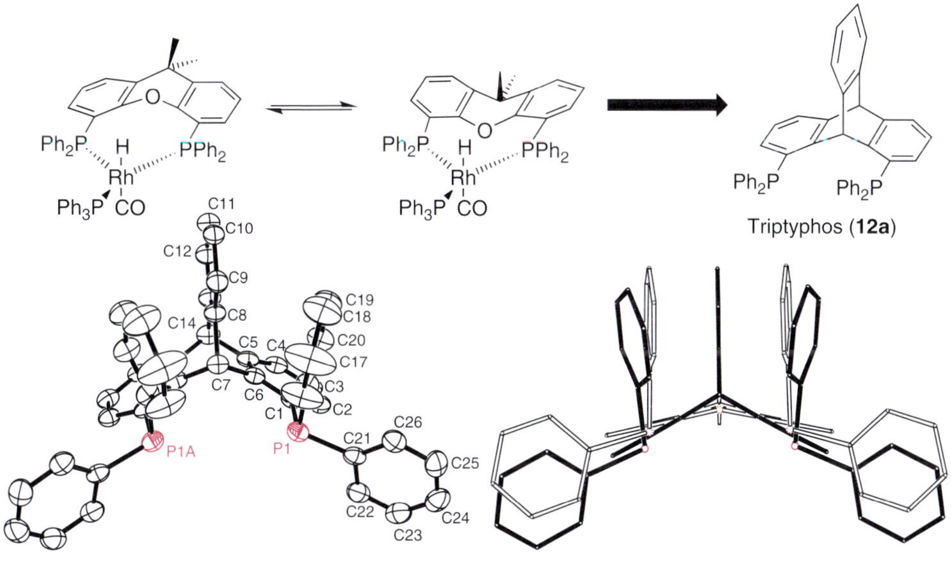

Scheme 8.8 (a) Butterfly flip of model complex [(Xantphos)Rh(CO)(PPh_3)H], (b) solid-state structure of Triptyphos **12a**, and (c) superposition of **12a** (black) and Xantphos (white).

As this facile adaptation of multiple conformers could potentially lead to different reaction pathways, thus lowering selectivity, we sought to design a ligand system that maintains the favorable features of Xantphos while suppressing backbone inversion and reducing free phenyl group rotation. This goal was realized by introducing a two-carbon bridge across the 9 and 10 positions of 9,10-dihydroanthracene. Scheme 8.8 displays one of the first representatives of this new ligand class, Triptyphos (TTP) **12a**. These new diphosphines showed very high activity combined with exceptional chemoselectivity and *n*-selectivity in the hydroformylation reaction (Scheme 8.7) [29, 30, 38]. Its diphosphonite and diphosphite congeners also displayed excellent performance in the Rh-catalyzed hydroformylation reaction [31–33, 39].

Scheme 8.8 depicts the solid-state structure of TTP **12a** as well as a superposition of **12a** and Xantphos to highlight the similarities and the differences. TTP features a smaller angle between the subtending planes of the phenyl rings bearing the PPh_2 substituents (Xantphos: 151°, **12a**: 121°). Compound **12a** also has a larger distance between the two P moities (4.34 Å) compared to Xantphos: (4.05–4.16 Å, depending on the solid-state structure) [30, 38, 40]. Meanwhile, a large body of synthetic, structural, mechanistic, and theoretical work has revealed (partly rather unexpected) details of Rh-hydroformylation catalysis employing ligands of the TTP-derived ligand family, which will be reported elsewhere.

8.3.2
Ni-Catalyzed Isomerization and Hydrocyanation with TTP-Type Phosphonite Ligands

Given the success of our diphosphonites featuring cycloalkane spacers as ligands for hydrocyanation and 2M3BN isomerization [15], and the good performance of the TTP-type diphosphonite ligands (Scheme 8.7, Figure 8.1) in hydroformylation, we considered it worthwhile to also investigate their activity in hydrocyanation catalysis and the mechanistically related 2M3BN isomerization reaction.

However, first attempts using the ethano-bridged ligand **13a** and benzeno-bridged ligand **13b** did not result in any conversion in hydrocyanation reactions or 2M3BN isomerization. The inactivity of these ligands can be explained by their coordination behavior toward Ni(0) precursors. In the presence of phosphonites **13a** and **13b**, $Ni(cod)_2$ (cod, cyclooctadiene) forms very stable catalytically inactive bischelate complexes $Ni(ligand)_2$ (Figure 8.2). Bischelate formation could not be counteracted by an increased amount of $Ni(cod)_2$. To tackle this challenge, more sterically demanding TTP-type phosphonites **13c–e** were synthesized (Figure 8.1) [31, 39]. Reaction of these bulky phosphonites with $Ni(cod)_2$ indeed resulted in the formation of mono-chelated complexes such as (**13c**)Ni(cod) (Figure 8.2). The modified ligands were then tested in Ni-catalyzed 2M3BN isomerization (Scheme 8.2, Equation II). The ligands were evaluated

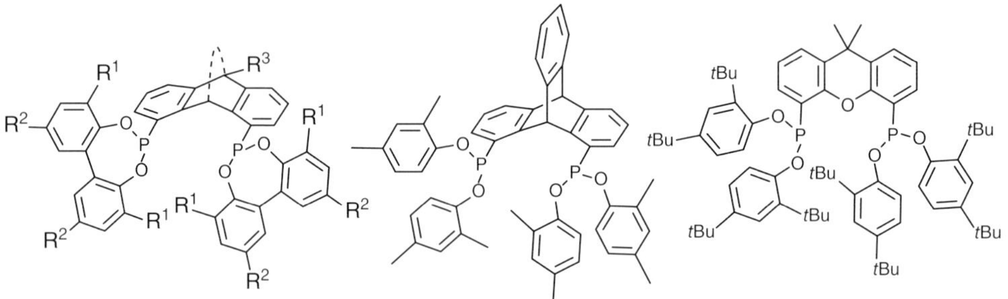

13e

15

13a $R^1 = R^2 = R^3 = H$, bridge = ethano
13b $R^1 = R^2 = H$, R^3 = *n*Oct, bridge = benzeno
13c $R^1 = R^2 = tBu$, $R^3 = H$, bridge = benzeno
13d R^1 = OMe, R^2 = Me, R^3 = H, bridge = cyclopentano

Figure 8.1 TTP-type phosphonite ligands **13a–e** and Xantphosphonite **15** [31, 39, 41].

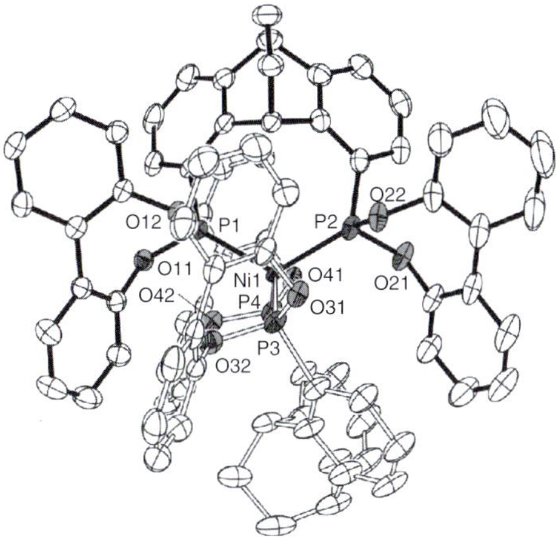

$(\mathbf{13a})_2$Ni

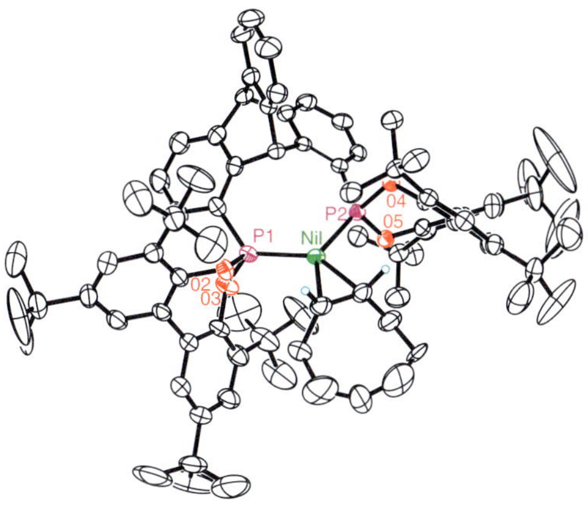

(**13c**) Ni(cod)

Figure 8.2 Solid-state structure of nickel complexes featuring bis-ligating and mono-chelating TTP-type phosphonite ligands [39, 42].

quantitatively in this reaction, using as little as 0.2 mol% catalyst loading and a moderate temperature of 100 °C [39]. While the extremely bulky phosphonite **13c** showed no activity, 24% conversion toward the desired t3PN was observed with ligand **13e** within 1.5 h. The reaction continued to full conversion (53% after 3 h, and 97% after 20 h). To date, to the best of our knowledge, ligand **13d** continues to be the most active phosphonite ligand reported for this transformation. Eighty percent t3PN was observed within 1.5 h (91% after 3 h, and 97% after 20 h), surpassing the most active Xantphosphonite derivative **15** (72% t3PN after 1.5 h at 90 °C, 2.5 mol% catalyst) [41] and our most active cycloalkane phosphonite ligand **7a** (58% t3PN after 1.5 h, 100 °C, 0.2 mol% catalyst) [15]. With these promising findings at hand, the ligands were also tested in the hydrocyanation of styrene, utilizing acetone cyanohydrin as a source for HCN. However, only low to moderate conversions were observed. Application of ligand **13e** yielded 4.5% 2-phenylpropionitrile (2PPN) (20 h, 1 mol% catalyst, 60 °C). Phosphonite **13d**, which performed best in the isomerization reaction, achieved 43% of the branched nitrile product within 16 h (0.8 mol% catalyst) [39].

8.3.3
Ni-Catalyzed Hydrocyanation Involving TTP-Type Phosphine Ligands

Catalytic hydrocyanation of styrene employing Xantphos-type ligands has been reported previously, but catalyst loading of 5 mol% was necessary to ensure good conversion [18, 43].

Before investigating potential applications of our TTP-type phosphine ligands **12** (Figure 8.3) in catalytic hydrocyanation and isomerization reactions, we

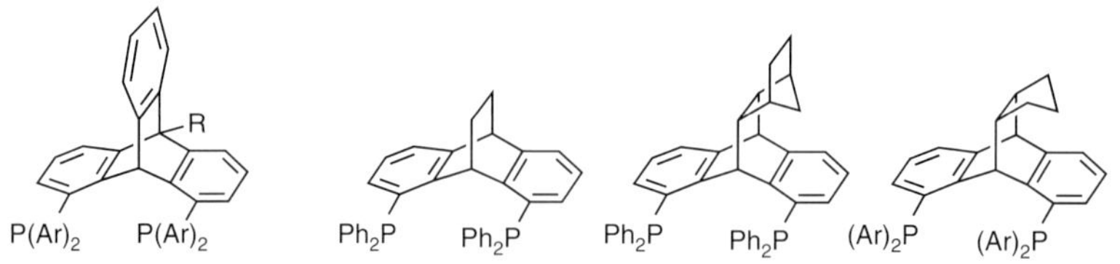

12a R = H, Ar = Ph (Triptyphos)
12b R = *n*Oct, Ar = Ph (Octyltriptyphos)
12c R = *n*Oct, Ar = 4-(Me)-C_6H_4

12d Rucaphos

12e Maophos

12f Ar = Ph (C5phos)
12g Ar = 4-(Me)-C_6H_4 (Me-C5phos)
12h Ar = 4-(OMe)-C_6H_4 (OMe-C5phos)
12i Ar = 4-(CF_3)-C_6H_4 (CF_3-C5phos)

Figure 8.3 TTP-type phosphine ligands. Rucaphos from Ruperto Carola (Latin name of Heidelberg University), Maophos from Martin August Ottfried Volland (a former member of our group) [29, 30, 38].

decided to study their coordination behavior. In contrast to the TTP-type phosphonites **13**, reaction of the phosphines **12** with Ni(cod)$_2$ yields mono-chelated complexes such as **16** (Scheme 8.9) [38]. The formation of bis-chelated complexes was observed only in the presence of an excess of **12**. The catalyst precursor **16** readily reacted with styrene featuring complex **17** [44].

We investigated the activity of the phosphine ligands in the catalytic hydrocyanation of styrene, employing acetone cyanohydrin as a source for HCN (Table 8.1). Best results were obtained in toluene at 60 °C.

We first evaluated a possible structural influence induced by the 9,10 bridge of the ligand's backbone using 1 mol% of catalyst. Starting with the presumably most rigid Octyltriptyphos **12b**, 82.9% conversion was achieved within 1.5 h. The reaction proceeded with similar results for Rucaphos **12d** (64.6%) and Maophos **12e** (65.5%). The probably most flexible C5phos **12f** (58.1%) showed the lowest activity (Table 8.1, entries 1, 4, 7, and 10). In all cases, >90% conversion as well as good iso-selectivity was achieved within 20 h. From this data, we concluded that the observed drop in activity was caused by an increase in flexibility of the ligand backbone.

To assess the impact of electronic modifications of our ligand system onto styrene hydrocyanation, we used substituted C5phos ligands **12g–i** (Figure 8.3), and compared the respective conversions after 1.5 h (1 mol% catalyst loading). Interestingly, a preference for the more electron-rich ligands Me-C5phos **12g** (91.5%) and OMe-C5phos **12h** (66.4%) was observed, compared to the nonsubstituted C5phos **12f** (58.1%, Table 8.1, entries 10–18). CF_3-C5phos showed no conversion at all in the reaction. The combined favorable steric and electronic features of ligand **12c** resulted in an increased activity of the system. With a catalyst loading of 0.5 mol%, **12c** reached 67.0% conversion within 1.5 h, surpassing Octyltriptyphos **12b** (59.1%, Table 8.1, entries 19–24).

Finally, we wondered whether the high selectivity for the branched hydrocyanation product 2PPN (2-phenylpropionitrile) was caused by an isomerization reaction similar to 2M3BN isomerization. Indeed, stoichiometric reaction of **16** with either 2PPN or 3PPN resulted in dehydrocyanation, producing styrene complex **17**. However, 3PPN could not be isomerized into 2PPN with **16** as a catalyst.

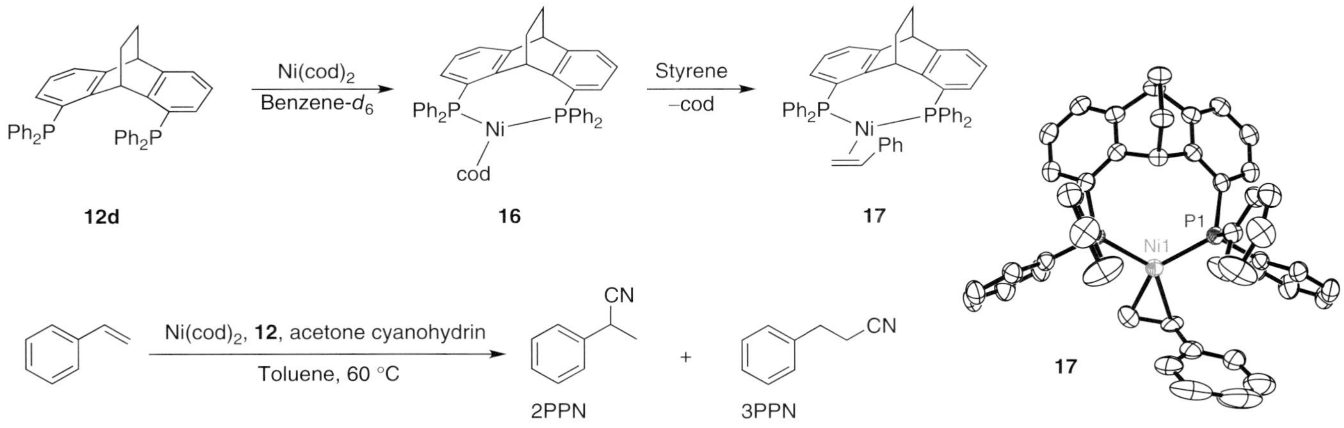

Scheme 8.9 Coordination properties of Rucaphos **12d** and hydrocyanation of styrene [44].

Table 8.1 Hydrocyanation of styrene using TTP-type phosphine ligands **12b–h** and $Ni(cod)_2$ as a catalyst [44].[a)]

Entry	Ligand	*t* (h)	Ni (mol%)	Conversion (%)	iso-Selectivity (%)	2PPN (%)	3PPN (%)
1	**12b**	1.5	1.00	82.9	98.5	81.7	1.3
2	**12b**	3.0	1.00	91.4	98.3	89.9	1.5
3	**12b**	20	1.00	98.7	98.4	97.1	1.6
4	**12d**	1.5	1.00	64.6	98.9	63.0	0.7
5	**12d**	3.0	1.00	75.6	98.9	74.8	0.8
6	**12d**	20	1.00	90.2	98.9	89.2	1.0
7	**12e**	1.5	1.00	65.5	99.6	65.3	0.2
8	**12e**	3.0	1.00	96.3	99.6	96.0	0.4
9	**12e**	20	1.00	100.0	99.6	99.6	0.4
10	**12f**	1.5	1.00	58.1	99.6	57.9	0.2
11	**12f**	3.0	1.00	86.2	99.5	85.8	0.4
12	**12f**	20	1.00	100.0	99.5	99.5	0.5
13	**12g**	1.5	1.00	91.5	99.8	91.3	0.2
14	**12g**	3.0	1.00	97.9	99.9	97.8	0.1
15	**12g**	20	1.00	99.7	99.9	99.6	0.1
16	**12h**	1.5	1.00	66.4	99.1	65.8	0.6
17	**12h**	3.0	1.00	91.0	99.0	90.0	0.9
18	**12h**	20	1.00	99.7	99.7	99.4	0.3
19	**12b**	1.5	0.50	59.1	99.0	58.5	0.6
20	**12b**	3.0	0.50	67.3	99.0	66.7	0.7
21	**12b**	20	0.50	77.7	99.1	77.0	0.7
22	**12c**	1.5	0.50	67.0	99.3	66.5	0.5
23	**12c**	3.0	0.50	71.2	99.3	70.7	0.5
24	**12c**	20	0.50	70.4	99.8	70.3	0.1

a) Reaction conditions: ligand/$Ni(cod)_2$ = 1.05, acetone cyanohydrin/styrene = 1.20, toluene, 60 °C, di(ethylene glycol)diethyl ether as internal standard for calibrated GC analysis.

Our ligand system attracted the interest of another group who evaluated the performance of TTP **12a** as a ligand in Ni-catalyzed hydrocyanation of 1,3-butadiene (Scheme 8.2, Equation I) [45]. Excellent activity and superior selectivity for the desired t3PN (up to 93%) were reported. The high selectivity for the linear nitrile product t3PN was thought to originate from a subsequent (fast) isomerization reaction, which is in good agreement with our own findings.

8.3.4
Applications and Mechanistic Studies of TTP-Type Phosphine Ligands in Ni-Catalyzed 2M3BN Isomerization

Given the success of the phosphine ligands **12** in styrene and 1,3-butadiene hydrocyanation, we decided to study their activity in the related isomerization of 2M3BN (Scheme 8.2, Equation II) [38]. Initial screening of the phosphine

Table 8.2 Ni-catalyzed 2M3BN isomerization with TTP-type phosphine ligands **12b–h** [38].[a)]

Entry	Ligand	*t* (h)	Ni (mol%)	Conversion (%)[b)]	2M3BN[b)](%)	t3PN[b)](%)	c3PN[b)](%)
1	**12b**	1.5	0.20	96.7	3.3	93.5	1.5
2	**12b**	3.0	0.20	96.7	3.3	91.9	3.0
3	**12b**	20.0	0.20	96.6	3.4	88.1	6.2
4	**12d**	1.5	0.20	71.2	28.8	69.1	0.4
5	**12d**	3.0	0.20	95.7	4.3	92.2	1.0
6	**12d**	20	0.20	96.8	3.3	88.4	5.5
7	**12e**	1.5	0.20	72.1	28.0	69.0	0.9
8	**12e**	3.0	0.20	96.0	4.0	92.3	1.5
9	**12e**	20	0.20	96.6	3.4	88.2	6.1
10	**12f**	1.5	0.20	55.6	44.4	52.0	1.3
11	**12f**	3.0	0.20	92.0	8.0	88.6	1.0
12	**12f**	20	0.20	96.8	3.3	90.1	4.2
13	**12g**	1.5	0.20	71.7	28.3	68.9	0.6
14	**12g**	3.0	0.20	96.8	3.2	93.5	1.6
15	**12g**	20	0.20	97.1	2.9	86.0	9.6
16	**12h**	1.5	0.20	68.0	32.0	65.5	0.9
17	**12h**	3.0	0.20	96.1	3.9	93.0	1.5
18	**12h**	20	0.20	97.2	2.9	87.6	7.6
19	**12i**	1.5	0.20	17.4	82.6	15.6	0.0
20	**12i**	3.0	0.20	31.4	68.6	29.5	0.1
21	**12i**	20.0	0.20	96.5	3.5	94.1	0.7
22	**12b**[c)]	1.5	0.011	46.5	53.5	45.5	0.4
23	**12b**[c)]	3.0	0.011	82.3	17.7	80.7	1.0
24	**12b**[c)]	20.0	0.011	96.1	3.9	87.2	8.0
22	**12c**[c)]	1.5	0.010	47.6	52.4	46.3	1.3
23	**12c**[c)]	3.0	0.010	80.0	20.0	78.2	1.8
24	**12c**[c)]	20.0	0.010	95.6	4.4	87.3	8.3

a) Reaction conditions: 10.0 μmol of $Ni(cod)_2$, 10.5 μmol of P-ligand, 100 μl of diethylene glycol diethyl ether (DEG), required amount of 2M3BN, 4 ml of toluene, 10 min preformation of catalyst, 100 °C.
b) Determined by calibrated GC analysis with DEG as an internal standard.
c) 120 °C.

ligands was performed with 0.2 mol% catalyst at 100 °C (Table 8.2). Interestingly, modification of steric and electronic properties of the phosphine ligands resulted in comparable relative activity as in styrene hydrocyanation. The more rigid ligand Octyltriptyphos **12b** depicted 96.7% conversion within 1.5 h, followed by Ruacaphos **12d** and Maophos **12e** with 71.2% and 72.1%, respectively (Table 8.2, entries 1–9). The presumably least rigid ligand backbone in the series, namely C5phos **12f** (55.6% within 1.5 h), required 3 h to reach full conversion (92%, Table 8.2, entries 10–12). Again, a preference for the more electron-rich ligands **12g** and **12h** (71.7% and 68%) was observed, while the electron-poorer CF_3-C5phos **12i** depicted only slow conversion (17.4%, Table 8.2, entries 13–19). The conversion almost doubled within 3 h (31.4%) and 96% conversion was achieved within 20 h, suggesting a deceleration of the reaction by electron-withdrawing substituents (Table 8.2, entries 14–21). However, NMR experiments showed the formation of multiple species after addition of 2M3BN to the catalyst (**12i**)Ni(cod), indicative of a decomposition process.

In the case of ligand **12b**, full conversion was reached well before 1.5 h. Further optimization (Table 8.2, entries 22–24) revealed that at 120 °C the catalyst concentration could be lowered to 0.011 mol% reaching 82.3% conversion within 3 h, which amounts to a TOF of up to 2850 h^{-1} and a TON of 9400. Ligand **12c** showed similar results. To the best of our knowledge, **12b** represents the most active ligand for 2M3BN isomerization reported so far.

In all cases, no formation of α,β-unsaturated nitriles (2PN/2M2BN) was observed. This excellent selectivity for 3-pentenenitriles is particularly remarkable, considering that formation of 2PN and 2M2BN is quite common among aryl diphosphine ligands such as dppf (1,1′-bis(diphenylphosphino)ferrocene, 12% "other nitriles") [24] or Xantphos (up to 50% 2PN) [41].

Technically employed ligands such as $P(Otolyl)_3$ often require a Lewis acidic cocatalyst such as $ZnCl_2$. When we added $ZnCl_2$ to our catalyst system, activity dropped below 10% conversion within 1.5 h. A similar effect has been reported for dppf [24].

The unsurpassed activity and robustness of our ligand system, as well as the unexpected slowing down of the reaction in the presence of Lewis acids, led us to believe that the 2M3BN isomerization might not occur via a cascade of dehydrocyanation and rehydrocyanation as suggested for catalytic systems employing monodentate phosphite ligands [46]. We therefore started a detailed study of the reaction mechanism.

Spectroscopic investigations using NMR and IR allowed us to gather some insight into the nature of the nickel complexes involved in the isomerization

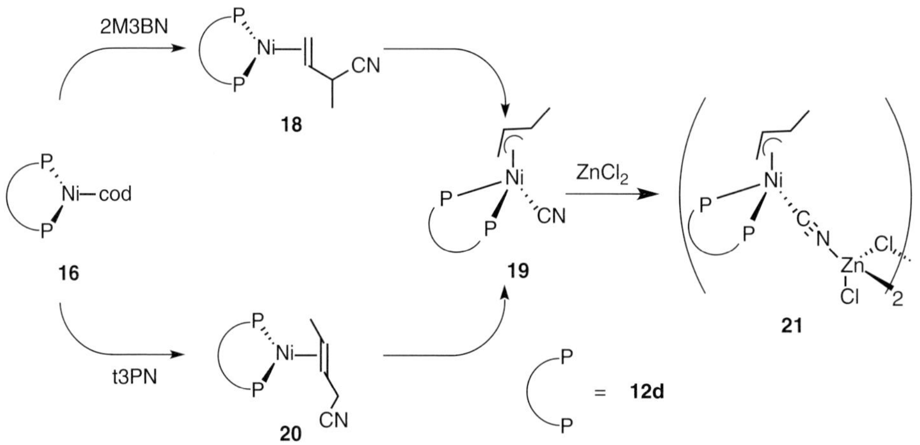

Scheme 8.10 Nickel complexes involved in 2M3BN isomerization [38, 44].

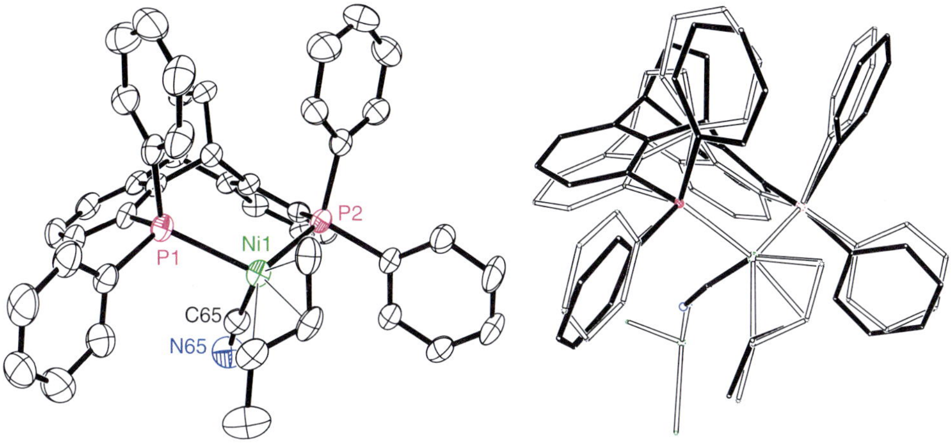

Figure 8.4 (a) Solid-state structure of resting state **19**. (b) Superposition of resting state **19** (black) and dimer complex **21** (white, only one unit shown for clarity) [38].

reaction [38, 44]. Precatalyst **16** reacted rapidly at room temperature with 2M3BN to form the methallyl complex **19**, which turned out to be the resting state of the reaction (Scheme 8.10). The solid-state structure of **19** (Figure 8.4) revealed a significant P–Ni bond length difference of 17 pm.

On the basis of DFT calculations, Li and Jones [27] had demonstrated that dissociation of one phosphorous donor ligand arm was feasible for (dippe)Ni(η^3-C_4H_7)CN. In ligands with a more flexible ligand backbone, such a dissociation process could potentially lead to catalyst decomposition. However, Because of the rigidity of our ligand backbone, the phosphorous donor cannot dissociate completely and remains in close proximity to the nickel center, thus possibly preventing catalyst decomposition.

Complex **19** could also be synthesized from **16** by addition of t3PN, thus demonstrating the reversibility of the isomerization reaction. At low temperatures, the intermediate alkene complexes **18** and **20** could be observed by *in situ* IR and NMR spectroscopy (Scheme 8.10).

Reaction of resting state **19** with $ZnCl_2$ results in the formation of the dimeric $ZnCl_2$-adduct complex **21** (Scheme 8.10), which displays structural features very similar to those of **19** (Figure 8.4).

To shed some light on the question how utilization of our diphosphine ligands results in such a high isomerization activity, we decided to repeat Druliner's experiment (Scheme 8.11). Druliner had found that isomerization of 2M3BN that was enriched with deuterium at the methyl group yielded t3PN with the deuterium equally located at the C2 and C5 positions. From this observation, it was concluded that isomerization occurred through a cascade of dehydrocyanation followed by rehydrocyanation [46]. However, when we performed this experiment with ligand **12b** in the absence of a Lewis acid, no deuterium scrambling was observed. To further exclude a mechanism featuring a dehydrocyanation step, we repeated the 2M3BN isomerization in the presence of deuterated 1,3-butadiene. In case of a dehydrocyanation mechanism, some exchange between labeled and unlabeled 1,3-butadiene should occur, thus yielding deuterated t3PN (Scheme 8.11). However, no trace of deuterium-enriched t3PN could be observed, most likely excluding a dehydrocyanation mechanism for our system.

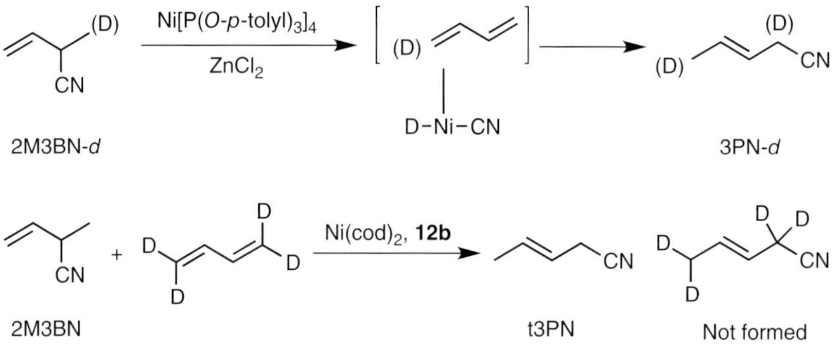

Scheme 8.11 Deuterium-labeling experiments. (a) Druliner's experiment and (b) control experiment [38, 44].

With regard to these results, we hypothesized that a reductive elimination from the resting state **19** might lead directly to the desired t3PN. Based on our experimental findings, we used quantum mechanical calculations (DFT) to model the catalytic cycle of 2M3BN isomerization featuring the Rucaphos ligand **12d** (Scheme 8.12) [38, 44]. The Gibbs free energy, ΔG, of the 2M3BN/t3PN transformation was calculated as $-16.6\,\text{kJ mol}^{-1}$. Starting from the trigonal-planar alkene complex **A**, oxidative addition was predicted to proceed through a late transition state **TS1** ($+51.1\,\text{kJ mol}^{-1}$), yielding the η^3-methallyl complex **B/19** ($-10.9\,\text{kJ mol}^{-1}$, see also Figure 8.4). A low activation barrier was found for the rotation of the allylic ligand (**TS2**, $+51.8\,\text{kJ mol}^{-1}$), which was identified as the most likely mechanistic pathway to lead to the rotamer of the resting state **C** ($+2.9\,\text{kJ mol}^{-1}$). From intermediate **C** onwards, two mechanistic pathways were feasible: reductive elimination via **TS4** ($+48.6\,\text{kJ mol}^{-1}$) leading to the product alkene complex **F** ($-10.4\,\text{kJ mol}^{-1}$), or a dehydrocyanation pathway. However, higher free energy intermediates were found for the latter pathway, as well as a very low barrier for an exergonic loss of 1,3-butadiene, which would be in contrast to our deuterium-labeling experiments (Scheme 8.11). We therefore believe the pathway outlined in Scheme 8.12 to be the most likely one.

The results described in this review, which are based upon combined experimental and theoretical efforts resulting in a new ligand class in the context of two industrially highly relevant, large-scale processes involving homogeneous catalysis with Rh and Ni, in our opinion have also opened many doors to further opportunities in molecular catalysis research also with other metals.

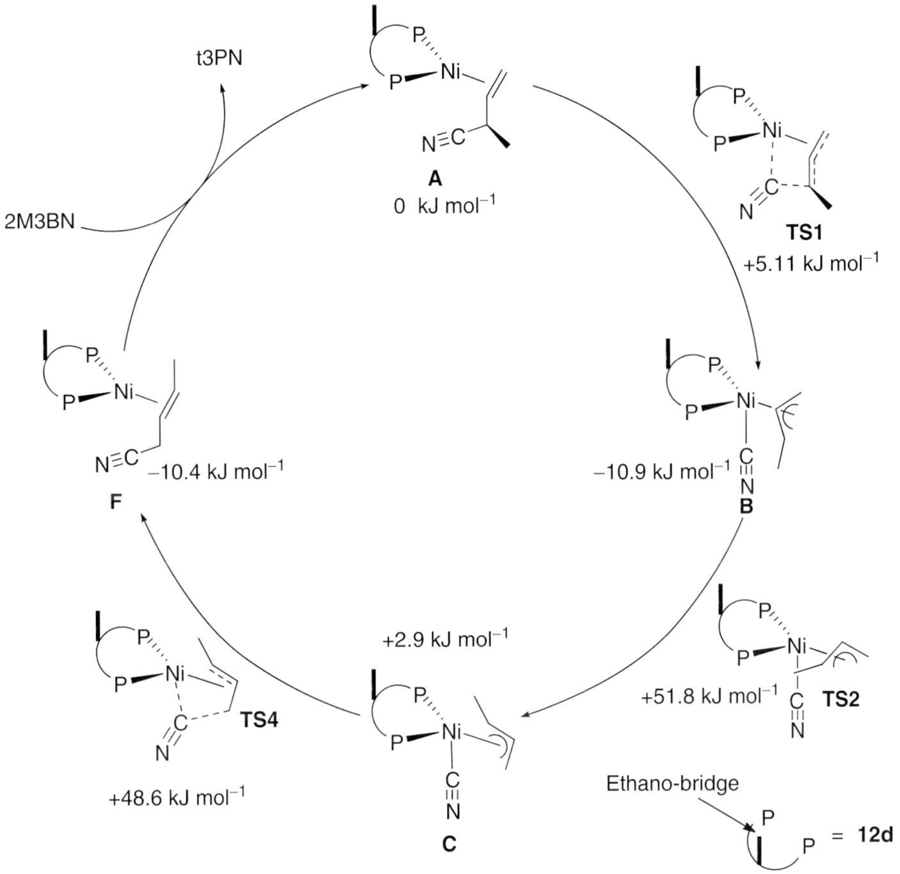

Scheme 8.12 Calculated catalytic cycle and Gibbs free energies (ΔG) of 2M3BN isomerization employing Rucaphos **12d** as a ligand [38, 44].

Acknowledgments

The work described here was supported by the Deutsche Forschungsgemeinschaft (SFB 623, "Molecular Catalysts: Structure and Functional Design"), the Graduate College 850 of the Deutsche Forschungsgemeinschaft, the University of Heidelberg, and the State of Baden-Württemberg. We also gratefully acknowledge generous support from BASF SE through donations of substances, instrumentation support, and fruitful discussions.

References

1. Miller, F.P., Vandome, A.F., and McBrewster, J. (2009) *Fischer-Tropsch Process*, Alphascript Publishing.
2. (a) Roelen, O. (1938/1952) Verfahren zur Darstellung von sauerstoffhaltigen Verbindungen DB Patent 849 548; (b) Cornils, B., Herrmann, W.A., and

Rasch, M. (1994) *Angew. Chem.*, **106**, 2219.
3. Platinum catalysis has been used for asymmetric hydroformylation of olefins, see e.g.: Stille, J.K., Hu, S., Brechot, P., Parrinello, G., and Hegedus, L.S. (1991) *Organometallics*, **10**, 1183.
4. For the Oxo-Reaction's 75th anniversary see also: Frey, G. and Dämbkes, G. (2013) *75 Years of Oxo-Synthesis*, Klartext-Verlag.
5. (a) Heck, R.F. and Breslow, D.S. (1961) *J. Am. Chem. Soc.*, **83**, 4023; (b) Heck, R.F. and Breslow, D.S. (1960) *Chem. Ind.*, 467.
6. (a) Ahlers, W., Paciello, R., Vogt, D., and Hofmann, P. (2002) Ligands for pnicogen chelate complexes with a metal of subgroup viii and use of the complexes as catalysts for hydroformylation, carbonylation, hydrocyanation or hydrogenation. WO Patent 02083695 A1; (b) Hofmann, P. (2004) Invited Lecture, 14th International Symposium on Homogeneous Catalysis, Munich, 2004 (Abstract IL7).
7. (a) Behr, A. (2008) *Angewandte Homogene Katalyse*, Wiley-VCH Verlag GmbH, Weinheim; (b) Arpe, H.J. (2007) *Industrielle Organische Chemie*, Wiley-VCH Verlag GmbH, Weinheim; (c) Steinborn, D. (2011) *Fundamentals in Organometallic Chemistry*, Wiley-VCH Verlag GmbH, Weinheim.
8. (a) Arthur, P. Jr.,, England, D.C., Pratt, B.C., and Whitman, G.M. (1954) *J. Am. Chem. Soc.*, **21**, 5364; (b) van Leeuwen, P.W.N.M. (2004) *Homogeneous Catalysis: Understanding the Art*, Wiley-VCH Verlag GmbH, Weinheim.
9. Tolman, C.A., McKinney, R.J., Seidel, W.C., Druliner, J.D., and Stevens, W.R. (1985) *Adv. Catal.*, **33**, 1.
10. Bini, L., Müller, C., and Vogt, D. (2010) *Chem. Commun.*, **46**, 8325.
11. Bini, L., Müller, C., and Vogt, D. (2010) *ChemCatChem*, **2**, 590.
12. RajanBabu, T.V. (2011) *Org. React.*, **75**, 1.
13. Tauchert, M.E. (2013) Hydrocyanation of alkenes, in *C-1 Building Blocks in Organic Chemistry* (ed P.W.N.M. van Leeuwen), Georg Thieme Verlag KG, 2014 in press.
14. Casalnuovo, A.L., RajanBabu, T.V., Ayers, T.A., and Warren, T.H. (1994) *J. Am. Chem. Soc.*, **116**, 9869.
15. Göthlich, A.P.V., Tensfeldt, M., Rothfuss, H., Tauchert, M.E., Haap, D., Rominger, F., and Hofmann, P. (2008) *Organometallics*, **27**, 2189.
16. Ahlers, W., Paciello, R., Roeper, M., Hofmann, P., Tensfeldt, M., Goethlich, A. WO 20011085739 A1 200111115 (2001). Hydroformylation method, bridged compounds of phosphorus, arsenic and antymony as catalyst comprising a complex of said compounds.
17. de Greef, M. and Breit, B. (2009) *Angew. Chem. Int. Ed.*, **48**, 551.
18. Goertz, W., Keim, W., Vogt, D., Englert, U., Boele, M.D.K., Veen, L.A., Kamer, P.C.J., and van Leeuwen, P.W.N.M. (1998) *J. Chem. Soc., Dalton Trans.*, 2981.
19. Falk, A., Göderz, A.L., and Schmalz, H.G. (2013) *Angew. Chem. Int. Ed.*, **52**, 1576.
20. Saha, B. and RajanBabu, T.V. (2006) *Org. Lett.*, **8**, 4657.
21. Mikhel, I.S., Garland, M., Hopewell, J., Mastroianni, S., McMulllin, C.L., Orpen, A.G., and Pringle, P.G. (2011) *Organometallics*, **30**, 974.
22. Fey, N., Garland, M., Hopewell, J.P., McMullin, C.L., Mastroianni, S., Orpen, A.G., and Pringle, P.G. (2012) *Angew. Chem. Int. Ed.*, **51**, 118.
23. Wilting, J., Müller, C., Hewat, A.C., Ellis, D.D., Tooke, D.M., Spek, A.L., and Vogt, D. (2005) *Organometallics*, **24**, 13.
24. Acosta-Ramírez, A., Muñoz-Hernández, M., Jones, W.D., and García, J.J. (2006) *J. Organomet. Chem.*, **691**, 3895.
25. Acosta-Ramírez, A., Flores-Gaspar, A., Muñoz-Hernández, M., Arévalo, A., Jones, W.D., and García, J.J. (2007) *Organometallics*, **26**, 1712.
26. Swartz, B.D., Reinartz, N.M., Brennessel, W.W., García, J.J., and Jones, W.D. (2008) *J. Am. Chem. Soc.*, **130**, 8548.
27. Li, T. and Jones, W.D. (2011) *Organometallics*, **30**, 547.
28. Acosta-Ramírez, A., Morales-Morales, D., Serrano-Becerra, J.M., Arévalo, A., Jones, W.D., and García, J.J. (2008) *J. Mol. Catal. A: Chem.*, **288**, 14.

29. Ahlers, W., Röper, M., Hofmann, P., Warth, D.C.M., and Paciello, R. (2001) WO 0158589.
30. Warth, D.C.M. (2004) Dissertation. Universität Heidelberg.
31. Kaiser, T.R. (2005) Dissertation. Universität Heidelberg.
32. Rosendahl, T. (2007) Dissertation. Universität Heidelberg.
33. Smith, S.E., Rosendahl, T., and Hofmann, P. (2011) *Organometallics*, **30**, 3643.
34. Hughes, O.R. and Unruh, J.D. (1981) *J. Mol. Catal.*, **12**, 71.
35. Kranenburg, M., van der Burgt, Y.E.M., Kamer, P.C.J., and van Leeuwen, P.W.N.M. (1995) *Organometallics*, **14**, 3081.
36. Carbó, J.J., Maseras, F., Bo, C., and van Leeuwen, P.W.N.M. (2001) *J. Am. Chem. Soc.*, **123**, 7630.
37. Landis, C.R. and Uddin, J. (2002) *J. Chem. Soc., Dalton Trans.*, 729.
38. Tauchert, M.E., Warth, D.C.M., Braun, S.M., Gruber, I., Ziesak, A., Rominger, F., and Hofmann, P. (2011) *Organometallics*, **30**, 2790.
39. Tauchert, M.E., Kaiser, T.R., Göthlich, A.P.V., Rominger, F., Warth, D.C.M., and Hofmann, P. (2010) *ChemCatChem*, **2**, 674.
40. Hillebrand, S., Bruckmann, J., Krüger, C., and Haenel, M.W. (1995) *Tetrahedron Lett.*, **36**, 75.
41. van der Vlugt, J.I., Hewat, A.C., Neto, S., Sablong, R., Mills, A.M., Lutz, M., Spek, A.L., Müller, C., and Vogt, D. (2004) *Adv. Synth. Catal.*, **346**, 993.
42. Gross, J.H., Nieth, N., Linden, H.B., Blumbach, U., Richter, F.J., Tauchert, M.E., Tompers, R., and Hofmann, P. (2006) *Anal. Bioanal. Chem.*, **386**, 52.
43. Kranenburg, M., Kamer, P.C.J., Leeuwen, P.W.N.M., Vogt, D., and Keim, W. (1995) *J. Chem. Soc., Chem. Commun.*, 2177.
44. Tauchert, M.E. (2009) Nickel-catalyzed hydrocyanation and pentenenitrile isomerization: ligand design and mechanistic studies. Dissertation. Universität Heidelberg.
45. Bini, L., Müller, C., Wilting, J., von Chrzanowski, L., Spek, A.L., and Vogt, D. (2007) *J. Am. Chem. Soc.*, **129**, 12622.
46. Druliner, J.D. (1984) *Organometallics*, **3**, 205.

9
Strongly Electron Donating Tridentate N-Heterocyclic Biscarbene Ligands for Rhodium and Iridium Catalysts

Doris Kunz and Eva Jürgens

9.1
Introduction

Although rhodium and iridium are high-priced precious metals, their complexes are indispensable for homogeneous catalysis. Prominent examples for their use on the industrial scale are, among others, asymmetric hydrogenation, hydroformylation, and the Monsanto acetic acid or the Cativa process [1].

Rh and Ir complexes bearing at least two N-heterocyclic carbene (NHC) moieties within a tridentate ligand framework, mostly in form of a pincer ligand, are the subject of this chapter. Special properties of each ligand type, its complexes, and catalytic examples will be discussed.

NHCs have proved to be valuable ligands for metal complexes in catalysis [2]. Despite their electronic similarity to phosphine ligands, they are in general the stronger ligands, due to their stronger σ-donor and a weaker π-acceptor character [3]. The second-generation Grubbs olefin metathesis catalyst is perhaps the most prominent example for an enhancement in catalytic activity by substitution of one phosphine with one NHC ligand [4]. Considering to the catalytic cycle that starts with a dissociation step of a phosphine ligand, it becomes clear that substitution of also this ligand by a second NHC ligand leads to a strong decrease in activity. In addition, the different steric properties of phosphines (cone-shaped coordination sphere) and NHCs (flat, fence-like coordination geometry) contribute to the higher activity of this the second generation Grubbs catalyst [5].

In general, introduction of more than one NHC ligand into metal complexes will have the following effects (positive and/or negative) on their catalytic activity:

- The additional electron-donating character stabilizes the Lewis acidic metal atom.
- The resulting higher electron density at the metal is favorable for oxidative addition steps and increases the nucleophilicity of the metal center.
- In reverse, the higher electron density stabilizes higher oxidation states and therefore retards reductive eliminations.

Molecular Catalysts: Structure and Functional Design, First Edition. Edited by Lutz H. Gade and Peter Hofmann.

- The stronger binding of NHCs can result in blocking of substrate binding if an NHC ligand dissociation step is part of the catalytic cycle.
- Substrates with a strong metal-to-ligand π-backbonding (e.g., CO, phosphines, olefins) will bind more strongly which could affect the further reactivity of the substrate.
- Substrates with a strong σ-donor character (e.g., alkyl, hydride) will be less strongly bound which could enhance their reactivity.
- Steric effects are harder to predict and can be both positive and negative. They strongly depend on the N-substituents of the NHC ligand, but also on the steric properties of the substrate itself.

In an octahedral coordination geometry, a tridentate ligand can be either meridionally or facially coordinating. Here we will focus mostly on meridionally coordinating ligands, the so-called pincer ligands.

The first pincer ligand was described by Shaw in 1976 [6]. Owing to the chelate effect, these ligands can form comparatively very stable complexes. Pincer ligands consist of three electron donating moieties. The central moiety is referred to as *backbone*. Each donating moiety can be charged or neutral. In most cases, the negative charge is centered at the backbone, for example, as a phenyl or amide moiety. For practical use, one way to abbreviate pincer ligands is to write only the three donating atoms in the same order as they appear in the ligand, for example, CCC or PNP for complexes with a C or N donor in the backbone and C and P donors in the lateral moiety. The overall charge of the ligand is omitted in this notation.

Many interesting reactions and catalytic applications are known especially for Pd and Ru pincer complexes [7]. Ir and Rh pincer complexes are less common, but for some catalytic applications that require high temperatures, such as alkane dehydrogenation, pincer complexes seem to be irreplaceable [8].

9.2 Ligand Systems

The first pincer ligand bearing NHC moieties **1** was presented by Peris and Crabtree in 2001 [9] (Figure 9.1). The bisimidazolium salt precursor of this neutral CNC ligand with a pyridine backbone **1** can be synthesized by a nucleophilic ipso substitution of 2,6-dichloro- or dibromopyridine. With this strategy, also

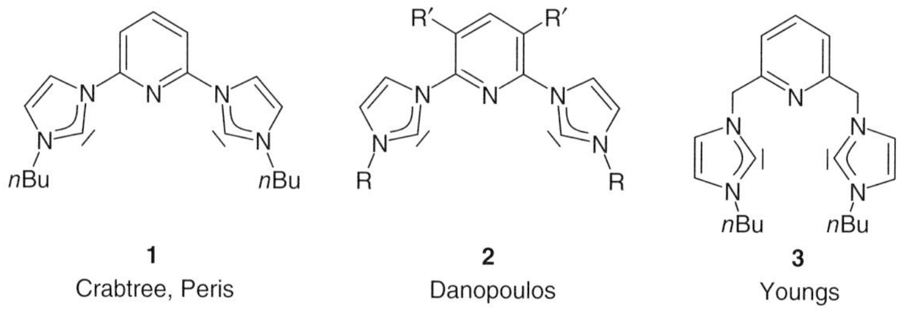

Figure 9.1 Neutral CNC pincer ligands bearing two NHC moieties.

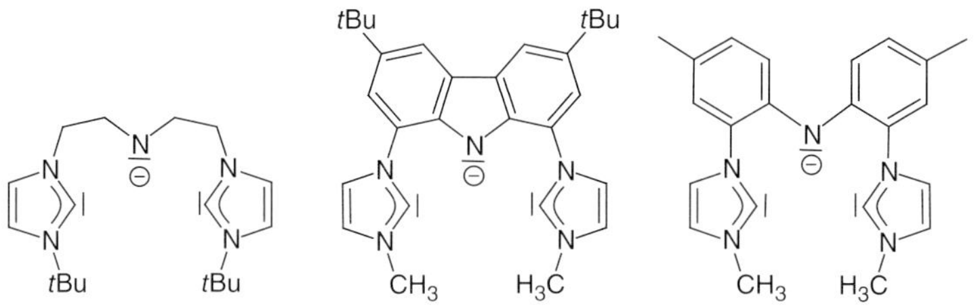

4 Douthwaite

5 bimca Kunz

6 Luo

Figure 9.2 Monoanionic CNC pincer complexes. Rh and Ir complexes haven been synthesized so far only with ligand **5**.

N-aryl-substituted imidazols can be introduced. Danopoulos [10] modified this ligand using 3,5-dimethylpyridine as backbone (R′ = CH_3), most likely to achieve a higher solubility with ligand **2**. The introduction of methylene groups between the pyridine and carbene moiety by Youngs [11] leads to a higher flexibility of the donor groups in **3**.

The first monoanionic pincer ligand with carbene moieties **4** was reported by Douthwaite *et al.* [12], who synthesized Pd complexes with a ligand consisting of two NHC moieties that are connected to a diethyleneamide backbone (Figure 9.2); however, neither Rh nor Ir complexes have been prepared so far.

We introduced bis(NHC) pincer ligands with carbazole backbone **5** to obtain a stable pyrrolid-like moiety and a full sp^2-hybridized arrangement [13]. The key step of the ligand synthesis is a Cu-catalyzed aminoarylation of diiodocarbazole. Subsequent N-alkylation of the imidazole moieties allows varying the N-substituents easily. However, aryl substituents cannot be introduced this way.

A more flexible ligand **6** was achieved by substitution of the carbazolide moiety with diphenylamide, reported by Luo in 2008 [14]. So far, neither Rh nor Ir complexes have been reported.

The monoanionic CCC pincer ligand 7 was introduced by Hollis [15] (Figure 9.3). Metallation of the ligand precursor was achieved by deprotonation with tetraamido zirconium(IV) complexes. Modifications were done by Chianese [16] who used stronger donating benzimidazolin-2-ylidene moieties in **8**, and by

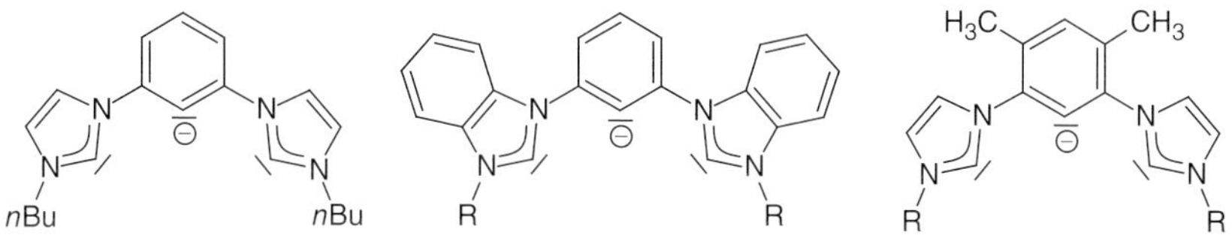

7 Hollis

8 Chianese

9 Braunstein

Figure 9.3 Examples for monoanionic CCC ligands containing two NHC moieties, first developed by Hollis.

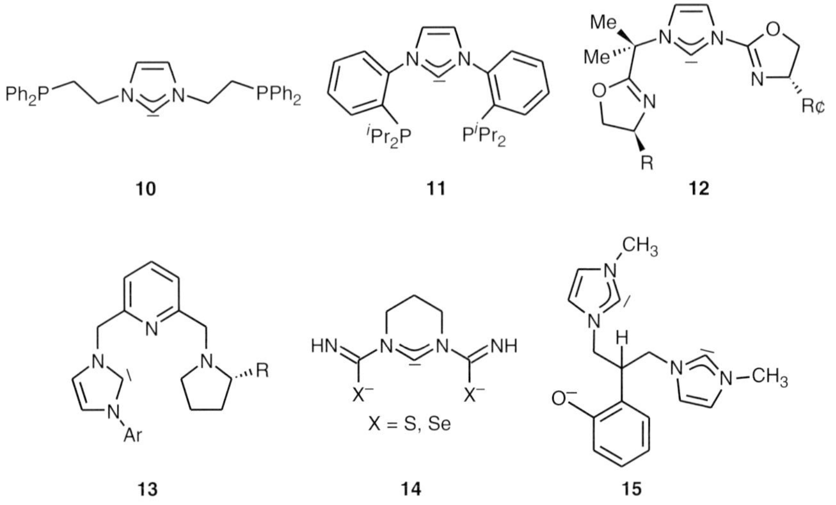

Figure 9.4 Selected pincer ligands with one carbene moiety and tripodal ligands.

Braunstein [17] who probed a 3,5-dimethyl-substituted backbone **9** to increase the solubility of the complexes.

Although going beyond the scope of this chapter, neutral pincer ligands with only one carbene moiety serving as backbone are of interest for Rh and Ir catalysis. Prominent examples are the ethylene-linked PCP ligand **10** of Lee [18], which is capable of switching from a facial to a meridional chelating mode [19] (Figure 9.4), and the phenyl analog **11** by Fryzuk [20]. Other examples are the chiral NCN ligands **12** of Gade [21] and **13** of Iglesias and Sánchez [22], as well as the very first example of a carbene-containing pincer ligand **14** by Matsumura [23] who synthesized SCS and SeCSe ligand precursors by reacting the doubly deprotonated hexahydropyrimidine-2-thione or imidazolidine-2-thione with isothiocyanate or isoselenocyanate in the key step.

Although there are several examples of tripodal NHC ligands by Baker, Meyer, Arnold, and others [24], only very few were used later for coordination to rhodium or iridium, for example, the monoanionic ligand **15** by Peris [25].

9.3
Synthesis and Reactivity of the Complexes

9.3.1
Synthesis of M(I) Complexes

One of the most common routes to synthesize carbene complexes is by ligand exchange using the free carbene ligand which is *in situ* generated by deprotonation. Usually, the metal precursor should keep its oxidation state. However, only very few Rh(I) and Ir(I) complexes were prepared this way. Danopoulos [10, 26] used the neutral *in situ* generated CNC ligand **2** and reacted it with [M(alkene)$_2$(μ-Cl)]$_2$ to obtain the monomeric Rh(I) and Ir(I) pincer complexes **16** and **17** with an intact M–Cl bond. For a successful synthesis of the Ir complex, he had to protect the pyridine backbone with methyl groups in 3 and 5 positions to avoid C–H activation at those positions and formation of the binuclear Ir(III) complex **18** or a mixed Ir(I)/Ir(III) complex.

The chlorido ligand of the Ir complex **16** can be exchanged by a hydrido ligand upon reaction with NaO*i*Pr, leading to complex **19**, or by neutral ligands L like pyridine, acetonitrile, ethylene, and CO upon chloride abstraction with KPF_6 or $NaBAr^F{}_4$ [10]. The CO stretching frequency of $\nu = 1980\,cm^{-1}$ in complex **20** indicates that the metal fragment possesses a weaker π-donor character than the analogous diphosphine pincer complex [Ir(P–N–P)(CO)] ($\nu = 1962\,cm^{-1}$) [27]. The respective Rh complex **21** (X = Cl) can be obtained without addition of halide abstractors [26].

The ease of oxidative addition reactions for the Rh(I)Cl complex **16** has been proven by the reaction with dichloromethane to yield complex **22** while the respective Rh(I)CO complex **21** does not show this reactivity.

A similar route was used by us to synthesize pincer complexes with the monoanionic CNC ligand **5** (bimca, 1,8-bis(imidazolin-2-yliden-1-yl)carbazolide). Three equivalents of the base (lithium diisopropylamide (LDA), MeLi) are necessary to obtain the *in situ* formed Li(bimca) complex **23** (Scheme 9.2). Salt metathesis and ligand exchange with [Rh(CO)$_2$(μ-Cl)]$_2$ afforded the Rh(CO)(bimca) complex **24** in good yields.

The respective Ir complex **25** could not be synthesized in pure form using [Ir(CO)(PPh$_3$)$_2$Cl] but it was possible with [Ir(CO)$_2$(acac)] (acac, acetylacetonate) as metal precursor [28a].

The very low CO stretching frequencies for **24** (1921 cm^{-1} (THF, tetrahydrofuran), 1916 cm^{-1} (KBr)) and **25** (1916 cm^{-1} (KBr)) are indicative of a strong π-backbonding character of the metal fragment to the CO ligand caused by the NHC moieties. In the analogous NNN-carbazole-based pincer complex **26** bearing imino donors, the stretching frequency is about 60 cm^{-1} higher (1980 cm^{-1}, KBr) [29]. This value lies in the same range as the values of the cationic complexes of Danopoulos [10, 26] bearing neutral CNC pincer ligands (1980 cm^{-1} (**20**) and 1968 (**21**)).

No CO exchange is observed for complex **24** in a ^{13}CO atmosphere, indicating the strong π-backbonding character of the metal center. In contrast to the cationic complex **21**, which was found to be unreactive (Scheme 9.1) under these conditions [26], the Rh complex **24** reacts with dichloromethane-d_2 to the oxidative addition product **27** within 2 h at 60 °C and slowly at ambient temperature (Scheme 9.2).

The iridium analog **25** shows identical reactivity in forming the dichloromethane activation product **29** as well as the η^1-allyl Ir complex **30** (A. Seyboldt, unpublished results) [28b].

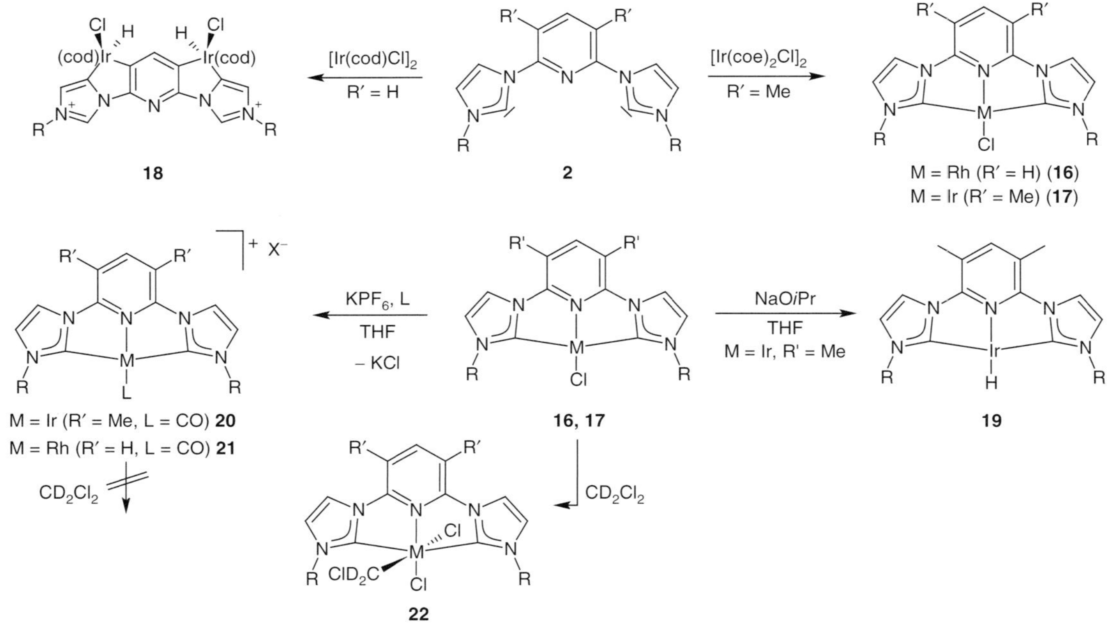

Scheme 9.1 Preparation and reactivity of Rh(I) and Ir(I) complexes bearing the pincer ligand **2**. (According to Danopoulos [10, 26].)

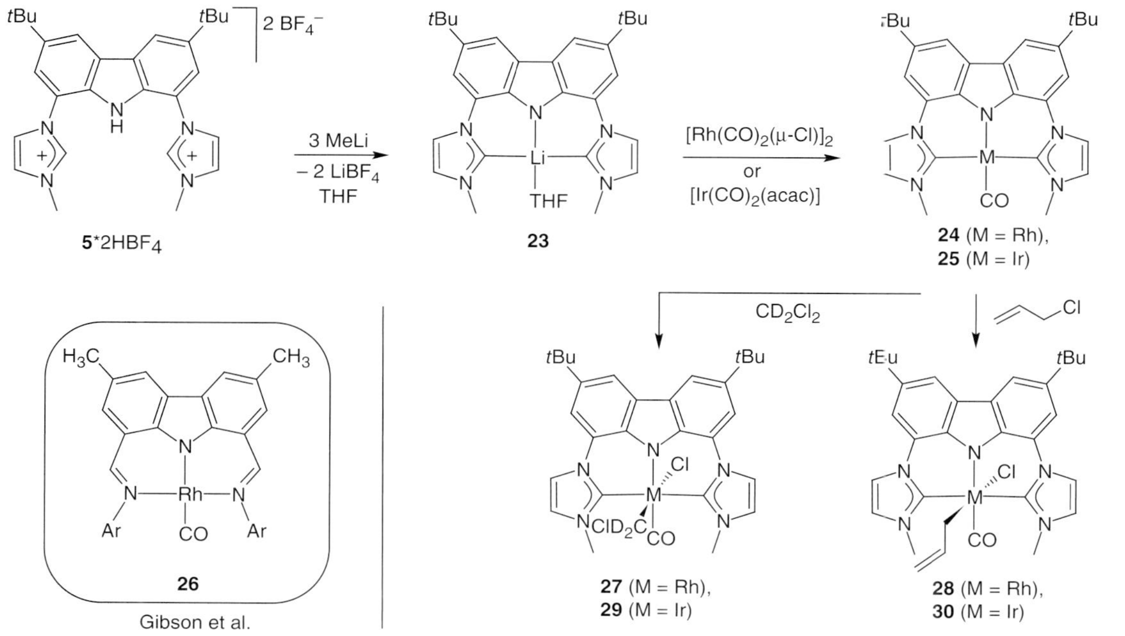

Scheme 9.2 Synthesis of the carbazole-based pincer complexes **24** and **25** (According to Kunz [13].), and their reactivity toward dichloromethane and allylchlorides.

9.3.2
Synthesis of M(III) Complexes

All other reported syntheses of pincer group 9 complexes bearing two NHC donor moieties involve the oxidation of the metal center. The first Rh pincer complex with two NHC donor moieties was reported by Peris and coworkers [30]. They reacted the imidazolium salt **1***2HBr [9] with $[Rh(COD)(\mu\text{-}Cl)]_2$ (COD, cyclooctadiene) and KBr in presence of triethylamine at 80 °C to obtain the Rh(III) tribromido complex **32**. Lower temperatures led to the dimetallic species **31**, which could be converted to the pincer complex **32** by heating and addition of KBr [30] (Scheme 9.3).

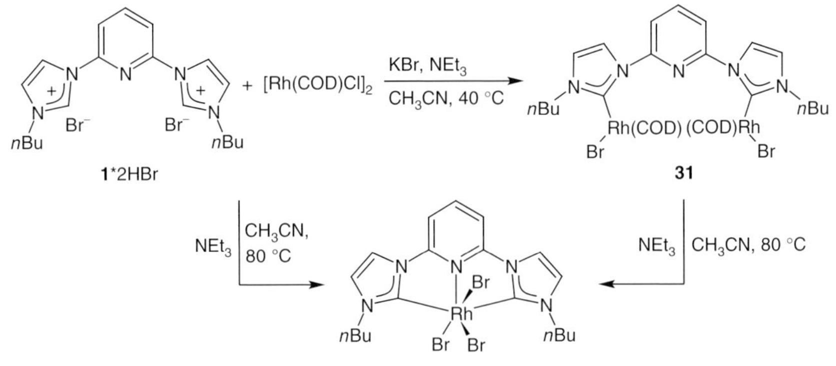

Scheme 9.3 Synthesis of Rh(III) pincer complexes bearing the neutral CNC ligand **1**. (According to Peris [30])

Youngs and coworkers used variation **3** of this ligand containing additional methylene groups between the NHC and pyridine moiety. They tried to synthesize a Rh complex by transmetallation from the dinuclear silver complex **33**. However, only the binuclear Rh complex **34** was obtained [11] (Scheme 9.4).

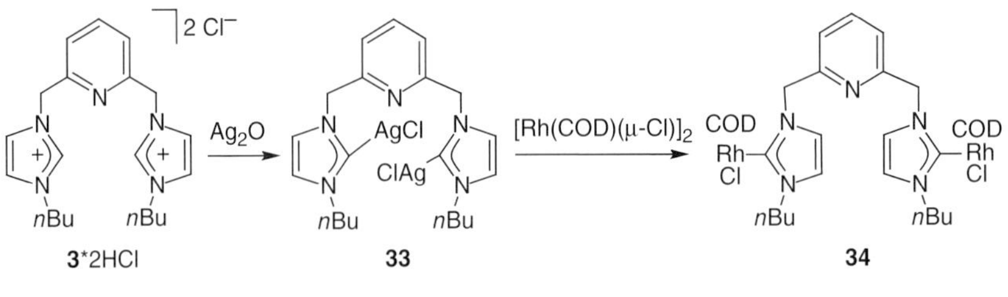

Scheme 9.4 The more flexible ligand system **3** leading only to formation of the dimetallic complex **34**.

Rh complexes with the monoanionic CCC pincer ligand **7** were first synthesized by Hollis *et al.* [15, 31] in a rather unusual way. They first synthesized the Zr complex **35** by amine elimination from $Zr(NMe_2)_4$ and subsequently transmetallated the ligand to $[Rh(COD)(\mu\text{-}Cl)]_2$ to obtain the Rh(III) complex **36**. Upon crystallization, the amine ligand was released and the iodo-bridged dimer **37** was formed. In the case of $[Ir(COD)(\mu\text{-}Cl)]_2$, the monometallic complex **38** was not observed but only dimer **39**.

Braunstein and coworkers reported on the synthesis of Ir complexes containing also this ligand type. Synthesis was not accomplished by transmetallation but with the use of a tertiary amine base (Scheme 9.5) [33]. The initially formed iodo-bridged Ir(III) dimer **39** could dissociate and be coordinated by solvents (acetonitrile, dimethylsulfoxide (DMSO)) to give complexes **38** upon heating or ultrasound activation. However, when they used a pincer ligand containing *N*-adamantyl substituents, one NHC moiety got bound via C5 to the metal center for steric reasons, thus forming the so-called abnormal carbene complex **40** (Scheme 9.6) [34]. To obtain better solubility, they introduced 3,5-dimethyl groups to the phenyl backbone and synthesized complex **41** bearing ligand **9** [17]. Complex **40** was transformed *in situ* into the proposed dihydrido complex **42** by addition of base at 1 atm of hydrogen. Gaining more flexibility by introducing methylene groups between NHC and backbone only led to the formation of dinuclear complexes with each of the Ir(I)(COD) fragments coordinated by one NHC moiety [35].

Heinekey and Goldberg [36] managed to synthesize different pincer Ir(III) complexes of ligand **43** by first generating a silver complex **44** upon reaction of the bisimidazolium salt ^{Br}**43***2HBr containing a 1-bromophenyl backbone with equimolar amounts of Ag_2O and adding the filtered solution to $[IrCl(C_2H_4)_4]$ prepared *in situ* by bubbling ethylene through a solution of $[Ir(\mu\text{-}Cl)(COE)_2]_2$ (Scheme 9.7). The reaction included oxidative addition of the bromophenyl moiety to obtain the desired Ir(III) complex **45** containing a mixture of the halide ligands (X = Cl, Br). Dehalogenation with sodium amalgam under hydrogen atmosphere led to the Ir(III) species **46**. The dihydrogen ligand could be substituted by pyridine or trimethylphosphine. Density functional theory (DFT) calculations and the experimental data indicate a thermodynamic preference for the dihydrido-dihydrogen binding mode of **46** (Ir(III)) over an Ir(V) tetrahydrido species, in contrast to the respective PCP pincer complex.

The monoanionic CCC ligand type was also used by Chianese and coworkers to prepare the Ir complexes **47a–f** with benzimidazolin-2-ylidene moieties to prevent the formation of abnormal carbenes. They treated the imidazolium salt **8***2HBr with 30 equiv of triethylamine as base and 0.5 equiv of $[(\mu\text{-}Cl)Ir(COD)]_2$ in acetonitrile at 80 °C to obtain the Ir(III) complexes **47** (Scheme 9.8). However, a clean reaction was obtained only in the case of **47** bearing *N*-mesityl (**a**) or *N*-2,6-di-isopropylphenyl (**b**) substituents at the NHC moiety. With *N*-3,5-dimethylphenyl (**c**), *N*-3,5-di-*tert*-butylphenyl (**d**), *N*-*tert*-butyl (**e**), and *N*-1-adamantyl (**f**) substituents, they had to use a slight excess of CsF as base to obtain the respective Ir(III) complexes **47** [16]. For complex **47b**, the group had to raise the temperature in an autoclave up to 150 °C [16b].

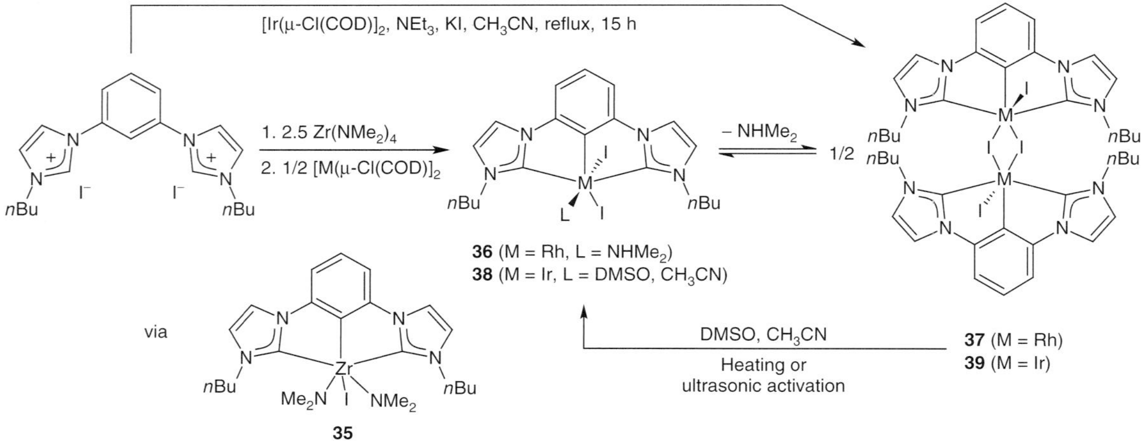

Scheme 9.5 Synthesis of monometallic complex **36** and the dimers **37** and **39** bearing a monoanionic CCC pincer ligand (According to Hollis [15, 31, 32].) and the preparation of iridium complex **38**. (According to Braunstein [33].)

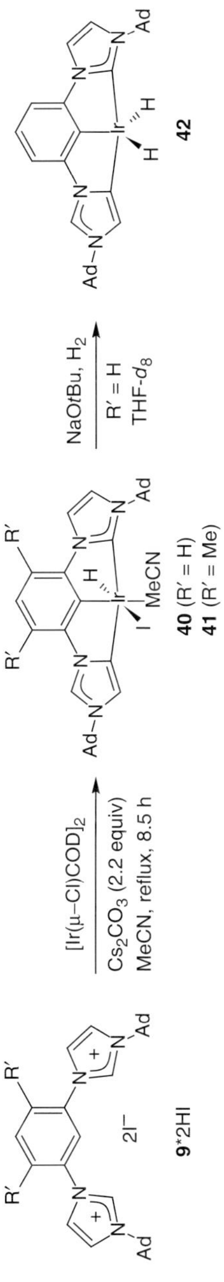

Scheme 9.6 Formation of abnormal carbene pincer complexes **40–42** due to steric reasons.

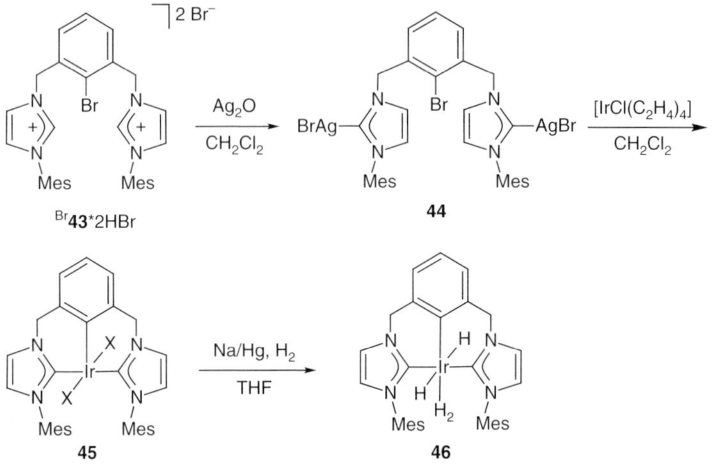

Scheme 9.7 Formation of flexible CCC pincer Ir(III) halide complexes and transformation to the dihydrido dihydrogen complex **46**. (According to Heinekey and Goldberg [36].)

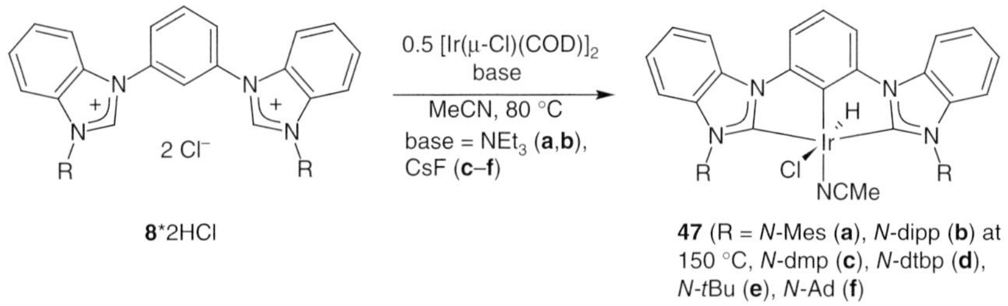

Scheme 9.8 Synthesis of CCC pincer Ir(III) complexes **47a–f** bearing benzimidazolinylidene moieties with various N-substituents. (According to Chianese [16].)

9.4
Catalytic Activities of the Rh Complexes

Although pincer complexes are in general ubiquitous in catalysis [7, 10], the reactions evaluated for Rh and Ir pincer complexes are still quite limited. Therefore, the few catalytic examples of those pincer complexes bearing two NHC moieties show the potential of these complexes rather than their limits, as the catalytic applications are highly diverse.

One key reaction for Rh and Ir (and Ru) catalysts is the transfer hydrogenation of ketones. This reaction, whose equilibrium is shifted to the product side by an excess of the alcohol that is used as hydrogen source (as solvent), is usually performed under basic conditions to ease binding of the alcohol (as alcoholate) to the complex. Peris and coworkers [30] showed that the Rh(III) complex **32** is a suitable catalyst precursor for this reaction (Figure 9.5). They performed this reaction at very low catalyst loading down to 0.006 mol% in a 0.1 M KOH ^{i}PrOH

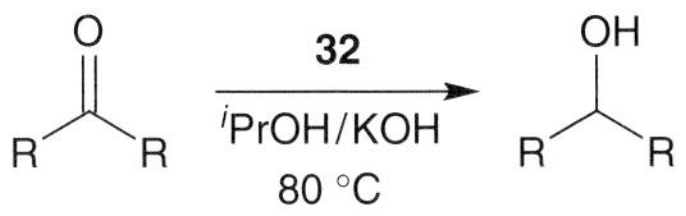

Substrate	Time (h)	Catalyst (mol%)	TON	TOF
Cyclohexanone	7	0.06	1322	189
Cyclohexanone	24	0.06	1322	55
Acetophenone	7	0.06	500	71
Acetophenone	24	0.06	1107	46
Acetophenone	6	0.006	3500	583
Benzophenone	7	0.06	1357	194
Benzophenone	24	0.06	1357	57
Benzophenone	7	0.006	6450	921

Figure 9.5 Results of Peris and coworkers for the transfer hydrogenation of ketones with complex **32** [30].

solution at 80 °C. For cyclohexanone and benzophenone, the turnover frequencies (TOFs) were identical after 7 and 24 h of reaction time. For acetophenone, they observed a still-unexplained activation period. Double bonds were not hydrogenated by this catalyst. The catalytic activity was similar to that found for other Rh(III) biscarbene complexes. While the group reported non-pincer Rh(I) species for hydrosilylation reactions in the same publication, they did not test complex **32** for this reaction.

However, Hollis showed that the Rh(III) complex **36** bearing a monoanionic pincer ligand was a suitable catalyst precursor for the hydrosilylation of acetylenes (Figure 9.6). At 80 °C, full conversion was monitored after 1 h for most internal and terminal acetylenes. The reaction proceeded much more slowly at room temperature (87% conversion, 12 h). The kinetic *Z*-product was preferred for terminal alkynes and the ratio of ~10 : 90 (E : Z) did not change over the reaction time, while sterically demanding groups or internal alkynes led to the thermodynamically favored *E*-product (~85 : 15). The catalyst precursor **36** has to be very pure, as additives can have a dramatic influence on the selectivity of this reaction. Addition of 1 equiv of LiCl or $(n\text{-Bu})_4NBr$ leads predominantly to the formation of the *E*-product together with an increase of the α-hydrosilylated acetylenes, whose content can be further increased by the addition of water.

The chemoselectivity was tested in a competition experiment containing dimethylphenylsilane, phenylacetylene, and styrene in a 1.1 : 1 : 1 ratio. The typical product distribution was found for the reaction with phenylacetylene, while styrene reacted slowly leading to 12% ethylbenzene. In comparison, the dinuclear Rh(I) NHC complex **37** led to a slight decrease in selectivity but comparable reactivities, and no ethylbenzene was found with this complex in the competition experiment. The results show that – in contrast to the findings by Peris – pincer Rh complexes are also be suitable catalyst precursors for the hydrogenation of C=C bonds. The results are consistent with the proposed mechanism for Rh catalysts by Ojima [37] that neutral Rh(I) complexes lead preferentially to *Z*-products while cationic Rh complexes lead to the *E*-isomer [38].

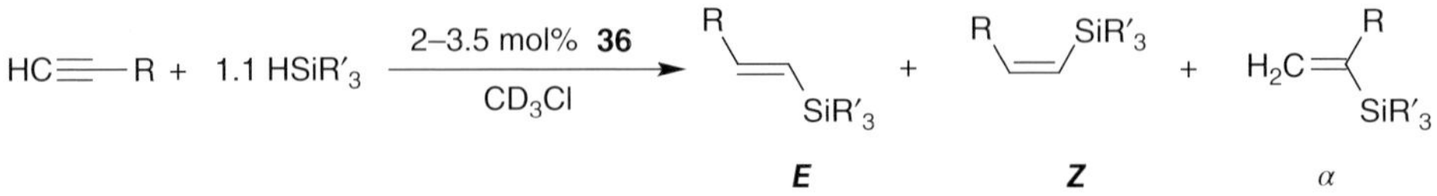

Alkyne	Temperature (°C)	Time (h)	Conversion (%)	*E*	*Z*	*α*
PhC≡CH	80	2	100	5	95	0
PhC≡CH	rt	12	87	5	95	0
PrC≡CH	80	1	100	8	89	3
CPh≡CPh	60	12	100	89	11	0

Figure 9.6 Hydrosilylation of terminal and internal acetylenes with the CCC pincer complex **36**. (According to Hollis [31].)

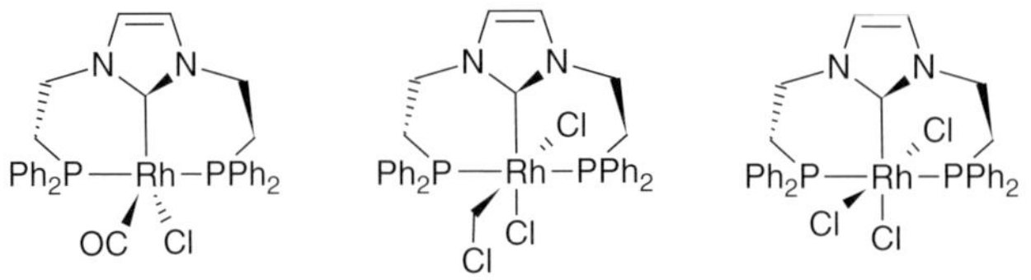

48 **49** **50**

Figure 9.7 Lee's PCP pincer complexes **48–50** that are catalytically active in hydrosilylation reactions [39].

Lee *et al.* [39] investigated various PCP-pincer complexes **48–50** of Rh(I) and Rh(III) bearing an NHC backbone as catalysts in the hydrosilylation of alkynes (Figure 9.7). The high *E*-selectivity of the hydrosilylation products indicates a cationic Rh species as active catalyst. The reaction is much faster than in the case of complex **36**. Already after 15 min at 60 °C in chloroform, quantitative yield was observed (Figure 9.8). The *E/Z* ratio was constant for 6 h and increased after this time slowly, while the amount of the α-addition product (~15%) stayed constant. An excess of the hydrosilane also favored the *E*-selectivity. The addition of NaI to suppress formation of a cationic complex did not change the *E/Z* ratio. Therefore, the authors attributed the high *E*-selectivity to the presence of an isomerization pathway. The *E*-selectivity depends also on the bulkiness of the hydrosilane. All three complexes **48–50** show similar selectivities and activities.

The dimeric pincer Rh and Ir complexes **37** and **39** reported later by Hollis [32] show catalytic activity in the intramolecular hydroamination of secondary aminoalkenes bearing geminal substituents in β-position to take advantage of the Thorpe–Ingold effect (Figure 9.9). The exo-trig cyclization yielded pyrrolidines and piperidines in very high yields within 16 h. Only small traces of isomerization products were observed. In the absence of β-substituents or using primary alkylamines, no hydroamination reaction was observed and the isomerization product was found exclusively. The reaction could be carried out in benzene or water

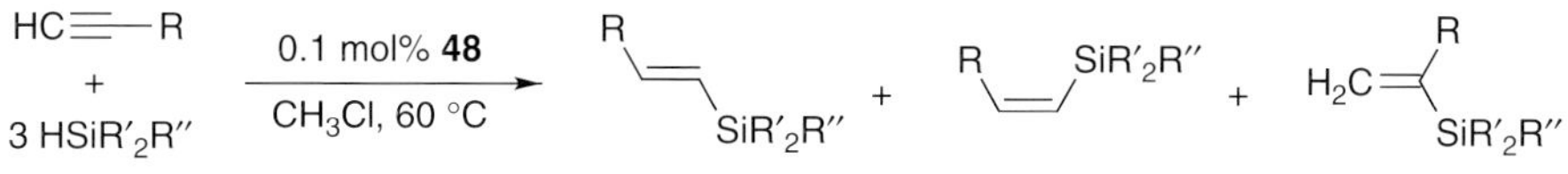

Time	*E*	*Z*	*α*
15 min	60	23	17
30 min	56	24	20
6 h	54	21	25
12 h	83	0	17
24 h	88	21	14

Figure 9.8 Results of hydrosilylation reactions with the PCP Rh(I) complex **48**.

under air without loss of activity. Control experiments with I_2 showed the cyclized iodination product, while without addition of catalyst no reaction occurred. The mechanism is consistent with an M–N bond formation, olefin coordination, and migratory insertion followed by either reductive elimination or π-coordination with amine attack [40].

The Rh(I) complexes **21** prepared by Danopoulos and **24** prepared by us are quite electron rich, as displayed by the relatively low (**21**) to very low (**24**) CO stretching frequencies due to strong metal–ligand backdonation. Therefore, they should have a strong tendency for oxidative addition reactions. Although Danopoulos found an easy exchange of the chlorido ligand in complex **16** by CO and formation of the cationic CO complex **21**, complex **16** does not seem suitable for hydrogen activation and thus does not catalyze the hydroformylation of olefins.

Complex **24** also does not react to a Rh(III) complex with hydrogen [13a] and is not active in the hydroformylation. Reaction with allyl chloride leads to the oxidative addition product **28**. Although the CO ligand is much more weakly bound in the Rh(III) complex and can be exchanged at room temperature, an η^3-coordination mode of the allyl ligand upon dissociation of CO is not observed. Interestingly, reaction of 3-chloro-1-butene as well as 1-chloro-2-butene leads to the identical product **51**, although an allylic S_N2' mechanism should lead to complex **52** bearing an allyl ligand with a terminal double bond (Figure 9.10).

In an exchange experiment, it was shown that the allyl ligand, bearing a terminal double bond, could be transferred from the Rh(III) complex **28** to another Rh(I) complex **24** (bearing different N-substituents) in a fast equilibrium reaction.

On the basis of these observations, complex **24** was tested as catalyst for the allylic alkylation of allyl carbonates. In general, Rh complexes led preferentially to the ipso alkylation products [41] while Ir complexes tended to produce the branched and Pd catalysts the linear as the major alkylation product [42]. The catalytic results of complex **24** showed a 6 : 1 ratio for the ipso product in case of

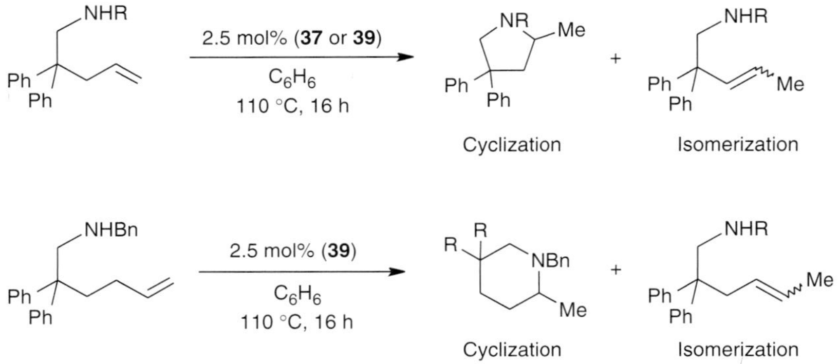

	Catalyst (Cyclization/Isomerization)	
R	**37 (Rh)**	**39 (Ir)**
Bn	>90/0	>98/0
n-Pr	75/<5	95/<5
Ph	—	>98/<2
H	0/75	0/50

	Catalyst (Cyclization/Isomerization)
R	**39 (Ir)**
Ph	90/0
H	0/100

Figure 9.9 Intramolecular hydroamination of geminal-substituted aminoalkenes with complexes **37** and **39**. (According to Hollis [32].)

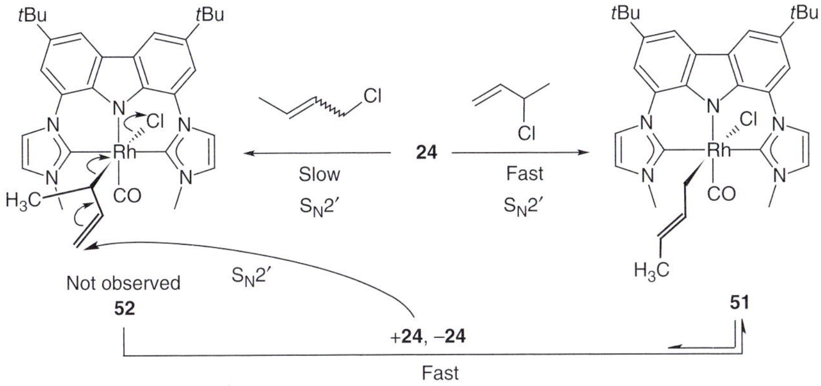

Figure 9.10 Catalytic isomerization of the proposed complex **52** to complex **51** by allyl transfer. (According to Kunz [13b].)

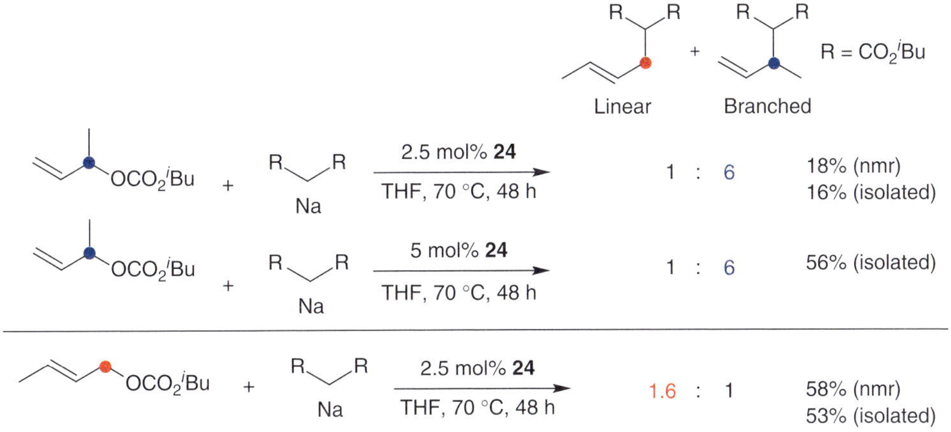

Figure 9.11 Allylic alkylation of allylcarbonates catalyzed by complex **24**. In each example, the respective ipso product is dominating, but in case of the linear carbonate a substantial isomerization occurred [13b].

a branched allylcarbonate (Figure 9.11) but only a 1.6 : 1 ratio of the ipso product if the reaction was carried out with the linear carbonate.

In case of the linear carbonate, an S_N2' reaction would lead to the Rh(III) intermediate **52** bearing the allyl ligand with a terminal double bond. This complex will undergo allyl transfer much faster than the Rh(III) allyl intermediate **51** with the thermodynamically more stable internal double bond. As the oxidative addition is the rate-limiting step in this reaction, a certain amount of Rh(I) complex **24** is present in the reaction mixture to catalyze the isomerization by allyl transfer, leading to a lower ipso-selectivity with linear allyl carbonates.

9.5 Catalytic Activities of the Ir Complexes

In contrast to the various applications of the pincer Rh complexes bearing two NHC moieties, the respective Ir complexes seem to be of interest only for C–H activation reactions so far.

In 2010, Chianese [16a] reported the catalytic activity of complexes **47** in the dehydrogenation of cyclooctene with both norbornene (NBE) and *tert*-butylethylene (TBE) as hydrogen acceptor as well as without acceptor. He tested various *N*-aryl substituents at the benzimidazolin-ylidene moieties and found that with the *N*-mesityl substituent (**47a**) he could achieve a turnover number (TON) of 10 at 150 °C for 24 h when using NBE as hydrogen acceptor (Figure 9.12). The complexes with *N*-3,5-dimethylphenyl (**47c**) and *N*-3,5-di-*tert*-butylphenyl (**47d**) substituents showed no catalytic activity under these conditions. The TONs were much smaller when using TBE as the hydrogen acceptor. In the acceptorless dehydrogenation of cyclooctene, which was performed under an argon stream and reflux for 20 h in cyclooctane, the *N*-mesitylene-substituted catalyst **47a** gave a TON of 68 and the 3,5-di-*tert*-butylphenyl-substituted complex **47d** gave a TON of 26 while complex **47c** showed no catalytic activity.

The same complexes were also tested in the catalytic borylation of arenes at 100 °C for 24 h using B_2pin_2 as borylating agent and 4% NaO*t*Bu as additive to promote dehydrochlorination of the catalysts **47** and thus reduction from Ir(III) to Ir(I). In this borylation reaction, complex **47a** also gave the best results with a conversion of 57% based on B_2pin_2 (^{1}H NMR) and 44% yield (^{1}H NMR). The electron-poor 1,3-bis(trifluoromethyl)benzene as substrate led to the best conversion (94%) and yield (74%). The borylation occured selectively in the meta position (Figure 9.13).

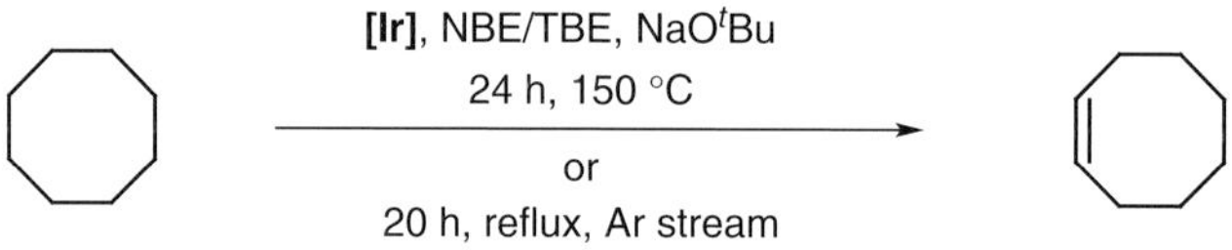

Catalyst	Acceptor	TON of cyclooctene
47a	NBE	10
47a	TBE	1
47c	NBE	0
47d	NBE	0
47d	TBE	2

Acceptorless

Catalyst	TON of cyclooctene
47a	68
47c	0
47d	26

1.0 mmol **[Ir]**, 2.0 mmol NaO^{t}Bu

Figure 9.12 Transfer dehydrogenation and acceptorless dehydrogenation of alkanes catalyzed by complexes **47**. (According to Chianese [16a].)

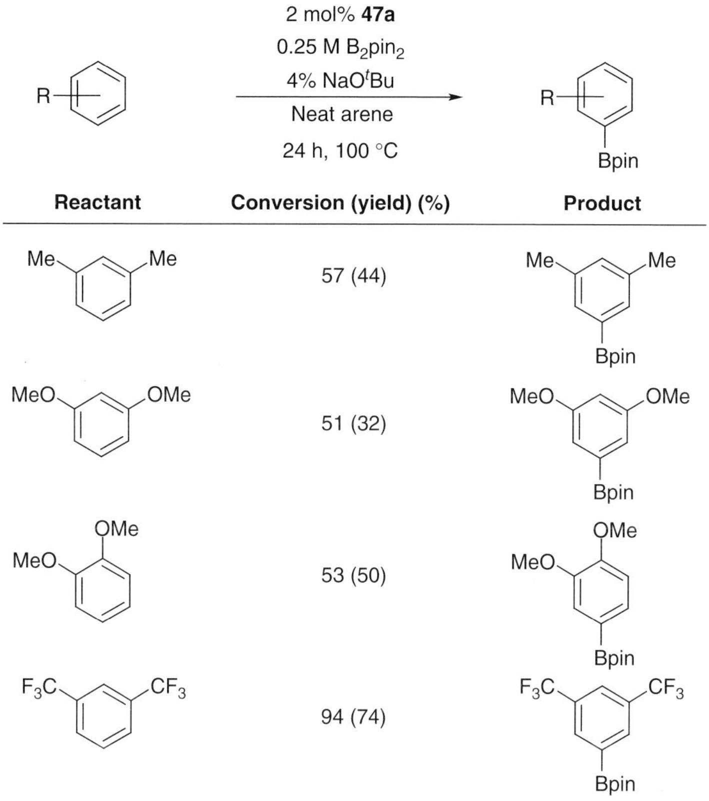

Figure 9.13 Borylation of arenes using complex **47a**. (According to Chianese [16a].)

The transfer dehydrogenation of *n*-octane was tested 2 years later including complexes bearing *N-tert*-butyl (**47e**) and *N*-adamantyl (**47f**) substituents. However, only complexes **47a** and **47d** showed catalytic activity in this reaction with small TONs of 12 and 10 under the same conditions used for the transfer dehydrogenation of cyclooctane [16b]. As already known from the reactivity of PCP Ir pincer complexes [43], Chianese observed only internal isomers of octene and therefore investigated the activity of complex **47a** in the isomerization of 1-hexene. Already after 15 min at 150 °C, 1-hexene was isomerized with a TON of 420 to a mixture of *trans*-2-hexene, *cis*-2-hexene, and 3-hexenes in a 67 : 29 : 4 ratio and after 60 min (TON 730) in a 65 : 26 : 8 ratio. Therefore the isomerization of terminal olefins is much faster than the transfer dehydrogenation. It was also concluded that the isomerization of terminal olefins is much faster than that of 2-hexenes to 3-hexenes. The isomerization of 1-octene was shown to proceed already at 100 °C with identical TON of almost 500 and nearly complete consumption of 1-octene after 24 h for **47a**, **47e**, and **47f**. In this case, the addition of NaO^tBu is required (Figure 9.14).

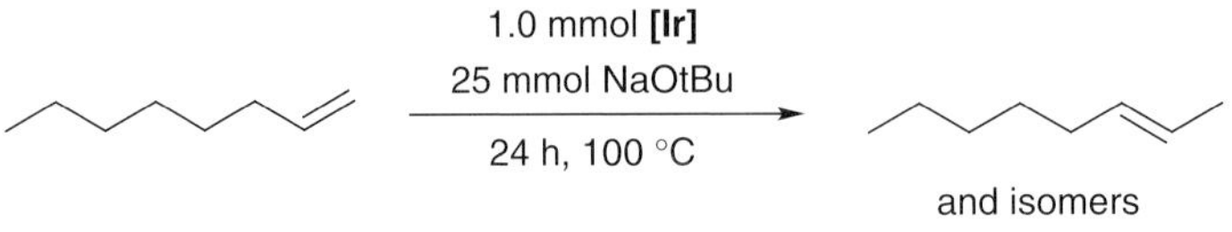

Catalyst	TON
47a	484
47e	487
47f	488

Figure 9.14 Isomerization of 1-*n*-octene with complexes **47**. (According to Chianese [16b].)

Braunstein [34b] reported a similar CCC pincer complex **38** bearing *N-n*-butylimidazolin-2-ylidene moieties, but the complex showed no activity in the catalytic dehydrogenation of cyclooctane. Upon change of the N-substituents to 1-adamantyl groups, the abnormal carbene pincer complex **40** was formed as a result of steric effects. This complex was converted to the dihydrogen complex **42** by adding NaO^{t}Bu and dihydrogen (Scheme 9.6) and tested in the transfer dehydrogenation of cyclooctane using TBE as hydrogen acceptor. In contrast to the results of Chianese, no activity was observed at 150 °C, but at 200 °C a TOF of 0.36 h^{-1} after 10 h was reported. The low activity of the catalyst compared to typical PCP-Ir systems [8], as well as to Chianese's CCC pincer complex **47a** [16], was supposed to stem from the low solubility of the complex in the nonpolar cyclooctane. Therefore, the group synthesized complex **41** bearing two additional methyl groups in 3 and 5 positions of the phenyl backbone; however, catalytic results have not been reported so far [17].

Goldberg and Heinekey [36] investigated the nature of complex **46** and its reactivity in detail. They reported that, upon removal of the H_2 atmosphere, an H/D exchange with benzene-d_6 under formation of **46**-d_4 was observed and, in case of toluene, the aromatic as well as the benzylic hydrogen atoms were exchanged except for those in the ortho position. The complex was not stable in alkanes and no H/D exchange was observed. Therefore, complex **46** could serve as a catalyst for H/D exchange for selective aromatic and benzylic hydrogen atoms.

Although all Ir catalysts presented so far contained Ir in its oxidation state +III, the active catalytic species is considered to be an Ir(I) complex formed *in situ*.

9.6 Discussion

Within the last years, a variety of Rh and Ir bis(NHC) pincer complexes could be synthesized. After publishing the first Rh bis(NHC) pincer complex **32**, Peris *et al.* showed that this complex class is highly valuable for catalytic applications and worth investigating. Although the activity of their complex in transfer hydrogenation of ketones is far less than that of the most active Ru PCP complexes [44], the activity was nevertheless comparable to that of other Rh NHC complexes at that time.

The Rh CCC pincer complex **36** tested by Hollis in the hydrosilylation of acetylenes shows a high activity and selectivity for the *Z*-product, which is also found for a bimetallic NHC complex bearing the same ligand in a bridging mode. Despite the fact that the selectivity does not yet match that of Tanke and Crabtree's [(triso)Ir(C_2H_4)$_2$] (triso, tris(diphenyloxosphosphoranyl)methanides) catalyst (in their case, the Rh analog was completely inactive) [37b], further optimization on the ligand system as well as on the complex itself could improve these initial results. Probably, the tedious synthesis of the ligand precursor, as well as the Rh complex itself, is a drawback for easy optimization studies. The ligand precursor synthesis allows only the introduction of alkyl groups as N-substituents so far. Synthesis of the respective monomeric Ir complexes has not been reported.

Similar reasons seem to count also for the promising catalytic experiments with the dimeric Rh and Ir complexes **37** and **39** which bear the same CCC pincer ligand in the intramolecular hydroamination of terminal olefins. Up to now, the air- and water-stable catalyst system is limited to the intramolecular reaction of β-disubstituted secondary amines.

With our Rh(CO) complex bearing the anionic CNC ligand **5**, we could show that Rh-catalyzed allyl transfer can be the source of isomerization during the allylic alkylation catalyzed by the same complex, especially when using linear allyl carbonates. The potential of this complex with regard to optimization of the catalyst, substrate scope, and asymmetric variations is not yet revealed, nor that of the respective Ir catalyst.

The first catalytic tests of Ir bis(NHC) pincer complexes **47** by Chianese *et al.* clearly showed a preference of the Ir complexes for C–H activation reactions. Even though the TONs for the acceptorless dehydrogenation of cyclooctane (TON 68, **4a**) are still lower than those reported for the known PCP pincer ligands [8] (TON 190), further optimization of the carbene pincer complex could improve the results. In his first study, Chianese concentrated on varying the N-substituents. However, other factors such as the carbene type could have an influence on the electronic structure of the catalyst. The observation that only NBE, but not TBE, is suitable as the acceptor olefin might be an indication that the olefin binds more strongly to the complex because of the better electron-donating properties of the NHC moieties. In turn, this could also lead to enhanced product inhibition. These findings are corroborated by the fact that complexes of type **47** are active in the isomerization of terminal alkenes to 2-alkenes after activation with sodium *tert*-butanolate already at 100 °C, while the PCP-Ir complexes need 150 °C to catalyze this reaction.

Danopoulos [10] reported the CNC pincer complex **17** and its transformation to the Ir-hydrido complex **19** upon reaction with NaO^{i}Pr, forming acetone as the oxidation product (Scheme 9.1). In light of this dehydrogenation reaction, complexes **17** and **19** could be potential catalysts for the transfer hydrogenation of ketones or other hydrogenation reactions.

9.7
Summary, Conclusion, and Outlook

In summary, we have shown that the use of pincer ligands bearing two NHC moieties in Rh and Ir catalysts is still at its beginning. One major drawback seems to be the laborious synthesis of the ligands and the complexes. In most cases, synthesis of the complex involves oxidation of the metal(+I) precursor leading to Rh(III) and Ir(III) complexes. Only under the conditions used by Danopoulos or by us do the metal centers keep their oxidation state of +I. The few examples of catalytic applications show already a high diversity and a large potential for the Rh complexes. Catalytic activities were reported for transfer hydrogenation of ketones, for regioselective hydrosilylation of acetylenes, and for the intramolecular hydroamination of aminoalkenes. The bis(NHC) pincer Rh(I) complex **24** is active in the regioselective allylic alkylation of allycarbonates and in the isomerization of terminal η^1-allyl complexes by allyl transfer. Although the pincer ligands do not seem to be suitable for hydroformylation catalysts, other reactions involving CO insertion should be tested. In addition it is important to take advantage of the electron richness in catalytic reactions involving oxidative addition steps.

Iridium complexes seem to be the catalysts of choice for the (transfer)dehydrogenation of alkanes. Other reactions involving a C–H activation step, such as H/D exchange of aromatic CH bonds or borylation of arenes, are also readily observed with the Ir complexes. No further catalytic applications with the Ir bis(NHC) pincer complexes have been reported so far, but, for example, transfer hydrogenations of ketones, CC coupling reactions, and the Cativa process for acetic acid synthesis are promising reactions. In addition, dehydrogenation or activation of X–H functionalities (X = C, N, B) could be another interesting option for further catalytic tests with Ir bis(NHC)-pincer complexes.

The application of Rh/Ir–bis(NHC) complexes has so far been limited to achiral reactions because of the lack of chiral pincer ligands. Improving the ligand synthesis also in this respect might accelerate the progress and success of catalytic applications with this ligand system. We are convinced that within the next few years the highly nucleophilic bis(NHC) pincer ligands will reveal their unique properties in Rh and Ir catalysis.

References

1. (a) Arpe, H.J. (2007) *Industrielle Organische Chemie*, Wiley-VCH Verlag GmbH, Weinheim; (b) Steinborn, D. (2007) *Grundlagen der Metallorganischen Komplexkatalyse*, Teubner, Wiesbaden; (c) Shimizu, H., Nagasaki, I., Matsumura, K., Sayo, N., and Saito, T. (2007) *Acc. Chem. Res.*, **40**, 1385–1393; (d) Sunley, G.J. and Watson, D.J. (2000) *Catal. Today*, **58**, 293–307.
2. (a) Regitz, M. (1996) *Angew. Chem.*, **108**, 791 *Angew. Chem. Int. Ed. Engl.* (1996), **35**, 725–728; (b) Arduengo, A.J. III, (1999) *Acc. Chem. Res.*, **32**, 913–921; (c) Bourissou, D., Guerret, O., Gabbaï, F.P., and Bertrand, G. (2000) *Chem. Rev.*, **100**, 39–91; (d) Herrmann, W.A. (2002) *Angew. Chem.*, **114**, 1342–1363; *Angew. Chem. Int. Ed.*, (2002), **41**, 1290–1309; (e) Hahn, F.E. and Jahnke, M.C. (2008) *Angew. Chem.*, **120**, 3166–3216; *Angew. Chem. Int. Ed.*, (2008), **47**, 3122–3172.
3. (a) Öfele, K., Herrmann, W.A., Mihalios, D., Elison, M., Herdtweck, E., Scherer, W., and Mink, J. (1993) *J. Organomet. Chem.*, **459**, 177–184; (b) Herrmann, W.A., Öfele, K., Elison, M., Kühn, F.E., and Roesky, P.W. (1994) *J. Organomet. Chem.*, **480**, C7–C9.
4. (a) Weskamp, T., Kohl, F.J., Hieringer, W., Gleich, D., and Herrmann, W.A. (1999) *Angew. Chem.*, **111**, 2573–2576;

Angew. Chem. Int. Ed. (1999), **38**, 2416–2419; (b) Bielawski, C. and Grubbs, R.H. (2000) *Angew. Chem.* **112**, 3025–3028; *Angew. Chem. Int. Ed.*, (2000), **39**, 2903–2906.

5. Straub, B.F. (2005) *Angew. Chem.*, **117**, 6129–6132; *Angew. Chem. Int. Ed.*, (2005), **44**, 5974–5975.
6. Moulton, C.J. and Shaw, B.L. (1976) *J. Chem. Soc., Dalton Trans.*, 1020–1024.
7. (a) Albrecht, M. and van Koten, G. (2001) *Angew. Chem.*, **113**, 3866–3898; *Angew. Chem. Int. Ed.*, (2001), **40**, 3750–3781; (b) van der Boom, M.E. and Milstein, D. (2003) *Chem. Rev.*, **103**, 1759–1792.
8. (a) Gupta, M., Hagen, C., Flesher, R.J., Kaska, W.C., and Jensen, C.M. (1996) *Chem. Commun.*, 2083–2084; (b) Gupta, M., Hagen, C., Kaska, W.C., Cramer, R.E., and Jensen, C.M. (1997) *J. Am. Chem. Soc.*, **119**, 840–841; (c) Jensen, C.M. (1999) *Chem. Commun.*, 2443–2449; (d) Haenel, M.W., Oevers, S., Angermund, K., Kaska, W.C., Fan, H.J., and Hall, M.B. (2001) *Angew. Chem.*, **113**, 3708–3712; *Angew. Chem. Int. Ed.* (2001), **40**, 3596–3600; (e) Morales-Morales, D., Redon, R., Yung, C., and Jensen, C.M. (2004) *Inorg. Chim. Acta*, **357**, 2953–2956; (f) Göttker-Schnetmann, I., White, P.S., and Brookhart, M. (2004) *J. Am. Chem. Soc.*, **126**, 1804–1811; (g) Kuklin, S.A., Sheloumov, A.M., Dolgushin, F.M., Ezernitkaya, M.G., Peregudov, A.S., Petrovskii, P.V., and Koridze, A.A. (2006) *Organometallics*, **25**, 5466–5476.
9. Peris, E., Loch, J.A., Mata, J., and Crabtree, R.H. (2001) *Chem. Commun.*, 201–202.
10. Danopoulos, A.A., Pugh, D., and Wright, J.A. (2008) *Angew. Chem.*, **120**, 9911–9913; *Angew. Chem. Int. Ed.*, (2008), **47**, 9765–9767.
11. Simons, R.S., Custer, P., Tessier, C.A., and Youngs, W.J. (2003) *Organometallics*, **22**, 1979–1982.
12. Douthwaite, R.E., Houghton, J., and Kariuki, B.M. (2004) *Chem. Commun.*, 698–699.
13. (a) Moser, M., Wucher, B., Kunz, D., and Rominger, F. (2007) *Organometallics*, **26**, 1024–1030; (b) Wucher, B., Moser, M., Schumacher, S.A., Rominger, F., and Kunz, D. (2009) *Angew. Chem.*, **121**, 4481–4485; *Angew. Chem. Int. Ed.*, (2009), **48**, 4417–4421.
14. Wie, W., Qin, Y., Luo, M., Xia, P., and Wong, M.S. (2008) *Organometallics*, **27**, 2268–2272.
15. (a) Rubio, R.J., Andavan, G.T.S., Bauer, E.B., Hollis, T.K., Cho, J., Tham, F.S., and Donnadieu, B. (2005) *J. Organomet. Chem.*, **690**, 5353–5364; (b) Vargas, V.C., Rubio, R.J., Hollis, T.K., and Salcido, M.E. (2003) *Org. Lett.*, **5**, 4847–4849.
16. (a) Chianese, A.R., Mo, A., Lampland, N.L., Swartz, R.L., and Bremer, P.R. (2010) *Organometallics*, **29**, 3019–3026; (b) Chianese, A.R., Shaner, S.E., Tendler, J.A., Pudalov, D.M., Shopov, D.Y., Kim, D., Rogers, S.L., and Mo, A. (2012) *Organometallics*, **31**, 7359–7367.
17. Zuo, W. and Braunstein, P. (2012) *Dalton Trans.*, **41**, 636–643.
18. Lee, H.M., Zeng, J.Y., Hu, C.H., and Lee, M.T. (2004) *Inorg. Chem.*, **43**, 6822–6829.
19. Chiu, P.L. and Lee, H.M. (2005) *Organometallics*, **24**, 1692–1702.
20. Shaw, B.K., Patrick, B.O., and Fryzuk, M.D. (2012) *Organometallics*, **31**, 783–786.
21. Schneider, N., Cesar, V., Bellemin-Laponnaz, S., and Gade, L.H. (2005) *Organometallics*, **24**, 4886–4888.
22. Boronat, M., Corma, A., González-Arellano, C., Iglesias, M., and Sánchez, F. (2010) *Organometallics*, **29**, 134–141.
23. (a) Matsumura, N., Kawano, J.-I., Fukunishi, N., and Inoue, H. (1995) *J. Am. Chem. Soc.*, **117**, 3623–3624; (b) Iwasaki, F., Manabe, N., Nishiyama, H., Takada, K., Yasui, M., Kusamiya, M., and Matsumura, N. (1997) *Bull. Chem. Soc. Jpn.*, **70**, 1267–1275.
24. Reviews: (a) Peris, E. and Crabtree, R.H. (2004) *Coord. Chem. Rev.*, **248**, 2239–2246; (b) Pugh, D. and Danopoulos, A.A. (2007) *Coord. Chem. Rev.*, **251**, 610–641; (c) Mata, J.A., Poyatos, M., and Peris, E. (2007) *Coord. Chem. Rev.*, **251**, 841–859. (d) Corberán, R., Mas-Marzá, E., and Peris, E. (2009) *Eur. J. Inorg. Chem.*,

2009, 1700–1716; (e) Bantreil, X., Broggi, J., and Nolan, S.P. (2009) *Annu. Rep. Prog. Chem., Sect. B*, **105**, 232–263; (f) Baker, M.V., Bosnich, M.J., Williams, C.C., Skelton, B.W., and White, A.H. (1999) *Aust. J. Chem.*, **52**, 823–825; (g) Hu, X.L., Castro-Rodriguez, I. and Meyer, K. (2004) *J. Am. Chem. Soc.*, **126**, 13464–13473; (h) Arnold, P.L. and Scarisbrick, A.C. (2004) *Organometallics*, **23**, 2519–2521.

25. Mas-Marzá, E., Poyatos, M., Sanaú, M., and Peris, E. (2004) *Organometallics*, **23**, 323–325.
26. Wright, J.A., Danopoulos, A.A., Motherwell, W.B., Carroll, R.J., Ellwood, S., and Saßmannshausen, J. (2006) *Eur. J. Inorg. Chem.*, **2006**, 4857–4865.
27. Klock, S.M., Heinekey, D.M., and Goldberg, K.I. (2006) *Organometallics*, **25**, 3007–3011.
28. (a) Wucher, B. (2011) *Studien zur Reaktivität von stark nucleophilen Rhodium- und Iridium-Pierce-Komplexen*, Dissertation, Eberhard Karls Universität Tübingen; (b) Seyboldt, A. *unpublished results.*
29. Gaunt, J.A., Gibson, V.C., Haynes, A., Spitzmesser, S.K., White, A.J.P., and Williams, D.J. (2004) *Organometallics*, **23**, 1015–1023.
30. Poyatos, M., Mas-Mazá, E., Mata, J.A., Sanaú, M., and Peris, E. (2003) *Eur. J. Inorg. Chem.*, **2003**, 1215–1221.
31. Andavan, G.T.S., Bauer, E.B., Letko, C.S., Hollis, T.K., and Tham, F.S. (2005) *J. Organomet. Chem.*, **690**, 5938–5947.
32. Bauer, E.B., Andavan, G.T.S., Hollis, T.K., Rubio, R.J., Cho, J., Kuchenbeiser, G.R., Helgert, T.R., Letko, C.S., and Tham, F.S. (2008) *Org. Lett.*, **10**, 1175–1178.
33. Raynal, M., Cazin, C.S.J., Vallée, C., Olivier-Bourbigou, H., and Braunstein, P. (2008) *Chem. Commun.*, 3983–3985.
34. (a) Zuo, W. and Braunstein, P. (2012) *Organometallics*, **31**, 2606–2615; (b) Raynal, M., Pattacini, R., Cazin, C.S.J., Vallée, C., Olivier-Bourbigou, H., and Braunstein, P. (2009) *Organometallics*, **28**, 4028–4047.
35. Raynal, M., Cazin, C.S.J., Vallée, C., Olivier-Bourbigou, H., and Braunstein, P. (2009) *Dalton Trans.*, 3824–3832.
36. Schultz, K.M., Goldberg, K.I., Gusev, D.G., and Heinekey, D.M. (2011) *Organometallics*, **30**, 1429–1437.
37. (a) Ojima, I., Clos, N., Donovan, R.J., and Ingallina, P. (1990) *Organometallics*, **9**, 3127–3133; (b) Tanke, R.S. and Crabtree, R.H. (1990) *J. Am. Chem. Soc.*, **112**, 7984–7989.
38. Jun, C.H. and Crabtree, R.H. (1993) *J. Organomet. Chem.*, **447**, 177–187.
39. Zeng, J.Y., Hsieh, M.-H., and Lee, H.M. (2005) *J. Organomet. Chem.*, **690**, 5662–5671.
40. (a) Hartwig, J.F. (2004) *Pure Appl. Chem.*, **76**, 507–516; (b) Takemiya, A. and Hartwig, J.F. (2006) *J. Am. Chem. Soc.*, **128**, 6042–6043.
41. (a) Leahy, D.K. and Evans, P.A. (2005) in *Modern Rhodium-Catalyzed Organic Reactions* (ed. P.A. Evans), Wiley-VCH Verlag GmbH, Weinheim, p. 191; (b) Kazmaier, U. and Stolz, D. (2006) *Angew. Chem.*, **118**, 3143–3146; *Angew. Chem. Int. Ed.*, (2006), **45**, 3072–3075.
42. (a) Kazmaier, U. and Pohlmann, M. (2004) in *Metal-Catalyzed Cross-Coupling Reactions*, vol. 2 (eds A. de Meijere and F. Diederich), Wiley-VCH Verlag GmbH, Weinheim, pp. 531–583; (b) Helmchen, G. and Pfaltz, A. (2000) *Acc. Chem. Res.*, **33**, 336–345.
43. Biswas, S., Huang, Z., Choliy, Y., Wang, D.Y., Brookhart, M., Krogh-Jespersen, K., and Goldman, A.S. (2012) *J. Am. Chem. Soc.*, **134**, 13276–13295.
44. Dani, P., Karlen, T., Gossage, R.A., Gladiali, S., and van Koten, G. (2000) *Angew. Chem.*, **112**, 759–761; *Angew. Chem. Int. Ed.*, (2000), **39**, 743–745.

10
NHCP Ligands for Catalysis

Peter Hofmann and Marcel Brill

10.1
Introduction

Since their first isolation by Arduengo [1] and the first reports of their potential as ligands by Öfele [2] and Herrmann [3], N-heterocyclic carbenes (NHCs) have initiated a spectacular ongoing success story associated with their application in molecular transition-metal catalysis. This development has been multiply documented in various books [4] and reviews [5] and has reached one of its culmination points through the chemistry of NHC-containing ruthenium-based so-called Grubbs-II precatalysts and their modified congeners. Their discovery and their broad, unprecedented application patterns for synthetic organic chemistry earned Grubbs (together with Schrock and Chauvin) the Nobel Prize of 2005. NHCs themselves have also found multiple valuable applications in organocatalysis [6].

Besides the now-established and more recent class of NHC ligands for $M-C^{NHC}$ coordination, there is another "old," well-established, important, and privileged family of ligands involving $M-P$ bonds, represented by a huge variety of P-based ligands, such as phosphines, phosphonites, phosphites, phosphoramidites, and so on, which have played and still play a dominant role in molecular transition-metal catalysis, both in the laboratory and in industrial applications [7]. The most important progress generated in the field of P-based ligand and catalysis research undoubtedly was the development of chiral, enantiopure representatives of this ligand class, a topic that was also pioneered by the works of Horner [8] and an area that has opened the field of efficient and practicable asymmetric transition-metal catalysis, for which Noyori and Knowles (together with Sharpless) were awarded the Nobel Prize in 2001.

As to the developments in NHC ligand research, more recently, functionalized NHC ligand systems have found appreciable interest [9]. Among them, molecular species with P-derived functional groups linked to a position of the NHC frame, which allows κ-P-, κ-C-chelating ligand coordination at transition-metal centers, seem particularly interesting. These types of ligands, which we will call *N*-phosphino-substituted *N*-heterocyclic carbene (NHCP) systems in the

Molecular Catalysts: Structure and Functional Design, First Edition. Edited by Lutz H. Gade and Peter Hofmann.

following for simplicity, should (i) add a novel ligand category to the large family of chelate ligands, being analogs of, for example, unsymmetrical chelating bisphosphines – achiral or chiral – and their relatives, but with electronically and sterically rather different properties, (ii) allow the elucidation and hopefully use of cooperative ligand effects of two of the most important ligand classes at transition-metal centers for catalysis, and (iii) be especially suitable for constructing chiral NHCP ligand systems for asymmetric catalysis through P-chirality at their phosphorus ligand arms. This chapter will describe recent developments in the field of P-functionalized chelating NHC ligand systems with a focus on their application in catalysis [10], at the same time summarizing some progress achieved in our own group aimed at the synthesis of new catalytically active metal complexes within the research program of SFB 623.

10.2
Recent Advances in Catalysis with NHCP Ligands

10.2.1
Cross-Coupling Catalysis and Related Reactions

In spite of the ubiquitous use of phosphines and the more recent and intensively explored NHCs as spectator ligands in homogeneous catalysis, it seems surprising that their hybrid bidentate or tridentate systems have evolved relatively late into an active field of research. The earliest example of a phosphino-functionalized NHC system was already synthesized in 1996 as an *N*-phosphinoethyl-substituted imidazolium salt in the laboratory of Herrmann [11], although no transition-metal catalysis was reported at that time. Five years later, the starting point of the more rapid development of these systems was set by Nolan *et al.* [12], who showed that a mixture of a similar *N*-phosphinoethyl-substituted imidazolium salt (**R-2·HBr**, R = Mes, see Figure 10.1), and $Pd(dba)_2$ (dba, dibenzylideneacetone) (0.5 mol%) efficiently catalyzes the Heck coupling of various aryl bromides with *n*-butyl acrylate in the presence of 2 equiv of Cs_2CO_3 giving high yields after just 15 min. No excess of ligand had to be employed, which is often required in phosphine-based catalyst systems, and the authors attributed this to the higher stability of NHC–metal bonds. Undoubtedly, this publication had a strong impact on the research community since this NHCP ligand became the most investigated of all known congeners and its performance has not only been studied in Heck–Mizoroki [13, 13a], Suzuki–Miyaura [13k], and Kumada–Corriu [13g,i,k] cross-coupling reactions but has also found applications in rhodium-catalyzed intramolecular hydroamination [14] and palladium-catalyzed allylations of aldehydes and α,β-unsaturated *N*-acylpyrroles [15] (see Section 10.2.2).

To date, most research on NHCP transition-metal catalysis has been devoted to cross-couplings [13] or related reactions such as hydroarylation of alkenes [18], direct arylation of alkynes [17], or oxidative homocoupling of terminal alkynes [19]. All NHCP systems used in these studies feature one or two

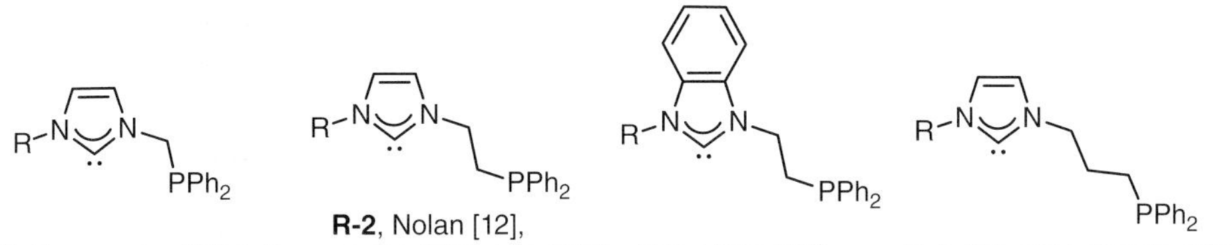

R-1, Danopoulos [16b] **R-2**, Nolan [12], Danopoulos [13a], Lee [13b], Poli and Labande [13g] **R-3**, Hahn [13j] **R-4**, Poli and Labande [13g]

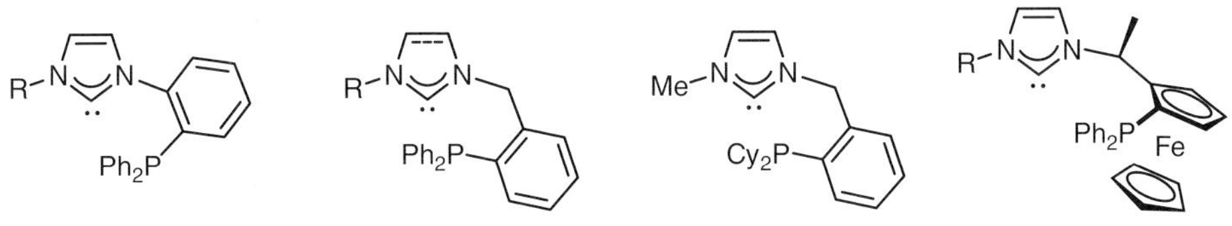

R-5, Zhou [18] **R-6**, Zhou [13d], Lee & Oberhauser [19] **7**, Lee [17] **R-8**, Shi [13h]

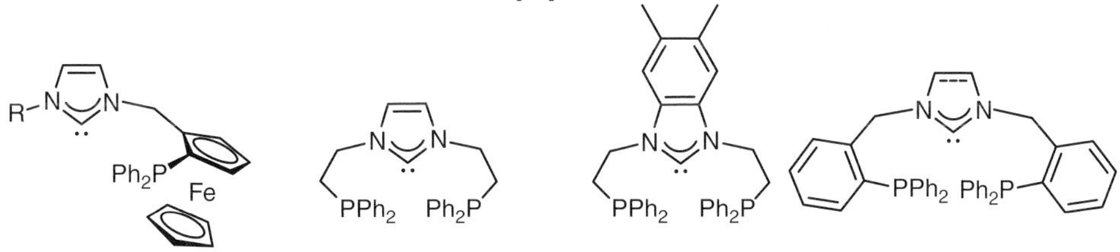

R-9, Labande and Poli [13n] **10**, Lee [13b] **11**, Hahn [13j] **12**, Zhou [13d]

Figure 10.1 NHCP ligands tested in cross-coupling and related reactions since 2001.

diphenylphosphino groups attached to the nitrogen atoms of the NHC scaffold by alkyl or aryl linkers of varying chain length, with the only exception being the di-cyclohexyl(aryl)phosphino-substituted ligand **7** from Lee *et al.* [17] (Figure 10.1).

Danopoulos and coworkers [13a] performed the Heck coupling of aryl bromides and methyl acrylates catalyzed by isolated palladium dimethyl complexes of NHCP **R-2** (R = Mes, DiPP) (DiPP, 2,6-diisopropylphenyl) and observed only moderate activities (0.5 mol% loading, low to good yields, 6 h), which were significantly lower than those previously reported by the Nolan group. Yet, as the authors pointed out, this is a commonly made observation when comparing the activities of *in situ* generated and defined precatalysts. The more bulky ligand **DiPP-2** showed enhanced catalytic activity over the *N*-mesityl counterpart, while the later developed methylene-bridged congener **DiPP-1** performed even better [13l]. However, the activities were still much inferior to the comparable *N*-pyridine- and picoline-functionalized NHC ligands, also studied in the group [13l]. Table 10.1 summarizes representative comparable results obtained in the Heck coupling of aryl bromides and acrylates using some of the NHCP ligands shown in Figure 10.1.

The closely related benzimidazolylidene ligands **R-3** (R = Et, *n*-Pr, *n*-Bu) and the pincer **11** introduced by Hahn *et al.* were tested in the Heck coupling of activated

Table 10.1 Representative comparable Heck–Mizoroki couplings of aryl bromides with alkyl acrylates mediated by NHCP palladium complexes.[a)]

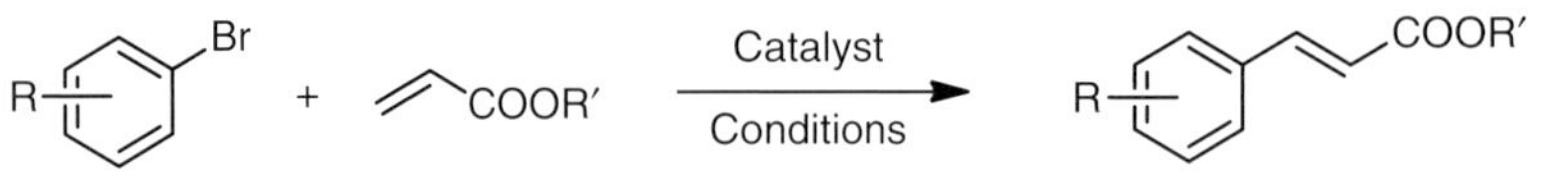

Entry	Catalyst	Pd (mol%)	Conditions	Time (h)	R	Yield[b)] (%)	TON	TOF (h^{-1})	References
1	[(**DiPP-1**)$PdMe_2$]	0.5	**A**	6	*p*-Ac	100	200	33	[13l]
2	[(**DiPP-1**)$PdMe_2$]	0.5	**B**	6	H	90	180	30	[13l]
3	**Mes-2·HBr**/$Pd(dba)_2$	0.5	**C**	0.25	*p*-CHO	100	200	800	[12]
4	**Mes-2·HBr**/$Pd(dba)_2$	0.5	**C**	1	H	100	200	200	[12]
5	[(**DiPP-2**)$PdMe_2$]	0.5	**A**	6	*p*-Ac	92	184	31	[13l]
6	[(**DiPP-2**)$PdMe_2$]	0.5	**B**	6	H	61	122	20	[13l]
7	[(***n*Bu-3**)$PdCl_2$]	0.1	**D**	24	*p*-Ac	100	1000	42	[13j]
8	[(**Et-3**)$PdCl_2$]	0.01	**D**	24	*p*-Ac	85	8500	355	[13j]
9	[(**11**)PdCl]Cl	0.1	**D**	24	*p*-Ac	62	620	26	[13j]
10	[(**11**)PdCl]Cl	0.01	**D**	24	*p*-Ac	52	5200	217	[13j]
11	[(**10**)Pd(MeCN)]$(BF_4)_2$	0.5	**B**[c)]	2	*p*-Ac	100	200	100	[13b]
12	[(**10**)Pd(MeCN)]$(BF_4)_2$	0.5	**B**[c)]	4	H	82	164	41	[13b]
13	**DiPP-6·HCl**/$Pd(dba)_2$	1	**E**	5	*m*-CHO	100	100	20	[13e]
14	**DiPP-6·HCl**/$Pd(dba)_2$	1	**F**	11	H	99	99	9	[13e]

a) Reaction conditions: unless noted otherwise, 1 equiv of aryl halide, 1.4 equiv of *n*-butyl acrylate, and 2 equiv of base were used. **A**: NEt_3, NMP (*N*-methyl-2-pyrrolidinone), 120 °C, methyl acrylate; **B**: Cs_2CO_3, DMAc, 120 °C, methyl acrylate; **C**: L/Pd (1 : 1), Cs_2CO_3, DMA (*N,N*-dimethylacetamide), 120 °C; **D**: NaOAc, DMAc, 110 °C; **E**: L/Pd (1 : 1) K_2CO_3, DMA 140 °C, 1.5 equiv acrylate; **F**: L/Pd (1 : 1), K_2CO_3, DMA 120 °C, 1.5 equiv acrylate.
b) Determined by GC.
c) 165 °C, 1.1 equiv NaOAc.

aryl bromides with styrene or *n*-butyl acrylate, which could be conducted at low catalyst loadings. In the presence of 0.01 mol% of defined precatalysts, couplings proceeded in good to excellent yields after 24 h at 110 °C (Table 10.1). Different N-substituents in precatalysts of the type [(**R-3**)$PdCl_2$] had little influence on the activities, and all bidentate ligands generally outperformed the cationic chloro complex of the pincer ligand [(**11**)PdCl]Cl. Yet, since these substituents differed only very little in their steric bulk, no firm conclusions about this influence could be drawn.

Lee tested the imidazolylidene analog **10** to Hahn's benzimidazolylidene PCP pincer **11** in the same coupling reactions. In comparison with Hahn's system, at similar catalyst loadings (0.5 mol%), the monocationic and biscationic complexes [(**10**)PdCl]Cl and [(**10**)Pd(MeCN)]$(BF_4)_2$ showed enhanced turnover numbers (TONs), leading to nearly full conversions after 1–2 h for activated aryl bromides, with the latter complex being slightly more efficient. However, very high temperatures (165 °C) were needed, and lower catalyst loadings were not reported.

Zhou and coworkers [13e] developed the triarylphosphino-NHC ligand **R-6** (R = Ph, Mes, DiPP), which was employed in the Heck coupling of styrene and *n*-butyl acrylate with a broad range of aryl bromides. The catalysts were generated *in situ* from various palladium precursors, a base, and the imidazolium salts **R-6·HCl**. Initial catalyst screening revealed that the increased steric bulk of the *N*-aryl substituent gave increasingly active catalysts. Using the optimized catalyst system **DiPP-6**/Pd(dba)$_2$/K_2CO_3 in *N*,*N*-dimethylacetamide (DMA) generally gave rise to high yields, even for deactivated aryl bromides, although in this case a temperature of 140 °C was needed to reach full conversions. Further, activities were fairly low in comparison with other NHCP systems (Table 10.1). Interestingly, the corresponding PCP pincer ligand **12** of these new triarylphosphino ligands showed far less activity than the bidentate ligands **R-6** and was excluded from further optimizations, which is in contrast to Hahn's results where only a slight drop in reactivity was observed for catalysts based on the pincer **11**.

The Suzuki–Miyaura coupling of arylboronic acids with aryl halides based on NHCP catalysts has also been studied in a number of publications. Lee and coworkers [13b,c,k] have been particularly interested in this field and have investigated the performance of various defined palladium and nickel complexes of ligand **R-2** (R = Mes, 1-naphthylmethyl (Naph), Bn, *p*-F-Ph, *m*-MeO-Ph) and pincer **10** for this purpose. Palladium dichloro complexes of the bidentate ligands [(**R-2**)$PdCl_2$] showed unsatisfactory activities under benchmark conditions, giving good conversions even for activated phenyl bromides only after 9 h at 0.5 mol% catalyst loading [13c]. The N-substituent played only a minor role in modulating catalyst activities. Quite untypically, in this case the PCP pincer **10** outperformed its bidentate analogs. The cationic complex [(**10**)PdCl]Cl, under the same conditions, coupled 4-bromoacetophenone and phenyl boronic acid quantitatively after just 2 h [13b]. However, this catalyst gave low conversions for deactivated aryl bromides even at prolonged reaction times (38% yield after 16 h). Table 10.2 summarizes representative comparable results obtained in the Suzuki coupling of phenylboronic acid with aryl bromides and in some cases aryl chlorides using some of the NHCP ligands shown in Figure 10.1.

A few years later, Lee's group was able to develop a highly efficient nickel catalyst. The air-stable bis-ligated nickel(II) complex *trans*-[[(**R-2**)$_2$Ni]Cl_2] (R = 1-naphthylmethyl (Naph), Bn, *p*-F-Ph, *m*-MeO-Ph) showed effective cross-coupling of phenylboronic acid with aryl chlorides (1–3 mol% catalyst loading, 0–6 mol% PPh_3, K_3PO_4, toluene, 80 °C), outperforming the reference system [(dppe)$NiCl_2$] (dppe, 1,2-bis(diphenylphosphino)ethane), especially in the absence of catalytic amounts of PPh_3. In contrast to the reference, the absence of PPh_3 did not lead to a substantial decrease in yields for the coupling of activated aryl chlorides. However, for deactivated substrates such as 4-chloranisole, at low catalyst loadings, the presence of PPh_3 led to enhanced reaction rates, although not quite as dramatic as for [(dppe)$NiCl_2$]. The steric influence of the N-substituent clearly had an impact on catalyst performance, as evidenced by the fact that the bulkiest ligand, containing an *N*-naphthylmethyl group, showed significantly enhanced activity in comparison to the *N*-benzyl or phenyl derivatives.

Table 10.2 Representative Suzuki–Miyaura couplings of aryl halides with phenylboronic acid with NHCP palladium and nickel complexes.a)

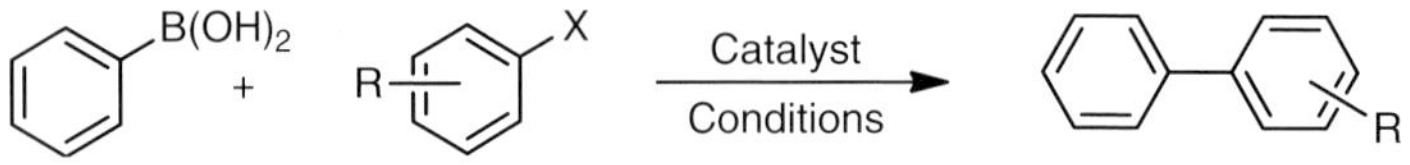

Entry	Catalyst	Pd (mol%)	Conditions	Time (h)	X	R	Yield[b] (%)	TON	TOF (h^{-1})	References
1	[(**Naph-2**)$PdCl_2$]	0.5	**A**	9	Br	*p*-Ac	83	166	18	[13c]
2	[(**Naph-2**)$PdCl_2$]	0.5	**A**	16	Br	H	85	170	11	[13c]
3	[(**10**)PdCl]Cl	0.5	**A**	2	Br	*p*-CHO	99	198	99	[13b]
4	[(**10**)PdCl]Cl	0.5	**A**	16	Br	H	90	180	11	[13b]
5	**DiPP-6·HCl**/ $[Pd(\eta^3\text{-}C_3H_5)Cl]_2$	0.1	**B**	6	Br	*p*-Ac	>99	1 000	167	[13d]
6	**DiPP-6·HCl**/ $[Pd(\eta^3\text{-}C_3H_5)Cl]_2$	0.1	**B**	6	Br	H	93	930	155	[13d]
7	**DiPP-6·HCl**/ $[Pd(\eta^3\text{-}C_3H_5)Cl]_2$	0.1	**B**	12	Cl	*p*-Ac	92	920	77	[13d]
8	[(**Naph-2**)$_2$Ni]Cl_2	1	**C**	2	Cl	*p*-Ac	97	97	49	[13k]
9	[(**Naph-2**)$_2$Ni]Cl_2	1	**C**	24	Cl	MeO	92	92	46	[13k]
10	***t*Bu-8·HCl**/$Pd(OAc)_2$	2	**A**c)	24	Cl	*p*-Ac	76	38	1.6	[13h]
11	***t*Bu-8·HCl**/$Pd(OAc)_2$	0.005	**D**	2	Br	H	>99	20 000	10 000	[13h]
12	***t*Bu-8·HCl**/$Pd(OAc)_2$	0.05	**D**d)	24	Br	MeO	92	1 480	77	[13h]

a) Reaction conditions: unless noted otherwise, 1 equiv of aryl bromide, 1.5 equiv of phenylboronic acid, 2 equiv of base, and 1,4-dioxane as solvent were used at 80 °C. **A**: base: Cs_2CO_3; **B**: base: K_3PO_4; **C**: base: 2.6 equiv $K_3PO_4 \cdot H_2O$, 2 mol% PPh_3, solvent: toluene; **D**: base: K_2CO_3, 110 °C, solvent: 95% aqueous ethanol.
b) Determined by GC.
c) L/Pd (1 : 1), 110 °C.
d) Solvent: 1,4-dioxane.

Zhou [13d] reported that his ligand family **R-6** and pincer **12** showed good activity in Suzuki coupling of various aryl bromides and activated aryl chlorides with some substituted arylboronic acids. Initial catalyst optimization gave the best results for mixtures of the imidazolium salts **R-6·HCl** and $[Pd(\eta^3\text{-}C_3H_5)Cl]_2$, while the most bulky ligand **DiPP-6** again showed the highest activity, as experienced earlier in Heck–Mizoroki couplings [13e]. This ligand was studied in further catalyst optimizations. Thus, at 0.1 mol% catalyst loading, couplings of activated aryl chlorides proceeded generally in high yields after 6–12 h.

The most active NHCP system developed for Suzuki–Miyaura coupling so far was introduced by Shi *et al.* They investigated the chiral ferrocenyl NHCP system **R-8**, which was developed by Chung and coworkers for asymmetric hydrogenation (see Section 10.3), because of its resemblance to the famous Josiphos ligand, which, next to its performance in asymmetric catalysis, is also known for its versatile application in cross-coupling reactions. A series of new derivatives of the precursor imidazolium salts featuring bulky substituents at the imidazolydene functionality were made (R = *t*Bu, 1-Ad, DiPP, 2,6-dimethylphenyl (DMP)) to gain insight into steric effects [13h]. Precatalysts were generally generated *in situ* from the precursor imidazolium salts **R-8·Cl** and $Pd(OAc)_2$. A palladium dichloro complex of ligand **DiMP-8** was also prepared, which performed similarly well. Initial ligand screening showed to some extent an influence of the size of the N-substituents, exhibiting the highest activity for ***t*Bu-8**, comparable activities for **Ad-8** and **DMP-8**, and significantly lower activity for the DiPP congener. A variety of substituted phenylboronic acids and a series of aryl bromides were then coupled with the ***t*Bu-8**-based system with high activities at 0.5 mol% catalyst loading even for deactivated substrates (e.g., >99% conversion for 4-bromoanisole after 2 h). At elevated temperatures (110 °C) and 2 mol% catalyst loading, the system also allowed the coupling of deactivated arylchlorides with high yields after 24 h. The most important feature of Shi's system was its robustness and efficiency at low catalyst loadings, giving the coupling products bromobenzene, 4-bromotoluene, and 3-toluene with arylboronic acid at 0.005 mol% (110 °C) in almost quantitative yield while generating TONs of 20 000 and turnover frequencies (TOFs) of up to 10 000. A few years later, Shi's group investigated the same system in the presence of KO*t*Bu as the base in the Buchwald–Hartwig reaction, where, as before, only a minor influence of the steric bulk of the imidazolylidene group was observed unless ortho-substituted aryl halides were used (more bulky ligands were less efficient). Again, aryl chlorides could be activated and were coupled with morpholine, although only modest activities were observed (e.g., *p*-CF_3-toluene: 1,4-dioxane, 110 °C, 24 h, 87%) [13m].

A similar ligand system based on the ferrocenyl skeleton **R-9** (R = Me, Mes), which possesses planar chirality, was introduced by Labande and Poli and was investigated toward the asymmetric Suzuki–Miyaura reaction. For optimization purposes, the standard coupling or aryl bromides with phenylboronic acid was investigated, in which similar activities to Shi's system were found using 0.1 mol% of the palladium dichloro complex of the more bulky ligand [(**Mes-9**)$PdCl_2$]. The good performance of the catalyst permitted them to perform the asymmetric coupling of naphthyl bromides and naphthylboronic acids at temperatures as low as 40 °C, even though catalyst loading had to be raised to 0.5 mol% in this case (Scheme 10.1). Although the observed enantioselectivities were in the low range

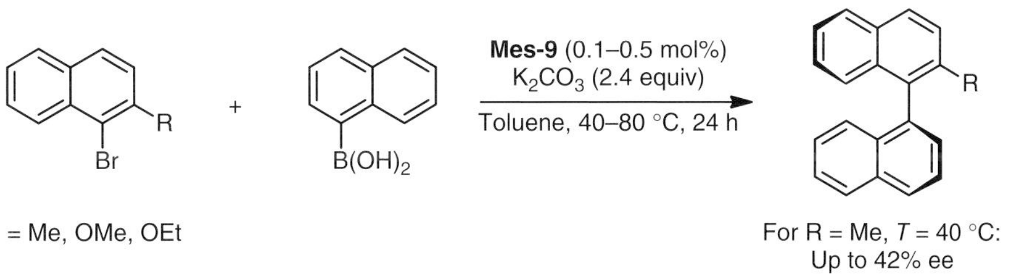

Scheme 10.1 First chiral NHCP system used in asymmetric Suzuki–Miyaura coupling.

of 23–42% ee, this was the first example of chiral NHC-palladium catalysis for this kind of transformation, which certainly leaves room for future improvements since activities were promising and no further ligand screening was investigated.

10.2.2 Miscellaneous Reactions

Labande, Poli, and coworkers [20] have developed achiral 1,1′-disubstituted ferrocenediyl NHCP ligands containing either an imidazolylidene or benzimidazolylidene as NHC donor. They investigated their efficiency to promote the hydrosilylation of arylketones using square-planar cationic rhodium complexes of the type [(NHCP)Rh(COD)]BF_4 (COD, cyclooctadiene) (**R-13** and **14**, Scheme 10.2). Surprisingly, the imidazolylidene system showed very poor activity in the hydrosilylation of acetophenone, while the benzimidazolylidene system led to a dramatic increase in activity. Electron-rich and electron-poor acetophenone derivatives were reduced to the corresponding alcohols in 2 h with 1.1 equiv diphenylsilane at 2 mol% of catalyst loading, followed by hydrolysis with MeOH/HCl. Unsubstituted acetophenone was reduced just as efficiently, requiring only 1 mol% catalyst. Future efforts will be toward employing the chiral ferrocenyl-NHCP systems **R-9** in the asymmetric variant of this reaction.

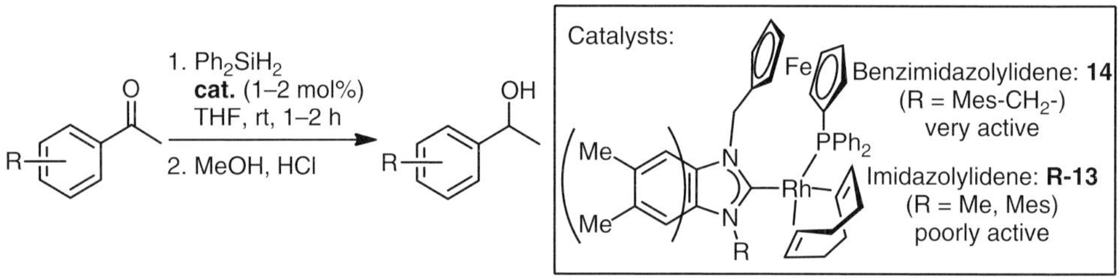

Scheme 10.2 Hydrosilylation of acetophenone derivatives with ferrocenyl-NHCP ligands.

Very effective catalysts for the hydrosilylation of alkynes were developed by Lee and coworkers. They prepared a series of rhodium(I) and rhodium (III) complexes containing their PCP pincer **10** (e.g., *mer*-[(**10**)Rh(CO)Cl], *mer*-[(**10**)$RhCl_3$]), all of which showed high activities in the hydrosilylation of phenylacetylene with $HSiMe_2Ph$, yielding the (*E*)-β-silylalkenyl products in >80% selectivity at 0.1–0.001 mol% catalyst loadings [21].

Lee *et al.* and Grotjahn *et al.* have reported two efficient ruthenium catalysts for the transfer hydrogenation of ketones [22]. While Lee used an interesting triply chloro-bridged dimeric NHCP ruthenium complex of his pincer ligand **10**, Grotjahn employed a new bidentate ethylene-bridged NHCP ruthenium complex [Cp(NHCP)RuCl] (**16**) featuring an *N*-H-substituted imidazolylidene group (Scheme 10.3). At similar catalyst loadings (0.1 and 0.05 mol%, respectively) and conditions (80 and 70 °C, respectively), high yields were obtained for the reduction of acetophenone [23].

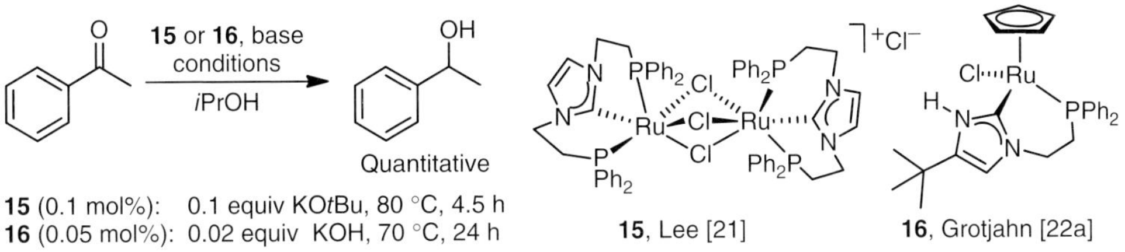

Scheme 10.3 Transfer hydrogenation of acetophenone mediated by NHCP ruthenium complexes.

Jarvo *et al.* have reported the use of isolated NHCP palladium allyl complexes of the type [(**R-2**)Pd(η^3-C_3H_5)]X (X = Cl, OAc, BF_4) as catalysts for nucleophilic allylation reactions. They investigated a series of electron-rich monodentate and bidentate ligands (bisphosphines, functionalized NHCs, monodentate NHCs) for their ability to promote nucleophilic attack of the corresponding palladium allyl complexes on electrophiles in, for example, the conjugate addition of allylboronic esters to a variety of α,β-unsaturated *N*-acylpyrroles [15a] as well as the allylation of aldehydes by allylstannanes (Scheme 10.4) [15b].

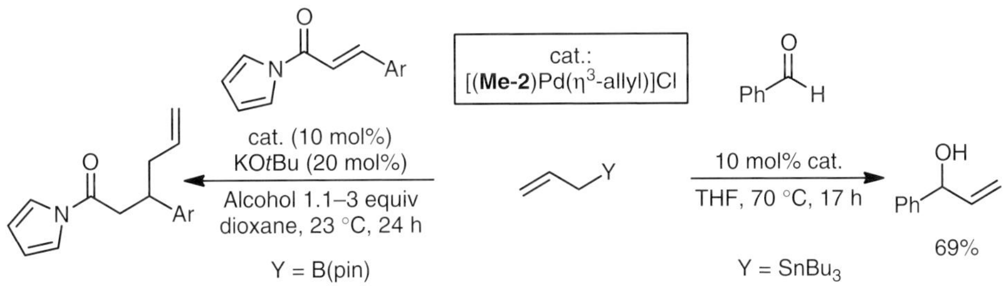

Scheme 10.4 Allylation of α,β-unsaturated *N*-acylpyrroles and aldehydes with allylboronic esters and allylstannanes catalyzed by an NHCP palladium allyl complex.

Initial studies on the allylation of aldehydes revealed that the chloride salts of the allyl complexes gave the highest activities (X = Cl), indicating the possibility of a nucleophilic η^1-allyl complex as intermediate in which the counterion would coordinate to palladium. Further, allyl complexes of monodentate NHCs or electron-rich bisphosphines proved to be poor catalysts, although the NHCP system proved to be inferior to a catalyst derived from a bidentate *N*-picoline-functionalized NHC. However, in the allylation of *N*-acyl-pyrroles, the NHCP-based catalyst exhibited the highest activities of all catalysts screened. Catalyst loadings in both reactions were relatively high, but yields ranged from good to excellent.

10.3 Recent Advances in Asymmetric Catalysis with Chiral NHCP Ligands

In the past decade, chiral NHC catalysis has made impressive progress, although the number of highly selective systems still remains very small in comparison to chiral phosphine-based catalysis – a fact that clearly results from the 30-year headstart of the latter systems. One of the main problems encountered in the development of new chiral NHC ligands results from the necessity to explore completely new concepts in ligand design for the electronically similar, yet topologically very distinct phosphine analogs, which NHCs represent. Effective chirality transfer from the usually remotely attached chiral entities (attached most commonly to the NHC backbone or as chiral N-substituents) to the reaction center remains a challenging task. Burgess and later Gade and Bellemin-Laponnaz [24, 25] have early on summarized the basic concepts of chiral NHC catalysis and have helped in characterizing efficient stereodirecting structural elements. Very recently, Shi [26] has published an excellent updated review on chiral NHC-based catalysis.

In spite of the overwhelming number of publications dedicated to the development of chiral bisphosphines [7], the development of the corresponding NHCP analogs has attracted much less attention. Partially, this is due to the fact that only one highly selective system has been developed to this point [27]. The majority of the newly developed ligands have been tested in asymmetric rhodium- and iridium-catalyzed hydrogenations (Figure 10.2) [27a, 28].

The first chiral NHCP employed in asymmetric catalysis was developed by Chung and coworkers in 2003 and featured a phosphino-functionalized ferrocenyl unit attached to the NHC group (**Me-8**, see Figure 10.1). This NHC-analog of Togni's very successful Josiphos ligand was tested in the rhodium-catalyzed asymmetric hydrogenation. Disappointingly, a very low selectivity of 13% ee was obtained in the hydrogenation of dimethyl itaconate with an *in situ* generated rhodium complex from $[Rh(COD)_2]BF_4$, imidazolium salt **R-8·HI**, and NaH as base. A preformed bis-ligated rhodium complex did not prove active at all. As mentioned in Section 10.2.1, this ligand was later investigated for its performance in cross-coupling catalysis [13m]. A derivative of this ligand has further been investigated in palladium-catalyzed allylic amination (see below) [29].

In 2004, Bolm and coworkers [28b] employed NHCP iridium complexes bearing a planar chiral [2,2]paracyclophane backbone **R-17** (R = Ph, Mes, DiPP) in the hydrogenation of various unfunctionalized and simple alkenes. Low to good activities and promising enantioselectivities of up to 82% ee (R = Ph) were shown in the hydrogenation of (*E*)-α-methylstilbene. Interestingly, an increase of steric bulk by changing the N-substituent from a phenyl to a mesityl or diisopropylphenyl group gave the product with inverse absolute configuration, albeit with low selectivities (29% and 22% ee, respectively). In contrast, the hydrogenations of functionalized alkenes such as dimethyl itaconates or cinnamic acid esters proceeded with the highest selectivities generally for the most bulky *N*-diisopropylphenyl-substituted system. Enantioselectivities ranged from low to good (up to 89% ee).

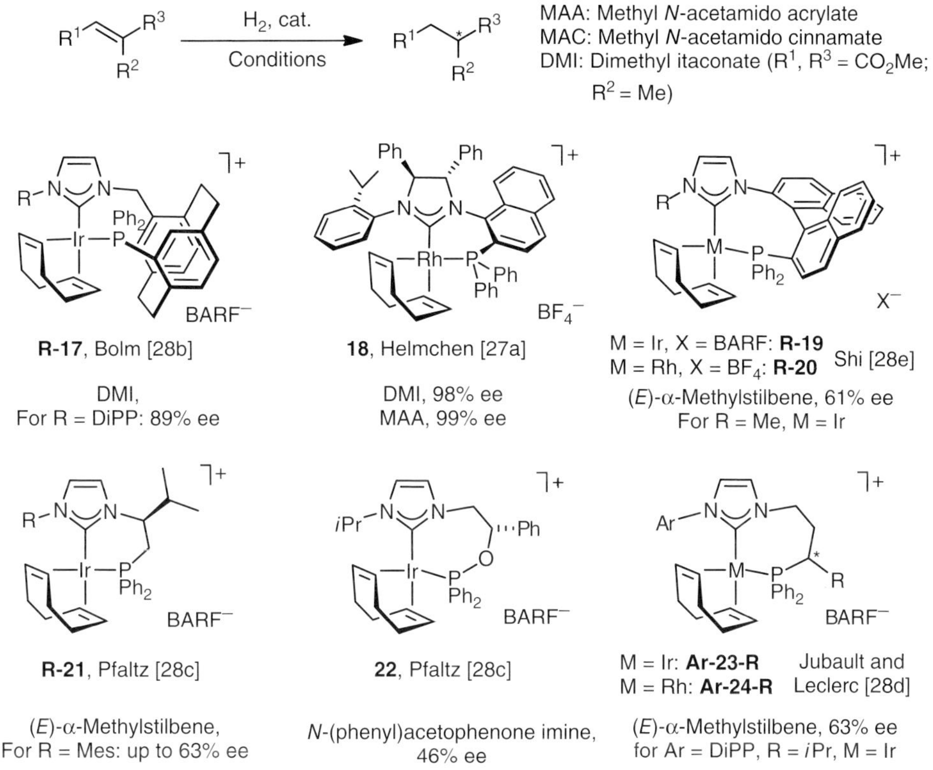

Figure 10.2 Chiral NHCP metal complexes used in asymmetric rhodium- and iridium-catalyzed hydrogenations (the highest selectivity exhibited for a certain type of substrate is given).

Helmchen and Bappert [27] introduced the NHCP rhodium complex **18** containing a stereogenic *N*-naphthyl axis as a very selective catalyst for asymmetric hydrogenation and 1,4-conjugate additions of phenylboronic acid to α,β-unsaturated esters. Although relatively high pressures and elevated temperatures (20–50 bar, up to 50 °C) were needed for the hydrogenation of itaconic and *N*-acetyldehydroamino acid esters in order to reach full conversion after 24 h, the selectivities were remarkable and have remained unmatched in the field of asymmetric NHCP-based catalysis since then [27a]. Thus, even at catalyst loadings of 0.1 mol%, dimethyl itaconate was obtained in 98% ee, while *N*-acetyldehydroamino acid esters were obtained in up to 99% ee at 1 mol% catalyst loading. The same precatalyst also exhibited the highest reported enantioselectivity in the conjugate addition of phenylboronic acid to certain linear α,β-unsaturated esters (>99% ee) (Scheme 10.5) [27b]. Under the same conditions, high yields and good to high enantioselectivities were also reported for the reaction of various arylboronic acids with cyclic enones.

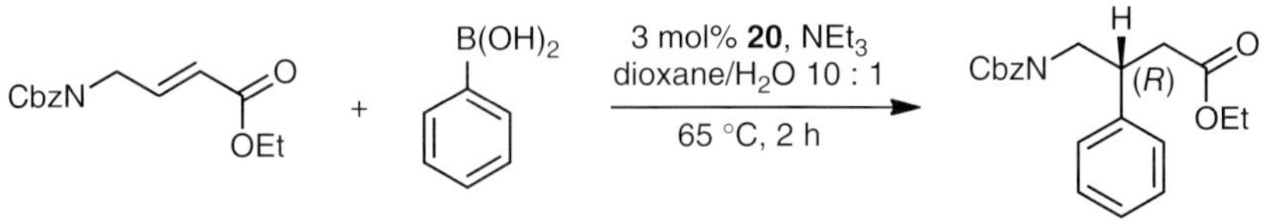

Scheme 10.5 Highly enantioselective conjugate addition of phenylboronic acid to a linear α,β-unsaturated ester using the chiral NHCP rhodium complex **18**.

Shi has just recently published a new axially chiral NHCP system and its rhodium and iridium diene complexes **R-19** (R = Me, *i*Pr) and **R-20** (R = Me). The new ligand is the NHCP analog of the 1,1′-binaphthalenyl-2,2′-diamine (BINAM) derived biscarbene ligands, which have been developed in this group and successfully applied in several asymmetric catalytic reactions [30]. The iridium complexes **R-19** were active in the hydrogenation of (*E*)-α-methyl-cinnamates, leading to full conversions at 25 bar hydrogen pressure after 24 h but in low selectivities (27–38% ee). Enantioselectivity for (*E*)-α-methylstilbene as substrate was slightly higher using catalyst **Me-19** (61% ee) but conversion was low (35%) even at 50 bar hydrogen pressure. The rhodium complex **R-20** was excluded from substrate screening since it showed even lower activities.

The iridium complexes **R-21** (R = Me, *i*Pr, Mes) and **22** containing NHCPs with an asymmetric center in the NHC and P connecting linker represent a development by Nanchen and Pfalz [28c]. It was hoped that effective chirality transfer from the backbone stereocenter to the diphenylphosphino group would lead to an edge-to-face orientation of the aromatic rings as observed, for example, in one of the most prominent chiral bisphosphines chiraphos [31]. Initial 2D NMR experiments on the square-planar iridium complexes revealed a fluxional behavior in which two boat-shaped conformers coexist, indicating that these six- and seven-membered metallacyclic systems might be lacking a certain degree of rigidity needed for effective chirality transfer. Indeed, modest to very low enantioinduction was observed in the hydrogenation of unfunctionalized alkenes; also, activities were unsatisfactory. Steric bulk seemed to have some influence on the selectivities as indicated by the fact that the highest enantiomeric excess in the hydrogenation of (*E*)-α-methylstilbene was achieved using the *N*-mesityl system (63% ee), while all other ligands yielded a nearly racemic product. Yet, for functionalized olefins this trend could not be observed. Notably, hydrogenation of functionalized alkenes and imines proceeded fast, albeit with low selectivities (up to 49% ee).

Similarly, the related systems **Ar-23-R** (Ar = Mes, DiPP; R = Me, Et, *i*Pr) developed recently by Jubault and Leclerc [28d] exhibited comparable selectivities and low activities (2 mol% catalyst, 50 bar hydrogen, elevated temperatures for full conversion, 38–63% ee) in the iridium-catalyzed hydrogenation of (*E*)-α-methylstilbene, once again indicating that an enhanced rigidity of the ligands might be required for future improvements. The rhodium complexes **Ar-24-R** (Ar = Mes, DiPP; R = Me, Et, *i*Pr) were applied in the hydrogenation of dehydroaminoacid esters, which also showed moderate activities and low enantioselectivities (17–60% ee for methyl *N*-acetamidoacrylate).

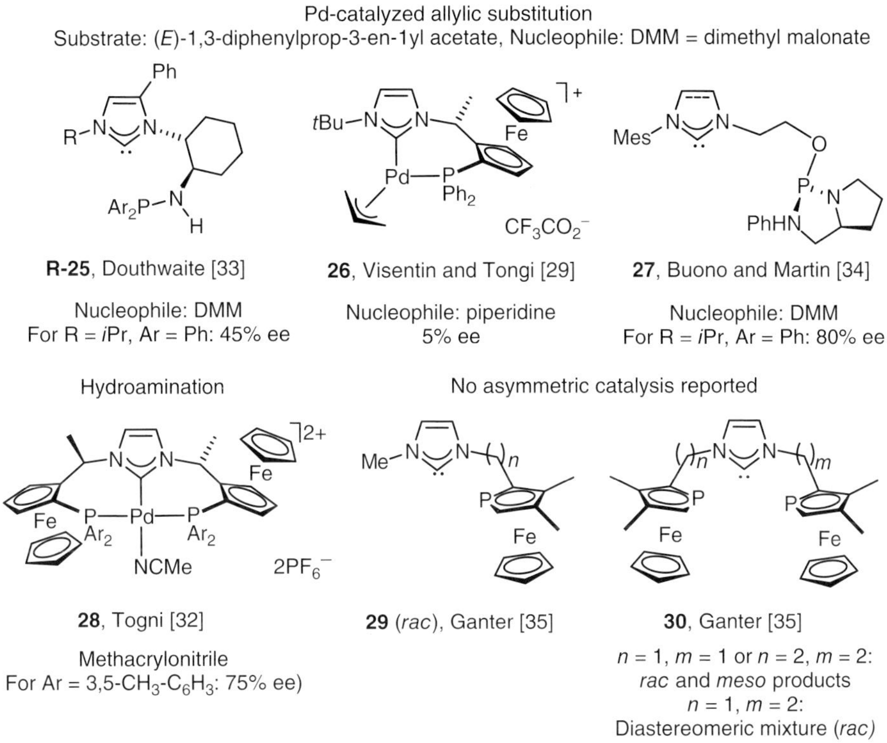

Figure 10.3 Chiral NHCP systems used in various catalytic reactions (the highest selectivity exhibited in catalysis for a certain type of substrate is given in brackets. If the free ligand is depicted, the catalyst was prepared *in situ*).

A few other chiral NHCP systems have been reported, and their application in various palladium-catalyzed transformations such as hydroamination of α,β-unsaturated nitriles [32], asymmetric Suzuki–Miyaura coupling (see Section 10.2.1) [13n], and most frequently allylic substitution reactions have been studied [29, 33, 34]. Figure 10.3 summarizes the chiral NHCP systems not used in asymmetric hydrogenations, which we will present in the following.

An interesting comparative study on the mechanism of palladium-catalyzed allylic amination with NHCP **R-8** (see Figure 10.1) and an isostructural aminophosphine ligand was performed by Visentin and Togni [29]. They investigated the amination of the typical benchmark substrate (*E*)-1,3-diphenylprop-3-en-1-yl acetate with piperidine and could show that a strong electronic asymmetry in the ligand was indeed required for induction of high enantioselectivities, as has been suggested in previous mechanistic studies with other dissymmetric ligands (Scheme 10.6) [36]. Therefore, NHCP systems that contain two donor functionalities of relatively equal σ-donor capacity seem to be unprivileged ligands for these types of reactions.

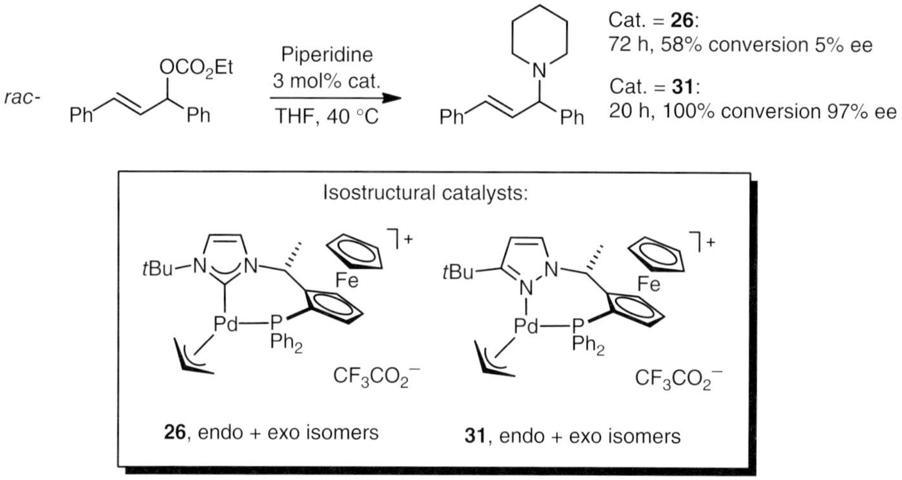

Scheme 10.6 Comparative study on isostructural NHC-phosphine and amino-phosphine chelate complexes in palladium-catalyzed allylic amination.

Douthwaite and Hodgson developed the NHCP system **R-25** (R = *i*Pr, $CHPh_2$, *n*Pr; Ar = Ph, Mes) featuring a *trans*-1,2-diaminocyclohexane backbone based on their experience with a previously developed NHC-imine analog that had been previously been employed successfully in palladium-catalyzed asymmetric alkylation (up to 92% ee) [33]. Disappointingly, only low enantiomeric excesses (8–45% ee) were obtained in the allylic alkylation of (*E*)-1,3-diphenylprop-3-en-1-yl acetate and dimethyl malonate (DMM) using *in situ* generated ligands from **R-25·HBr** and $KN(SiMe_3)_2$ in combination with $[Pd(\eta^3\text{-allyl})Cl]_2$. On the bright side, reaction rates were significantly improved over the previous NHC-imine systems, giving the product after 5 h in nearly quantitative yield even a 0 °C. The authors attributed this to the higher π-acidity present in these systems, which possibly relieves charge buildup at the palladium center in the transition state.

In line with Visentin and Togni's conclusion of NHC-phosphine hybrids being rather disadvantageous for asymmetric palladium-catalyzed allylic substitution reactions due to the similarity in σ-donor ability of both donor moieties, Martin and Buono developed NHCP **27**. The ligand features a very electron-poor diazaphospholidine P-donor , which therefore should possess a rather strong electronic asymmetry in its donor atoms. Fortunately, it showed the highest reported enantiomeric excesses for NHCP-based palladium-catalyzed asymmetric substitutions (up to 80% ee in the alkylation of (*E*)-1,3-diphenylprop-3-en-1-yl acetate with DMM at 25 °C). Yet, since the activities exhibited were fairly low (full conversion after 48 h at 5 mol% catalyst loading), and since the study was rather focused on the investigation of cooperative effects in the corresponding mixed monodentate NHC and phospholidine systems, no further derivatives were made to improve results with the bidentate ligands.

Based on the successful results obtained in the application of the C_1-symmetric triphosphine Pigiphos in asymmetric hydroamination of cyanoolefins [37, 38], Gishig and Togni [39] have developed its phosphino-NHC analog in which the NHC unit replaces the internal P-donor (formerly an R_2PCy unit) of the original Pigiphos ligand. The new biscationic C_2-symmetric PCP pincer complex **28** was used as a Lewis acidic catalyst in the asymmetric hydroamination of acrylonitrile derivatives. It was reasoned that a more defined chiral environment should be present in this square-planar complex compared to the parent Pigiphos system. Yet, only moderate selectivities were obtained using a series of piperidine derivatives as nucleophiles (up to 63% and 75% ee for the addition of thiomorpholine to methacrylonile at room temperature and −80 °C, respectively). A similar chiral PCP pincer system **30** featuring phospha-ferrocene donors was introduced by Ganter in 2008. The modular synthesis also allowed easy preparation of the bidentate phospha-ferrocene NHC hybrid **29** [35]. However, no application in asymmetric catalysis was reported for either ligand system.

10.4 Recent Advances in NHCP Chemistry Featuring Bulky, Electron-Rich, Small-Bite-Angle Ligands

Based upon earlier experiences with chelating, bulky, electron-rich, small-bite-angle di-*tert*-butylphosphinomethane (dtbpm) and di-*tert*-butylphosphinoethane (dtbpe) ligands and their transition-metal complexes in homogeneous catalysis and bond-activation chemistry [40–42], we became interested in two generic types of NHCP ligand systems **A** and **B**, which we wanted to carry bulky and electron-rich substituents both at the NHC moiety (R^1) and at their P-functionality (R^1, R^2), respectively (Figure 10.4).

N-Phosphinomethyl-substituted NHCP ligands **B** were expected to yield five-membered chelate rings, probably more or less strain free, if both donor centers C and P coordinate to one metal center, and *N*-phosphino-substituted ligands **A** should form four-membered chelate rings if their higher ring strain allows sufficient stability at all. Therefore, transition-metal complexes of ligands **A** may as well be associated with hemilabile properties which can be beneficial or even crucial in catalysis due to stabilization of the reactive intermediates. On the other hand, ligands of type **B** should lead to strongly bound chelates, which should contrast the two diphenylphosphino-substituted analogs known in the literature [13l, 16]. Both types of NHCPs, of course, also could have the option of bridging two

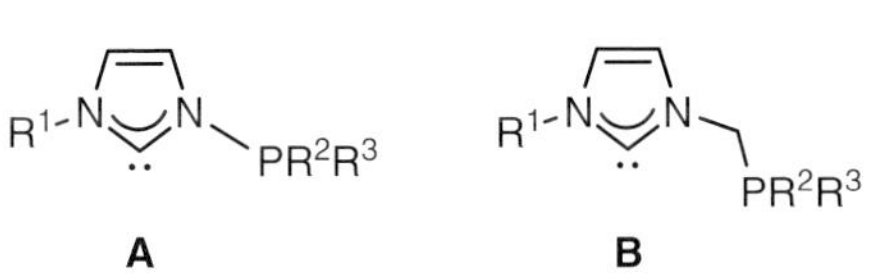

Figure 10.4 Small-bite-angle ligands investigated in our group.

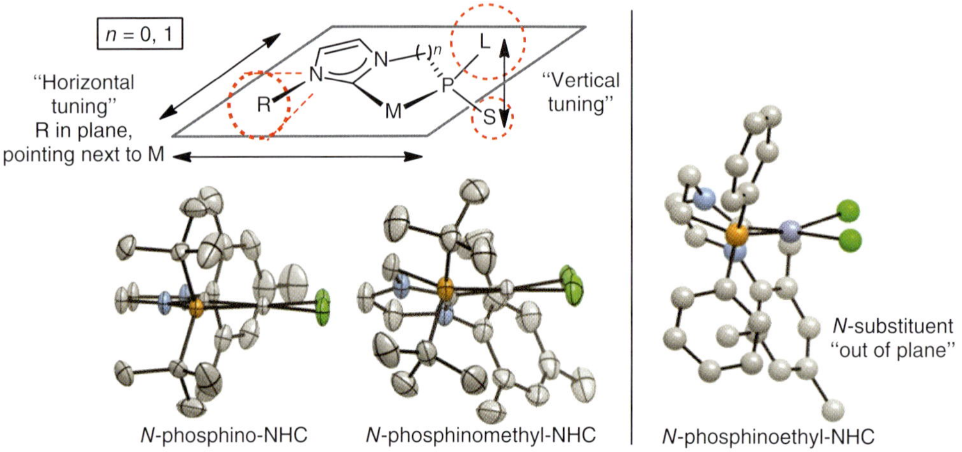

Figure 10.5 Steric properties of NHCP metal complexes exemplified by a simple graphical representation of *N*-phosphino- and *N*-phosphinomethyl-NHCs (above) and solid state structures of *N*-phosphino-NHC [46], *N*-phosphinomethyl-NHC [45e], and *N*-phosphinoethyl-NHC [13a] platinum- or palladium dihalide model complexes (below, hydrogen atoms omitted for clarity).

metal centers. Large, sterically demanding groups R^2, R^3 should stabilize chelating structures through the Thorpe–Ingold effect [43].

We reasoned that the bulky dialkyl-phosphino substitution, which is rarely found in NHCP chemistry [10g, 17, 44], and the small bite angles would lead to rigid metallacycles enabling the development of more defined molecular environments than with most systems known in the literature. One key feature that distinguishes their metal complexes from other NHCP systems – for example, the commonly used ethylene-bridged NHCPs or the comparably electron-rich biscarbene ligands – results from this enhanced rigidity present in our systems. Therefore, the NHC plane is only slightly tilted (five-membered ring system) or nearly coplanar (four-membered ring system) with respect to the coordination plane, as evidenced in various model complexes made in our group [45, 46]. Consequently, the N-substituents are oriented toward the metal center, presumably leading to a greater steric influence than in biscarbene or most known NHCP systems, while the distinct topology of the P-donor further allows steric refinement perpendicular to the coordination plane ("vertical tuning," Figure 10.5). The strong influence of the N-substituent will be particularly important for modulation of catalyst activities in, for example, cross-coupling or asymmetric catalysis.

10.4.1 Ligand Synthesis of *N*-Phosphino- and *N*-Phosphinomethyl NHCs

Ligands of type **A** were synthesized via two different routes (Scheme 10.7). Route (a), which was developed in our group [43a, 44], proceeded via selective

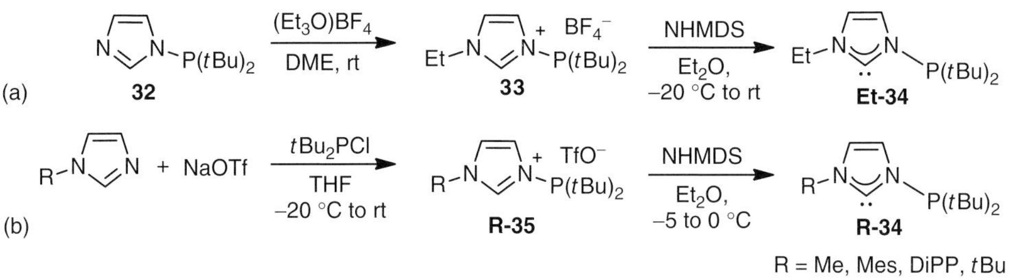

Scheme 10.7 Synthesis of *N*-phosphino NHC ligands by (a) Hofmann *et al.* and (b) Kostyuk *et al.* (R denotes only the N-substituents investigated in our group).

quaternization of the easily accessible phosphino-imidazole **32** at the N-atom with ethyl Meerwein salt. At low temperature, imidazolium salt **33** was then converted to the free carbene **Et-34** with NHMDS (sodium hexamethyldisilazide) which was isolated and characterized by X-ray crystallography. Kostyuk *et al.* [47] have quaternized N-substituted imidazoles in the presence of NaOTf with $t\mathrm{Bu_2PCl}$ for the same purpose. Thereby, a large number of *N*-phosphino imidazolium salts **R-35** could be obtained that are not limited by the availability of strong alkylating agents which usually only transfer small alkyl groups. The free carbenes were generated similarly, but could even be purified via Kugelrohr distillation. Recently, the same group introduced a very elegant third possibility of generating *N*-phosphino NHCs by directly converting *in situ* deprotonated N-substituted imidazoles with secondary chlorophosphines into the corresponding carbenes at −60 °C [48]. This procedure sadly is not general since it is successful only for very large N-substituents (R = Ad, *t*Bu) in combination with sterically bulky chlorophosphines such as di-*tert*-butylchlorophosphine. Otherwise, phosphorylation occurs at the carbene carbon, leading to 2-phosphino-substituted *N*-alkyl or *N*-aryl imidazoles [44].

The synthesis of *N*-phoshinomethyl-substituted NHC systems initially followed a six-step protocol starting from di-*tert*-butylchlorophosphine which was hydrolyzed to di-*tert*-butylphosphine oxide (Scheme 10.8) [49]. This compound was hydroxymethylated using formaldehyde giving **36** followed by tosylation to

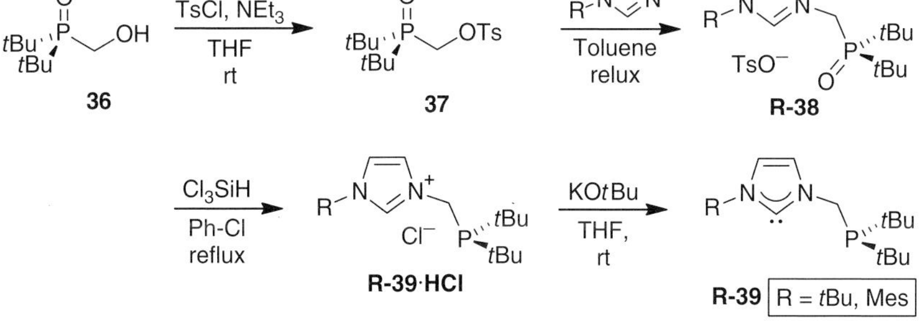

Scheme 10.8 Synthesis of *N*-phosphinomethyl NHC ligands.

yield the tertiary phosphine oxide **37**, which was coupled with N-substituted imidazoles (R = *t*Bu, Mes) at high temperatures. After reduction of the obtained imidazolium tosylates **R-38** using a large excess of $HSiCl_3$, the imidazolium salts **R-39·HCl** was obtained in which the complete substitution of the tosylate counterion with chloride had occurred. Generation and isolation of the free carbenes **R-39** proceeded smoothly, which in the case of the *N-tert*-butyl system were also characterized via X-ray crystal structure analysis. Besides the remarkably stable *N*-phosphino carbene systems, only for one other known NHCP ligand the isolation and X-ray characterization has been reported [50].

Recently, we have developed an alternative synthesis that enables the formation of bulky and electron-rich *N*-phosphinomethyl-substituted NHC metal complexes in just three steps (Scheme 10.9) [51]. Further, this procedure does not rely on the isolation of air-sensitive trialkylphosphines. The coupling of borane-protected di-*tert*-butylphosphine and the chloromethylated imidazolium salts **R-40** (R = tBu, Mes) with *n*-butyl lithium generated the air-stable protected imidazolium salts **R-41**. In case of an *N*-mesityl substitution, this compound could be converted into the palladium allyl complex **Mes-42** in a high-yielding one-pot procedure. This feature could encourage the development of new trialkylphosphine–NHC hybrid ligands.

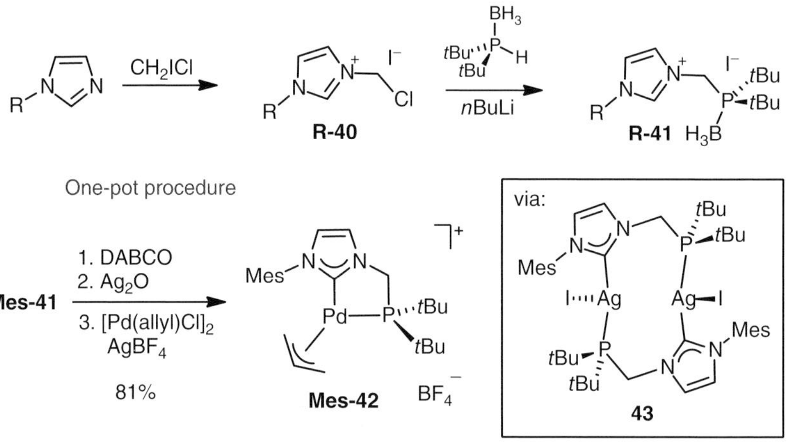

Scheme 10.9 Efficient three-step synthesis of NHCP palladium allyl complexes.

10.4.2
N-Phosphino-NHC Transition-Metal Complexes

We were delighted to find that the NHCP ligands with a direct N–P linkage **R-34** form stable chelating complexes with metals of the nickel triade and ruthenium, featuring very small bite angles between 65° and 70° as determined by X-ray crystal structure analysis of all new compounds made (Scheme 10.10) [44]. The NHCP palladium dimethyl complexes **45** show reactivity toward electron-poor

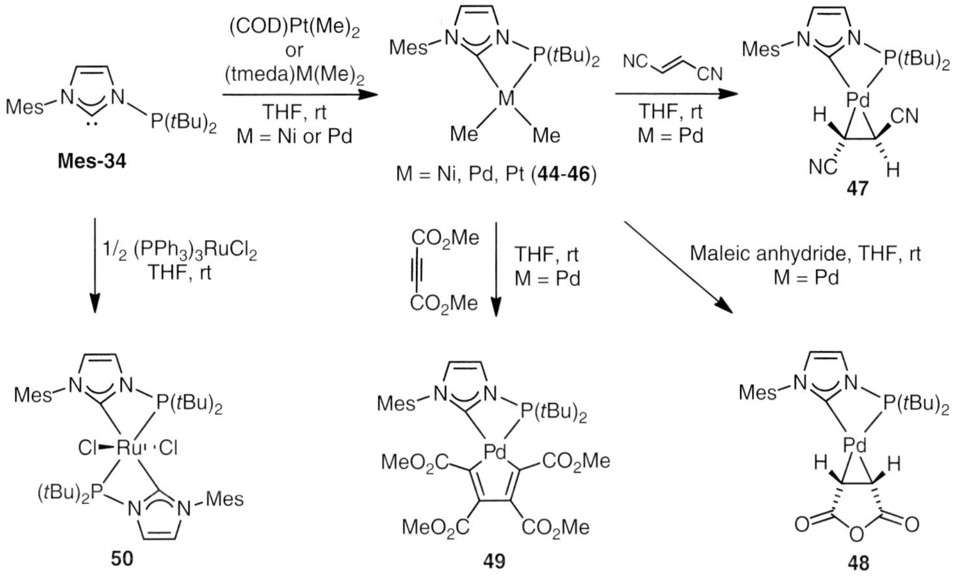

Scheme 10.10 Synthesis and reactivity of *N*-phosphino NHC metal complexes of the nickel triade.

olefins, yielding the corresponding olefin complexes **47** and **48** under reductive elimination of ethane at room temperature. Adding an electron-poor alkyne such as dimethyl acetylene dicarboxylate cleanly led to the formation of the palladacycle **49**. The bis-ligated ruthenium complex **50** was obtained upon treatment of $(PPh_3)_2RuCl_2$ with 2 equiv of NHCP ligand and featured a remarkably small bite angle of 65°.

In the light of the unusual reactivity exhibited by tailor-made d^{10}-ML_2 fragments of the similarly electron-rich, small-bite-angle ligand dtbpm in bond activation reactions of all sorts [39], these systems seem particularly interesting, especially since they feature even smaller bite angles. The reactivity of these metal(0) fragments is currently being explored in our lab.

Our interest in bis(carbene) copper(I) coordination chemistry [52] and our efforts dedicated to the elucidation and isolation of reactive intermediates in copper(I)-catalyzed cyclopropanation [53] have led us to the investigation of the coordination properties of the new NHCP ligands toward cationic and halide-containing copper(I) precursors. Further, since attractive $d^{10}-d^{10}$ interactions in binuclear copper(I) complexes are still a subject of debate, unlike, for example, in gold(I) chemistry, we reasoned that by studying the new and structurally easily variable *N*-phosphino NHCs in this context we could gain valuable insights into this so-called "cuprophilicity" phenomenon.

Adding the isolated NHCP ligands **R-34** (R = Me, Mes, *t*Bu) to copper(I) precursors led to the formation of mostly bridging systems displaying interesting structural motifs in the solid state, as determined by X-ray crystal structure

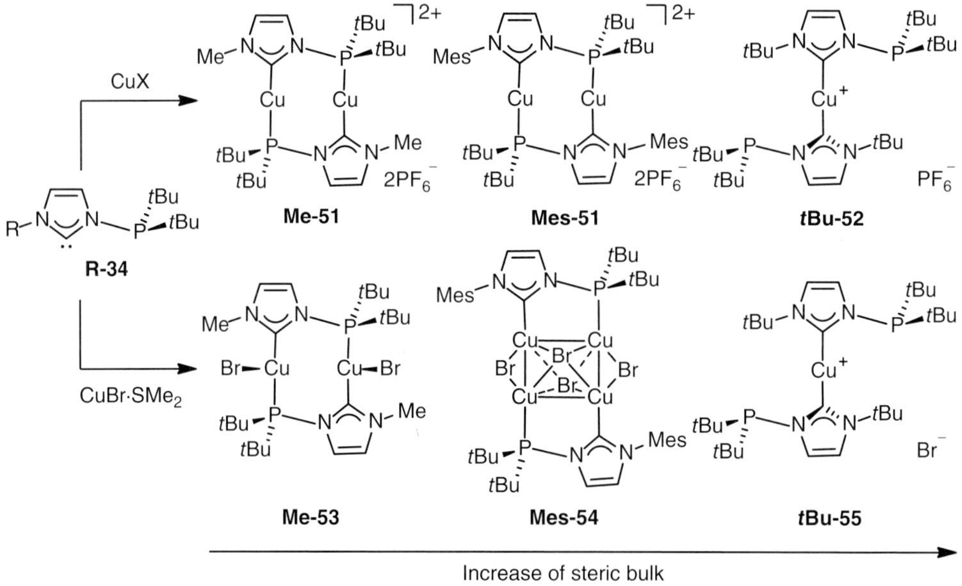

Scheme 10.11 Coordination chemistry of *N*-phosphino-NHC copper(I) complexes.

analysis (Scheme 10.11) [54]. It was obvious that the structure obtained depended strongly on the N-substituent of the ligand. Thus, in cationic systems, ligands with lower steric bulk (R = Me, Mes) led to the formation of the dimeric 10-membered ring metallacycles **R-51** in which the ligands selectively coordinated trans to each other, very likely caused by the steric nature of the NHC moiety. We are certain that a cis coordination is clearly disfavored because the N-substituents would inevitably face each other in these rather rigid systems. Enlarging the N-substituent to a *tert*-butyl group by using ligand ***t*Bu-34** led to the formation of the mononuclear bis-ligated complex **52** in which only the NHC groups coordinate to the copper center. The NHC planes are oriented nearly perpendicular to each other. Kostyuk and Biffis [55] have recently also prepared cationic complexes of all coinage metals and, in agreement with our predictions, have observed the formation of only the dinuclear species of type **R-51**, since they used only *N*-phosphino azolylidene ligands bearing rather small N-substituents (R = Me, Et, *i*Pr).

Using a copper(I) source such as $CuBr{\cdot}SMe_2$ and reacting it with **Mes-34** revealed another dinuclear, yet rarely observed structural motif, which likely is formed driven by steric effects. We presume that by going from the *N*-methyl NHCP **Me-34** to the more bulky *N*-mesityl ligand favors the formation of cluster **Mes-54**, since the ligands are then at a greater distance from each other. A further increase of steric bulk to the NHCP ***t*Bu-34** prevented the formation of a bis-ligated species. Instead, the mono nuclear complex ***t*Bu-55** [56] was formed. Extraordinarily, the bromide does not coordinate to the copper center, which has never been observed previously. The only other known complex of the type

$[(NHC)_2Cu]X$ (with X = halide) showed the coordination of an iodide to the copper center to form a Y-shaped geometry [57].

The assumption of a sterically enforced cluster formation was corroborated later in silver(I) iodide coordination chemistry ligated by our methylene-bridged NHCP ligands (Scheme 10.12) [48]. Thus, at lower steric bulk, the *N*-mesityl ligand **Mes-39** in this case still permitted the formation of a dinuclear complex, likely due to the enhanced flexibility of the system in comparison to the rigid *N*-phosphino systems, which allows the N-substituents to twist out of the coordination plane. For greater steric bulk, an *N-tert*-butyl substitution yielded the tetranuclear iodide cluster **56**.

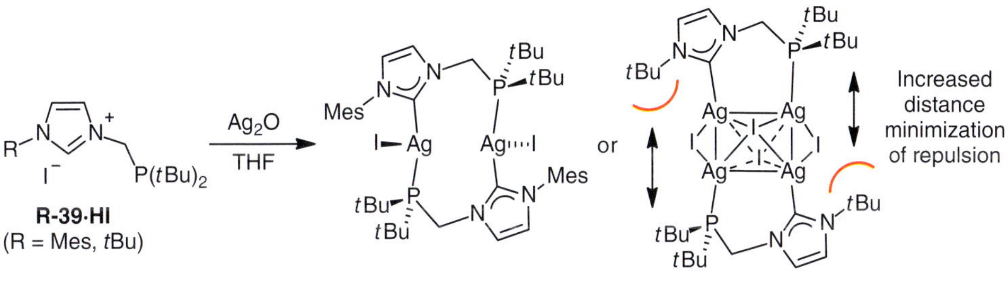

Scheme 10.12 Coordination chemistry of silver(I) *N*-phosphinomethyl-NHC complexes.

10.4.3 *N*-Phosphinomethyl-NHC Ruthenium Alkylidene Complexes

We and others have investigated bisphosphine ruthenium alkylidene complexes of the type $[(P-P)RuCl_2(alkylidene)]$ as analogs of the Grubbs first-generation catalyst in ring-opening metathesis polymerizations (ROMPs) [35, 58, 59]. In our systems, featuring the very bulky, small-bite-angle ligand dtbpm, the addition of halide abstractors such as methyl triflate resulted in the formation of highly active chloro-bridged biscationic dimers which were found to polymerize cyclooctene very efficiently at very low catalyst loadings [41b,c]. These results kindled our interest in the synthesis of the corresponding NHCP systems with our new small-bite-angle systems.

We were successful in synthesizing novel chelating NHCP analogs of the Grubbs II catalyst based on two different routes [49]. Route (a) involves the direct conversions of either the NHCP ligand or its silver adduct obtained from the reaction of the corresponding imidazolium salt **R-39·HI** and silver oxide, with Grubbs I type precursors (Scheme 10.13). In case of the very bulky NHCP ***t*Bu-39**, transmetallation from its silver adduct to Grubbs I catalyst was the key to success, while the less bulky **Mes-39** reacted readily with Umicore M1 catalyst. Direct conversions of isolated **Mes-39** with Grubbs I catalyst led to the formation of unidentifiable side products while ***t*Bu-39** did not react at all.

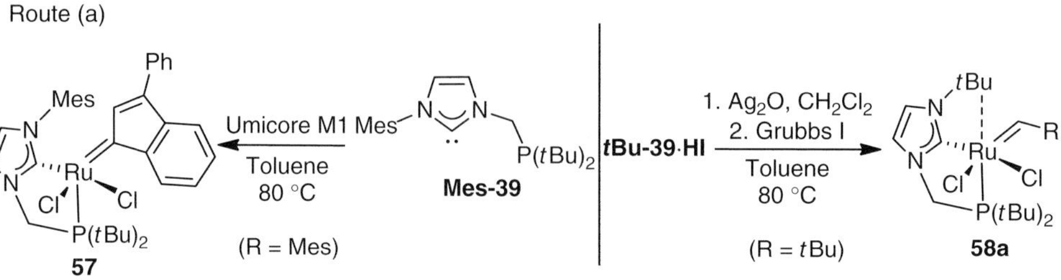

Scheme 10.13 Direct synthesis of NHCP ruthenium alkylidene complexes. The dashed line in complex **58a** does not indicate an agostic interaction but implies the steric shielding of the free coordination site as a consequence of the distinct topology of the NHC group.

Route (b) involves the introduction of the alkylidene ligand with various diazo compounds at a later stage of the synthesis (Scheme 10.14). The first step was based on the reaction of the isolated NHCP ***t*Bu-39** with either $[Ru(COD)Cl_2]_n$ in the presence of H_2 pressure or $[Ru(p\text{-cymene})Cl]_2$ at elevated temperatures to give a mixture of the chloro-bridged diastereomeric dimers **59a/b** as determined by NMR spectroscopy at room temperature. Of the numerous possible diastereoisomers conceivable for this compound, we were able to show via density functional theory (DFT) calculations and variable-temperature NMR experiments that these are the only feasible structures (P-donors located in the apical positions of the square-pyramidal substructures and their mutual trans configuration with respect to the Ru_2Cl_2 plane) and that they dynamically interconvert, possibly

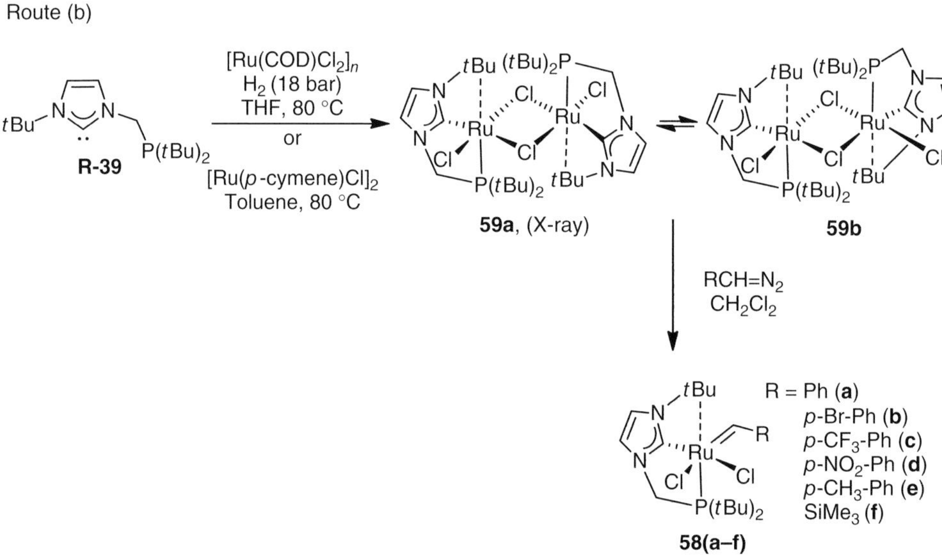

Scheme 10.14 Synthesis of NHCP ruthenium alkylidene complexes by introducing the alkylidene ligand with diazo compounds.

via a triply chloro-bridged intermediate or transition state. Crystals suitable for X-ray analysis showed only the isomer **59a**, which, according to DFT calculations (Turbomole 5.7, [BP86/SV(P)], gas phase), was found to be 36 kJ mol^{-1} more stable than the isomer **59b**. The conversions of these dimers with diazo compounds yielded the alkylidene complexes **58a–f**, which were all characterized by X-ray crystal structure analysis.

Reactions of the new NHCP ruthenium alkylidene complexes **57** and **58a–f** toward halide abstractors and application of the formed cationic species in ROMP catalysis will be published in due course. Additionally, the preparation of the corresponding *N*-phosphino-NHC systems is also under way.

References

1. (a) Arduengo, A.J., Harlow, R.L., and Kline, M. (1991) *J. Am. Chem. Soc.*, **113**, 361; (b) Arduengo, A.J., Kline, M., Calabrese, J.C., and Davidson, F. (1991) *J. Am. Chem. Soc.*, **113**, 9704.
2. Öfele, K.J. (1968) *J. Organomet. Chem.*, **12**, 42.
3. Herrmann, W.A., Elison, M., Fischer, J., Köcher, C., and Artus, G.R.J. (1995) *Angew. Chem.*, **107**, 2602; *Angew. Chem., Int. Ed. Engl.*, 1995, **34**, 2371.
4. (a) Nolan, S.P. (ed) (2006) *N-Heterocyclic Carbenes in Synthesis*, Wiley-VCH Verlag GmbH, Weinheim; (b) Glorius, F. (ed) (2007) *N-Heterocyclic Carbenes in Transition Metal Catalysis*, Springer-Verlag, Berlin; (c) Díez-González, S.S. (ed) (2010) *N-Heterocyclic Carbenes: From Laboratory Curiosities to Efficient Synthetic Tools*, Royal Society of Chemistry, Cambridge; (d) Cazin, C.S.J. (ed) (2010) *N-Heterocyclic Carbenes in Transition Metal Catalysis and Organocatalysis. Catalysis by Metal Complexes*, Springer-Verlag, Heidelberg.
5. (a) Herrmann, W.A. (2002) *Angew. Chem.*, **114**, 1342; *Angew. Chem. Int. Ed.* 2002, **41**, 1290; (b) Peris, E. and Crabtree, R.H. (2004) *Coord. Chem. Rev.*, **248**, 2239; (c) Kantchev, E.A.B., O'Brien, C.J., and Organ, M.G. (2007) *Angew. Chem.*, **119**, 2824; *Angew. Chem. Int. Ed.* **2007**, 46, 2768; (d) Hahn, F.E. and Jahnke, M.C. (2008) *Angew. Chem.*, **120**, 3166; *Angew. Chem. Int. Ed.* (2008) **47**, 3122; (e) Díez-González, S., Marion, N., and Nolan, S.P. (2009) *Chem. Rev.*, **109**, 3612; (f) Vougioukalakis, G.C. and Grubbs, R.H. (2010) *Chem. Rev.*, **110**, 1746; (g) Dröge, T. and Glorius, F. (2010) *Angew. Chem.*, **122**, 7049; *Angew. Chem. Int. Ed.* (2010) **49**, 6940; (h) Valente, C., Çalimsiz, S., Hoi, K.H., Mallik, D., Sayah, M., and Organ, M.G. (2012) *Angew. Chem.*, **124**, 3370; *Angew. Chem. Int. Ed.* (2012) **51**, 3314.
6. Selected reviews: (a) Marion, N., Díez-González, S., and Nolan, S.P. (2007) *Angew. Chem.*, **119**, 3046; *Angew. Chem. Int. Ed.* (2007) **46**, 2988; (b) Enders, D., Niemeier, O., and Henseler, A. (2007) *Chem. Rev.*, **107**, 5606; (c) Nair, V., Vellalath, S., and Babu, B.P. (2008) *Chem. Soc. Rev.*, **37**, 2691; (d) Grossmann, A. and Enders, D. (2012) *Angew. Chem.*, **124**, 320; *Angew. Chem. Int. Ed.* (2012) **51**, 314.
7. (a) Börner, A. (ed) (2008) *Phosphorus Ligands in Asymmetric Catalysis. Synthesis and Applications*, Wiley-VCH Verlag GmbH, Weinheim; (b) Kamer, P.C.J. and Van Leeuwen, P.W.N.M. (eds) (2012) *Phosphorus(III) Ligands in Homogeneous Catalysis. Design and Synthesis*, John Wiley & Sons, Inc., New York.
8. Horner, L., Siegel, H., and Buthe, H. (1968) *Angew. Chem.*, **80**, 1034; *Angew. Chem., Int. Ed. Engl.* (1968) **7**, 942.
9. (a) Kühl, O. (2007) *Chem. Soc. Rev.*, **36**, 592; (b) Lee, H.M., Lee, C.-C., and Cheng, P.-Y. (2007) *Curr. Org. Chem.*, **11**, 1491; (c) Normand, A.T. and Cavell, K.J. (2008) *Eur. J. Inorg. Chem.*, 2781; (d) Kühl, O. (ed) (2010) *Functionalised N-Heterocyclic Carbene Complexes*, John Wiley & Sons, Inc, New York;

For polydentate NHC-systems, see: (e) Edwards, P.G. and Hahn, F.E. (2011) *Dalton Trans.*, **40**, 10278.

10. Some interesting works in NHCP coordination chemistry have been published, which we are not able to discuss within the context of this book chapter: For ruthenium complexes, see: (a) Benhamou, L., Wolf, J., César, V., Labande, A.S., Poli, R., Lugan, N.L., and Lavigne, G. (2009) *Organometallics*, **28**, 6981; (b) Cabeza, J.A., Damonte, M., García-Álvarez, P., Kennedy, A.R., and Pérez-Carreño, E. (2011) *Organometallics*, **30**, 826; (c) Cabeza, J.A., Damonte, M., García-Álvarez, P., Hérnández-Cruz, M.G., and Kennedy, A.R. (2012) *Organometallics*, **31**, 327; For metal complexes of group 9, see: (d) Stylianides, N., Danopoulos, A.A., and Tsoureas, N. (2005) *J. Organomet. Chem.*, **690**, 5948; (e) Wolf, J., Labande, A., Daran, J.-C., and Poli, R. (2008) *Eur. J. Inorg. Chem.*, **2008**, 3024; (f) Labande, A., Debono, N., Sournia-Saquet, A., Daran, J.-C., and Poli, R. (2013) *Dalton Trans.*, **42**, 6531; (g) Liu, X. and Braunstein, P. (2013) *Inorg. Chem.*, **52**, 7367; (h) For a dinuclear iron complex bridged by a NHCP, see: Song, L.-C., Gai, B., Feng, Z.-H., Du, Z.-Q., Xie, Z.-J., Sun, X.-J., and Song, H.-B. (2013) *Organometallics*, **32**, 3673; For chiral NHCP complexes not applied in catalysis, see: (i) Lang, H., Vittal, J.J., and Leung, P.-H. (1998) *J. Chem. Soc., Dalton Trans.*, 2109; (j) Abdellah, I., Lepetit, C., Canac, Y., Duhayon, C., and Chauvin, R. (2010) *Chem. Eur. J.*, **16**, 13095; For a phosphine-tethered acyclic diaminocarbene ligand, see: (k) Yu, I., Wallis, C.J., Patrick, B.O., Diaconescu, P.L., and Mehrkhodavandi, P. (2010) *Organometallics*, **29**, 6065; For interesting structural work on PCP pincers: (l) Kaufhold, O., Stasch, A., Pape, T., Hepp, A., Edwards, P.G., Newman, P.D., and Hahn, F.E. (2009) *J. Am. Chem. Soc.*, **131**, 306; (m) Steinke, T., Shaw, B.K., Jong, H., Patrick, B.O., and Fryzuk, M.D. (2009) *Organometallics*, **28**, 2830; (n) Steinke, T., Shaw, B.K., Jong, H., Patrick, B.O., Fryzuk, M.D., and Green, J.C. (2009) *J. Am. Chem. Soc.*, **131**, 10461; (o) Shaw, B.K., Patrick, B.O., and Fryzuk, M.D. (2012) *Organometallics*, **31**, 783; (p) Vabre, B., Canac, Y., Duhayon, C., Chauvin, R., and Zargarian, D. (2012) *Chem. Commun.*, **48**, 10446; For a more general review on imidazole-based phosphorus compounds and their use as ligands, see: (q) Gaillard, S. and Renaud, J.-L. (2013) *Dalton Trans.*, **42**, 7255.
11. Herrmann, W.A., Köcher, C., Gookn, L.J., and Artus, G.R.J. (1996) *Chem. Eur. J.*, **2**, 1627.
12. Yang, C., Lee, H.M., and Nolan, S.P. (2001) *Org. Lett.*, **3**, 1511.
13. (a) Tsoureas, N., Danopoulos, A.A., Tulloch, A.A.D., and Light, M.E. (2003) *Organometallics*, **22**, 4750; (b) Lee, H.M., Zeng, J.Y., Hu, C.-H., and Lee, C.-H. (2004) *Inorg. Chem.*, **43**, 6822; (c) Lee, H.M., Chiu, P.L., and Zeng, J.Y. (2004) *Inorg. Chim. Acta*, **357**, 4313; (d) Wang, A.-E., Zhong, J., Xie, J.-H., Li, K., and Zhou, Q.-L. (2004) *Adv. Synth. Catal.*, **346**, 595; (e) Wang, A.-E., Xie, J.-H., Wang, L.-X., and Zhou, Q.-L. (2005) *Tetrahedron*, **61**, 259; (f) Zhong, J., Xie, J.-H., Wang, A.-E., Zhang, W., and Zhou, Q.-L. (2006) *Synlett*, **8**, 1193; (g) Wolf, J., Labande, A., Daran, J.-C., and Poli, R. (2006) *J. Organomet. Chem.*, **691**, 433; (h) Shi, J.-C., Yang, P.-Y., Tong, Q., Wu, Y., and Peng, Y. (2006) *J. Mol. Catal. A: Chem.*, **259**, 7; (i) Wolf, J., Labande, A., Natella, M., Daran, J.C., and Poli, R. (2006) *J. Mol. Catal. A: Chem.*, **259**, 205; (j) Hahn, F.E., Jahnke, M.C., and Pape, T. (2006) *Organometallics*, **25**, 5927; (k) Lee, C.-C., Ke, W.-C., Chan, K.-T., Lai, C.-L., Hu, C.-H., and Lee, H.M. (2007) *Chem. Eur. J.*, **13**, 582; (l) Fiddy, S.G., Evans, J., Neisius, T., Newton, M.A., Tsoureas, N., Tulloch, A.A.D., and Danopoulos, A.A. (2007) *Chem. Eur. J.*, **13**, 3652; (m) Shi, J.-c., Yang, P., Tong, Q., and Jia, L. (2008) *Dalton Trans.*, 938; (n) Debono, N., Labande, A., Manoury, E., Daran, J.-C., and Poli, R. (2010) *Organometallics*, **29**, 1879; For a non-chelating NHCP imidazolium salt (not shown in Figure 1) used in Suzuki-coupling, see: (o) Milde, B., Schaarschmidt, D., Rüffer, T., and Lang, H. (2012) *Dalton Trans.*, **41**, 5377.

14. Field, L.B., Messerle, B.A., Vuong, K.Q., and Turner, P. (2005) *Organometallics*, **24**, 4241.
15. (a) Barczak, N.T., Grote, R.E., and Jarvo, E.R. (2007) *Organometallics*, **26**, 4863; (b) Shaghafi, M.B., Kohn, B.L., and Jarvo, E.R. (2008) *Org. Lett.*, **10**, 4743.
16. *N*-substituent of the NHCP = DiPP: (a) Stylianides, N., Danopoulos, A.A., and Tsoureas, N. (2005) *J. Organomet. Chem.*, **690**, 5948. (b) Danopoulos, A.A., Tsourears, N., Macgregor, S.A., and Smith, C. (2007) *Organometallics*, **26**, 253 ; see also Ref. [13l]; *N*-substituent = *i*Pr: (c) Song, G., Wang, X., Li, Y., and Li, X. (2008) *Organometallics*, **27**, 1187; (d)*N*-substituent = H: Hahn, F.E., Naziruddin, A.R., Hepp, A., and Pape, T. (2010) *Organometallics*, **29**, 5283.
17. Ho, C.-C., Chatterjee, S., Wu, T.-L., Chan, K.-T., Chang, Y.-W., Hsiao, T.-H., and Lee, H.M. (2009) *Organometallics*, **28**, 2837.
18. Zhong, J., Xie, J.-H., Wang, A.-E., and Zhou, Q.-L. (2006) *Synlett*, **8**, 1193.
19. Hsiao, T.-H., Wu, T.-L., Chatterjee, S., Chiu, C.-Y., Lee, H.M., Bettucci, L., Bianchini, C., and Oberhauser, W. (2009) *J. Organomet. Chem.*, **692**, 4014.
20. (a) Labande, A., Daran, J.-C., Manoury, E., and Poli, R. (2007) *Eur. J. Inorg. Chem.*, **2007**, 1205; (b) Gülcemal, S., Labande, A., Daran, J.-C., Çetinkaya, B., and Poli, R. (2009) *Eur. J. Inorg. Chem.*, **2009**, 1806.
21. Zeng, J.Y., Hsieh, M.-H., and Lee, H.M. (2005) *J. Organomet. Chem.*, **690**, 5662.
22. (a) Miranda-Soto, V., Grotjahn, D.B., DiPasquale, A.G., and Rheingold, A.L. (2008) *J. Am. Chem. Soc.*, **130**, 13200; (b) Miranda-Soto, V., Grotjahn, D.B., Cooksy, A.L., Golen, J.A., Moore, C.E., and Rheingold, A.L. (2011) *Angew. Chem. Int. Ed.*, **50**, 631.
23. Chiu, P.L. and Lee, H.M. (2005) *Organometallics*, **24**, 1692.
24. Perry, M.C. and Burgess, K. (2003) *Tetrahedron: Asymmetry*, **14**, 951.
25. César, V., Bellemin-Laponnaz, S., and Gade, L.H. (2004) *Chem. Soc. Rev.*, **33**, 619.
26. Wang, F., Liu, L.-J., Wang, W., Li, S., and Shi, M. (2012) *Coord. Chem. Rev.*, **256**, 804.
27. (a) Bappert, E. and Helmchen, G. (2004) *Synlett*, **10**, 1789; (b) Becht, J.-M., Bappert, E., and Helmchen, G. (2005) *Adv. Synth. Catal.*, **347**, 1495.
28. (a) Seo, H., Park, H.-J., Kim, B.Y., Lee, J.H., Son, S.U., and Chung, Y.K. (2003) *Organometallics*, **22**, 618; (b) Focken, T., Raabe, G., and Bolm, C. (2004) *Tetrahedron: Asymmetry*, **15**, 1693; (c) Nanchen, S. and Pfaltz, A. (2006) *Helv. Chim. Acta*, **89**, 1559; (d) Passays, J., Ayad, T., Ratovelomanana-Vidal, V., Gaumont, A.-C., Jubault, P., and Leclerc, E. (2011) *Tetrahedron: Asymmetry*, **22**, 562; (e) Gu, P., Zhang, J., Xu, Q., and Shi, M. (2013) *Dalton Trans.*, **42**, 13599.
29. Visentin, F. and Togni, A. (2007) *Organometallics*, **26**, 3746.
30. See for example: (a) Zhang, R., Xu, Q., Zhang, X., Zhang, T., and Shi, M. (2010) *Tetrahedron: Asymmetry*, **21**, 1928; (b) Wang, W.F., Zhang, T., Wang, F.J., and Shi, M. (2011) *Tetrahedron*, **67**, 1523; (c) Liu, Z., Gu, P., and Shi, M. (2011) *Chem. Eur. J.*, **17**, 5796.
31. Fryzuk, M.D. and Bosnich, B. (1977) *J. Am. Chem. Soc.*, **99**, 6262.
32. Gischig, S. and Togni, A. (2005) *Eur. J. Inorg. Chem.*, **2005**, 4745.
33. Hodgson, R. and Douthwaite, R.E. (2005) *J. Organomet. Chem.*, **690**, 5822.
34. Toselli, N., Martin, D., and Buono, G. (2008) *Org. Lett.*, **10**, 1453.
35. Willms, H., Frank, W., and Ganter, C. (2008) *Chem. Eur. J.*, **14**, 2719.
36. (a) Steinhagen, H., Reggelin, M., and Helmchen, G. (1997) *Angew. Chem., Int. Ed. Engl.*, **36**, 2108; (b) Junker, J., Reif, B., Steinhagen, H., Junger, B., Felli, I.C., Reggelin, M., and Griesinger, C. (2000) *Chem.Eur. J.*, **6**, 3281.
37. Barbaro, P., Bianchini, C., and Togni, A. (1997) *Organometallics*, **16**, 3004.
38. Fadini, L. and Togni, A. (2003) *Chem. Commun.*, 30.
39. (a) Gishig, S. and Togni, A. (2004) *Organometallics*, **23**, 2479; (b) Gischig, S. and Togni, A. (2005) *Organometallics*, **24**, 203.
40. For bond-activation chemistry, see: (a) Hofmann, P., Heiss, H., Müller, G., and

Lachmann, J. (1990) *Angew. Chem.*, **102**, 935; *Angew. Chem. Int. Ed.*, (1990), **29**, 880; (b) Hofmann, P. and Unfried, G. (1992) *Chem. Ber.*, **125**, 659; (c) P. Hofmann Silicon chemistry in (Eds. N. Auner, J. Weis) *Organosilane Chemistry at Tailor-Made Platinum Centers*, Wiley-VCH Verlag GmbH: Weinheim, 1993.

41. (a) For application in ROMP catalysis, see: Hansen, S.M., Metz, M., Rominger, F., and Hofmann, P. (1999) *Chem. Eur. J.*, **5**, 557; (b) Hansen, S.M., Volland, M.A.O., Rominger, F., Eisenträger, F., and Hofmann, P. (1999) *Angew. Chem.*, **111**, 1360; *Angew. Chem. Int. Ed.*, (1999), **38**, 1273; (c) Volland, M.A.O., Hansen, S.M., Rominger, F., and Hofmann, P. (2004) *Organometallics*, **23**, 800.
42. (a) For application in nickel-catalyzed polymerization of olefins, see: Kristen, M.O., Hofmann, P., and Eisenträger, F. (2002) Metallic compounds and the use thereof in the polymerisation of olefins. WO 0202573A1 (BASF SE); (b) Schultz, M., Eisenträger, F., Regius, C., Rominger, F., Hanno-Igels, P., Jakob, P., Gruber, I., and Hofmann, P. (2012) *Organometallics*, **31**, 207; For palladium-catalyzed copolymerization of ethylene/CO, see:(c) Lippert, F., Höhn, A., and Hofmann, P. (1995) Process for preparing polyketones. WO 9503353 A1 (BASF SE); (d) Lippert, F., Höhn, A., and Hofmann, P. (1996) Carbon monoxide/olefin-copolymers. WO 9637537 A1 (BASF SE).
43. (a) Recent studies on the thermodynamic component to the *gem*-dimethyl effect in carbocyclic compounds: Ringer, A.L. and Magers, D.H. (2007) *J. Org. Chem.*, **72**, 2533; (b) Bachrach, S.M. (2008) *J. Org. Chem.*, **73**, 2466.
44. (a) Naziruddin, A.R., Hepp, A., Pape, T., and Hahn, F.E. (2011) *Organometallics*, **30**, 5859.
45. (a) Herrlich, U. (neé Blumbach) (2007) Neue NHCP-Chelatliganden für die homogene Katalyse. Dissertation, Universität Heidelberg; (b) Eisenhauser, D. (2007) Synthese neutraler und kationischer Rhodium- und Iridiumkomplexe mit chelatisierenden Bisphosphan-, Biscarben- und Phosphinocarben-Liganden. Dissertation, Universität Heidelberg; (c) Schmitt, M. (2008) Neue Rutheniumchelatkomplexe mit Anwendungspotential in der molekularen Katalyse. Universität Heidelberg; (d) Hanno-Igels, P. (2011) Chirale NHCP-Liganden für die Katalyse. Dissertation, Universität Heidelberg; (e) Nägele, P. (2010) Neue NHCP-Übergangsmetallkomplexe. Diplomarbeit, Universität Heidelberg.
46. Nägele, P., Herrlich, U. (neé Blumbach), Rominger, F., and Hofmann, P. (2013) *Organometallics*, **32**, 181.
47. Marchenko, A.P., Koidan, H.N., Huryeva, A.N., Zarudnitskii, E.V., Yurchenko, A.A., and Kostyuk, A.N. (2010) *J. Org. Chem.*, **75**, 7141.
48. Marchenko, A.P., Koidan, H.N., Pervak, I.I., Huryeva, A.N., Zarudnitskii, E.V., Tolmachev, A.A., and Kostyuk, A.N. (2012) *Tetrahedron Lett.*, **53**, 494.
49. Salem, H., Schmitt, M., Herrlich, U. (née Blumbach), Kühnel, E., Brill, M., Nägele, P., Rominger, F., Bogado, A.L., and Hofmann, P. (2013) *Organometallics*, **32**, 29.
50. Danopoulos, A.A., Winston, S., Gelbrich, T., Hursthouse, M.B., and Tooze, R.P. (2002) *Chem. Commun.*, 482.
51. Brill, M., Kühnel, E., Scriban, C., Rominger, F., and Hofmann, P. (2013) *Dalton Trans.*, **42**, 12861.
52. Shishkov, I.V., Rominger, F., and Hofmann, P. (2009) *Dalton Trans.*, 1428.
53. (a) Straub, B.F., Rominger, F., and Hofmann, P. (2001) *Angew. Chem.*, **113**, 1328; *Angew. Chem. Int. Ed.* (2001), **40**, 1288; (b) Straub, B.F., Gruber, I., Rominger, F., and Hofmann, P. (2003) *J. Organomet. Chem.*, **684**, 124; (c) Shishkov, I.V., Rominger, F., and Hofmann, P. (2009) *Organometallics*, **28**, 1049.
54. Kühnel, E., Shishkov, I.V., Rominger, F., Oeser, T., and Hofmann, P. (2012) *Organometallics*, **31**, 8000.
55. Marchenko, A.P., Koidan, H.N., Hurieva, A.N., Gutov, O.V., Kostyuk, A.N., Tubaro, C., Lollo, S., Lanza, A., Nestola, F., and Biffis, A. (2013) *Organometallics*, **32**, 718.

56. Kühnel, E. (2013) Neue Münzmetall-NHCP-Komplexe: Synthese, Struktur und Eigenschaften. Dissertation, Universität Heidelberg.
57. Venkatachalam, G., Heckenroth, M., Neels, A., and Albrecht, M. (2009) *Helv. Chim. Acta*, 1034.
58. Amoroso, D. and Fogg, D.E. (2000) *Macromolecules*, **33**, 2815.
59. Six, C., Beck, K., Wegner, A., and Leitner, W. (2000) *Organometallics*, **19**, 4639.

Part III
Catalysts in Synthesis

Günter Helmchen

Stereoselective homogeneous catalysis has become an integral issue of organic synthesis. Today, a multistep synthesis of a natural product would not be considered state of the art if stereoselectivity were not essentially controlled by catalysts. This concerns preparation of the starting material with the help of asymmetric catalysis, that is, selective buildup of one of a pair of enantiomers, as well as control of diastereoselectivity in further steps.

In addition to transition-metal catalysis, biocatalysis and organocatalysis are very important aspects of stereoselective catalysis. Recently, dual catalysis, involving combinations of several types of catalysis, for example, transition-metal catalysis and biocatalysis, has become of interest. Multicomponent and domino reactions are desirable because of the often high atom economy and efficiency. All these aspects have been pursued within the Heidelberg Collaborative Research Network (Sonderforschungsbereich, SFB 623), though most of the effort has been devoted to transition-metal catalysis.

Efforts in the latter area have been traditionally directed at the development of new chiral ligands, typically chelating diamines and diphosphines or, recently, N-heterocyclic carbenes. Screening of ligands for a particular reaction is the traditional mode of operation and, doubtless, still successful. However, within the research efforts presented in this monograph, there has always been attempt to elucidate reaction mechanism by experiment, mainly spectroscopic measurement, and quantum chemical calculation and to try to develop catalysts on a rational basis as far as is possible today.

Analytical evaluation is of great importance for stereoselective catalysis. Much of the success in recent years has its origin in new tools such HPLC and GC with modified celluloses, cyclodextrins, or siloxane-bound amino acid derivatives as chiral stationary phases. As described in Chapter 21, it has been shown that upon use of polymer-bound chiral catalysts or selectors, the kinetic parameters of stereoselective reactions can be determined in a parallel mode with high precision.

Finally, at a university, the educational aspects must not be forgotten. Stereoselective catalysis needs knowledge and skills in a variety of directions. First

Molecular Catalysts: Structure and Functional Design, First Edition. Edited by Lutz H. Gade and Peter Hofmann.

and foremost, the principles of stereochemistry are required, both in the traditional field of carbon compounds and the more demanding coordination chemistry. Catalysis is, by necessity, interdisciplinary in the sense that it amalgamates inorganic and organic chemistry to the extent that the boundaries between these classical subdisciplines vanish, except perhaps in undergraduate teaching. Interdisciplinary interaction among professors and students is a particularly pleasing side effect of the SFB.

Looking in more detail on the work presented in this monograph, several areas of stereoselective catalysis have been particularly intensely explored. One is the allylic substitution reaction, of which several aspects were investigated with considerable success. The Menche group investigated substrate-controlled diastereoselective combinations of a Michael addition followed by an intramolecular allylic substitution. This concept could be realized in several versions with a high degree of diastereoselectivity. A variety of structures of interest for natural products synthesis, particularly in the area of polyketides, can be generated on the basis of this approach and its several extensions.

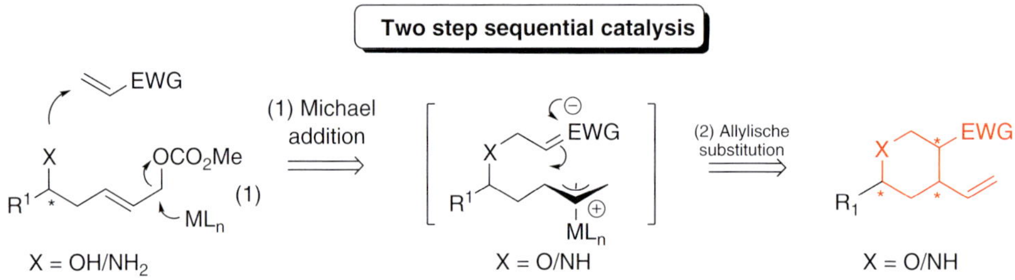

Multicomponent sequential transition-metal-catalyzed reactions were accomplished by the group of Müller. Their approach involves cycloisomerizations or combinations of a Heck reaction and an insertion or allylic substitution reaction, yielding an intermediate from which by β-H-elimination an aldehyde is generated, which reacts with an amine to give an imine, which in turn is reduced or otherwise modified. This conceptually challenging approach allows the experimentally remarkably facile generation of a variety of compounds of interest in medicinal chemistry.

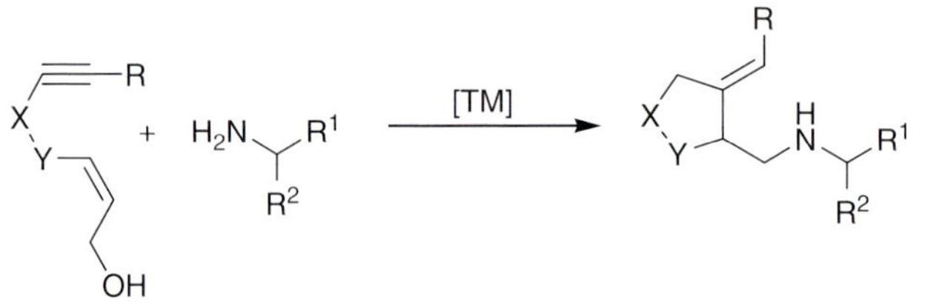

Neutral group 4 metal complexes appear to possess a relatively broad scope for catalytic hydroaminations. They have been employed for the intramolecular as well as the intermolecular hydroaminations of alkenes, alkynes, and allenes. Catalytic hydroaminations (and hydrohydrazinations) of alkynes have been exploited as key steps in catalytic multicomponent reactions, giving rise to highly functionalized substrates, in particular to several types of N-heterocycles. Chapter 13 by Gade focuses, *inter alia*, on two case histories involving hydrohydrazinations which exemplify key challenges and the way they may be addressed in practice.

Work of the Helmchen group was focused on the iridium-catalyzed asymmetric allylic substitution, which was introduced by the group and developed into a broadly applicable tool for organic synthesis:

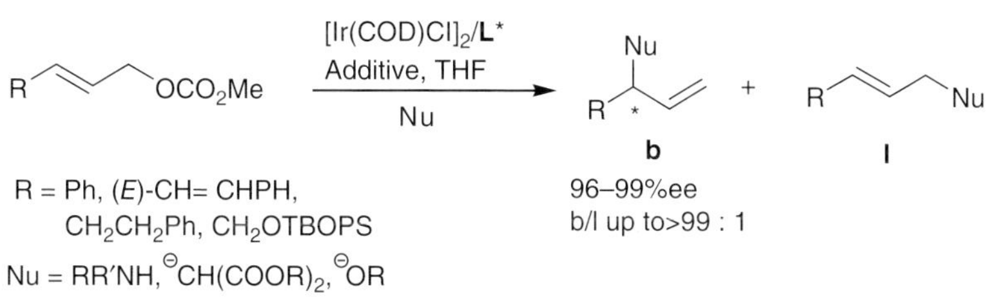

Phosphoramidites are effective ligands for these reactions. Particularly good results with respect to enantio- and regioselectivity were achieved by *in situ* C–H activation of their iridium complexes or by use of isolated (π-allyl)Ir complexes, which have become easily available.

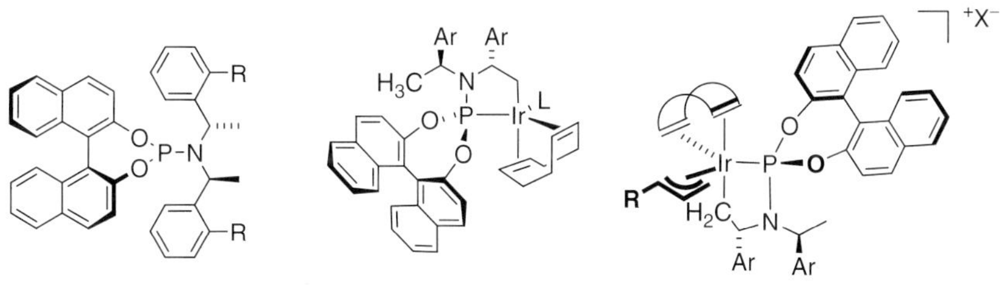

L *In situ* activated catalyst

Recent progress has allowed the method to be used for relatively short syntheses of biologically active compounds with up to 99% ee.

Development of new chiral ligands is of crucial importance for asymmetric catalysis. The Gade group has established several new classes of multidentate ligands. The modular makeup of these ligands allows structures to be varied conveniently. Furthermore, attachment to solid supports has added to their practical significance. The new ligands have been applied in numerous reactions with excellent success. Examples are catalytic hydrogenations, allylic and other oxidations, as well as carbon–carbon and carbon–heteroatom bond connections. Among the latter polymerization, cyclopropanation, hydrosilylation reactions, and introduction of fluorine-containing groups should be mentioned. Mechanistic investigation of complexes of the potentially tridentate trisox and phospholane ligands has shown that complicated equilibria between complexes with bi- and tridentate bonding exist.

11
Ir-Catalyzed Asymmetric Allylic Substitution Reactions – Fundamentals and Applications in Natural Products Synthesis

Günter Helmchen

11.1
Introduction

Over the last few years, the Ir-catalyzed allylic substitution has been investigated in several laboratories and was found to be well suited for applications in organic synthesis. Using this reaction, branched chiral allylic derivatives can be prepared with high selectivity from simple achiral monosubstituted allylic substrates (Scheme 11.1). The reaction has been carried out with C, N, O, and S nucleophiles. The very broad range of nucleophiles is impressive. Several reviews have appeared on the subject [1]. At present, the best catalysts are prepared from $[Ir(cod)Cl]_2$ (cod, cycloocta-1,5-diene) [2, 3] or $[Ir(dbcot)Cl]_2$ (dbcot, dibenzocyclooctatetraene) [4] and a chiral phosphoramidite by base-induced C–H activation. Reliable experimental procedures have been developed for the reaction [5].

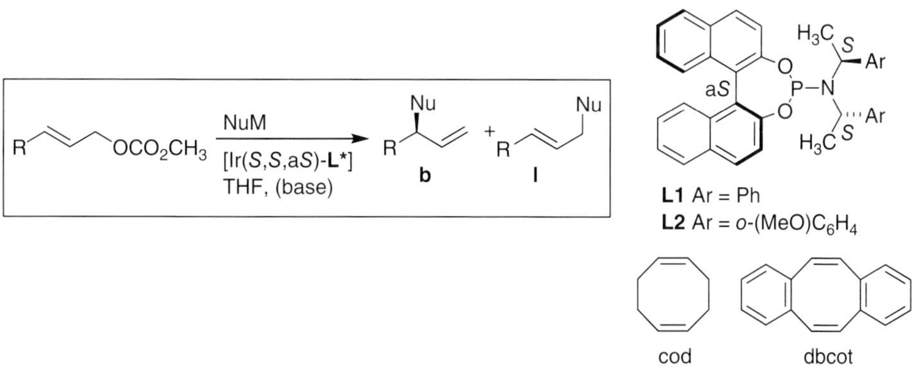

Scheme 11.1 Allylic substitutions with Ir catalysts derived from phosphoramidites.

Molecular Catalysts: Structure and Functional Design, First Edition. Edited by Lutz H. Gade and Peter Hofmann.

11.2
Background on Reaction Mechanism

The following aspects of the Ir-catalyzed allylic substitution have been elucidated by mechanistic investigations: (i) determination of the resting state, (ii) isolation and characterization of catalytically active (π-allyl)Ir complexes, and (iii) quantum chemical calculations. Furthermore, a very simple and broadly applicable method for the preparation of (π-allyl)Ir complexes has been developed. Their structural characterization and evaluation as catalysts have given valuable mechanistic insight into the catalytic process [6]. In our work, the particularly effective phosphoramidite **L2**, which was introduced by Alexakis *et al.* [7], was mainly used. Independent work was carried out simultaneously by Hartwig *et al.* [8, 9] using the ligand **L1**. They also succeeded in preparing and characterizing (π-allyl)Ir complexes using a comparatively complicated preparative procedure different from ours.

Our current view on the mechanism of the Ir-catalyzed allylic substitution is summarized in Scheme 11.2. The complexes **C1**, which are obtained by simply mixing [{Ir(diene)Cl}$_2$] and a phosphoramidite, are catalytically inactive. Catalyst preparation involves activation with a base, which is required in order to generate the cyclometallated complexes **C2** (18 valence electrons (VEs)) [3]. An additional base is not required if the nucleophile is sufficiently basic, for example, in case of benzyl or propyl amine. In our laboratory, the non-nucleophilic base TBD (1,5,7-*triaza*-bicyclo[4.4.0]dec-5-ene) is usually employed because it effects the

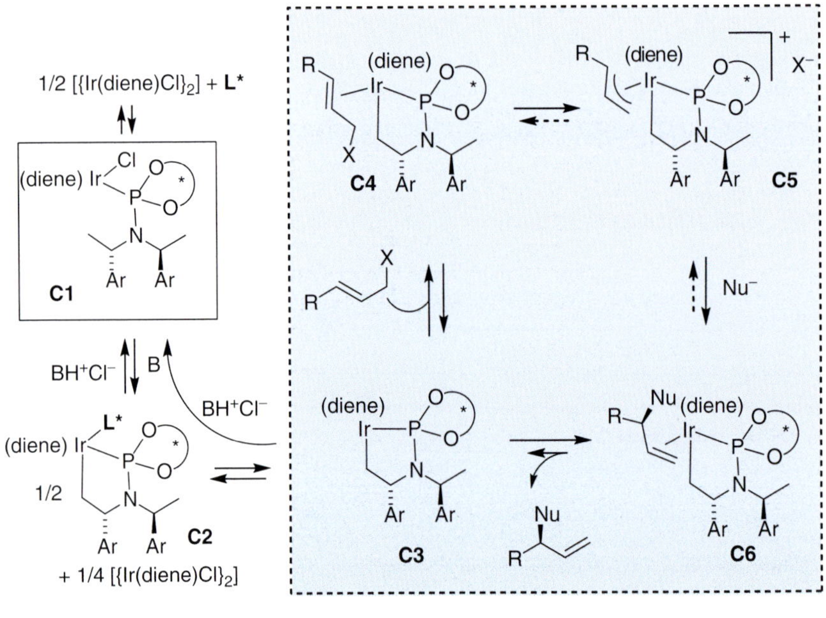

Scheme 11.2 Mechanism of the Ir-catalyzed allylic substitution with catalysts prepared *in situ* from **C1** by reaction with a base B. (According to [10].) (Wiley-VCH.)

activation within a few minutes. The complexes of type **C2** are not themselves catalytically active because they are coordinatively saturated. With respect to rate of an allylic substitution, a 1 : 1 ratio of **L*** and Ir is optimal. The ligand **L*** of **C2** can be easily replaced by other ligands, but an excess of **L*** causes lowering of the rate. Therefore, it is assumed that the active species in the catalytic cycle is the intermediate **C3** with 16 VE.

The method for preparation of the (π-allyl)Ir complexes **C5** mentioned above is a convenient one-pot procedure (Scheme 11.3). The complexes are stable against oxygen and water. Trace impurities can even be removed by conventional column chromatography.

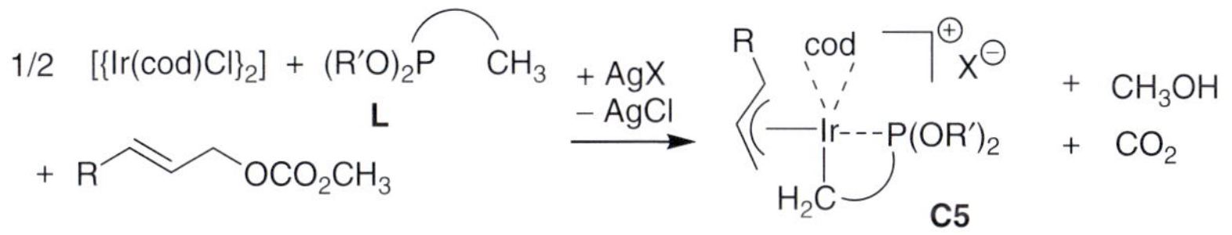

Scheme 11.3 One-pot procedure for the preparation of (π-allyl)Ir complexes via cyclometallation.

The structures of the allyl complexes were determined by X-ray crystal structure analysis. An example is shown in Figure 11.1. It is apparent that the allyl ligand sits in a narrow cleft, which is formed by the cod and the aryl groups A and B. In the outward direction the cleft is open, and the substituent R can be varied quite freely.

On first view, the situation with respect to the π-allyl complexes appears simple. However, this is deceptive. There are no less than 16 possibilities, which are listed in Figure 11.2 [6]. This is a demonstration how complicated the situation is when one has to deal with octahedral complexes. According to density functional theory (DFT) calculations and using cyclooctatetraene as ancillary ligand [6] and mechanistic investigations [11], there is a kinetic as well as thermodynamic preference for the formation of the experimentally observed stereoisomer.

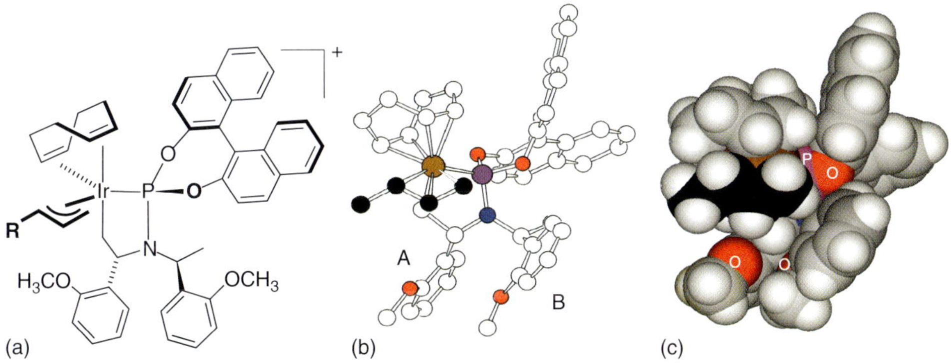

Figure 11.1 X-ray crystal structure of a typical methylallyl ($R = CH_3$) complex. (a) Standard chemical formula. (b) Ball-and-stick model. (c) Corey–Pauling–Koltun (CPK) model.

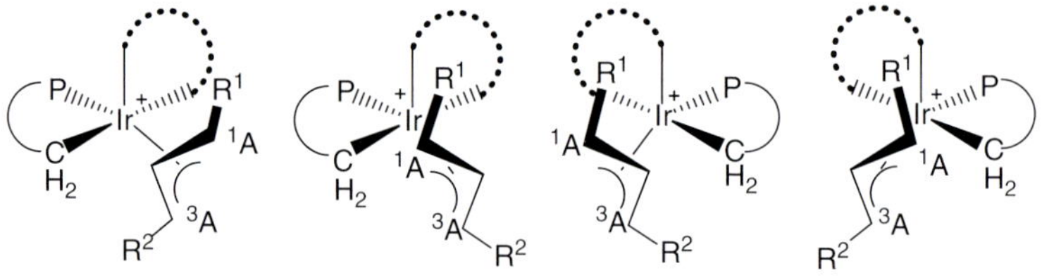

	Δ-P-endo	Δ-P-exo	Λ-P-endo	Λ-P-exo
$R^1 = CH_3$, $R^2 = H$	0.0	3.2	3.9	4.6
$R^1 = H$, $R^2 = CH_3$	3.0	4.7	6.6	7.0

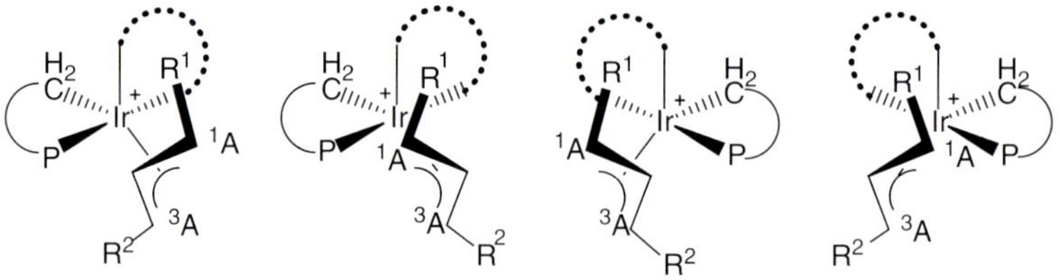

	Δ-C-endo	Δ-C-exo	Λ-C-endo	Λ-C-exo
$R^1 = CH_3$, $R^2 = H$	4.1	3.9	7.2	7.0
$R^1 = H$, $R^2 = CH_3$	6.3	6.7	9.1	9.1

Figure 11.2 Configurations of the (π-allyl)Ir complexes **C5**. •••, cot; P and C series: one allylic carbon is positioned trans to P and the CH_2 group of the C,P ligand, respectively; exo and endo: the $C2_A$–H bond of the allyl ligand is pointing toward cot or away from cot, respectively; energies in kilocalories per mole.

11.3 Dibenzocyclooctatetraene (dbcot) as Ancillary Ligand

Crabtree introduced dbcot into catalysis research [12] as an inhibitor of homogeneous catalytic hydrogenations. It is an inhibitor there because removal of this ligand by hydrogenation is more difficult than that of cod. On the other hand, it was postulated early that the ancillary ligand is retained in the Ir-catalyzed allylic substitution. Accordingly, it was of interest to compare the rates of reactions with the two dienes. This required a convenient procedure for preparation of dbcot, which we have worked out [13]. Indeed, the rate of the reaction was not significantly affected upon exchange of cod by dbcot. Examples of aminations and alkylations are given in Scheme 11.4. It was surprising that the level of regioselectivity was found to be generally improved in comparison with reactions catalyzed with cod complexes [4, 10]. This is due to the fact that dbcot is a stronger electron acceptor than cod [14].

Another very pleasant surprise was the observation that reactions catalyzed by dbcot complexes could be run under an atmosphere of air (Scheme 11.5),

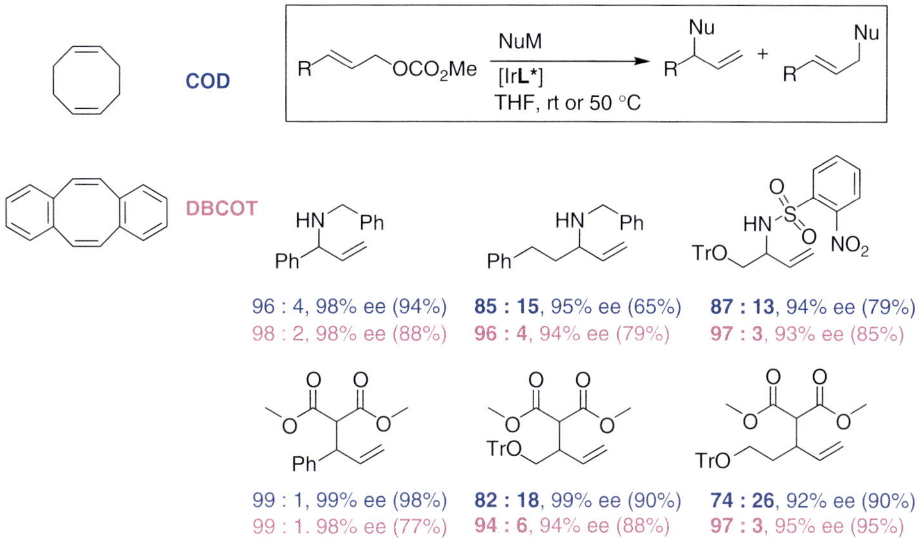

Scheme 11.4 Allylic substitutions with dbcot or cod as ancillary ligand.

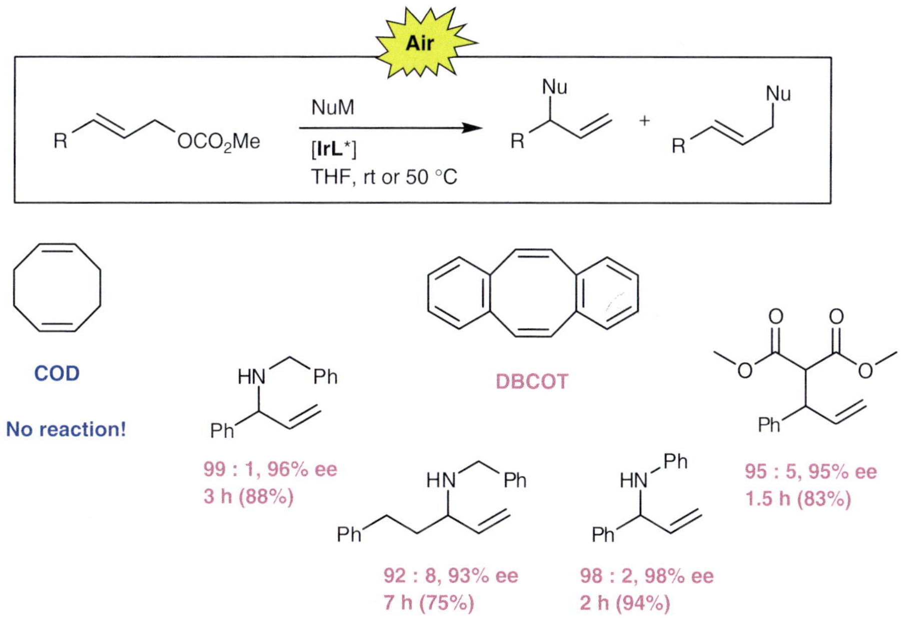

Scheme 11.5 Ir-catalyzed allylic substitutions under aerobic conditions.

which is a very desirable aspect for technical application. The generally high robustness of the dbcot complexes is due particularly to the strong bond between dbcot and iridium as well as high bond strength of the C, Ir-σ-bond.

The robustness of the dbcot complexes allowed us to realize a long-standing problem in the field, namely the asymmetric hydroxylation to give branched allylic alcohols, which are valuable intermediates in stereoselective syntheses (Scheme 11.6) [15]. In prior work, the allylic substitution with water or hydroxyl anion, using a catalyst that was generated *in situ* from $[Ir(cod)Cl]_2$, a chiral phosphoramidite ligand, and a base, failed and was only indirectly accomplished via reaction with a silanoate [16]. With (π-allyl)Ir complexes containing dbcot as ancillary ligand, the reaction with potassium bicarbonate in an aqueous solvent system under air proceeded smoothly to give branched allylic alcohols [17]. The mechanism indicated in Scheme 11.6 is based on previous work by Gais *et al.* [18] on Pd-catalyzed allylic hydroxylation.

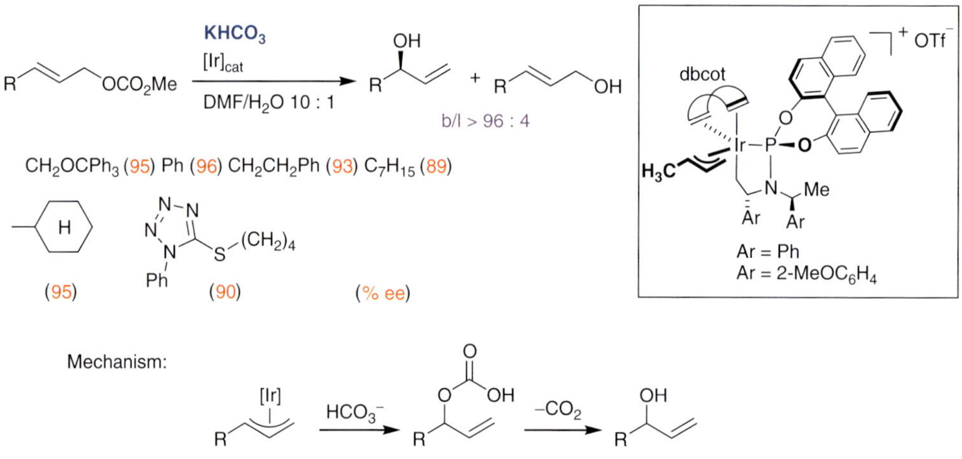

Scheme 11.6 Asymmetric allylic hydroxylation using a dbcot complex as catalyst.

Despite the effort involved in the synthesis of dbdot, its complexes are gaining increasing popularity [19]. An example was recently published by Han *et al.* [20], which nicely illustrates the enhanced thermal stability of these complexes (Scheme 11.7). This reaction is a decarboxylative allylic substitution (Tsuji reaction), which can also be carried out with C and N nucleophiles.

11.4 Applications in Organic Synthesis

The methods described above are very well suited for applications in syntheses of biologically active compounds, in particular N-heterocycles such as alkaloids. Generally, the Ir-catalyzed allylic substitution is particularly well suited for

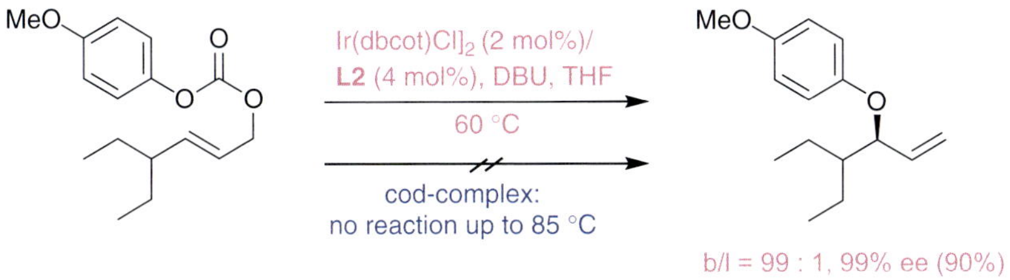

Scheme 11.7 Decarboxylative allylic etherifications illustrating the enhanced thermal stability of dbcot complexes.

cyclization reactions. In Scheme 11.8, a number of routes to get precursors for cyclization reactions are shown. A generally very good combination is provided by the allylic substitution and ring-closing metathesis (RCM), as illustrated on the left-hand side of Scheme 11.8. Similarly, the combination of an allylic and a propargylic unit offers many possibilities, for example, cyclization by a Pauson–Khand reaction. Hydroformylation of allylic amines (Nu = NR′) can, in domino fashion, lead to pyrrolidines (hydroaminomethylation). On the right-hand side, hydroboration followed by Suzuki coupling is described. Subsequent conjugate addition leads to mono- or bicyclic structures. In the following, an example from each of the categories is presented.

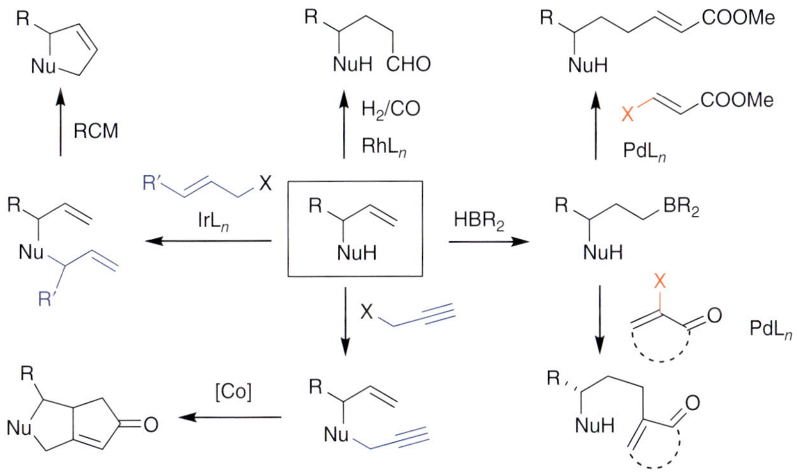

Scheme 11.8 General routes to biologically active compounds based on the Ir-catalyzed allylic substitution.

11.4.1
Allylic Substitution in Combination with Ring Closing Metathesis

The combination of an allylic substitution and an RCM [21] constitutes a powerful approach for syntheses of carbo- and heterocycles; this approach was

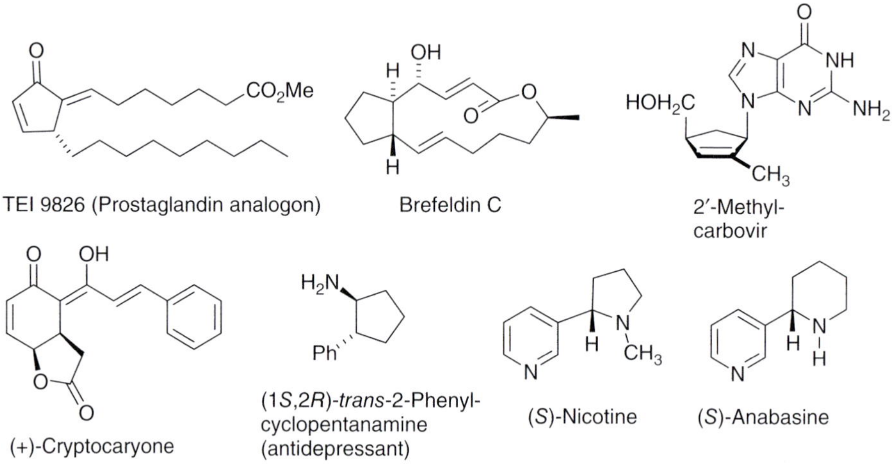

Figure 11.3 Examples of biologically active compounds prepared via ring closing metathesis.

pioneered by Evans using Rh-catalyzed allylic alkylations and etherifications [22]. We found it particularly well suited for the Ir-catalyzed variant. Formulas of biologically active compounds synthesized in this way in our group are listed in Figure 11.3.

In Scheme 11.9, a synthesis of (*S*)-nicotine is described as example [23]. In this case, the Ir-catalyzed allylic substitution proceeded with an exceptionally high degree of both regio- and enantioselectivity. The most problematic step in this synthesis was the reduction of the double bond. Transition-metal-catalyzed hydrogenation was not suitable because of partial racemization. In general,

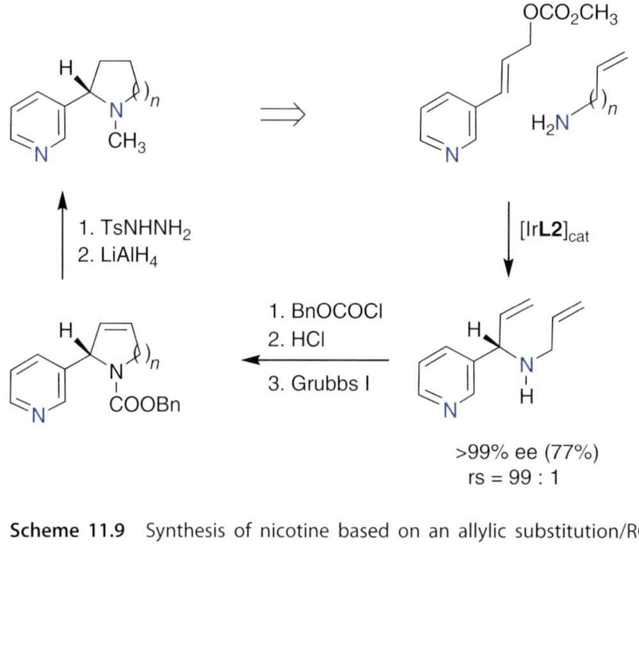

Scheme 11.9 Synthesis of nicotine based on an allylic substitution/RCM approach.

isomerization of the double bond into an allylic position occurs to some extent with all transition-metal catalysts. Reduction with diimide, generated from $TsNHNH_2$, was successfully used as a solution of this problem.

As example of an application of the RCM strategy to carbocyclic targets, the syntheses of the anticancer agents infecto- and cryptocaryone are described in Scheme 11.10 as a retrosynthetic scheme [19a].

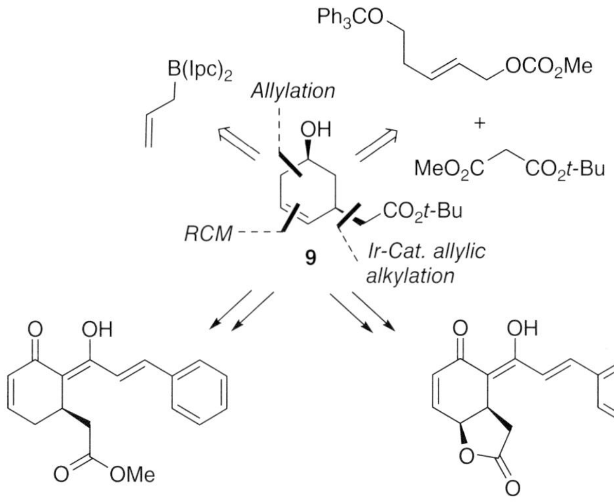

Scheme 11.10 Retrosynthetic scheme of syntheses of infecto- and cryptocaryone.

11.4.2
Domino-Hydroformylation–Cyclization (Hydroaminomethylation)

The hydroformylation of allylamines is a versatile reaction because it allows realization of domino reactions to give pyrrolidines or piperidines in one pot [24]. Our initial attempts were quite cumbersome, because a variety of products can be formed (Scheme 11.11). After some experimentation, it was found that the substituents at nitrogen are controlling the product distribution of the hydroformylation [25]. Thus, secondary amines, R^1 = alkyl, undergo the hydroaminomethylation reaction yielding pyrrolidines C via enamines B, which are reduced under the reaction conditions; amides and sulfonamides, that is, R^2 = COR′ or SO_2R', yield hemiaminals A, which are useful starting materials in synthesis, for example, in acyliminium ion chemistry. Finally, primary amines, R^2 = H, give rise to imines E.

The formation of the products A–D can be rationalized with the help of the mechanistic proposals presented in Scheme 11.12.

Given our interest in a synthesis of nicotine, we wondered whether the hydroaminomethylation could be applied as described in Scheme 11.13. While the Ir-catalyzed allylic amination with methylamine again gave an excellent

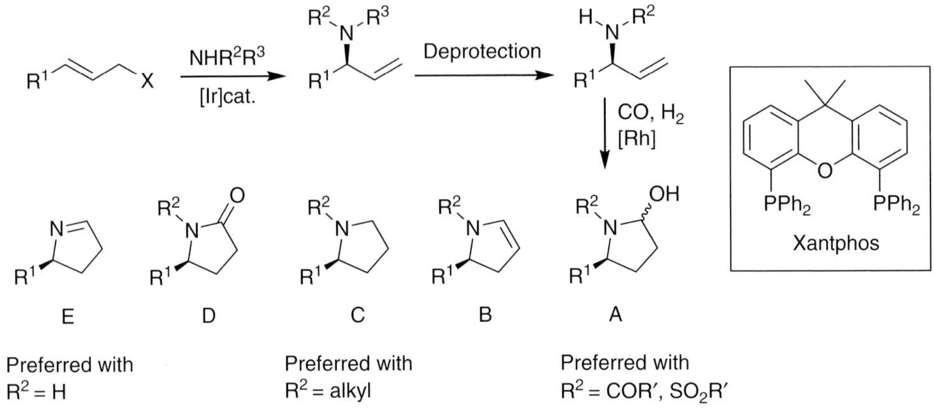

Scheme 11.11 Products formed upon hydroformylation of allylamines.

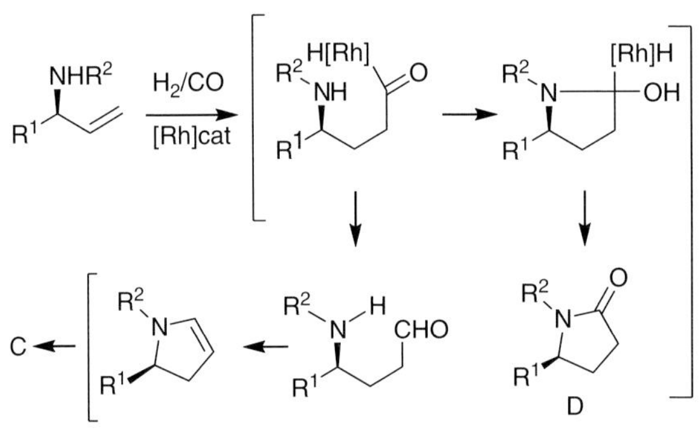

Scheme 11.12 Mechanistic aspects of the hydroformylation of allylamines.

result, the subsequent hydroformylation caused problems. With Xantphos (cf. Scheme 11.11) as ligand, mainly the lactam of type D was formed. This problem was solved by using the more bulky ligand Biphephos, which also allowed lowering of the reaction temperature and increasing the partial pressure of hydrogen. Note that all steps of this synthesis are controlled by transition-metal catalysis.

11.4.3 The Allylic Substitution in Combination with the Suzuki–Miyaura Reaction

The Suzuki–Miyaura reaction [26] is a very reliable method for connecting the building blocks. It was of interest to apply it to allylamines in that capacity. We found that hydroboration of the vinyl group can be reliably accomplished with 9-borabicyclo[3.3.1]nonane (9-BBN), even in the presence of further disubstituted double bonds, and the subsequent Suzuki–Miyaura coupling works well [27]. Several targets were addressed with these key reactions [28]. An example is

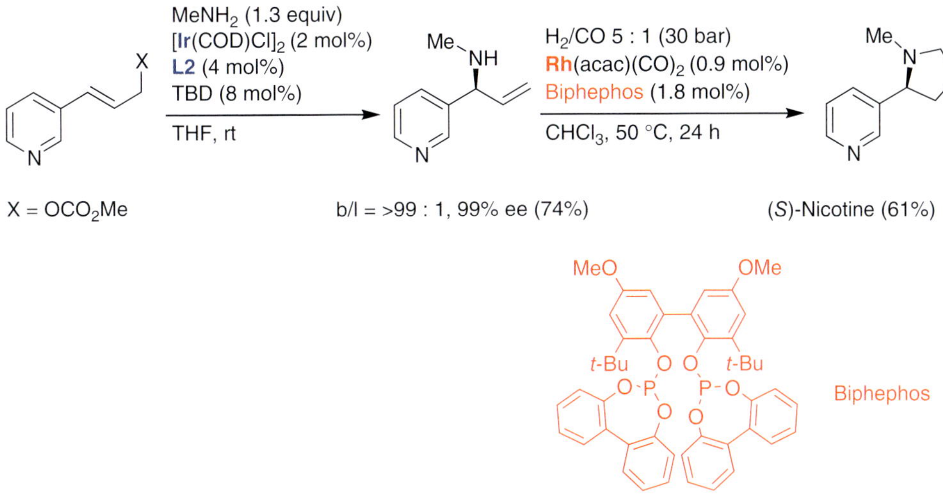

Scheme 11.13 Synthesis of (*S*)-nicotine by combination of Ir and Rh catalysis.

the synthesis of pumiliotoxin C (alternative designation: *cis*-**195A**), a very potent noncompetitive blocker of nicotinic acetylcholine receptors [29]. It was our ambition to control the installation of each chirality center by asymmetric catalysis. This was indeed possible by starting with an Ir-catalyzed allylic substitution, connection of the product to cyclohexenone by hydroboration/Suzuki–Miyaura coupling, and asymmetric Cu-catalyzed conjugate addition of $AlMe_3$, using a method developed by Alexakis and coworkers [30]. Most remarkably, the Ir- and the Cu-catalyzed reaction were controlled by diastereomers of the isoconstitutional phosphoramidite ligands **L2a** and **L2b**. In the final step, a further chirality center was introduced by a diastereoselective, that is, substrate-controlled, reductive amination (Schemes 11.14 and 11.15).

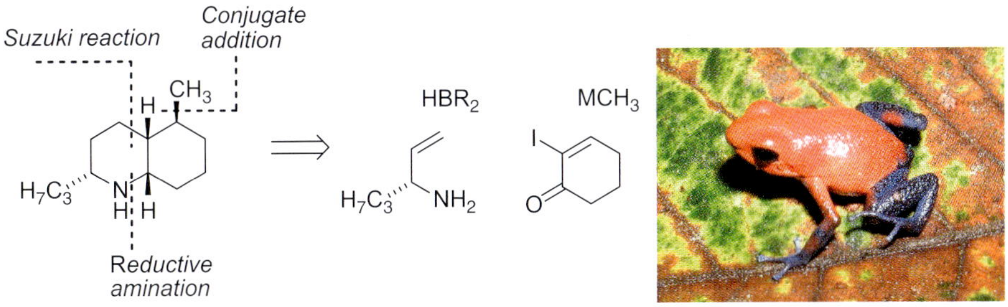

Scheme 11.14 (a, b) Retrosynthetic analysis of the synthesis of pumiliotoxin C and its biological source, the strawberry poison dart frog (*Dendrobates pumilio*). (The picture of *Dendrobates pumilio* is reproduced from Wikimedia commons (public domain)). (Source: Wikimedia Commons, Dendrobates pumilio.jpg, CC BY-SA 3.0.)

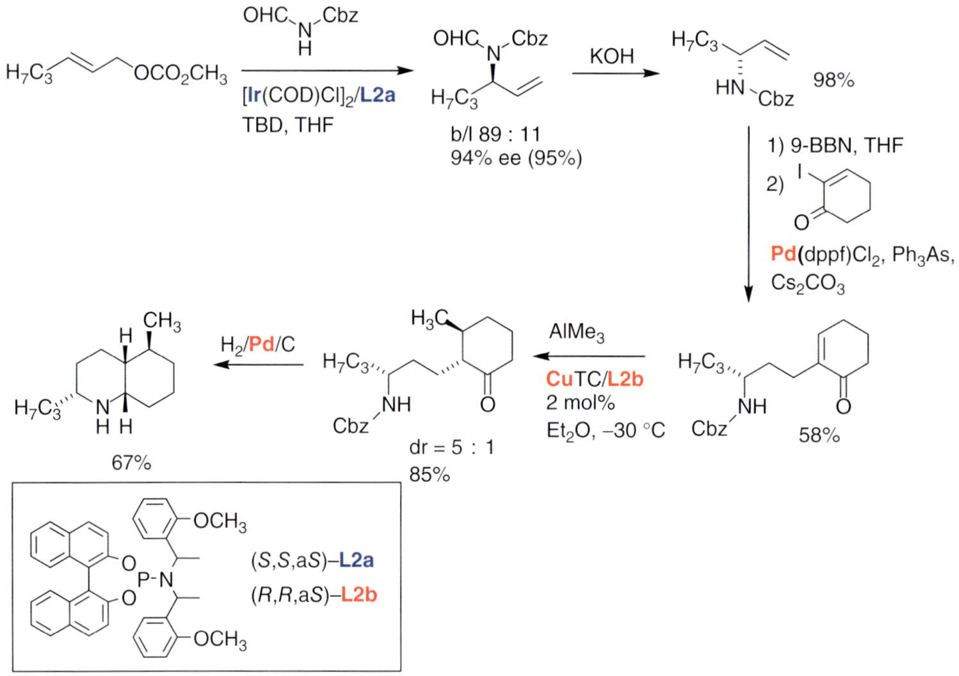

Scheme 11.15 Synthesis of pumiliotoxin C using multiple catalyst-controlled asymmetric reactions.

11.4.4 Reactions of Enines Derived from Allylic Substitution Products

Enines derived from allylic substitution products by propargylation offer many possibilities in organic synthesis. We have pursued Au-catalyzed cycloadditions with carbonyl compounds [31], enyne metathesis reactions in combination with Diels–Alder reactions [32], and a Pauson–Khand reaction as part of a synthesis of kainic acid [33]. The latter synthesis is described in Scheme 11.16 as a retrosynthetic scheme. The Ir-catalyzed allylic amination under *in situ* conditions proceeded with excellent enantioselectivity. Overall, our synthesis required 12 steps and gave a total yield of 12%.

11.5 Conclusions

The asymmetric Ir-catalyzed allylic substitution reaction has recently been developed into a broadly applicable synthetic method. In particular, isolated (π-allyl)Ir complexes derived from phosphoramidites are effective single-species catalysts. As ancillary ligand, the standard cod has been supplemented with

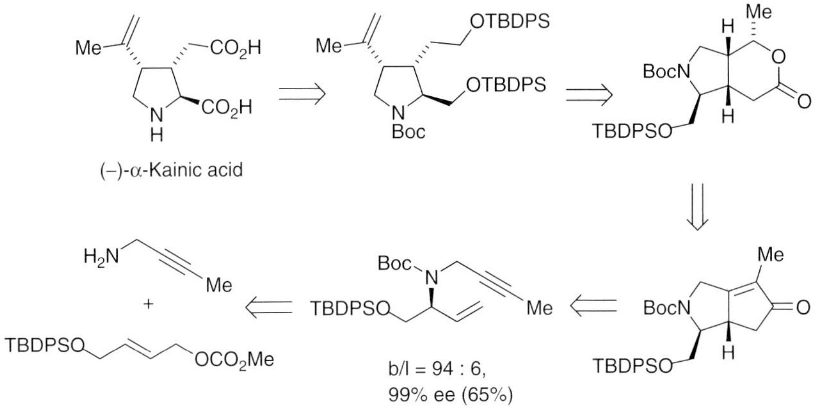

Scheme 11.16 Retrosynthetic analysis of the synthesis of (−)-kainic acid [TBDPS = Si(*t*-Bu)Me_2].

dbcot, which yields catalysts with superior thermal stability and enhanced induction of regioselectivity. The reaction is compatible with a variety of C, N, O, and S nucleophiles and a very broad range of solvents as well as water and air. Branched substitution products, in particular chiral allylamines and their derivatives, are available with excellent regio- and enantioselectivity. Applications in syntheses of natural products have been systematically explored. Particular efficient synthetic strategies result by combinations of the Ir-catalyzed allylic substitution with RCM, hydroboration/Suzuki–Miyaura coupling, or Rh-catalyzed hydroformylation.

Acknowledgments

The work described here was supported by the Deutsche Forschungsgemeinschaft (SFB 623), the Graduiertenkolleg 850, and, concerning applications in natural products synthesis, by the Dr. Rainer Wild-Stiftung. Donation of enantiopure 1-arylethylamines by Prof. Dr. K. Ditrich, BASF SE, is gratefully acknowledged.

References

1. Reviews: (a) Helmchen, G., Dahnz, A., Dübon, P., Schelwies, M., and Weihofen, R. (2007) *Chem. Commun.*, 675. (b) Helmchen, G. (2009) in *Iridium Complexes in Organic Synthesis* (eds L.A. Oro and C. Claver), Wiley-VCH Verlag GmbH, Weinheim, pp. 211–250. (c) Hartwig, J.F. and Stanley, L.M. (2010) *Acc. Chem. Res.*, **43**, 1461. (d) Takeuchi, R. and Kezuka, S. (2006) *Synthesis*, **20**, 3349. (e) Miyabe, H. and Takemoto, Y. (2005) *Synlett*, 1641. (f) Liu, W.-B., Xia, J.-B., and You, S.-L. (2012) *Top. Organomet. Chem.*, **38**, 155. (g) Tosatti, P., Nelson, A., and Marsden, S.P. (2012) *Org. Biomol. Chem.*, **10**, 3147.
2. Welter, C., Koch, O., Lipowsky, G., and Helmchen, G. (2004) *Chem. Commun.*, 896.

3. Kiener, C.A., Shu, C., Incarvito, C., and Hartwig, J.F. (2003) *J. Am. Chem. Soc.*, **125**, 14272.
4. Spiess, S., Welter, C., Franck, G., Taquet, J.-P., and Helmchen, G. (2008) *Angew. Chem. Int. Ed.*, **47**, 7652.
5. Weihofen, R., Tverskoy, O., and Helmchen, G. (2006) *Angew. Chem. Int. Ed.*, **45**, 5546.
6. (a) Spiess, S., Raskatov, J.A., Gnamm, C., Brödner, K., and Helmchen, G. (2009) *Chem. Eur. J.*, **15**, 11087. (b) Raskatov, J.A., Spiess, S., Gnamm, C., Brödner, K., Rominger, F., and Helmchen, G. (2010) *Chem. Eur. J.*, **16**, 6601.
7. (a) Polet, D., Alexakis, A., Tissot-Croset, K., Corminboeuf, C., and Ditrich, K. (2006) *Chem. Eur. J.*, **12**, 3596. (b) Tissot-Croset, K., Polet, D., and Alexakis, A. (2004) *Angew. Chem. Int. Ed.*, **43**, 2426.
8. Marković, D. and Hartwig, J.F. (2007) *J. Am. Chem. Soc.*, **129**, 11680.
9. Madrahimov, S.T., Marković, D., and Hartwig, J.F. (2009) *J. Am. Chem. Soc.*, **131**, 7228.
10. Raskatov, J.A., Jäkel, M., Straub, B.F., Rominger, F., and Helmchen, G. (2012) *Chem. Eur. J.*, **18**, 14314.
11. Madrahimov, S.T. and Hartwig, J.F. (2012) *J. Am. Chem. Soc.*, **134**, 8136.
12. Anton, D.R. and Crabtree, R.H. (1983) *Organometallics*, **2**, 621.
13. Franck, G., Brill, M., and Helmchen, G. (2012) *Org. Synth.*, **89**, 55.
14. Defieber, C., Grützmacher, H., and Carreira, E.M. (2008) *Angew. Chem. Int. Ed.*, **47**, 4482.
15. (a) Hodgson, D.M. and Humphreys, P.G. (2007) in *Science of Synthesis*, vol. 36 (ed. J.P. Clayden), Thieme, Stuttgart, p. 583. (b) Guzman-Martinez, A. and Hoveyda, A.H. (2010) *J. Am. Chem. Soc.*, **132**, 10634.
16. Lyothier, I., Defieber, C., and Carreira, E.M. (2006) *Angew. Chem. Int. Ed.*, **45**, 6204.
17. Gärtner, M., Seehafer, K., and Helmchen, G. (2011) *J. Am. Chem. Soc.*, **133**, 2072.
18. (a) Lüssem, B.J. and Gais, H.-J. (2003) *J. Am. Chem. Soc.*, **125**, 6066. (b) Gais, H.-J., Bondarev, O., and Hetzer, R. (2005) *Tetrahedron Lett.*, **46**, 6279. (c) Takahata, H., Suto, Y., Kato, E., Yoshimura, Y., and Ouchi, H. (2007) *Adv. Synth. Catal.*, **349**, 685.
19. (a) Frank, G., Brödner, K., and Helmchen, G. (2010) *Org. Lett.*, **12**, 3886. (b) Tosatti, P., Horn, J., Campbell, A.J., House, D., Nelson, A., and Marsden, S.P. (2010) *Adv. Synth. Catal.*, **352**, 3153. (c) Ye, K.-Y., He, H., Liu, W.-B., Dai, L.-X., Helmchen, G., and You, S.-L. (2011) *J. Am. Chem. Soc.*, **133**, 19006. (d) Ye, K.-Y., Dai, L.-X., and You, S.-L. (2012) *Org. Biol. Chem.*, **10**, 593.
20. Kim, D., Reddy, S., Singh, O.V., Lee, J.S., Kong, S.B., and Han, H. (2013) *Org. Lett.*, **15**, 512.
21. Grubbs, R.H. (ed.) (2003) *Handbook of Metathesis*, John Wiley & Sons, Inc., New York.
22. Review: Leahy, D.K. and Evans, P.A. (2005) in *Modern Rhodium-Catalyzed Organic Reactions* (ed. P.A. Evans), John Wiley & Sons, Inc., New York, p. 191.
23. Welter, C., Moreno, R.M., Streiff, S., and Helmchen, G. (2005) *Org. Biomol. Chem.*, **3**, 3266.
24. Reviews: (a) Eilbracht, P. and Schmidt, A.M. (2006) *Top. Organomet. Chem.*, **18**, 65. (b) Eilbracht, P., Bärfacker, L., Buss, C., Hollmann, C., Kitsos-Rzychon, B.E., Kranemann, C.L., Roggenbruck, R., and Schmidt, A. (1999) *Chem. Rev.*, **99**, 3329. (c) Eguchi, M., Tzamarioudaki, M., and Ojima, I. (1995) *J. Org. Chem.*, **60**, 7078. (d) Vidal, E.S. and Ojima, I. (1998) *J. Org. Chem.*, **63**, 7999. (e) Spangenberg, T., Airiau, E., Bui The Thuong, M., Donnard, M., Billet, M., and Mann, A. (2008) *Synlett*, 2859.
25. Dübon, P., Farwick, A., and Helmchen, G. (2009) *Synlett*, 1413.
26. Reviews: (a) Miyaura, N. and Suzuki, A. (1995) *Chem. Rev.*, **95**, 2457. (b) Kotha, S., Lahiri, K., and Kashinath, D. (2002) *Tetrahedron*, **58**, 9633.
27. Gärtner, M., Weihofen, R., and Helmchen, G. (2011) *Chem. Eur. J.*, **17**, 7605.
28. Satyanarayana, G., Pflästerer, D., and Helmchen, G. (2011) *Eur. J. Org. Chem.*, **2011**, 6877.
29. Gärtner, M., Qu, J., and Helmchen, G. (2012) *J. Org. Chem.*, **77**, 1186–1190.

30. (a) Alexakis, A., Albrow, V., Biswas, K., d'Augustin, M., Prieto, O., and Woodward, S. (2005) *Chem. Commun.*, 2843. (b) Vuagnoux-d'Augustin, M. and Alexakis, A. (2007) *Chem. Eur. J.*, **13**, 9647.
31. (a) Schelwies, M., Moser, R., Dempwolff, A., Rominger, F., and Helmchen, G. (2009) *Chem. Eur. J.*, **15**, 10888. (b) Schelwies, M., Dempwolff, A., Rominger, F., and Helmchen, G. (2007) *Angew. Chem.*, **119**, 5694; *Angew. Chem. Int. Ed.*, (2007), **46**, 5598.
32. Schelwies, M., Farwick, A., Rominger, F., and Helmchen, G. (2010) *J. Org. Chem.*, **75**, 7917.
33. (a) Farwick, A. and Helmchen, G. (2010) *Org. Lett.*, **12**, 1108. (b) Farwick, A., Engelhart, J.U., Tverskoy, O., Welter, C., Umlauf, Q.A., Rominger, F., Kerr, W.J., and Helmchen, G. (2011) *Adv. Synth. Catal.*, **353**, 349.

12
Sequential Catalysis Involving Metal-Catalyzed Cycloisomerizations and Cyclizations

Thomas J. J. Müller

12.1
Introduction

The advent of transition-metal catalysis has considerably revolutionized strategies and concepts in synthetic organic chemistry. In particular, reaction sequences consisting of multiple iterations of organometallic elementary steps have been shown to considerably enhance structural complexity. These peculiar transition-metal-catalyzed processes have become known as *domino*, *tandem*, or *cascade* reactions [1]. Although, all three terms have slightly different meanings, they have been used synonymously during the past two decades because they all are one-pot transformations [2]. Simple starting materials lead to an enormous increase in structural complexity, causing a rapid development of this steadily expanding field [3]. Transition-metal-catalyzed cascade reactions fall into two major categories: domino reactions that involve multiple iterations of the same elementary step [3k–m], or consisting of reaction sequences involving two or more fundamentally different elementary processes [4].

Metal-catalyzed domino reactions are characterized by organometallic specimens that are regenerated in multiple insertion events and yield after an elimination step the complex, often polycyclic organic product. Predominantly in palladium chemistry, unimolecular reactions proceeding in an intramolecular fashion, for example, by cyclic carbopalladation [5], furnish complex polycyclic structures starting from linear polyunsaturated substrates.

However, metal-catalyzed cascade reactions performed parallelly or sequentially [4a, e] are even more challenging if the particular catalyst enables the performance of multiple different catalytic processes. Conceptually, parallel catalysis is significantly more difficult to develop than sequential catalysis. Especially, the latter offers the possibility of altering and triggering reaction conditions from step to step with additives. In the sense of Fogg's and De Santo's terminology, sequential catalysis can be considered as bi- or multicatalytic one-pot processes, assisted tandem catalysis, or auto tandem catalysis [1]. As a consequence, the challenge is the identification of a single precatalyst that can be elaborated into multiple catalytic reactions in a sequential one-pot manner. In addition, these types of sequences are

Molecular Catalysts: Structure and Functional Design, First Edition. Edited by Lutz H. Gade and Peter Hofmann.

setting the stage for new reactions in diversity-oriented syntheses [6] of complex molecular scaffolds.

Two prerequisites for successfully performing sequentially catalyzed processes can be easily identified. As in domino reactions, the initial catalyzed process has to generate or retain a functional group that is suitably reactive for a subsequent transformation. And the catalyst or catalyst precursor has to be present in the reaction medium during the whole sequence without further addition of the transition-metal source after the initial step. Triggering additives might be added from step to step.

In the past decade, cycloisomerizations and cyclizations have received particular attention as processes with a high degree of atom economy. Linear precursors are cyclized, and an increase of molecular complexity is induced in the initial step. On the other hand, the effective catalyst system still stands available for further catalytic transformations.

Sequential transition-metal-catalyzed processes are dominated by palladium- [7], rhodium- [8], and ruthenium- [9] catalyzed sequences. With respect to Fogg's and de Santos' classification, only cyclizing metal-catalyzed one-pot processes [1] in the sense of auto and assisted tandem catalysis will be discussed and highlighted in this overview.

12.2 Sequences Initiated by Cycloisomerizations

12.2.1 Sequentially Pd-Catalyzed Sequences Initiated by Cycloisomerizations

In the 1980s, the group of Trost [10, 11] discovered and developed intramolecular palladium-catalyzed Alder-ene reactions. Changing the mechanism from the classical pericyclic process to a reaction that proceeds via transition organometallic intermediates has enriched the understanding and realization of the concept of atom-economic processes [12]. Therefore, the cycloisomerization of 1,*n*-enynes has remained an intensively pursued field of synthetic methodologies ever since [13]. In addition, major advances in enantioselective transition-metal-catalyzed cycloisomerizations of enynes have also been explored [14].

Two mechanistic pathways for Pd-catalyzed enyne cycloisomerizations have been proposed. While the palladacyclization followed by β-hydride elimination is likely to be operative for Pd(II) catalysts, Pd(0) complexes in combination with catalytic amounts of carboxylic acids enter a hydropalladation–cyclic carbopalladation–β-hydride elimination cycle. This latter mechanism is particularly suited for the development of sequentially catalyzed processes.

The alkyl-Pd species formed in the cyclic carbopalladation can be subjected to a subsequent transmetallation with organometallic substrates. As a consequence, a reductive elimination concludes the second Pd-mediated step by releasing the Pd(0) species.

Kibayashi and coworkers [15] have introduced a sequential cyclic carbopalladation–Stille vinylations of enyne compounds. Upon treating the enyne **1** and vinyl tributylstannane with catalytic amounts of system $Pd_2(dba)_3{\cdot}CHCl_3$/AcOH (dba, dibenzylideneacetone), the allyl-substituted methylene cyclopentane **2** was formed in 53% yield (Scheme 12.1). The expected Alder-ene product **3** was not detected. This sequence was extended to heteroatom-tethered enynes **4** and vinyl tin compounds **5** to furnish heterocycles **6** in moderate to excellent yields.

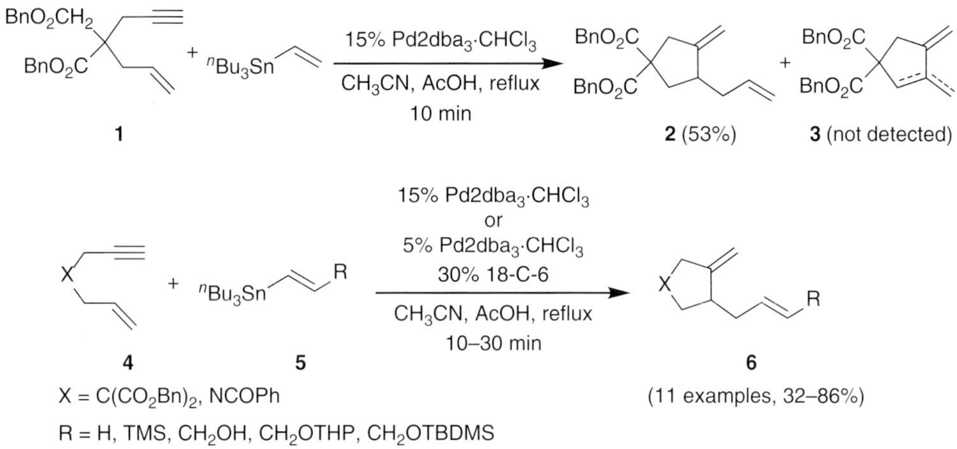

Scheme 12.1 Enyne cycloisomerization–Stille sequence.

By the choice of the enyne substrates, the obtained cyclized products might contain functionalities other than dienes. The unusual catalyst combination of a Pd precatalyst and formic acid enables sequential catalysis initiated by cycloisomerization in a very peculiar way. Kressierer and Müller [16] demonstrated in several cases that the palladium(0)-catalyzed Alder-ene reaction of α-alkynyl ω-allyl alcohols **7** furnishes cyclic γ,δ-enals **8** as a consequence of the *in situ* enol-aldehyde tautomerism (Scheme 12.2).

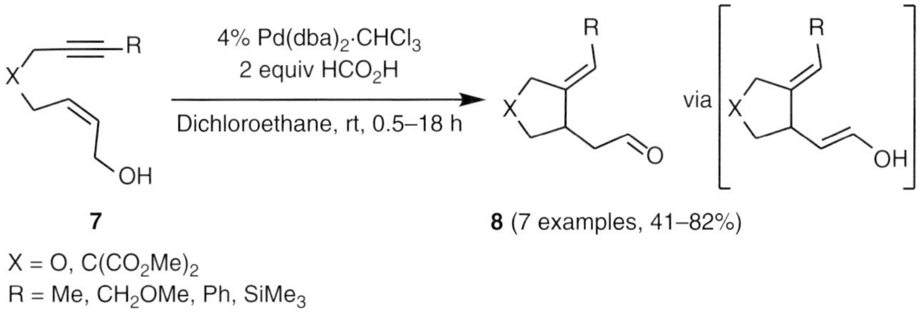

Scheme 12.2 Pd(0)-catalyzed Alder-ene cycloisomerization of alkynyl allyl alcohols to give cyclic γ,δ-enals.

The newly generated aldehyde functionality was reacted in consecutive carbonyl transformations, such as cycloisomerization–Wittig [16] and cycloisomerization–Leuckart–Wallach [17] one-pot sequences to give five-membered carbo- and heterocycles with densely functionalized side chains.

Kressierer and Müller [18] could also show that yne allyl alcohols are rapidly cycloisomerized by a Pd_2dba_3–*N*-acetyl phenyl alanine catalyst system to give silylsubstituted heterocyclic enals in excellent yields. This intriguing catalyst system is capable of performing a sequentially Pd-catalyzed processes as well as an acid-catalyzed iminium aldol condensation.

As showcases for sequentially catalyzed one-pot processes, sequential cycloisomerization–reductive amination (Scheme 12.3) [18] and cycloisomerization–Knoevenagel condensation sequences (Scheme 12.4) [19] were established. In the former case, this intriguing catalyst system catalyzes a subsequent reductive amination with secondary amines in the same pot, under the same conditions, and in the presence of one atmosphere of hydrogen. Therefore, alkyne allyl alcohols **7** are readily transformed in moderate to good yields into β-amino ethyl alkylidene tetrahydrofurans (THFs) or pyrrolidines **9** in

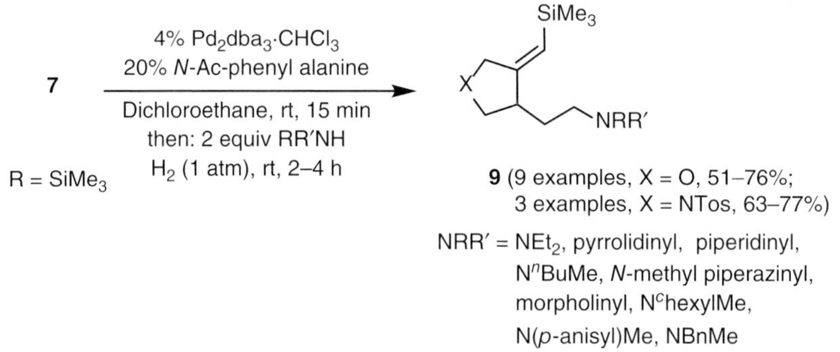

Scheme 12.3 Pd(0)-catalyzed Alder-ene cycloisomerization–reductive amination sequence to β-amino ethyl alkylidene tetrahydrofurans and pyrrolidines **9**.

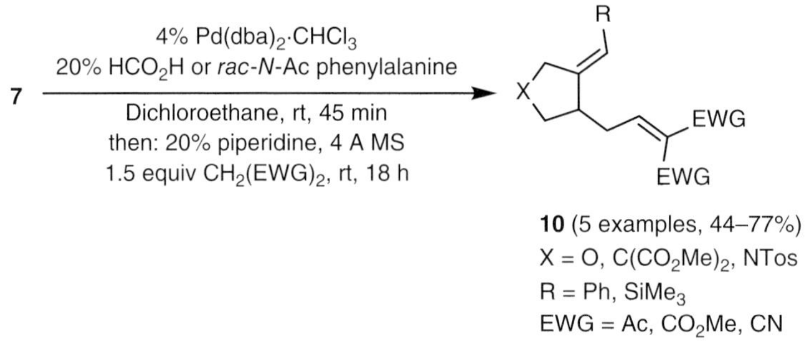

Scheme 12.4 Pd(0)-catalyzed Alder-ene cycloisomerization–Knoevenagel to five-membered carbo- and heterocycles **10** with super-Michael acceptor side chains.

the sense of a Pd-catalyzed cycloisomerization–reductive amination sequence with remarkable chemoselectivity.

In the latter case, catalytic amounts of acid, formic acid or *N*-acetyl phenyl alanine, catalyze in the second step together with a catalytic amount of piperidine the concluding Knoevenagel condensation, furnishing five-membered carbo- and heterocycles **10** with super-Michael acceptors in the side chain. In addition, this newly installed Michael acceptor functionality undergoes a Michael addition with an excess of 1,3-cyclohexadiones as CH-acidic components in good to excellent yields in the sense of a cycloisomerization–Knoevenagel–Michael addition sequence (Scheme 12.5). Again, the presence of the acid catalyst and piperidine triggers this third step of the one-pot sequence to give the pseudo-three-component products **11**.

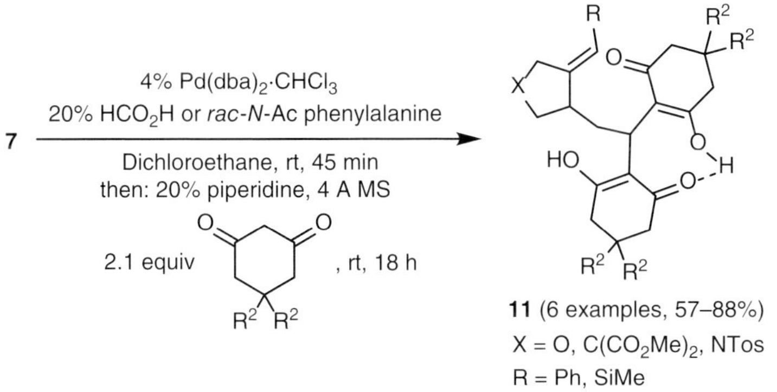

Scheme 12.5 Palladium(0)-catalyzed Alder-ene cycloisomerization-Knoevenagel–Michael addition sequence.

12.2.2
Sequentially Rh-Catalyzed Sequences Initiated by Cycloisomerizations

Rh-catalyzed cycloisomerizations of enynes and allenynes and subsequent one-pot sequences have also been developed in the past decade [8]. Stimulated by the success of enantioselective rhodium-catalyzed hydrogenation, and taking advantage of a plethora of chiral chelating diphosphane ligands, enantioselective intramolecular Rh-catalyzed Alder-ene reactions have evolved. In 2002, Zhang's group [20] was the first to report an efficient and highly enantioselective synthesis of functionalized α-methylene-γ-butyrolactones via a rhodium(I)-catalyzed intramolecular Alder ene reaction (Scheme 12.6). While *S*-BINAP (2,2′-bis(diphenylphosphano)-1,1′-binaphthyl) as a chiral chelating ligand furnishes the butyrolactons **(*S*)-12**, *R*-BINAP delivers the enantiomers **(*R*)-12**. In contrast to Pd-catalyzed cycloisomerization, the 1,4-diene isomer is formed exclusively. Interestingly, allyl ethers ($R^2 = OR$), allyl acetates ($R^2 = OAc$), and allyl alcohols ($R^2 = OH$) are equally well accepted as substrates. However, as

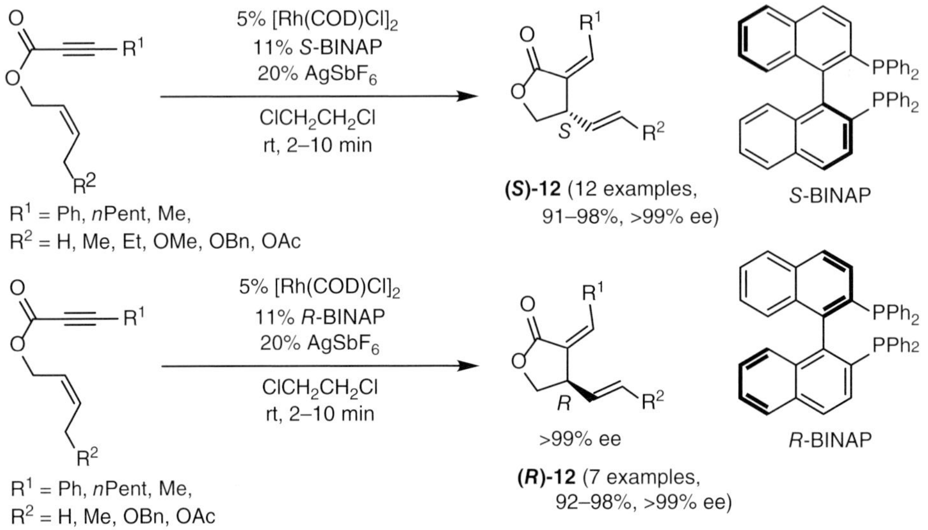

Scheme 12.6 Enantioselective synthesis of functionalized (*S*)- and (*R*)-α-methylene-γ-butyrolactones **12** via Rh(I)-catalyzed cycloisomerization.

shown before in Pd catalysis, the resulting elusive enol immediately tautomerizes to the corresponding aldehyde. This aspect has been utilized in a concise formal synthesis of (+)-pilocarpine.

The Rh–BINAP-catalyzed intramolecular Alder-ene cycloisomerization is very rapid and, therefore, sequentially Rh-catalyzed sequences for the efficient enantioselective generation of five-membered carbo- and heterocycles were envisioned. Körber *et al.* [21] reported the enantioselective rhodium-catalyzed cycloisomerization of alkyl and (hetero)aryl alkynyl allyl alcohols for the generation of aldehyde-bearing chiral 4-alkyl 3-alkylidene THFs and tetrahydrofuranones, which were converted into α,β-unsaturated carbonyl side chains in a one-pot manner via a concluding Wittig olefination in good yields.

Based on the initial Alder-ene cycloisomerization of alkyl and (hetero)aryl-substituted alkynyl allyl alcohols **13**, the rhodium(I)–BINAP complex subsequently can be employed in the reduction of the primary products with hydrogen, furnishing 2,7-dioxabicyclo[3.2.1]octanes **14** in the sense of a sequentially Rh-catalyzed one-pot process (Scheme 12.7) [22].

According to mechanistic studies, the hydrogenation of the aldehyde precedes the double-bond isomerization to the cyclic enolether, which is rapidly transformed by a Lewis acid-catalyzed acetalization. In this sequence, the cationic rhodium complexes obviously catalyze four steps. The three last steps all proceed in a domino manner and a rhodium–hydride complex is the presumed catalytically active species.

Brummond's group [23] and independently the group of Shibata [24] discovered that rhodium(I) complexes are excellent catalysts in the formal Alder-ene cycloisomerization of allenynes to give cross-conjugated trienes under mild conditions

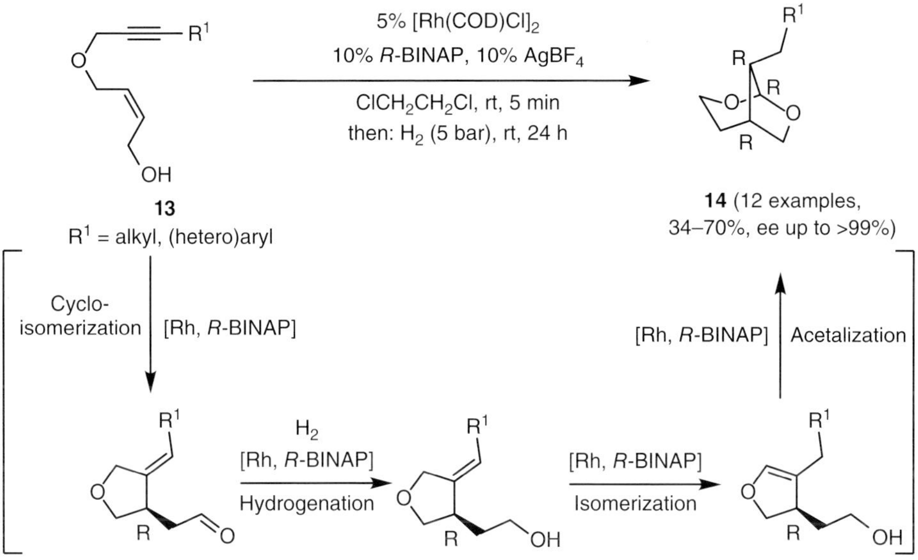

Scheme 12.7 Enantioselective sequentially Rh(I)-BINAP-catalyzed cycloisomerization–hydrogenation–isomerization–acetalization sequence.

and in good yields. In the past decade, Brummond considerably developed and extended this Alder-ene type cycloisomerization of allenynes to the formation of δ- and ε-lactams [25] and to yne-allenamides as substrates [26], and applied it as a key reaction toward the synthesis of the angiogenesis inhibitor ovalicin [27]. The obtained cross-conjugated trienes [28] are excellent building blocks for complex molecular frameworks. Therefore, consecutive one-pot transformations initiated by the rhodium(I)-catalyzed Alder-ene reactions were devised, such as Alder-ene–Diels–Alder [29], and Alder-ene–Diels–Alder–Diels–Alder sequences [30], furnishing the complex polycyclic products **15** and **16** (Scheme 12.8).

However, in the latter case, a cationic rhodium(I) complex had to be added to catalyze the intramolecular Diels–Alder reaction with the triene moiety to give a diene, ready for an intermolecular [4 + 2] cycloaddition. Strictly speaking, this case is definitely a borderline one with respect to the rigorous definition given in the introduction.

Okamoto *et al.* [31] reported an enantioselective Rh–BINAP-catalyzed allyl ether isomerization–cycloisomerization domino sequence of phenol- or naphthol-linked 1,7-enynes **17** to give dihydrobenzofurans and dihydronaphthofurans **18** (Scheme 12.9).

Iridium complexes, such as $[IrCl(COD)]_2$, (COD, cycloocta-1,5-diene) also catalyze intramolecular Alder-ene type cycloisomerizations 1,6-enynes to give cyclic 1,4-dienes. Shibata *et al.* [32] demonstrated the increased efficiency of this catalyst system in an ionic liquid over toluene. Iridium(I)-catalyzed Alder-ene cycloisomerizations of an alkynoyl cyclohexadienyl ester or *N*-tosyl amide have been shown to furnish spirocyclic products [33].

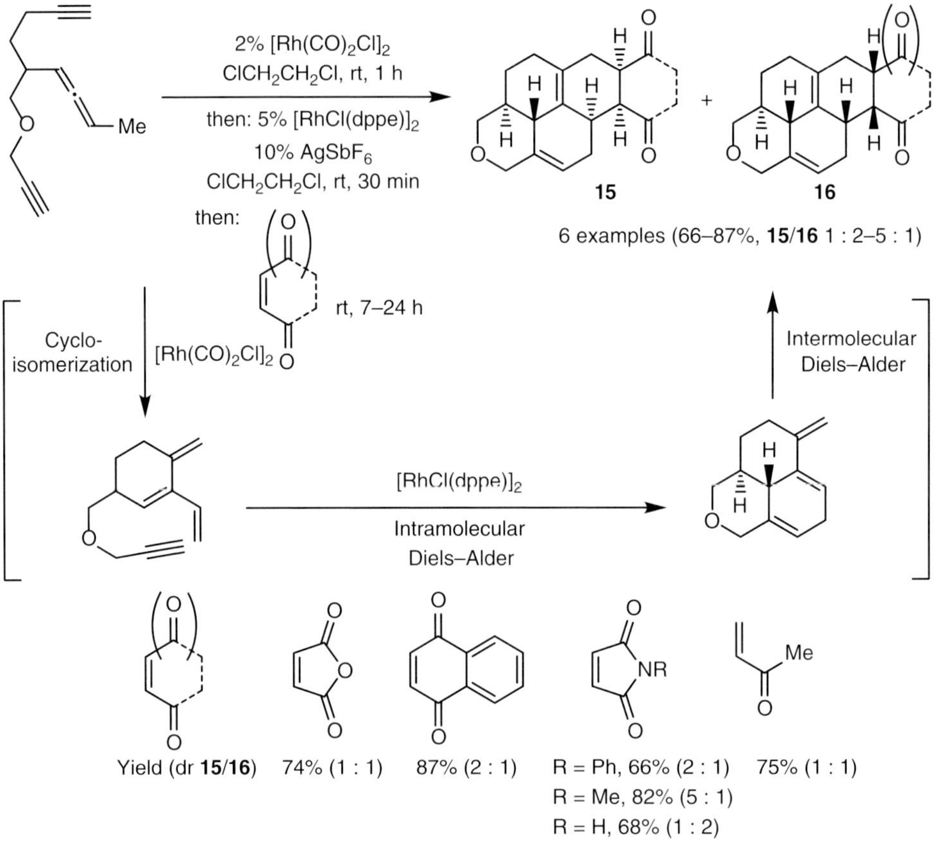

Scheme 12.8 One-pot Alder-ene–Diels–Alder–Diels–Alder reactions initiated by Rh-catalyzed cycloisomerization of an allenyne.

Kummeter *et al.* [34] successfully employed this type of Alder-ene cycloisomerization as an entry to a sequentially Ir-catalyzed cycloisomerization–Murahashi reaction sequence in a one-pot manner, where the intermediate aldehyde arising from the Alder-ene tautomerization step was condensed with cyano acetic esters without addition of acids or bases, furnishing five-membered heterocycles **19** with super-Michael acceptors in the side chain (Scheme 12.10).

12.3
Sequences Initiated by Ring-Closing Olefin Metathesis

Olefin metathesis [35] regioselectively ligates two olefinic sp^2-hybridized carbon atoms in an inter- or intramolecular sense and represents the most unusual carbon–carbon bond coupling for the synthesis of complex molecules. The latter variation is called *ring-closing olefin metathesis* [36] and has received considerable

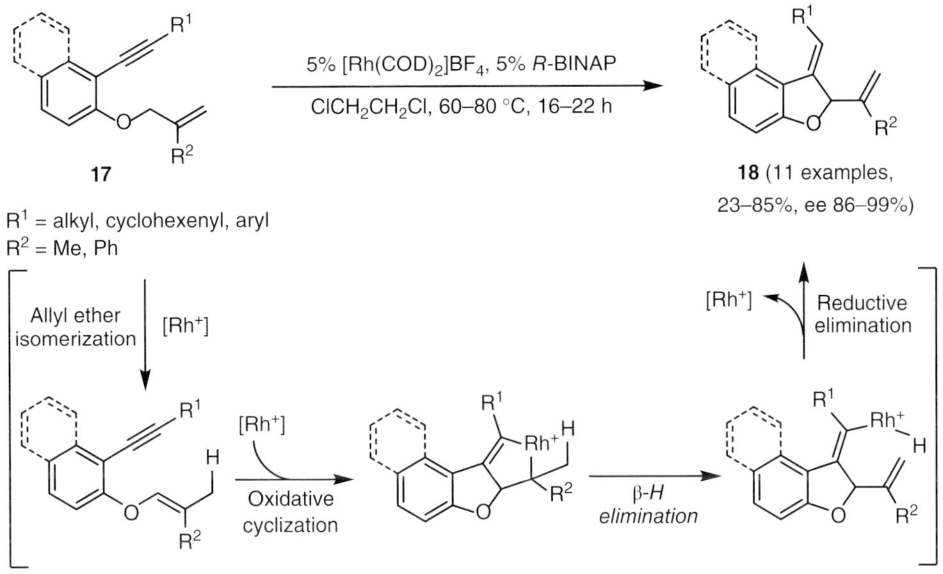

Scheme 12.9 Enantioselective sequentially Rh(I)-BINAP-catalyzed allyl ether isomerization–cycloisomerization reaction.

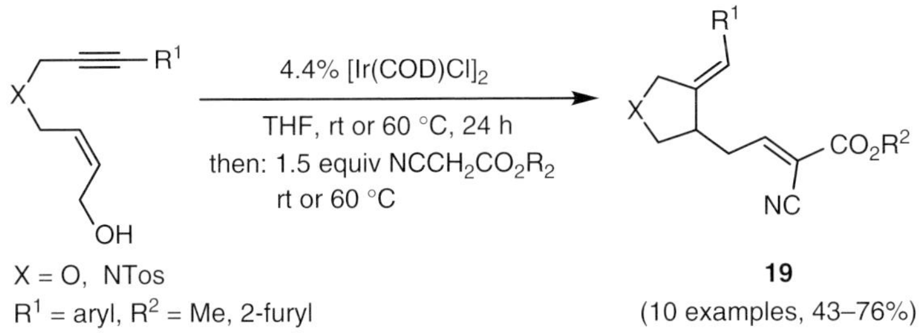

Scheme 12.10 Sequentially Ir-catalyzed Alder-ene–Murahashi reaction sequence.

attention with respect to macrocycle syntheses and total syntheses of complex natural products [37]. The application of Grubbs' carbene complexes in sequentially Ru-catalyzed processes based upon ring-closing metathesis (RCM) has been reviewed [38]. Therefore, only recent progress will be highlighted.

12.3.1
Ring-Closing Metathesis–Isomerization Sequences

Cyclic allyl ethers generated by RCM can be isomerized into 2,3-dihydropyrans by a ruthenium complex and eventually by conditioning ruthenium carbene complex precursors [39]. This combination of sequentially Ru-catalyzed ring closure and subsequent isomerization has set the stage for rapid syntheses of complex heterocyclic frameworks by cyclization strategies.

Schmidt and Biernat have successfully applied the sequentially Ru-catalyzed RCM–isomerization reaction in the synthesis of six- to eight-membered oxacycles, which can be regarded as glycals of 3-deoxyheptoses with variable ring size (Scheme 12.11). Interestingly, in this study, the same precursor was transformed in a divergent manner to furnish the corresponding substrate for

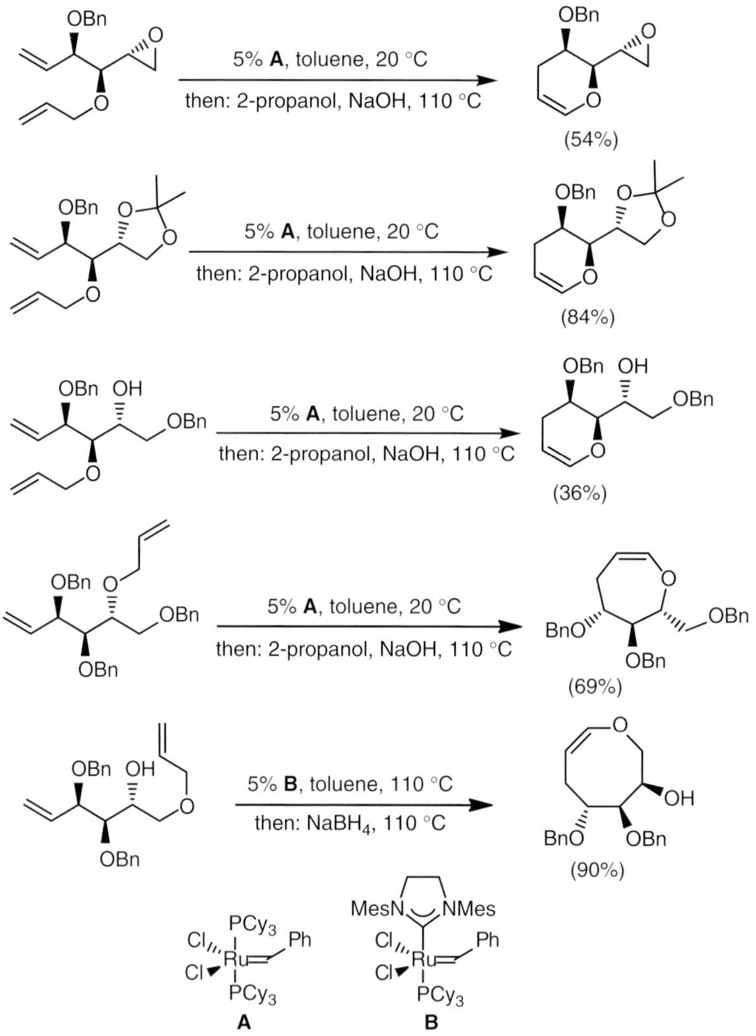

Scheme 12.11 Sequentially Ru-catalyzed RCM–isomerization reaction in the synthesis of six- to eight-membered oxacycles.

the RCM–isomerization sequence. The same RCM–isomerization sequence was also applied in an efficient approach to 6-deoxyglycals [40].

Likewise, Schmidt and Kunz [41] very recently reported that α,β-unsaturated δ-valerolactones **20** can be readily prepared by the RCM–isomerization sequence of butenoates of allyl alcohols (Scheme 12.12), an efficient alternative to the RCM of acrylates derived from homoallylic alcohols, that proceeds with lower catalyst loading and at higher substrate concentrations.

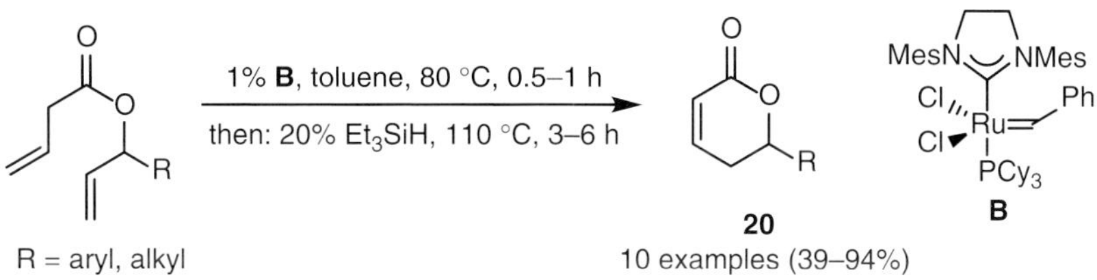

Scheme 12.12 Sequentially Ru-catalyzed RCM–isomerization reaction in the synthesis of six- to eight-membered oxacycles.

The sequential combination of RCM and double-bond isomerization was recently employed by You's group [42] for the design of an RCM–isomerization–Pictet–Spengler cascade for a highly efficient enantioselective synthesis of tetrahydro-β-carbolines **21**, where a Hoveyda–Grubbs II complex and a chiral phosphoric acid (CPA-H) cooperatively acted as a bicatalytic system (Scheme 12.13).

Previously, Ascic *et al.* [43] reported the same reaction with the Hoveyda–Grubbs II catalyst and trifluoroacetic acid as a bicatalytic system, but as a racemic version [43]. The intermediate acyliminium ion readily attacks also other tethered nucleophiles, such as trimethoxybenzene, benzothiophene, thiophene, *O,O*-dimethyl catechol, and furan, intramolecularly. When tryptophane or the corresponding alanine derivatives are employed, reasonable levels of diastereoselectivity can be observed for the terminal Pictet–Spengler acyliminium cyclization. Even a tethered alcohol functionality diastereoselectively traps the chiral acyliminium intermediate, furnishing bicyclic *O,N*-acetals.

Very recently, Petersen and Nielsen [44] could not only show that the nucleophilic trapping of the acyliminium ion can be conducted in an intramolecular manner but also extend the sequentially Ru-catalyzed RCM–isomerization sequence to exploit the heteroaromatic hydroxyl tautomer for an aldol condensation to give pyrrolone derivatives **22** (Scheme 12.14). The combination of RCM–isomerization–intermolecular nucleophilic trapping and the acid-catalyzed aldol condensation has been elegantly illustrated in the total synthesis of the antileukemic indole alkaloid violacein.

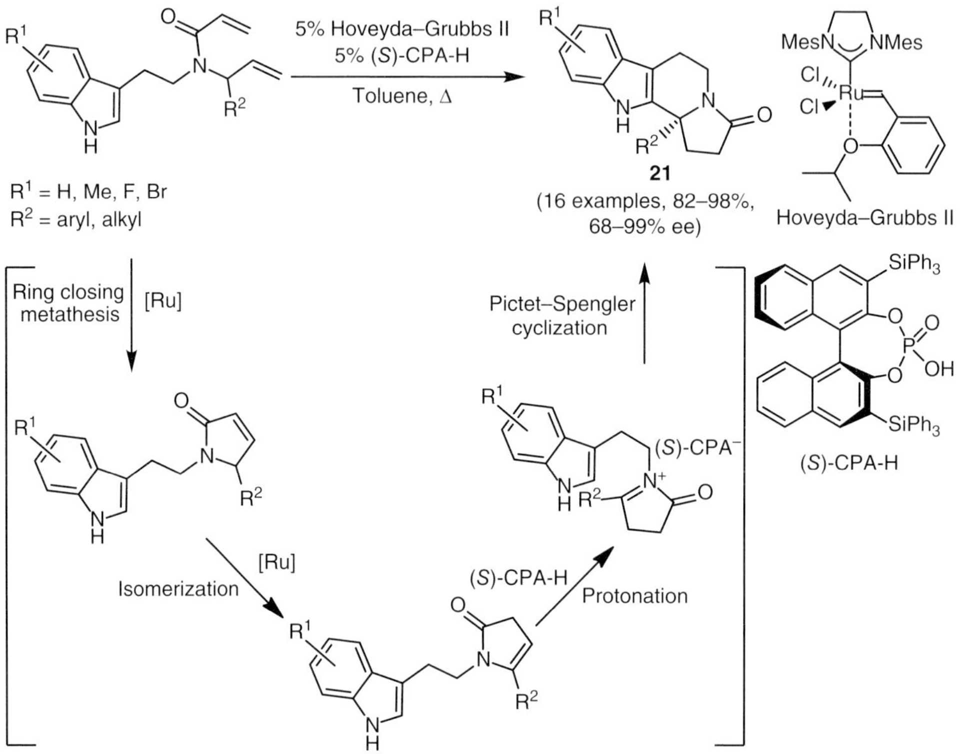

Scheme 12.13 Sequentially Ru-catalyzed RCM–isomerization–Pictet–Spengler synthesis of enantioenriched tetrahydro-β-carbolines **21**.

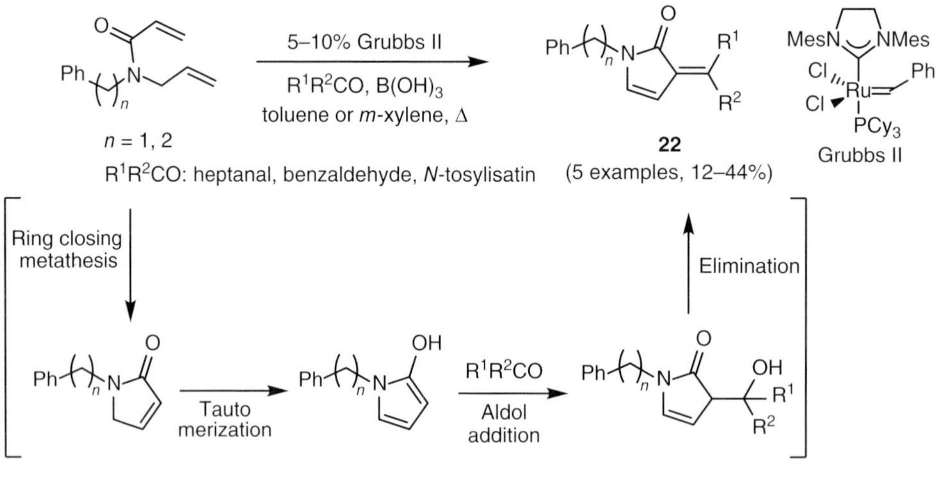

Scheme 12.14 RCM–isomerization–aldol condensation synthesis of pyrrolone derivatives **22**.

12.3.2
Ring-Closing Metathesis–Oxidation Sequences

Ruthenium–carbene complexes capable of metathesis reactions can be transformed into oxidation catalysts by the addition of oxidation agents. This principle has been applied in sequentially Ru-catalyzed metathesis–dihydroxylation reactions [45] which can be elegantly employed in the syntheses of complex and densely functionalized scaffolds.

Independently, the groups of Schmidt in Germany and Arisawa and Shuto in Japan have developed sequentially Ru-catalyzed RCM–allylic oxidation syntheses of coumarins and 2-quinolinones. Schmidt and Krehl [46] demonstrated that allyloxy styrenes can be transformed by a sequentially Ru-catalyzed RCM–allylic oxidation into coumarins **23** by Grubbs first-generation catalysts (Scheme 12.15). And Kato *et al.* [47] reported an RCM–allylic oxidation sequence of 2-vinyl-*N*-allylanilines to furnish 2-quinolines **24** in good yield (Scheme 12.16).

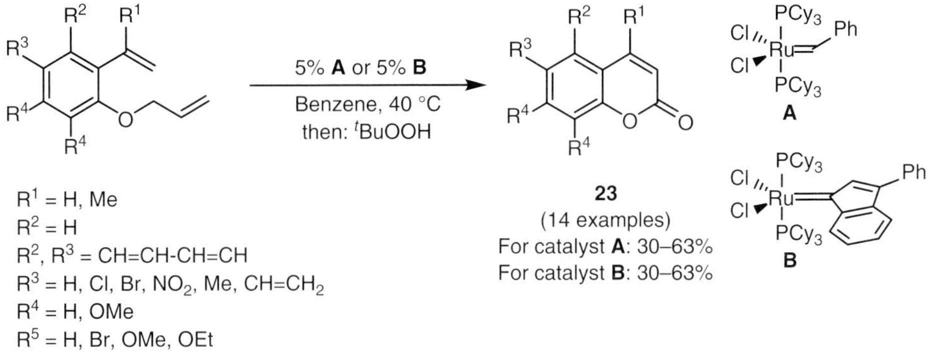

Scheme 12.15 Sequentially Ru-catalyzed RCM–allylic oxidation synthesis of coumarins **23**.

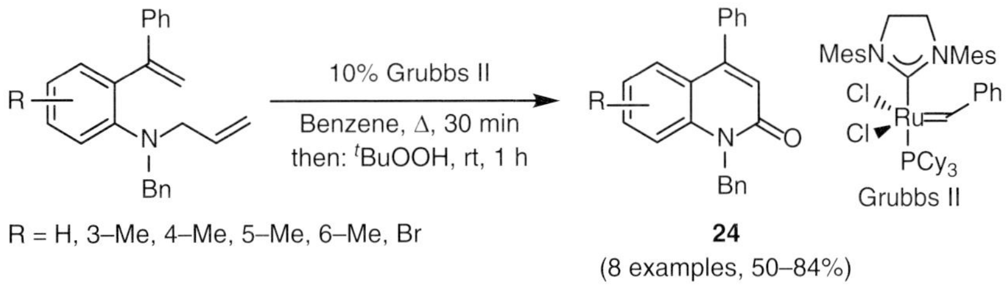

Scheme 12.16 Sequentially Ru-catalyzed RCM–allylic oxidation synthesis of quinolinones **24**.

12.4
Sequences Initiated by Alkynylation and Carbopalladative Insertions

Most Pd-catalyzed sequential, consecutive, and domino processes [48] are based on the insertion multiple carbon–carbon bonds as the underlying mechanistic elementary step. Carbopalladation of alkynes is particularly interesting because triple bonds are excellent ligands for transition metals, which, after insertion, do not terminate by β-hydride eliminations as in Heck vinylations. Possible terminations are either Pd-mediated nucleophilic trapping or transmetallation. Recent advances in Pd-catalyzed cascade cyclizations are briefly highlighted. Catalyst systems enabling Sonogashira coupling are among the most versatile to trigger a subsequent Pd- or Pd–Cu-catalyzed cyclization.

Three different N-heterocyclic carbene (NHC)–Pd–pyridine complexes (**A**, **B**, and **C**), where NHC is 1,2,4-trimethyltriazolyldiylidene, 1,3-dimethylimidazolylidene, and 1,4-dimethyltriazolylidene, have been reported to efficiently catalyze a domino sequence of alkyne coupling of an *ortho*-iodo phenol or an *ortho*-iodo benzylalcohol and the subsequent cycloisomerization of the intermediate tolane derivative to afford benzofuran **25** or benzoisofuran **26** (Scheme 12.17) [49].

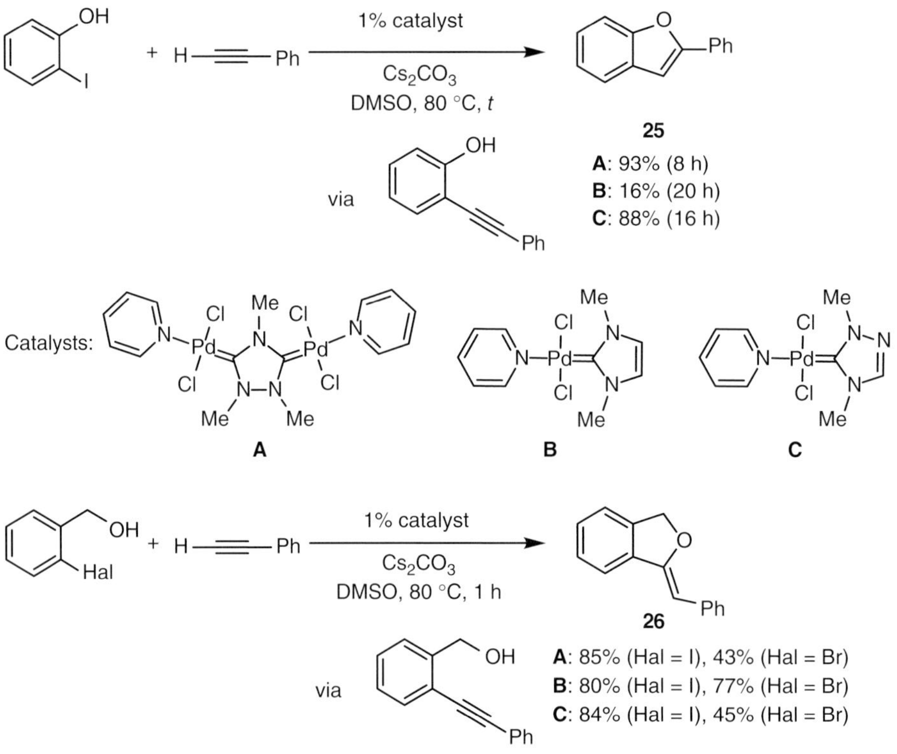

Scheme 12.17 Sequentially Pd-catalyzed alkynylation–hydroxyalkylation synthesis of benzofuran **25** and isobenzofuran **26**.

Previously, Olivi *et al.* [50] had shown that a consecutive amine propargylation–Sonogashira sequence proceeds very efficiently in the sense of a three-component coupling–cyclization to furnish the benzofuran **27** or the aniline **28** upon applying the standard Sonogashira catalyst system (Scheme 12.18).

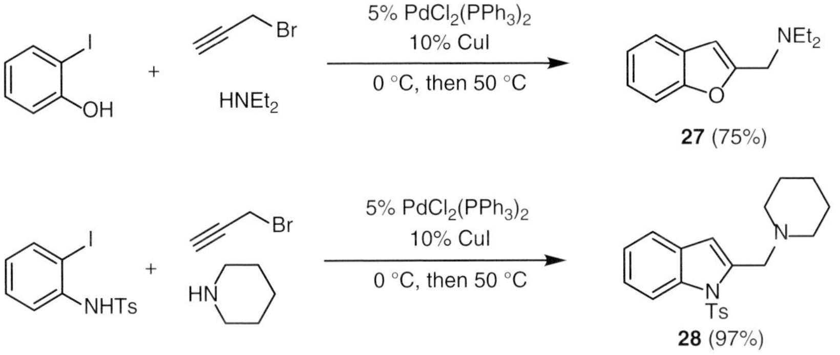

Scheme 12.18 Consecutive amine propargylation–Sonogashira coupling–cyclization synthesis of benzofuran **27** and indole **28**.

The group of Pal [51] reported an interesting sequence to thiophene-annelated α-pyrones **29** and **30** in good yields based upon the Sonogashira catalyst system (Scheme 12.19). Although the mechanistic rationale for this pseudo-three-component synthesis remains obscure because an oxidation step has to be formally involved, the cooperativity of the palladium–copper system is remarkable.

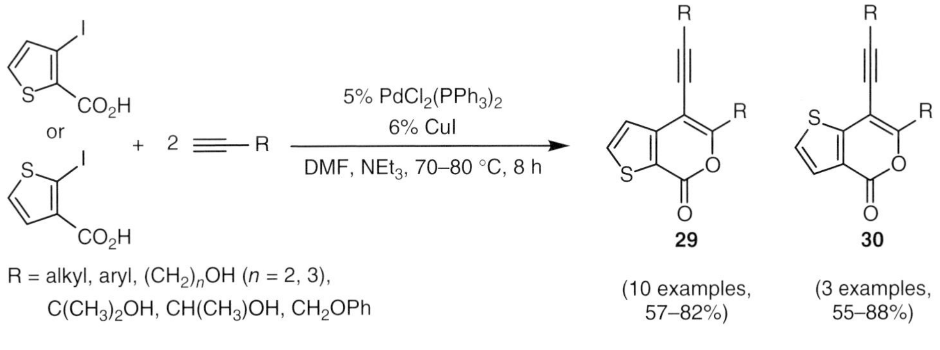

Scheme 12.19 Sequentially Pd-catalyzed alkynylation–cyclization–alkynylation synthesis of 4-alkynylthieno[2,3-*c*]pyran-7-ones **29** and 7-alkynylthieno[3,2-*c*]pyran-4-ones **30**.

Pinto *et al.* [52] have reported a consecutive three-component synthesis of oxindoles that proceeds in the sense of a Sonogashira alkynylation–carbopalladation–CH activation–cyclization sequence, furnishing 3-(diarylmethylene)oxindoles **31** in moderate to good yields (Scheme 12.20).

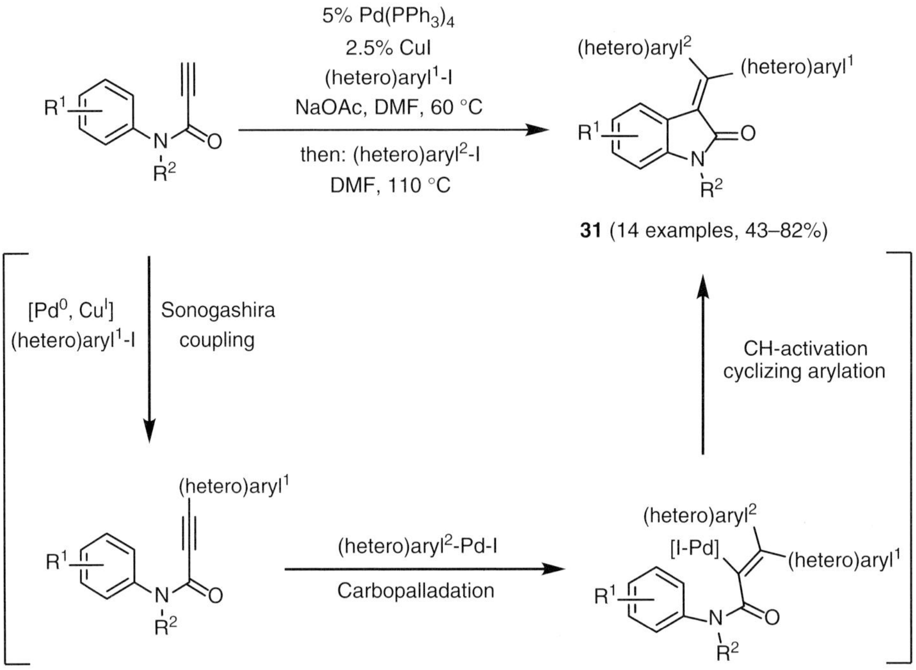

Scheme 12.20 Sequentially Pd/Cu-catalyzed Sonogashira carbopalladation–CH activation–cyclization synthesis of 3-(diarylmethylene)indolin-2-ones **31**.

The same authors disclosed a convergent three-component access to the indolone scaffold by a sequentially Pd-catalyzed N-arylation–carbopalladation–CH activation–cyclization process [53]. Without further addition of catalyst or ligands after the *N*-arylation of the secondary propiolamide, a second aryliodide concludes the synthesis of 3-(diarylmethylene)indolin-2-ones **32** in moderate to good yields via an intermolecular carbopalladation–CH activation–arylation (Scheme 12.21).

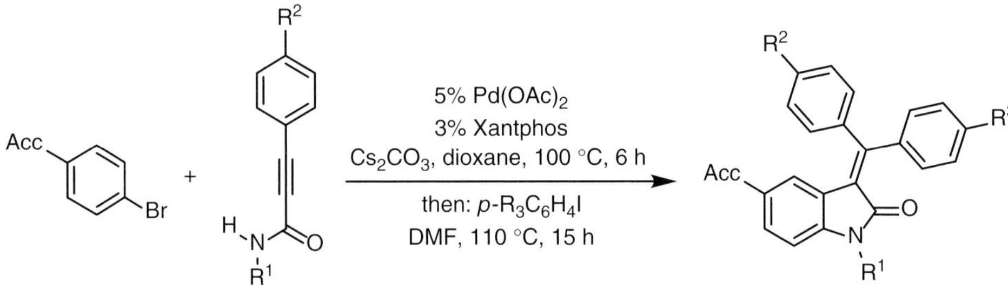

32 (8 examples, 32–67%)

Scheme 12.21 Sequentially Pd-catalyzed *N*-arylation–carbopalladation–CH activation–cyclization synthesis of 3-(diarylmethylene)indolin-2-ones **32**.

ortho-Iodo anilines are perfect starting materials for generating suitable substrates for domino reactions, in particular in heterocycle synthesis [3a–c,e, 54–56] and for functional materials [57]. For instance, the domino insertion–coupling–isomerization–intramolecular Diels–Alder sequence of alkynoyl *ortho*-iodo phenolesters or alkynoyl *ortho*-iodo anilides and propargyl allyl ethers furnishes solution- and solid-state luminescent (tetrahydroisobenzofuran) spiro-benzofuranones or spiro-indolones in moderate to excellent yields [58]. Based upon this concept, the enyne indolone intermediate **33** was exploited for a general consecutive one-pot three-component approach to solid-state, red fluorescent push–pull indolones **34** in the sense of an insertion–alkynylation–amination sequence (Scheme 12.22) [59]. In a related study, Balalaie and coworkers [60] reported an elegant entry to alkynoyl *ortho*-iodo anilides by an Ugi four-component condensation.

For secondary *ortho*-iodo alkynoyl anilides ($R^1 = H$), Schönhaber *et al.* [61] observed that the intermediate **33** ($R^1 = H$) is consumed in a transition-metal-catalyzed cycloisomerization, furnishing the new class of proto- and metallochromic luminescent 2,4-diarylpyrano[2,3-*b*]indoles **35** in moderate yields (Scheme 12.23). The overall reaction proceeds as a sequentially Pd–Cu-catalyzed insertion–coupling–cycloisomerization domino reaction.

12.5 Sequences Intercepted by Cyclizations

Alkynones [62] and 1,3-diaryl propenones (chalcones) [63] are reactive three-carbon building blocks that can be transformed into five-, six-, and seven-membered heterocycles with bifunctional nucleophiles in the sense of Michael addition and cyclocondensation. Therefore, a catalytic access to ynones [64] and enones [65] by Sonogashira alkynylation turned out to be a versatile entry to consecutive multicomponent syntheses of heterocycles. This reactivity-based concept has been considerably developed by the Müller group in the past decade. In several reviews, the major advancements of this one-pot methodology have been summarized [66]. Therefore, only very recent conceptual achievements with respect to sequentially Pd-catalyzed processes [7] will be considered here.

A consecutive three-component synthesis of pyrazoles [67] starting from (hetero)aroyl chlorides, terminal alkynes, and hydrazines represents an excellent entry to more sophisticated one-pot transformations on the regioselectively substituted heterocyclic core. For instance, the catenation of an electrophilic halogenation step led to a one-pot four-component synthesis of 4-halo pyrazoles **36** (halo = chloro or bromo) in good to excellent yield (Scheme 12.24) [68].

Indeed, by virtue of the *in situ* generated 4-halo pyrazole **36** and the presence of the catalytic metal complexes, a sequentially Pd-catalyzed Suzuki coupling was conceived. Upon addition of catalytic amounts of triphenylphosphane, the proper

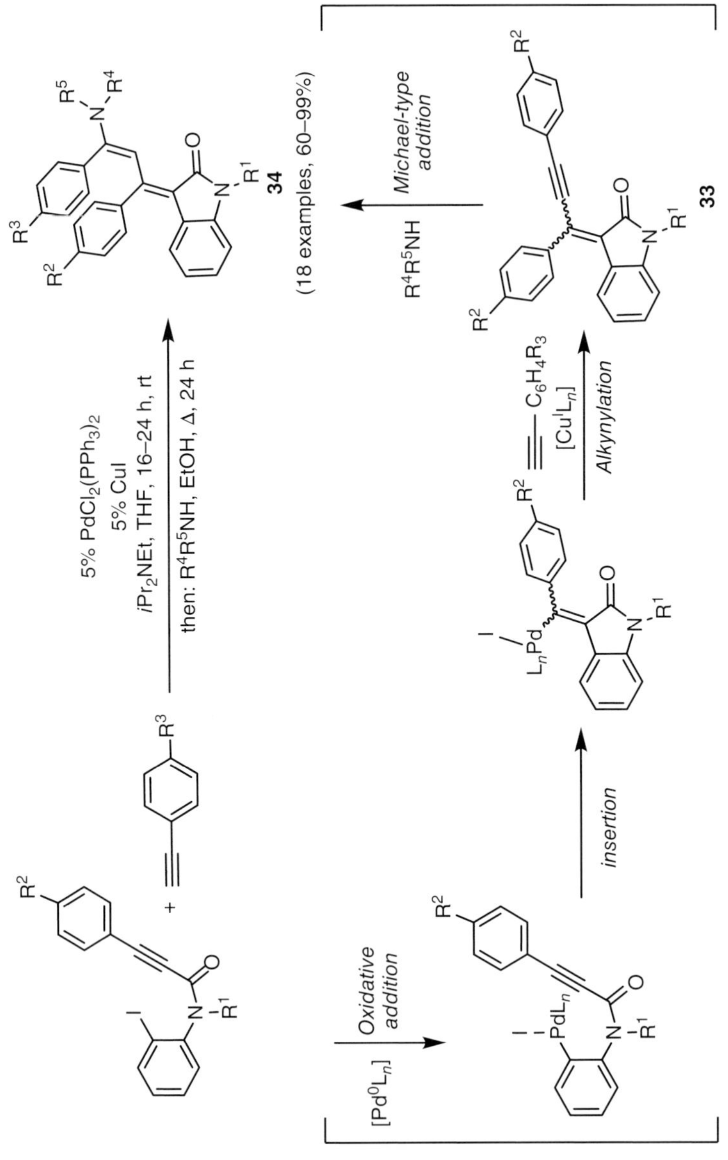

Scheme 12.22 Three-component synthesis of push-pull indolones **34**.

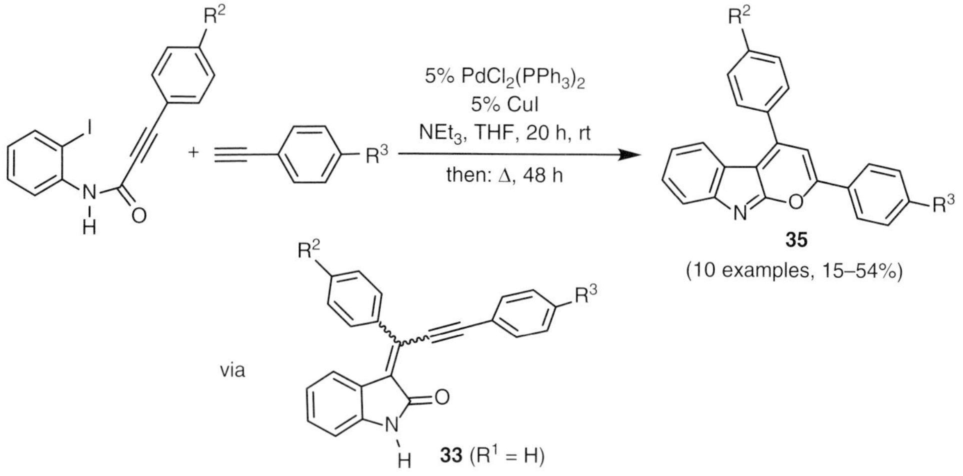

35
(10 examples, 15–54%)

33 (R^1 = H)

Scheme 12.23 Sequentially Pd/Cu-catalyzed insertion–coupling–cycloisomerization domino synthesis of 2,4-diarylpyrano[2,3-*b*]indoles **35**.

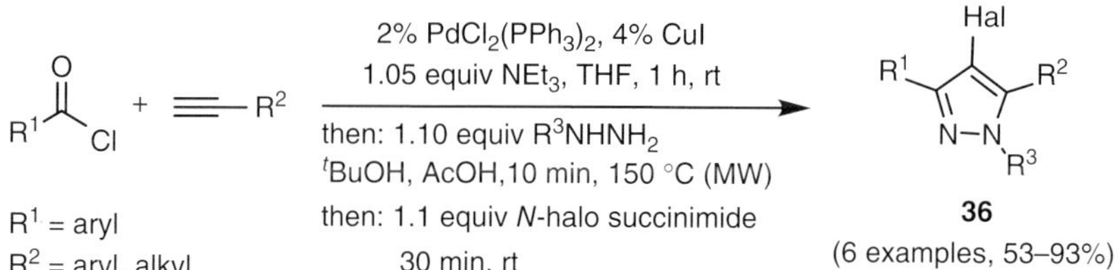

R^1 = aryl
R^2 = aryl, alkyl

36
(6 examples, 53–93%)
Hal = Cl, Br

Scheme 12.24 Four-component synthesis of 4-halo pyrazoles **36**.

oxidation stage of the Pd source is ensured, giving rise to the formation of 1,3,4,5-tetrasubstituted pyrazoles **37** in a one-pot four-step sequence consisting of Sonogashira coupling, addition–cyclocondensation, bromination, and Suzuki coupling (Scheme 12.25) [68]. All title compounds displayed intense blue fluorescence and high quantum yields.

Interestingly, under the peculiar conditions of the alkynylation of acid chlorides, the reaction medium is easily conditioned to Brønsted acidic reactions. Hence, Müller and coworkers [69] applied tetrahydropyranyl (THP)-protected propargyl alcohols as alkynes and sodium chloride or iodide, respectively, in a three-component coupling–acetal cleavage–hydrogen halide Michael addition–cyclocondensation sequence yielding the class of 3-halofurans. Furthermore, the conditions are compatible with subsequent Suzuki reactions by simply adding the required boronic acids and excess of Na_2CO_3 to the reaction mixture with the iodo furan, leading to the formation of trisubstituted furans [69b].

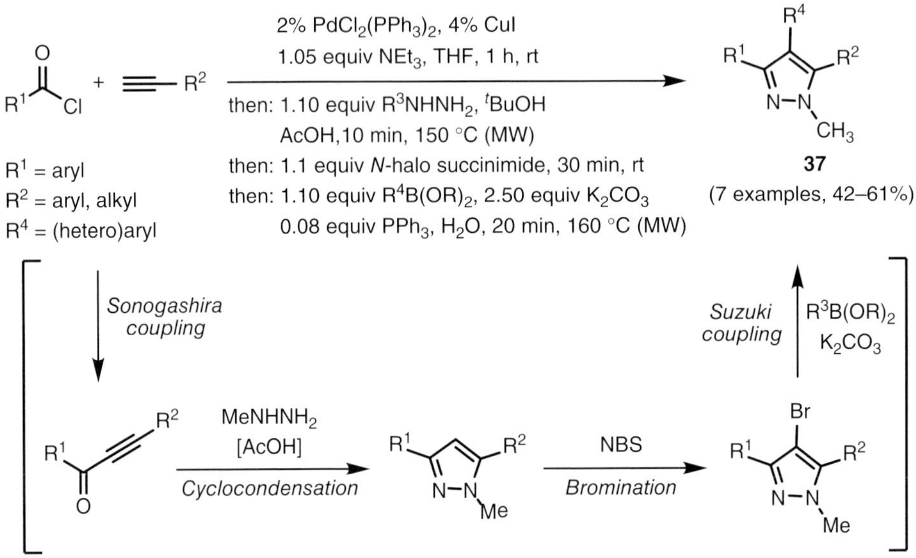

Scheme 12.25 Sequentially Pd/Cu-catalyzed Sonogashira alkynylation–cyclocondensation–bromination–Suzuki arylation synthesis of 1,3,4,5-tetrasubstituted pyrazoles **37**.

For the three-component synthesis of the related 4-iodo pyrroles, the proficient choice of the right nitrogen protecting group plays a crucial role. The Boc (*tert*-butyloxycarbonyl) group is a versatile carbamate protecting group for the pyrrole nitrogen atom, useful for many transformations on the pyrrole core and easily removable. Upon reacting (hetero)aroyl chlorides and *N*-Boc-protected propargyl amine under modified Sonogashira conditions, the intermediate alkynone was transformed by addition–cyclocondensation with sodium iodide and *p*-toluenesulfonic acid (PTSA), furnishing *N*-Boc-4-iodo-2-substituted pyrroles **38** in good yields (Scheme 12.26) [70].

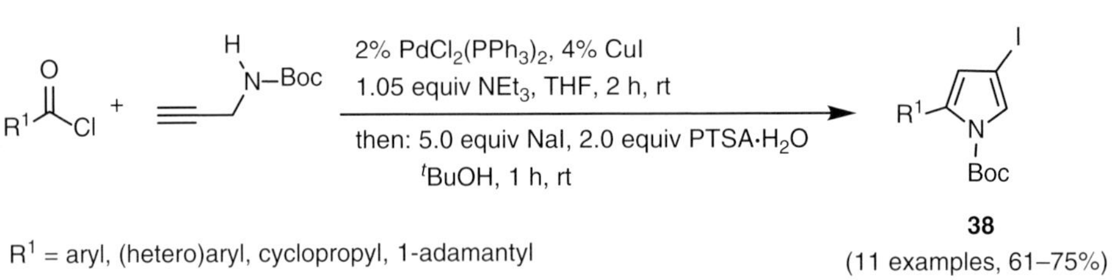

Scheme 12.26 Three-component synthesis of *N*-Boc-4-iodo-2-substituted pyrroles **38**.

Since *N*-Boc-protected 4-iodo pyrroles **38** are highly useful synthetic building blocks that can be transformed in the sense of a sequential Pd/Cu-catalyzed process, assuming that the catalyst system is still active after the coupling–addition–cyclocondensation sequence. Upon addition of another terminal alkyne to the reaction mixture, *N*-Boc-2-aryl-4-alkynyl pyrroles **39** were obtained in good yields in the sense of a one-pot four-step coupling–addition–cyclocondensation–coupling sequence (Scheme 12.27) [70]. The Boc group remains uncleaved in the complete sequence.

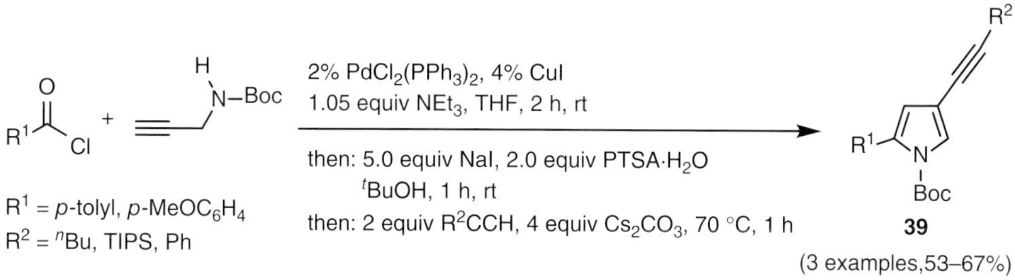

39
(3 examples,53–67%)

Scheme 12.27 Sequentially Pd–Cu-catalyzed three-component synthesis of *N*-Boc-2-aryl-4-alkynyl pyrroles **39**.

Very recently, Murata *et al.* [71] have communicated a remarkable sequential Pd–Cu-catalyzed sequence in a domino manner furnishing hetero α,α′-dimers of heterocycles **40** in good to excellent yields (Scheme 12.28). The domino sequence

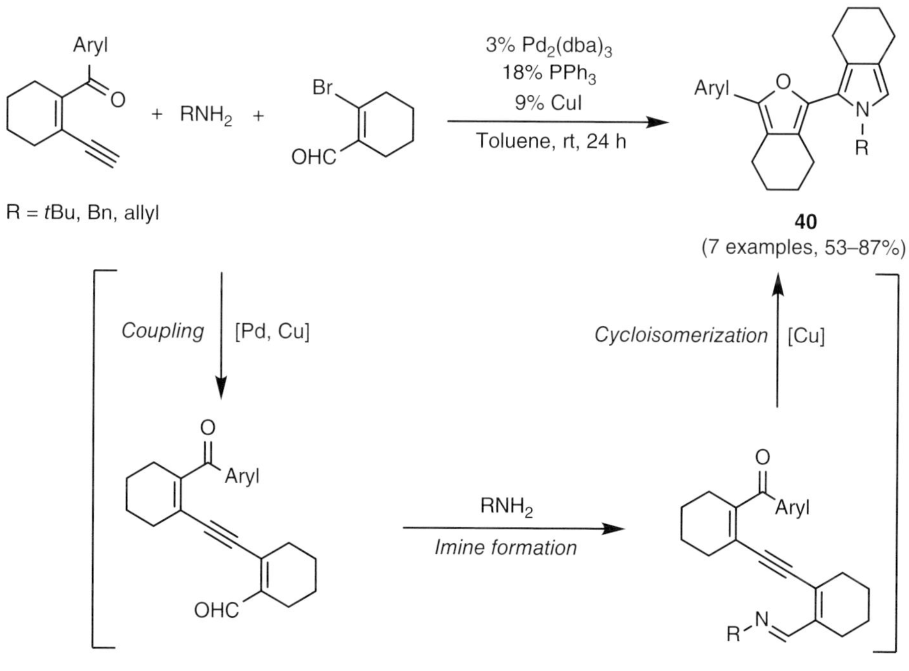

Scheme 12.28 Sequentially Pd–Cu-catalyzed three-component synthesis of hetero α,α′-dimers of heterocycles **40**.

commences with the Sonogashira coupling of a conformationally fixed enyne carbonyl compound with 2-bromo cyclohexen-1-carbaldehyde, followed by imine formation and concludes with double cycloisomerization.

The same sequentially catalyzed domino reaction was successfully applied to very efficient syntheses of 2-(5′-thienyl)furan and -pyrrole (**41**) and the monodisperse heterooligomer **42** (Scheme 12.29).

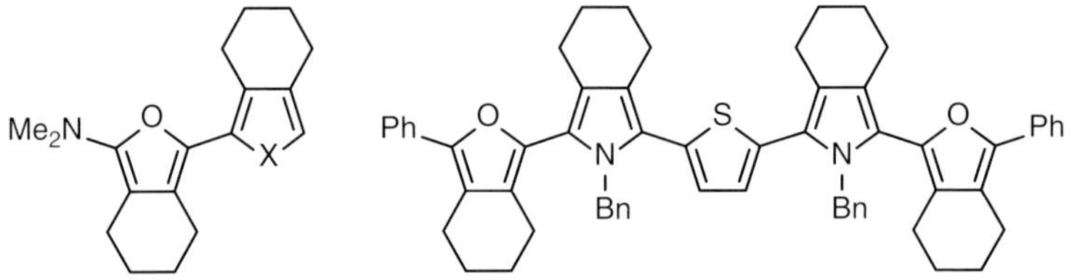

41
(90%, X = O)
(84%, X NBn)

42
(53%)

Scheme 12.29 Sequentially Pd–Cu-catalyzed synthesis of 2-(5′-thienyl)furan and -pyrrole (**41**) and the monodisperse heterooligomer **42**.

12.6 Conclusion

Sequential transition-metal-catalyzed processes based upon cycloisomerization and cyclization are receiving increasing attention because they combine the increase of structural complexity (cyclizing reactions) with an atom economically benign use of the employed catalyst sources (sequential catalysis). This reactivity-based concept has been predominantly illustrated for sequentially Pd-, Rh-, and Ru-catalyzed processes, and currently the literature is dominated by the three transition metals. However, many other metal complexes can be envisioned to be equally fruitful in further developments and, without any doubt, even more unusual sequences are waiting to be discovered. Inevitably, only the opening chapter of this never-ending story has been told.

Acknowledgment

The author gratefully acknowledges the continuous support by the Fonds der Chemischen Industrie.

Abbreviations

Ac	acetyl
AcOH	acetic acid
BINAP	2,2′-bis(diphenylphosphano)-1,1′-binaphthyl

Bn	benzyl
Boc	*tert*-butyloxycarbonyl
Bu	butyl
COD	cycloocta-1,5-diene
Cy	cyclohexyl
dba	dibenzylideneacetone
DMF	*N*,*N*-dimethylformamide
DMSO	dimethylsulfoxide
dppe	bis(diphenylphosphanyl) ethane
Et	ethyl
EtOH	ethanol
EWG	electron-withdrawing group
Me	methyl
Ms	mesityl
MW	microwave (reactor)
NEt_3	triethylamine
NBS	*N*-bromosuccinimide
NHC	N-heterocyclic carbine
PTSA	*p*-toluenesulfonic acid
THF	tetrahydrofuran
THP	tetrahydropyranyl
TMS	trimethylsilyl
Tos	*p*-tolylsulfonyl
Xantphos	4,5-bis(diphenylphosphano)-9,9-dimethylxanthene

References

1. For a recent review with a suggestion for a taxonomy of domino, tandem and cascade metal catalyzed reactions see Fogg, D.E. and de Santos, E.N. (2004) *Coord. Chem. Rev.*, **248**, 2365–2379.
2. For recent reviews on one-pot catalysis, see e.g.(a) Patil, N.T., Shinde, V.S., and Gajula, B. (2012) *Org. Biomol. Chem.*, **10**, 211–224. (b) Burke, A.J., Marinho, V.R., and Furtado, O.M.R. (2010) *Curr. Org. Synth.*, **7**, 94–119. (c) Lee, J.M., Na, Y., Han, H., and Chang, S. (2004) *Chem. Soc. Rev.*, **33**, 302–312.
3. (a) Kirsch, G., Hesse, S., and Comel, A. (2004) *Curr. Org. Synth.*, **1**, 47–63. (b) Nakamura, I. and Yamamoto, Y. (2004) *Chem. Rev.*, **104**, 2127–2198. (c) Balme, G., Bossharth, E., and Monteiro, N. (2003) *Eur. J. Org. Chem.*, **2003**, 4101–4111. (d) de Meijere, A. and Schelper, M. (2003) *Actual. Chim.*, 51–56. (e) Battistuzzi, G., Cacchi, S., and Fabrizi, G. (2002) *Eur. J. Org. Chem.*, **2002**, 2671–2681. (f) Poli, G., Giambastiani, G., and Heumann, A. (2000) *Tetrahedron*, **56**, 5959–5989. (g) Grigg, R. and Sridharan, V. (1999) *J. Organomet. Chem.*, **576**, 65–87. (h) de Meijere, A. and Bräse, S. (1999) *J. Organomet. Chem.*, **576**, 88–110. (i) Molander, G.A. and Harris, C.R. (1996) *Chem. Rev.*, **96**, 307–338. (j) Malacria, M. (1996) *Chem. Rev.*, **96**, 289–306. (k) Tietze, L.F. (1996) *Chem. Rev.*, **96**, 115–136. (l) Tietze, L.F. and Beifuss, U. (1993) *Angew. Chem.*, **105**, 137–170; *Angew. Chem., Int. Ed. Engl.*, (1993), **32**, 131–163. (m) Tietze, L.F. (1990) *J. Heterocycl. Chem.*, **27**, 47–69.
4. For reviews, see (a) Wasilke, J.-C., Obrey, S.J., Baker, R.T., and Bazan, G.C. (2005) *Chem. Rev.*, **105**, 1001–1020. (b)

Ajamian, A. and Gleason, J.L. (2004) *Angew. Chem.*, **116**, 3842–3848; *Angew. Chem. Int. Ed.*, (2004), **43**, 3754–3760. For examples, see e. g.(c) Yamamoto, Y., Nakagai, Y.-i., and Itoh, K. (2004) *Chem. Eur. J.*, **10**, 231–236. (d) Cossy, J., Bargiggia, F., and BouzBouz, S. (2003) *Org. Lett.*, **5**, 459–462. (e) Thadani, A.N. and Rawal, V.H. (2002) *Org. Lett.*, **4**, 4317–4320. (f) Son, S.U., Park, K.H., and Chung, Y.K. (2002) *J. Am. Chem. Soc.*, **124**, 6838–6839. (g) Louie, J., Bielawski, C.W., and Grubbs, R.H. (2001) *J. Am. Chem. Soc.*, **123**, 11312–11313. (h) Evans, P.A. and Robinson, J.E. (2001) *J. Am. Chem. Soc.*, **123**, 4609–4610.

5. For reviews on carbopalladation see (a) Cacchi, S. (1999) *J. Organomet. Chem.*, **576**, 42–64. (b) Negishi, E., Copéret, C., Ma, S., Liuo, S.-Y., and Liu, F. (1996) *Chem. Rev.*, **96**, 365–393. (c) Negishi, E., Wang, G., and Zhu, G. (2006) *Top. Organomet. Chem.*, **19**, 1–48.
6. For reviews on diversity oriented syntheses, see (a) Schreiber, S.L. and Burke, M.D. (2004) *Angew. Chem.*, **116**, 48–60; *Angew. Chem. Int. Ed.*, (2004), **43**, 46–58. (b) Burke, M.D., Berger, E.M., and Schreiber, S.L. (2003) *Science*, **302**, 613–618. (c) Arya, P., Chou, D.T.H., and Baek, M.G. (2001) *Angew. Chem.*, **113**, 351–358; *Angew. Chem. Int. Ed.*, (2001), **40**, 339–346. (d) Cox, B., Denyer, J.C., Binnie, A., Donnelly, M.C., Evans, B., Green, D.V.S., Lewis, J.A., Mander, T.H., Merritt, A.T., Valler, M.J., and Watson, S.P. (2000) *Prog. Med. Chem.*, **37**, 83–133. (e) Schreiber, S.L. (2000) *Science*, **287**, 1964–1969.
7. Müller, T.J.J. (2006) *Top. Organomet. Chem.*, **19**, 149–205.
8. Körber, N. and Müller, T.J.J. (2006) *Chem. Today Chim. Oggi*, **24**, 22–29.
9. Bruneau, C., Dérien, S., and Dixneuf, P.H. (2006) *Top. Organomet. Chem.*, **19**, 295–326.
10. (a) Trost, B.M. and Krische, M.J. (1998) *Synlett*, 1–16. (b) Trost, B.M. (1990) *Acc. Chem. Res.*, **23**, 34–42.
11. For leading reviews on Pd catalyzed cycloisomerizations, see (a) Trost, B.M. (1991) *Janssen Chim. Acta*, **9**, 3–10. (b) Lloyd-Jones, G.C. (2003) *Org. Biomol. Chem.*, **1**, 215–236. (c) Aubert, C., Buisine, O., and Malacria, M. (2002) *Chem. Rev.*, **102**, 813–834.
12. (a) Trost, B.M. (1995) *Angew. Chem.*, **107**, 285–307; *Angew. Chem., Int. Ed. Engl.*, (1995), **34**, 259–281. (b) Trost, B.M. (1991) *Science*, **34**, 1471–1477.
13. For recent reviews on transition metal catalyzed enyne cycloisomerizations, see e. g. (a) Pradal, A., Toullec, P.Y., and Michelet, V. (2011) *Synthesis*, **10**, 1501–1514. (b) Michelet, V., Toullec, P.Y., and Genet, J.-P. (2008) *Angew. Chem.*, **120**, 4338–4386; *Angew. Chem. Int. Ed.*, (2008), **47**, 4268–4315. (c) Widenhoefer, R.A. (2002) *Acc. Chem. Res.*, **53**, 905–913.
14. (a) Marinetti, A., Jullien, H., and Voituriez, A. (2012) *Chem. Soc. Rev.*, **41**, 4884–4908. (b) Tietze, L.F., Ila, H., and Bell, H.P. (2004) *Chem. Rev.*, **104**, 3453–3516.
15. Yamada, H., Aoyagi, S., and Kibayashi, C. (1997) *Tetrahedron Lett.*, **38**, 3027–3030.
16. Kressierer, C.J. and Müller, T.J.J. (2004) *Synlett*, 655–658.
17. Kressierer, C.J. and Müller, T.J.J. (2004) *Tetrahedron Lett.*, **45**, 2155–2158.
18. Kressierer, C.J. and Müller, T.J.J. (2005) *Org. Lett.*, **7**, 2237–2240.
19. Kressierer, C.J. and Müller, T.J.J. (2005) *Synlett*, 1721–1725.
20. Lei, A., He, M., and Zhang, X. (2002) *J. Am. Chem. Soc.*, **124**, 8198–8199.
21. Körber, N., Rominger, F., and Müller, T.J.J. (2010) *Synlett*, 782–786.
22. Körber, N., Rominger, F., and Müller, T.J.J. (2009) *Adv. Synth. Catal.*, **351**, 2921–2935.
23. Brummond, K.M., Chen, H., Sill, P., and You, L. (2002) *J. Am. Chem. Soc.*, **124**, 15186–15187.
24. Shibata, T., Takesue, Y., Kadowaki, S., and Takagi, K. (2003) *Synlett*, 268–270.
25. Brummond, K.M., Painter, T.O., Probst, D.A., and Mitasev, B. (2007) *Org. Lett.*, **9**, 347–349.
26. Brummond, K.M. and Yan, B. (2008) *Synlett*, 2303–2308.
27. Brummond, K.M. and McCabe, J.M. (2006) *Tetrahedron*, **62**, 10541–10554.
28. Brummond, K.M. and Mitasev, B. (2004) *Org. Lett.*, **6**, 2245–2248.

29. Mitasev, B., Yan, B., and Brummond, K.M. (2006) *Heterocycles*, **70**, 367–388.
30. Brummond, K.M. and You, L. (2005) *Tetrahedron*, **61**, 6180–6185.
31. Okamoto, R., Okazaki, E., Noguchi, K., and Tanaka, K. (2011) *Org. Lett.*, **13**, 4894–4897.
32. Shibata, T., Yamasaki, M., Kadowaki, S., and Takagi, K. (2004) *Synlett*, 2812–2814.
33. Sim, S.H., Lee, S.I., Park, J.H., and Chung, Y.K. (2010) *Adv. Synth. Catal.*, **352**, 317–322.
34. Kummeter, M., Ruff, C.M., and Müller, T.J.J. (2007) *Synlett*, 717–720.
35. For reviews on olefin metathesis, see e. g. (a) Grubbs, R.H., Miller, S.J., and Fu, G.C. (1995) *Acc. Chem. Res.*, **28**, 446–452. (b) Hashmi, A.S.K. (1997) *J. Prakt. Chem.*, **339**, 195–199. (c) Schuster, M. and Blechert, S. (1997) *Angew. Chem.*, **109**, 2124–2144; *Angew. Chem. Int. Ed.*, (1997), **36**, 2036–2056. (d) Grubbs, R.H. and Chang, S. (1998) *Tetrahedron*, **54**, 4413–4450. (e) Fürstner, A. (1998) *Top. Organomet. Chem.*, **1**, 37–72. (f) Fürstner, A. (2000) *Angew. Chem.*, **112**, 3141–3172; *Angew. Chem. Int. Ed.*, (2000), **39**, 3012–3043. (g) Grubbs, R.H. (2004) *Tetrahedron*, **60**, 7117–7140.
36. For reviews on ring closing metathesis, see e. g. (a) van Otterlo, W.A.L. and de Koning, C.B. (2009) *Chem. Rev.*, **109**, 3743–3782. (b) Conrad, J.C. and Fogg, D.E. (2006) *Curr. Org. Chem.*, **10**, 185–202.
37. For reviews on metathesis in organic synthesis, see e. g. (a) Nicolaou, K.C., Bulger, P.G., and Sarlah, D. (2005) *Angew. Chem.*, **117**, 4564–4601; *Angew. Chem. Int. Ed.*, (2005), **44**, 4490–4527. (b) Deiters, A. and Martin, S.F. (2004) *Chem. Rev.*, **104**, 2199–2238.
38. (a) Schmidt, B. (2004) *Eur. J. Org. Chem.*, **2004**, 1865–1880. (b) Schmidt, B. (2006) *J. Mol. Catal. A*, **254**, 53–57. (c) Dragutan, V. and Dragutan, I. (2006) *J. Organomet. Chem.*, **691**, 5129–5147. (d) Schmidt, B. (2006) *Pure Appl. Chem.*, **78**, 469–476. (e) Krompiec, S., Krompiec, M., Penczek, R., and Ignasiak, H. (2008) *Coord. Chem. Rev.*, **252**, 1819–1841. (f) Alcaide, B., Almendros, P., and Luna, A. (2009) *Chem. Rev.*, **109**, 3817–3858.
39. (a) Sutton, A.E., Seigal, B.A., Finnegan, D.F., and Snapper, M.L. (2002) *J. Am. Chem. Soc.*, **124**, 13390–13391. (b) Schmidt, B. (2004) *J. Org. Chem.*, **69**, 7672–7687.
40. Schmidt, B. and Biernat, A. (2008) *Chem. Eur. J.*, **14**, 6135–6141.
41. Schmidt, B. and Kunz, O. (2012) *Synlett*, 851–854.
42. Cai, Q., Liang, X.-W., Wang, S.-G., Zhang, J.-W., Zhang, X., and You, S.-L. (2012) *Org. Lett.*, **14**, 5022–5025.
43. Ascic, E., Jensen, J.F., and Nielsen, T.E. (2011) *Angew. Chem.*, **123**, 5294–5297; *Angew. Chem. Int. Ed.*, (2011), **50**, 5188–5191.
44. Petersen, M.T. and Nielsen, T.E. (2013) *Org. Lett.*, **15**, 1986–1989.
45. For representative examples, see (a) Beligny, S., Eibauer, S., Maechling, S., and Blechert, S. (2006) *Angew. Chem.*, **118**, 1933–1937; *Angew. Chem. Int. Ed.*, (2006), **45**, 1900–1903. (b) Neisius, N.M. and Plietker, B. (2008) *J. Org. Chem.*, **73**, 3218–3227, and references cited therein.
46. Schmidt, B. and Krehl, S. (2011) *Chem. Commun.*, **47**, 5879–5881.
47. Kato, H., Ishigame, T., Oshima, N., Hoshiya, N., Shimawaki, K., Arisawa, M., and Shuto, S. (2011) *Adv. Synth. Catal.*, **353**, 2676–2680.
48. (a) Vlaar, T., Ruijter, E., and Orru, R.V.A. (2011) *Adv. Synth. Catal.*, **353**, 809–841. (b) Negishi, E.-I. and Anastasia, L. (2003) *Chem. Rev.*, **103**, 1979–2018. (c) Negishi, E.-I. (2002) in *Handbook of Organopalladium Chemistry for Organic Synthesis* (ed. E.-I. Negishi), John Wiley & Sons, Inc., New York.
49. Zanardi, A., Mata, J.A., and Peris, E. (2009) *Organometallics*, **28**, 4335–4339.
50. Olivi, N., Spruyt, P., Peyrat, J.-F., Alami, M., and Brion, J.-D. (2004) *Tetrahedron Lett.*, **45**, 2607–2610.
51. Raju, S., Batchu, V.R., Swamy, N.K., Dev, R.V., Sreekanth, B.R., Babu, J.M., Vyas, K., Kumar, P.R., Mukkanti, K., Annamalai, P., and Pal, M. (2006) *Tetrahedron*, **62**, 9554–9570.
52. Pinto, A., Neuville, L., and Zhu, J. (2007) *Angew. Chem.*, **119**, 3355–3359; *Angew. Chem. Int. Ed.*, (2007), **46**, 3291–3295.

53. Pinto, A., Neuville, L., and Zhu, J. (2009) *Tetrahedron Lett.*, **50**, 3602–3605.
54. (a) Zeni, G. and Larock, R.C. (2004) *Chem. Rev.*, **104**, 2285–2309. (b) Lie, J.J. and Gribble, G.W. (2000) *Palladium in Heterocyclic Chemistry*, Pergamon Press, New York.
55. Balme, G., Bouyssi, D., and Monteiro, N. (2006) *Pure Appl. Chem.*, **78**, 231–239.
56. For a review on domino syntheses of complex heterocycles, see e. g. Padwa, A. and Bur, S.K. (2007) *Tetrahedron*, **63**, 5341–5378.
57. For a recent account, see e. g. Müller, T.J.J. and D'Souza, D.M. (2008) *Pure Appl. Chem.*, **80**, 609–620.
58. (a) D'Souza, D.M., Rominger, F., and Müller, T.J.J. (2005) *Angew. Chem.*, **117**, 156–161; *Angew. Chem. Int. Ed.*, (2005), **44**, 153–158. (b) D'Souza, D.M., Kiel, A., Herten, D.P., and Müller, T.J.J. (2008) *Chem. Eur. J.*, **14**, 529–547.
59. D'Souza, D.M., Muschelknautz, C., Rominger, F., and Müller, T.J.J. (2010) *Org. Lett.*, **12**, 3364–3367.
60. Bararjanian, M., Balalaie, S., Rominger, F., Movassagh, B., and Bijanzadeh, H.R. (2010) *J. Org. Chem.*, **75**, 2806–2812.
61. Schönhaber, J., Frank, W., and Müller, T.J.J. (2010) *Org. Lett.*, **12**, 4122–4125.
62. Bol'shedvorskaya, R.L. and Vereshchagin, L.I. (1973) *Russ. Chem. Rev.*, **42**, 225–240.
63. For a review on the chemistry of 1,3-diaryl propenones, see e. g. Thebtaranonth, C. and Thebtaranonth, Y. (1989) in *The Chemistry of Enones*, vol. 29 (eds S. Patai and Z. Rappoport), John Wiley & Sons, Ltd, Chichester, New York, Brisbane, Toronto, Singapore, pp. 199–280.
64. (a) D'Souza, D.M. and Müller, T.J.J. (2008) *Nat. Protoc.*, **3**, 1660–1665. (b) Karpov, A.S. and Müller, T.J.J. (2003) *Org. Lett.*, **5**, 3451–3454.
65. (a) Müller, T.J.J., Ansorge, M., and Aktah, D. (2000) *Angew. Chem.*, **112**, 1323–1326; *Angew. Chem. Int. Ed.*, (2000), **39**, 1253–1256. (b) Braun, R.U., Ansorge, M., and Müller, T.J.J. (2006) *Chem. Eur. J.*, **12**, 9081–9094. (c) Schramm, O.G., Dediu, N., and Müller, T.J.J. (2006) *Adv. Synth. Catal.*, **348**, 2565–2570. (d) Liao, W.-W. and Müller, T.J.J. (2006) *Synlett*, 3469–3473.
66. (a) Müller, T.J.J. (2010) *Top. Heterocycl. Chem.*, **25**, 25–94. (b) Willy, B. and Müller, T.J.J. (2009) *Curr. Org. Chem.*, **13**, 1777–1790. (c) Willy, B. and Müller, T.J.J. (2008) *ARKIVOC*, **2008** (Pt. I), 195–208.
67. (a) Willy, B. and Müller, T.J.J. (2008) *Eur. J. Org. Chem.*, **2008**, 4157–4168. (b) Liu, H.L., Jiang, H.F., Zhang, M., Yao, W.J., Zhu, Q.H., and Tang, Z. (2008) *Tetrahedron Lett.*, **49**, 3805–3809.
68. Willy, B. and Müller, T.J.J. (2011) *Org. Lett.*, **13**, 2082–2085.
69. (a) Karpov, A.S., Merkul, E., Oeser, T., and Müller, T.J.J. (2006) *Eur. J. Org. Chem.*, **2006**, 2991–3000. (b) Karpov, A.S., Merkul, E., Oeser, T., and Müller, T.J.J. (2005) *Chem. Commun.*, 2581–2583.
70. Merkul, E., Boersch, C., Frank, W., and Müller, T.J.J. (2009) *Org. Lett.*, **11**, 2269–2272.
71. Murata, T., Murai, M., Ikeda, Y., Miki, K., and Ohe, K. (2012) *Org. Lett.*, **14**, 2296–2299.

13
C–N-Coupling Reactions in Catalytic One-Pot Syntheses Using Molecular Group 4 Catalysts

Lutz H. Gade and Solveig A. Scholl

13.1
Introduction

The development of catalytic methods for the hydroamination of nonactivated alkenes, allenes, and alkynes has received considerable attention in recent years [1]. These highly atom-economical processes allow direct access to industrially and biologically relevant classes of compounds such as amines, enamines, and imines from cheap and readily available starting materials. This has recently led to an ever-increasing range of molecular compounds that have been identified as catalysts for these transformations (Scheme 13.1). Whereas rare-earth catalysts have been found to be mainly active in intramolecular hydroamination, other catalysts – in particular those of the late transition metals – are frequently limited to the addition of weakly basic substrates (aniline, sulfonamides, carboxamides, etc.) to alkenes, alkynes, and allenes.

13.2
Group 4 Metal Catalysts for the Hydroamination and Hydrohydrazination of C–C Multiple Bonds as well as Complex Reaction Sequences Based Thereon

Neutral group 4 metal complexes appear to possess a relatively broad scope for catalytic hydroaminations. They have been employed for the intramolecular hydroamination of alkynes [2], allenes [3], and alkenes [4] as well as the intermolecular hydroaminations of alkynes [5] and allenes [6]. Primary aryl- and alkylamines readily react, but secondary amines have posed a greater challenge for this type of transformation with neutral catalysts [7]. For the reactions of the latter, cationic Zr and Ti complexes have been employed in intramolecular cyclizations of aminoalkenes [8]. Very recent work suggests that substrates that are difficult to hydroaminate may favor hydroaminoalkylations instead (Scheme 13.2) [9].

Catalytic hydroaminations (and hydrohydrazinations, see below) of alkynes have been exploited as key steps in catalytic multicomponent reactions by

Molecular Catalysts: Structure and Functional Design, First Edition. Edited by Lutz H. Gade and Peter Hofmann.

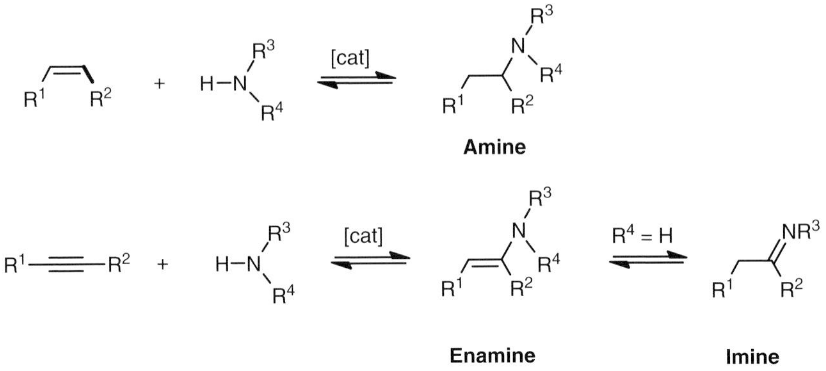

Scheme 13.1 Hydroamination of alkenes and alkynes.

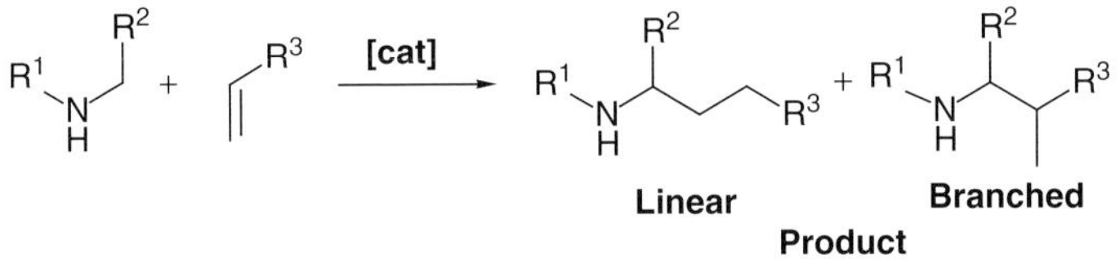

Scheme 13.2 Hydroaminoalkylation of a terminal alkene [9c].

Odom and coworkers [10]. The titanium-catalyzed reaction cascades give rise to highly functionalized substrates, in particular to several types of N-heterocycles (Scheme 13.3).

Catalytic hydrohydrazinations, in which the amine substrate is replaced by hydrazine derivatives, are much less developed to date [11]. However, it is now well established that the hydrohydrazination of alkynes provides an atom-economical access to hydrazones [12], which are potentially valuable reagents for further transformations. Homogeneous catalysts employed for this reaction are titanium- or zinc-based, the latter allowing the tandem combination of hydrohydrazination and subsequent Fischer indole cyclization (Scheme 13.4).

Hydrazinediido complexes have been identified as active species in the titanium-catalyzed hydrohydrazination [12] and iminohydrazination of alkynes and the subsequent transformation of the hydrazones into indoles or tryptamine derivatives (Scheme 13.5) [13]. The postulated reaction mechanism of the group 4 metal-catalyzed reaction is based on the extensively studied mechanistic scheme for hydroaminations of alkynes, in which imido complexes are the key active species.

In contrast to the considerable amount of published work on the structures and reactivity of imido group 4 metal compounds, there has been comparatively little published work on their corresponding hydrazides, in particular hydrazinediides, until recently. Despite a report of a hydrazinediidotitanium complex by Wiberg *et al.* as early as 1978 [14], the chemistry of this class of complexes has only begun to be developed further during the past decade, mainly through the work

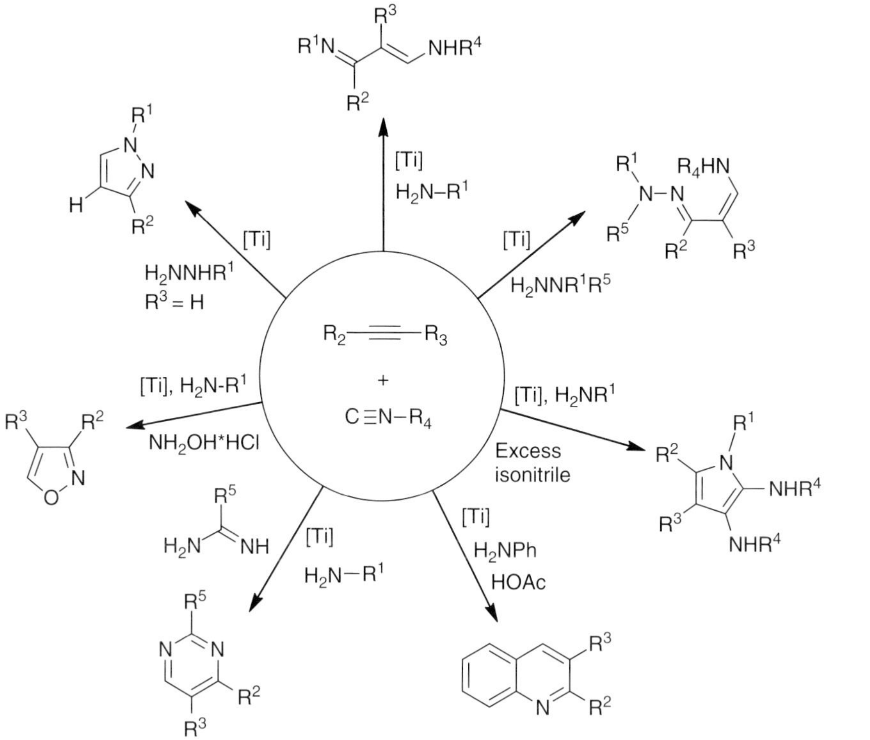

Scheme 13.3 Titanium-catalyzed multicomponent reaction involving a hydroamination/hydrohydrazination step.

of Mountford [15] and Odom [10]. Access to the hydrazido and hydrazinediido complexes of the heavier group 4 metal homologs has been gained in the pioneering study by Bergman and coworkers [16] and in subsequent work from our group (Figure 13.1) [17].

13.3
Case Histories

The following sections will focus on two case histories from our own activities in the field that exemplify the key challenges and the way they may be addressed in practice. First, the focus will be on a new class of titanium hydrazinediido half-sandwich complexes, which have displayed remarkable activity in the hydrohydrazination of terminal alkynes and of carbodiimides. This will be followed by the presentation of a catalytic cascade of N–N and C–H scissions and C–C and C–N coupling steps in the coordination sphere of zirconium, which directly converts alkynes and hydrazines to indoles along a non-Fischer-type reaction pathway.

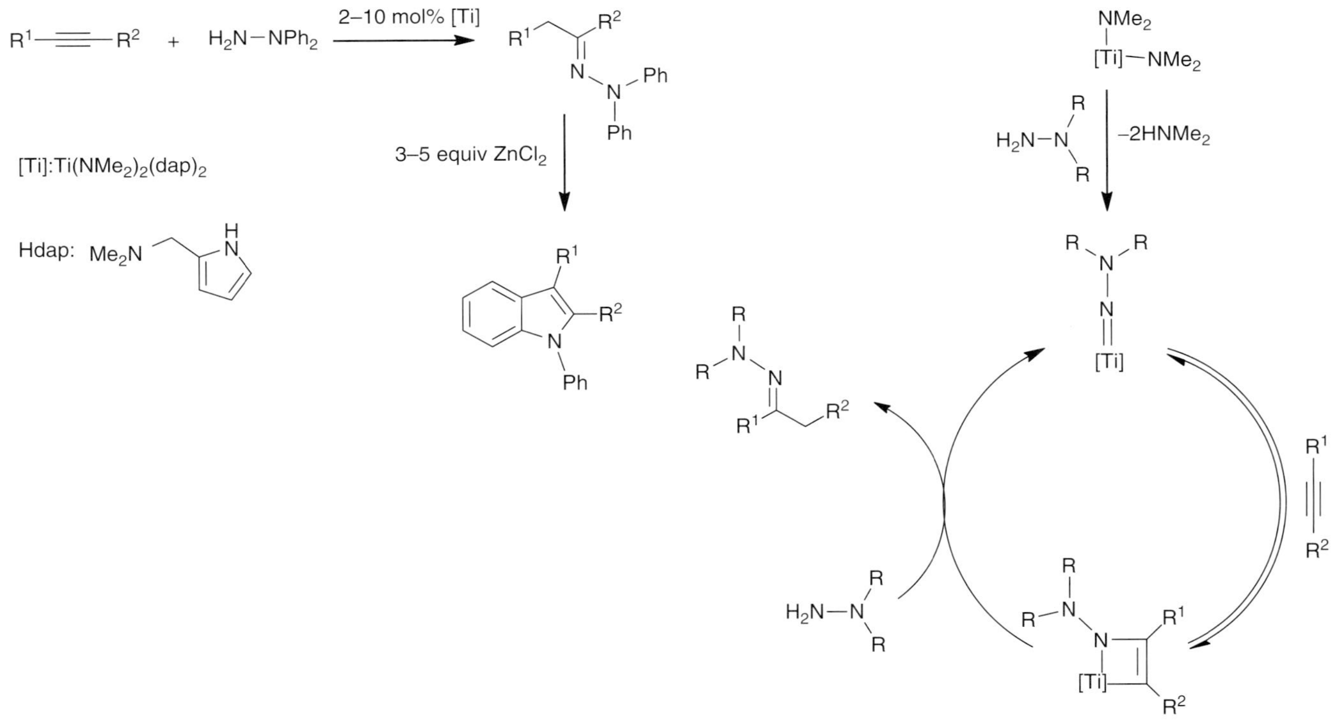

Scheme 13.4 (a) Titanium-catalyzed hydrohydrazination of alkynes and subsequent $ZnCl_2$-catalyzed Fischer indole synthesis. (b) Postulated catalytic cycle.

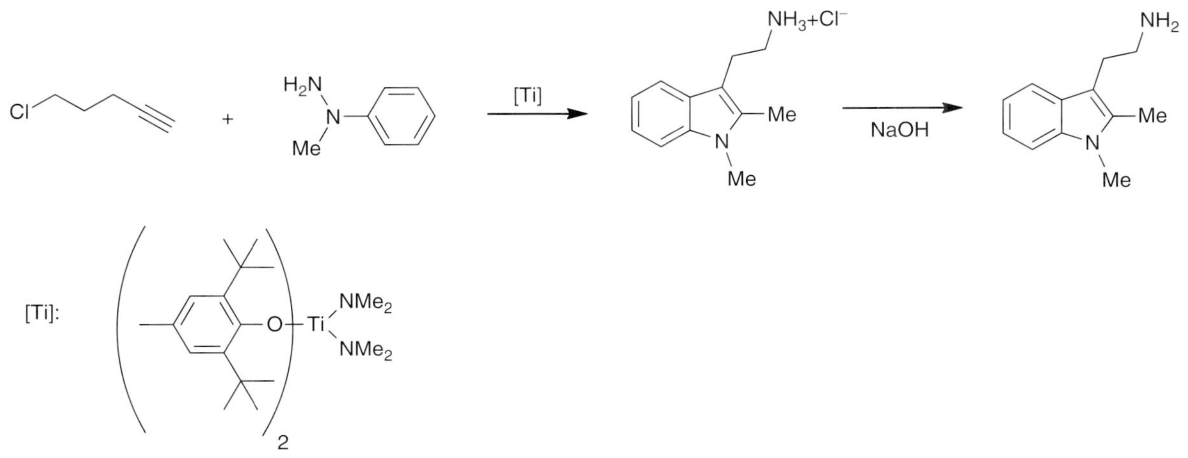

Scheme 13.5 Tryptamine synthesis reported by Beller *et al.* The HCl generated in this process catalyzes the Fischer indole cyclization, which makes the addition of a Lewis acid such as $ZnCl_2$ unnecessary [13].

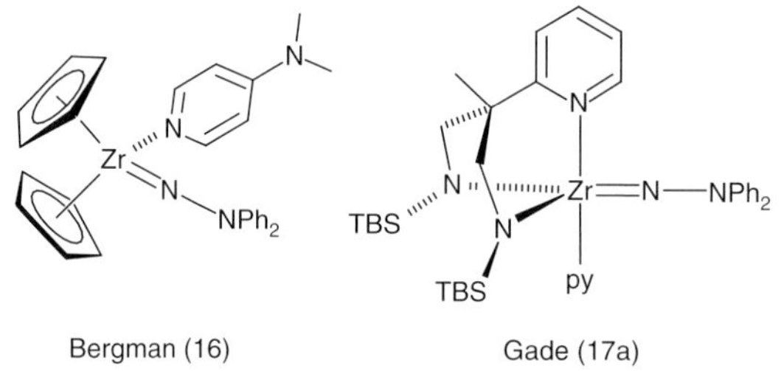

Figure 13.1 (a,b) Two structurally characterized hydrazinediido-zirconium complexes.

13.3.1 Highly Active Titanium Catalysts for the Hydrohydrazination of Terminal Alkynes and Aminoguanylation of Carbodiimides

A new class of titanium hydrazinediido half-sandwich complexes [Cp*Ti(N_{xyl}N)(=NNR_2)(L)] (**A**) was found to display remarkable activities in catalytic hydrohydrazinations of alkynes at ambient temperatures (Scheme 13.6). A key feature has been the use of 2-aminopyrrolinato spectator ligands in which the amidinato binding unit is exocyclic with respect to the heterocycle [18]. The possibility of amidinates to convert from κ^2 to κ^1 coordination and thus to temporarily liberate a coordination site at the metal may be crucial in the development of active molecular catalysts.

For a catalyst loading of 5 mol%, complete conversions of the terminal alkynes and diynes with Markovnikov regioselectivities of over 99% were observed within 1 h. Despite the unusually mild reaction conditions and the excellent selectivity, the substrate scope proved to be limited. This became apparent in the attempted conversion of 1,2-diethynylbenzene. No reaction was observed in this case, which was probably due to steric constraints of the two ethynyl units that are in close proximity. This is consistent with the facile double hydrohydrazination of the constitutional isomer 1,3-ethynylbenzene as well as 1,7-octadiyne (Scheme 13.6) [19].

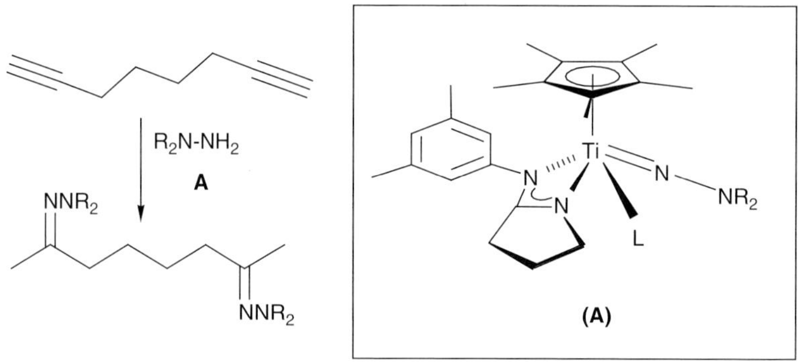

Scheme 13.6 Titanium half-sandwich complexes **A** that were found to hydrohydrazinate a series of terminal alkynes at ambient temperature [19].

The same type of catalyst has been used for the synthesis of a wide range of aminoguanidine derivatives. Aminoguanidines (**I**) exhibit interesting physiological behavior as dopamine β-oxidase inhibitors and antihypertensives. Moreover, anti-HIV activity has been found for *N*-glycosyl-*N′*-(4-arylthiazol)-aminoguanidines (**II**), and several *N*-hydroxy-*N′*-aminoguanidines (**III**) display antitumor activity [20].

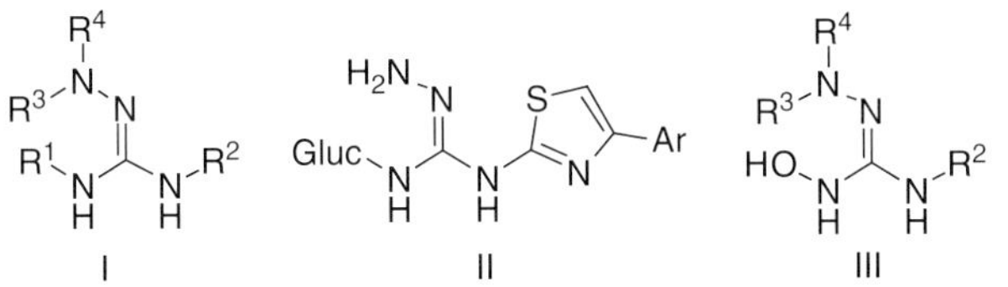

While the catalytic amination of carbodiimides ("guanylation") has been explored for many transition-metal, lanthanide, and even group 1 and 2 complexes as catalysts [21], there was only one previous example of catalytic aminoguanylation employing alkylaluminum precatalysts, reported by Koller and Bergman in 2010 [22].

The aminoguanylation of a series of carbodiimides was carried out at 80 or 105 °C employing titanium hydrazinediido complexes as catalysts or the imido complex [Cp*Ti(N_{xyl}N)(=NR_2)(L)] as precatalyst (Scheme 13.7). The use of the appropriate hydrazinediido complexes or their *in situ* generation from the catalyst precursor [Cp*Ti(N_{xyl}N)(=NR_2)(L)] gave similar yields. Generally, 5 mol% of the corresponding catalyst (or precatalyst) was used, but the catalyst loading could be reduced to 2 mol% by employing higher temperatures or longer reaction times [23].

Among the different carbodiimides tested, ^{i}PrNCN^{i}Pr displayed the highest reactivity, followed by CyNCNCy (DCC, *N*,*N′*-dicyclohexylcarbodiimide). On the other hand, ^{t}BuNCNEt, TolNCNTol, and especially 3-*N*,*N*-dimethylaminopropyl-ethyl-carbodiimide were generally less reactive and demonstrated the limitations of the system. Finally, ^{t}BuNCN^{t}Bu did not react with any of the hydrazines under the conditions outlined above, probably because of steric hindrance. For yields and activity, the trend $Ar_2NNH_2 > MePhNNH_2 > Me_2NNH_2$ was observed, which is the opposite of the trend found by Koller and Bergman. This may be rationalized by assuming that for the aluminum-catalyzed reaction, nucleophilicity of the hydrazine is important whereas the titanium-catalyzed reaction is controlled by NH acidity of the hydrazine.

13.3.2
A Zirconium-Catalyzed Non-Fischer-Type Pathway to Indoles

The transformations of the M=N−NR_2 unit in group 4 metal hydrazinediides combine potential addition of unsaturated molecules at the highly polar M=N bond and the facile fragmentation of the N−N-bond. Both processes may occur sequentially or almost concomitantly and may lead to the combined formation

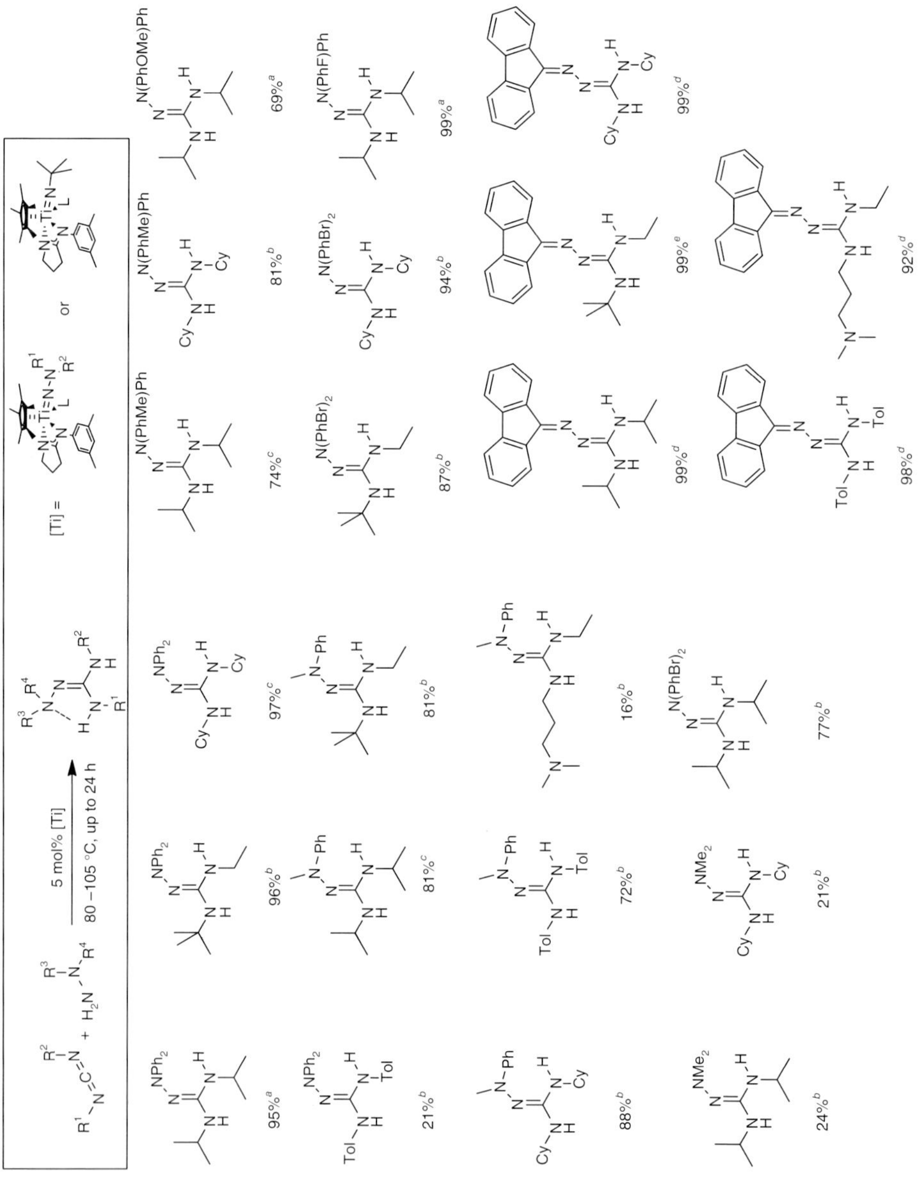

Scheme 13.7 Catalytic aminoguanylations of a range of carbodiimides using [$Cp^*Ti(N_{xyl}N)(=NR_2)(L)$] as precatalyst [23].

and scission of several chemical bonds in one process. This pattern of reactivity was first demonstrated by Bergman and Walsh in 1991 [16] for *in situ* generated $[Cp_2Zr(N_2Ph_2)]$ (Scheme 13.8).

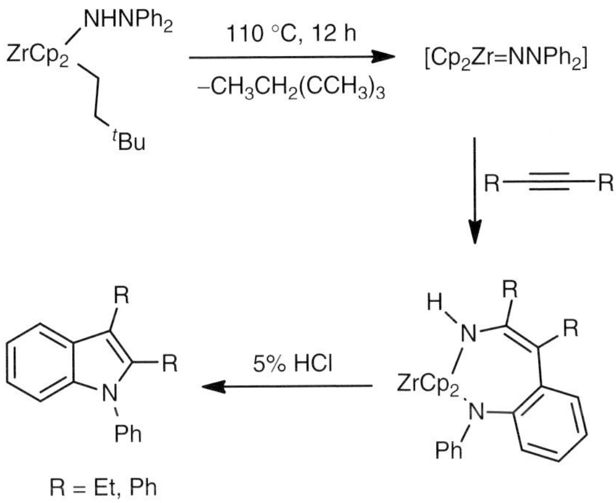

Scheme 13.8 Reaction of *in situ* generated $[Cp_2Zr(=NNPh_2)]$ with an alkyne giving a metallacycle which was converted to an indole derivative upon aqueous acidic work-up [16].

Our own investigations into the chemistry of the heavier group 4 analogs have primarily focused on the hydrazinediido-zirconium complex $[Zr(N_2{}^{TBS}N_{py})(NNPh_2)(py)]$ (Figure 13.1) [17]. The latter has displayed similar stoichiometric and catalytic reactivity toward unsaturated substrates as found for titanium. However, its reactivity toward alkynes appears to be characterized by the absence of a hydrohydrazination step in favor of N–N bond cleavage before the formation of a hydrazone and a reaction pathway similar to that depicted in Scheme 13.8 [24].

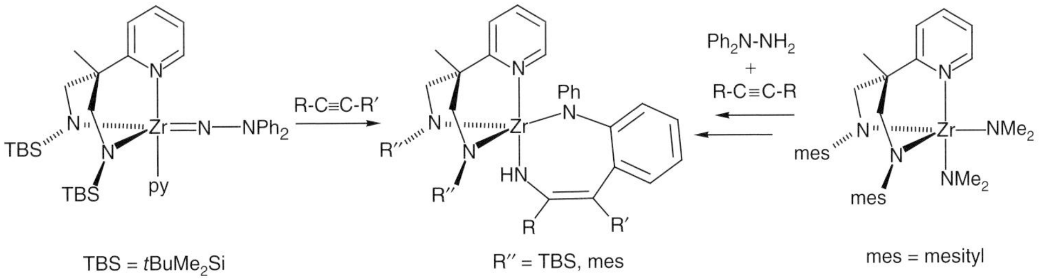

Scheme 13.9 Stoichiometric formation of the seven-membered diazazirconacycles by reaction of an isolated or an *in situ* generated hydrazinediido complex with an alkyne [24].

Reaction of 1 equiv of the zirconium hydrazinediide $[Zr(N_2{}^{TBS}N_{py})(NNPh_2)(py)]$ with disubstituted alkynes led to the formation of seven-membered diazazirconacycles in high yields (Scheme 13.9). Similar reactivity was observed

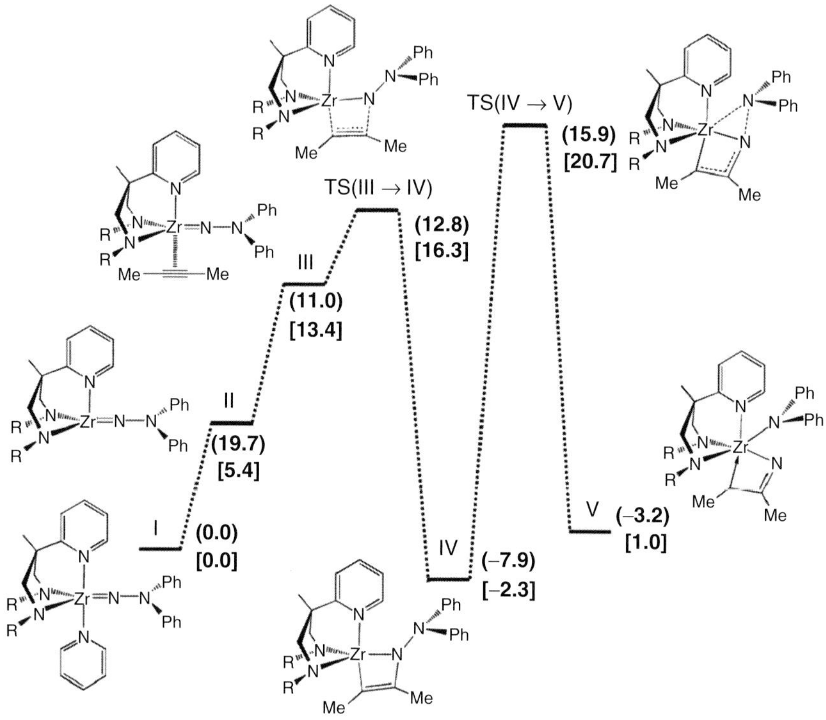

Figure 13.2 Reaction profile (density functional theory (DFT), B3PW91) of the cycloaddition of 2-butyne to the hydrazinediido unit (**I** → **IV**) and the subsequent (rate-determining) N–N bond scission to give the diphenylamido(azaallyl) intermediate **V** (R = TBS). Energies (kcal · mol^{-1}) are given in parentheses, free energies in brackets [24].

when reacting the bis(amido) complex $[Zr(N_2{}^{mes}N_{py})(NMe_2)_2]$ with 1 M equiv of the corresponding alkyne and diphenylhydrazine, respectively. The formation of the seven-membered zirconacycles implied a key coupling step involving the alkyne and one of the aryl rings of the diphenylhydrazinediido ligand.

First, the [2 + 2]-cycloaddition of 2-butyne to the {Zr=N} bond, a reaction step that has been studied in detail for Ti hydrazides, was modeled (Figure 13.2: conversion of **I** → **IV**) and found to occur without prohibitive energetic barriers, giving rise to the metallacyclic intermediate **IV**. In a subsequent rate-determining step, the N–N-bond is broken and the "azaallyl" species **V** is formed.

From the intermediate **V**, no low-activation-barrier pathways to the reaction products were found in a systematic search of the active conformational and configurations space. However, its rearrangement to the energetically only slightly (by $\Delta G = 2.6\,\text{kcal mol}^{-1}$) disfavored constitutional isomer, the dimethylazirinido

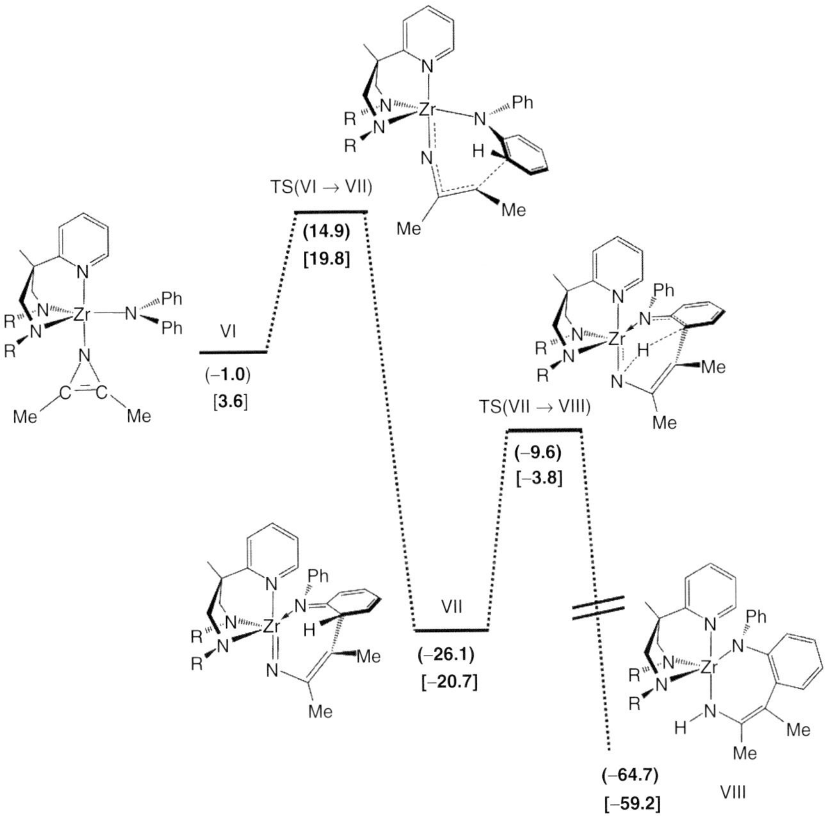

Figure 13.3 Reaction profile (DFT, B3PW91) of the transformation the 2,3-dimethylazirinido species **VI** to the metallacycle **VIII** (R = TBS). Energies ($kcal\,mol^{-1}$) are given in parentheses, free energies in brackets [24].

complex **VI** (Figure 13.3), allows the subsequent transformation without prohibitive activation barriers. The latter is converted to the seven-membered metallacyclic complex **VIII** via nucleophilic attack of a carbon atom in ortho position of a phenyl ring of the NPh_2 fragment at one of the (electrophilic) carbon atoms of the metallated azirine ring to form the intermediate **VII** ($\Delta G = -20.7\,kcal\,mol^{-1}$) via the relatively low activation barrier **TS(VI–VII)** ($\Delta G^{\dagger} = 19.8\,kcal\,mol^{-1}$).

As already mentioned, Bergman and coworkers [16] showed that their diazazirconacycles could be hydrolyzed with 5% HCl to yield *N*-phenylindoles. As hydrazides and amides have similar pK_A values, and hydrazine had been shown to replace amido ligands, we tried to protonate the zirconacycles by reacting them in toluene with a slight excess of diphenyl hydrazine. Reaction of the metallacyclic complexes with two molar equivalents of diphenylhydrazine gave

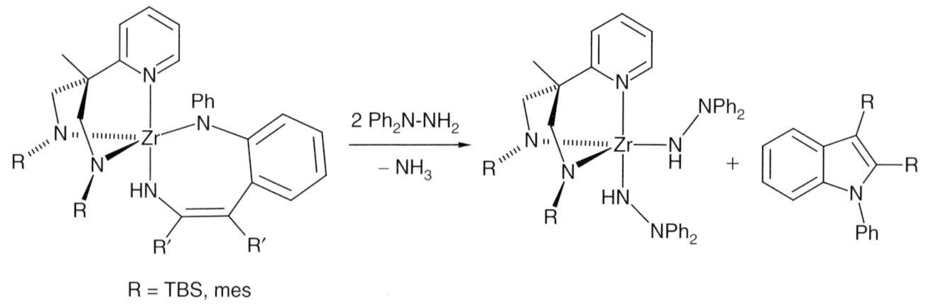

Scheme 13.10 Hydrazinolysis of the metallacyclic zirconium compounds with diphenylhydrazine, giving the bis(hydrazido) complexes and the corresponding indoles [24].

the bis{hydrazido(1-)} complexes as well as 1 equiv of the corresponding indole derivative (Scheme 13.10) [24].

As bis(hydrazido) complexes may be precursors in the generation of hydrazinediido compounds [25], the reaction sequence leading to the indoles was thought to be potentially part of a *catalytic* cycle for the direct generation of substituted indoles from alkynes and hydrazines. Based on the combined experimental and computational evidence presented in this work, we propose the mechanistic cycle depicted in Scheme 13.11 for the catalytic formation

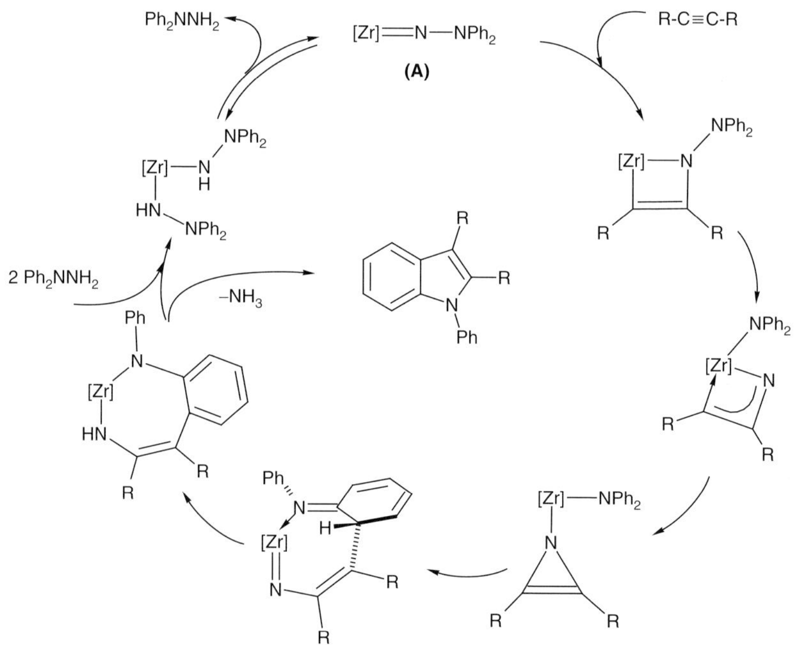

Scheme 13.11 Proposal of a mechanistic cycle for the catalytic formation of indoles from 1,2-diarylhydrazines and alkynes [24].

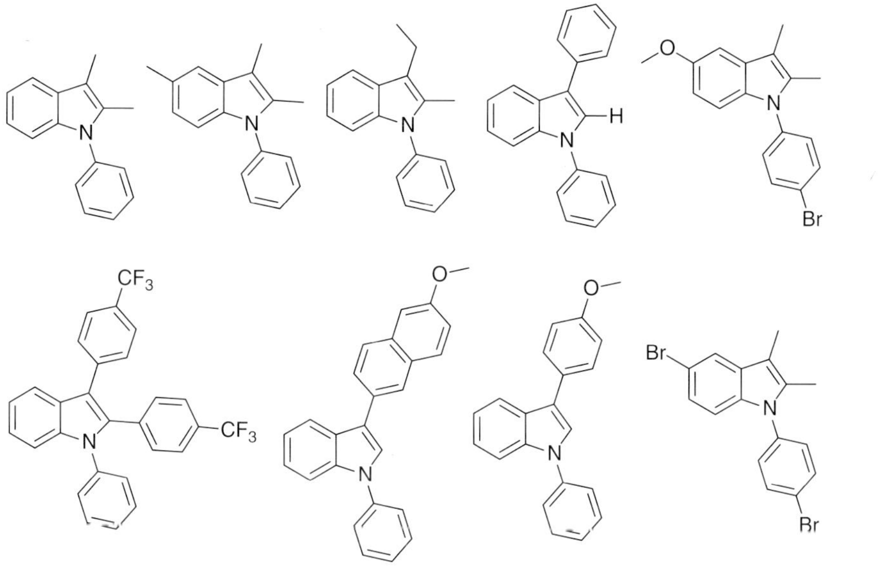

Figure 13.4 Selected targets in the catalytic indole synthesis of various diarylhydrazines with alkynes [24b].

of the N-substituted indoles from 1,2-diarylhydrazines and alkynes. The key intermediate, the hydrazinediido complex **A**, is formed by dissociation of the axial pyridine ligand in a pre-equilibrium, as has been found experimentally.

The catalytic synthesis of indoles on a preparative scale was achieved by the reaction of 1.2 equiv of several diarylhydrazines and various substituted alkynes (1 equiv) in the presence of 10 mol% of the catalyst precursor $[Zr(N_2{}^{xyl}N_{py})(NMe_2)_2]$ (Figure 13.4).

In particular, 1,7-annulated indole derivatives, such as the prototypical Cilansetron **A**, display high activity as 5-HT_3 receptor antagonists and are being clinically tested for the treatment of symptoms in the gastrointestinal system such as the irritable bowel syndrome (IBS) [26]. To extend our previous synthetic strategy based on the reactivity of hydrazinediido zirconium complexes to annulated polycyclic indole derivatives, we aimed to prepare such hydrazides derived from cyclic hydrazines (**B**).

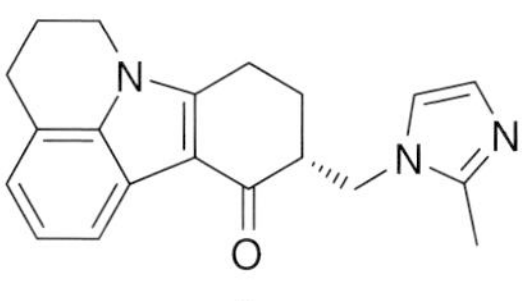

A

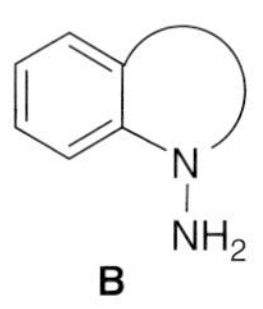

B

The amidozirconium complex $[Zr(N_2{}^{xyl}N_{py})(NMe_2)_2]$ was tested as catalyst for the formation of 1,7-annulated indoles, starting from the

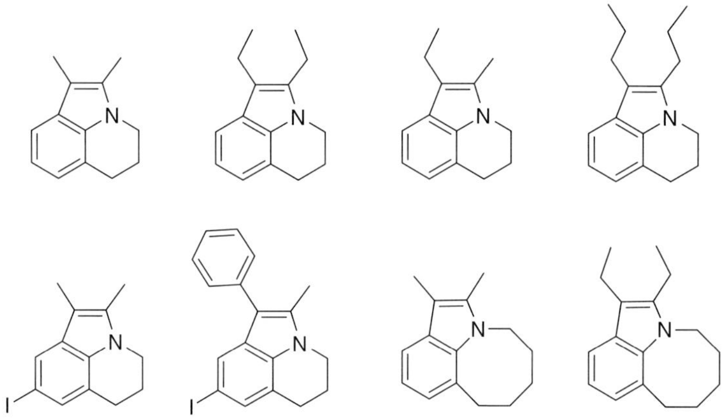

Figure 13.5 Selected examples of the zirconium-catalyzed synthesis of annulated heterocycles [27].

corresponding hydrazines and disubstituted alkynes (Figure 13.5) [27]. The reactions were carried out at 80 °C in toluene using a catalyst loading of 10 mol%.

In the few examples of zirconium-catalyzed domino reactions reported in the literature to date [15], the metal primarily appears to act as a Lewis acid, though mechanistic details have not been established. The reaction of hydrazines with alkynes reported in this chapter proceeds along a complex pathway in which the metal stabilizes a range of reactive intermediates, distinct from those found for titanium-catalyzed hydrohydrazinations. This capability merits the development of further applications of the heavier group 4 metals in the synthesis of N-heterocycles [16].

References

1. (a) Müller, T.E., Hultsch, K.C., Yus, M., Foubelo, F., and Tada, M. (2008) *Chem. Rev.*, **108**, 3795; (b) Severin, R. and Doye, S. (2007) *Chem. Soc. Rev.*, **36**, 1407; (c) Pohlki, F. and Doye, S. (2003) *Chem. Soc. Rev.*, **32**, 104.
2. (a) Bytschkov, I. and Doye, S. (2001) *Tetrahedron Lett.*, **43**, 3715; (b) Bytschkov, I., Sibeneicher, H., and Doye, S. (2003) *Eur. J. Org. Chem.*, **2003**, 2888; (c) Li, C., Thomson, R.K., Gillon, B., Patrick, B.O., and Schafer, L.L. (2003) *Chem. Commun.*, 2462.
3. (a) Ackermann, L. and Bergman, R.G. (2002) *Org. Lett.*, **4**, 1475; (b) Ackermann, L. and Bergman, R.G. (2003) *J. Am. Chem. Soc.*, **125**, 11956; (c) Hoover, J.M., Petersen, J.R., Pikul, J.H., and Johnson, A.R. (2004) *Organometallics*, **23**, 4614; (d) Ayinla, R.O. and Schafer, L.L. (2006) *Inorg. Chim. Acta*, **359**, 3097.
4. (a) Bexrud, J.A., Beard, J.D., Leitch, D.C., and Schafer, L.L. (2005) *Org. Lett.*, **7**, 1959; (b) Kim, H., Lee, P.H., and Livinghouse, T. (2005) *Chem. Commun.*, 5205; (c) Müller, C., Loos, C., Schulenberg, N., and Doye, S. (2006)

Eur. J. Org. Chem., **2006**, 2499; (d) Thomson, R.K., Bexrud, J.A., and Schafer, L.L. (2006) *Organometallics*, **25**, 4069; (e) Watson, D.A., Chiu, M., and Bergman, R.G. (2006) *Organometallics*, **25**, 4731; (f) Lee, A.V. and Schafer, L.L. (2006) *Organometallics*, **25**, 5249; (g) Kim, H., Kim, Y.K., Shim, J.H., Kim, M., Han, M., Livinghouse, T., and Lee, P.H. (2006) *Adv. Synth. Catal.*, **348**, 2609; (h) Wood, M.C., Leitch, D.C., Yeung, C.S., Kozak, J.A., and Schafer, L.L. (2007) *Angew. Chem. Int. Ed.*, **46**, 354; (i) Stubbert, B.D. and Marks, T.J. (2007) *J. Am. Chem. Soc.*, **129**, 6149; (j) Marcsekova, K. and Doye, S. (2007) *Synlett*, 2564; (k) Bexrud, J.A., Li, C., and Schafer, L.L. (2007) *Organometallics*, **26**, 6366; (l) Müller, C., Saak, W., and Doye, S. (2008) *Eur. J. Org. Chem.*, **2008**, 2731; (m) Majumder, S. and Odom, A.L. (2008) *Organometallics*, **27**, 1174; (n) Gott, A.L., Clarke, A.J., Clarkson, G.J., and Scott, P. (2008) *Chem. Commun.*, 1422.

5. A Selected examples: (a) Haak, E. and Doye, S. (2000) DE 19913522; *Chem. Abstr.*, (2000), **133**, 266588; (b) Haak, E., Bytschkov, I., and Doye, S. (1999) *Angew. Chem. Int. Ed.*, **38**, 3389; (c) Haak, E., Siebeneicher, H., and Doye, S. (2000) *Org. Lett.*, **2**, 1935; (d) Heutling, A. and Doye, S. (2002) *J. Org. Chem.*, **67**, 1961; (e) Pohlki, F., Heutling, A., Bytschkov, I., Hotopp, T., and Doye, S. (2002) *Synlett*, 799; (f) Heutling, A., Pohlki, F., and Doye, S. (2004) *Chem. Eur. J.*, **10**, 3059; (g) Tillack, A., Garcia Castro, I., Hartung, C.G., and Beller, M. (2002) *Angew. Chem. Int. Ed.*, **41**, 2541; (h) Tillack, A., Jiao, H., Garcia Castro, I., Hartung, C.G., and Beller, M. (2004) *Chem. Eur. J.*, **10**, 2409; (i) Tillack, A., Khedkar, V., Jiao, H., and Beller, M. (2005) *Eur. J. Org. Chem.*, **2005**, 5001; (j) Zhang, Z. and Schafer, L.L. (2003) *Org. Lett.*, **5**, 4733; (k) Cao, C., Ciszewski, J.T., and Odom, A.L. (2001) *Organometallics*, **20**, 5011; (l) Shi, Y., Hall, C., Ciszewski, J.T., Cao, C., and Odom, A.L. (2003) *Chem. Commun.*, 586; (m) Swartz, D.L. II, and Odom, A.L. (2006) *Organometallics*, **25**, 6125; (n) Esteruelas, M.A., López, A.M., Mateo, A.C., and Onate, E. (2005) *Organometallics*, **24**, 5084; (o) Buil, M.L., Esteruelas, M.A., López, A.M., and Mateo, A.C. (2006) *Organometallics*, **25**, 4079; (p) Buil, M.L., Esteruelas, M.A., López, A.M., Mateo, A.C., and Onate, E. (2007) *Organometallics*, **26**, 554.
6. Johnson, J.S. and Bergman, R.G. (2001) *J. Am. Chem. Soc.*, **123**, 2923.
7. Stubbert, B.D. and Marks, T.J. (2007) *J. Am. Chem. Soc.*, **129**, 6149.
8. (a) Knight, P.D., Munslow, I., O'Shaughnessy, P.N., and Scott, P. (2004) *Chem. Commun.*, 894; (b) Gribkov, D.V. and Hultzsch, K.C. (2004) *Angew. Chem. Int. Ed.*, **43**, 5542.
9. (a) Bexrud, J.A., Eisenberger, P., Leitch, D.C., Payne, P.R., and Schafer, L.L. (2009) *J. Am. Chem. Soc.*, **131**, 2116; (b) Roesky, P.W. (2009) *Angew Chem.*, **121**, 4988; *Angew. Chem. Int. Ed.*, (2009), **48**, 4892; (c) Preuss, T., Saak, W., and Doye, S. (2013) *Chem. Eur. J.*, **19**, 3833.
10. (a) Cao, C., Shi, Y., and Odom, A.L. (2003) *J. Am. Chem. Soc.*, **125**, 2880; (b) Banerjee, S., Shi, Y., Cao, C., and Odom, A.L. (2005) *J. Organomet. Chem.*, **690**, 5066; (c) Majumder, S., Gipson, K.R., and Odom, A.L. (2009) *Org. Lett.*, **11**, 4720; (d) Barnea, E., Majumder, S., Staples, R.J., and Odom, A.L. (2009) *Organometallics*, **28**, 3876; (e) Majumder, S., Gipson, K.R., Staples, R.J., and Odom, A.L. (2009) *Adv. Synth. Catal.*, **351**, 2013; (f) Majumder, S. and Odom, A.L. (2010) *Tetrahedron*, **66**, 3152; (g) Dissanayake, A. and Odom, A.L. (2012) *Tetrahedron*, **68**, 807; (h) Dissanayake, A. and Odom, A.L. (2012) *Chem. Commun.*, **48**, 440.
11. Odom, A.L. (2005) *Dalton Trans.*, 225.
12. (a) Alex, K., Tillack, A., Schwarz, N., and Beller, M. (2008) *Angew. Chem.*, **120**, 2377; *Angew. Chem. Int. Ed.*, (2008), **47**, 2304; (b) Alex, K., Tillack, A., Schwarz, N., and Beller, M. (2008) *Org. Lett.*, **10**, 2377; (c) Alex, K., Tillack, A., Schwarz, N., and Beller, M. (2008) *Tetrahedron Lett.*, **49**, 4607; (d) Li, Y., Shi, Y., and Odom, A.L. (2004) *J. Am. Chem. Soc.*, **126**, 1794; (e) Ackermann,

L. and Born, R. (2004) *Tetrahedron Lett.*, **45**, 9541; (f) Tillack, A., Jiao, H., Castro, I.G., Hartung, C.G., and Beller, M. (2004) *Chem. Eur. J.*, **10**, 2409.

13. (a) Schwarz, N., Alex, K., Sayyed, I.A., Khedkar, V., Tillack, A., and Beller, M. (2007) *Synlett*, 1091; (b) Tillack, A., Khedkar, V., and Beller, M. (2004) *Tetrahedron Lett.*, **45**, 8875; (c) Barnea, E. and Odom, A.L. (2008) *Dalton Trans.*, 4050; (d) Sayyed, I.A., Alex, K., Tillack, A., Schwarz, N., Michalik, D., and Beller, M. (2007) *Eur. J. Org. Chem.*, **2007**, 4525; (e) Review:Krueger, K., Tillack, A., and Beller, M. (2008) *Adv. Synth. Catal.*, **350**, 2153.
14. Wiberg, N., Haring, H.W., Huttner, G., and Friedrich, P. (1978) *Chem. Ber.*, **111**, 2708.
15. (a) Selby, J.D., Manley, C.D., Feliz, M., Schwarz, A.D., Clot, E., and Mountford, P. (2007) *Chem. Commun.*, 4937; (b) Clulow, A.J., Selby, J.D., Cushion, M.G., Schwarz, A.D., and Mountford, P. (2008) *Inorg. Chem.*, **47**, 12049; (c) Tiong, P.-J., Schofield, A.D., Selby, J.D., Nova, A., Clot, E., and Mountford, P. (2010) *Chem. Commun.*, **46**, 85; (d) Schofield, A.D., Nova, A., Selby, J.D., Manley, C.D., Schwarz, A.D., Clot, E., and Mountford, P. (2010) *J. Am. Chem. Soc.*, **132**, 10484; (e) Schofield, A.D., Nova, A., Selby, J.D., Schwarz, A.D., Clot, E., and Mountford, P. (2011) *Chem. Eur. J.*, **17**, 265; (f) Schofield, A.D., Nova, A., Selby, J.D., Schwarz, A.D., Clot, E., and Mountford, P. (2011) *Chem. Eur. J.*, **17**, 265; (g) Schwarz, A.D., Onn, C.S., and Mountford, P. (2012) *Angew. Chem. Int. Ed.*, **51**, 12298; (h) Tiong, P.J., Groom, L.R., Clot, E., and Mountford, P. (2013) *Chem. Eur. J.*, **19**, 4198.
16. Walsh, P.J., Carney, M.J., and Bergman, R.G. (1991) *J. Am. Chem. Soc.*, **113**, 6343.
17. (a) Herrmann, H., Lloret Fillol, J., Wadepohl, H., and Gade, L.H. (2007) *Angew. Chem.*, **119**, 8578; *Angew. Chem. Int. Ed.*, (2007), **46**, 8426; (b) Herrmann, H., Lloret Fillol, J., Gehrmann, T., Enders, M., Wadepohl, H., and Gade, L.H. (2008) *Chem. Eur J.*, **14**, 8131; (c) Gehrmann, T., Lloret Fillol, J., Wadepohl, H., and Gade, L.H. (2009) *Angew. Chem.*, **121**, 2186; *Angew. Chem. Int. Ed.*, (2009), **48**, 2152.
18. (a) Ward, B.D., Risler, H., Bellemin-Laponnaz, S., Wadepohl, H., and Gade, L.H. (2006) *Inorg. Chem.*, **45**, 7777; (b) Weitershaus, K., Ward, B.D., Kubiak, R., Müller, C., Doye, S., Wadepohl, H., and Gade, L.H. (2009) *Dalton Trans.*, 4586.
19. Weitershaus, K., Wadepohl, H., and Gade, L.H. (2009) *Organometallics*, **28**, 3381.
20. (a) Augstein, J., Green, S.M., Monro, A.M., Wrigley, T.L., Katritzky, A.R., and Tiddy, G.J.T. (1967) *J. Med. Chem.*, **10**, 391; (b) Chen, H., Zhao, Y., and Cao, L. (2010) *J. Chin. Chem. Soc.*, **57**, 1085; (c) Basu, A., Sinha, B.N., Saiko, P., Graser, G., and Szekeres, T. (2011) *Bioorg. Med. Chem. Lett.*, **21**, 3324.
21. Selected references:(a) Zhu, X., Du, Z., Xu, F., and Shen, Q. (2009) *J. Org. Chem.*, **74**, 6347; (b) Zhang, W.-X., Nishiura, M., and Hou, Z. (2007) *Chem. Eur. J.*, **13**, 4037; (c) Li, Q., Wang, S., Zhou, S., Yang, G., Zhu, X., and Liu, Y. (2007) *J. Org. Chem.*, **72**, 6763; (d) Alonso-Moreno, C., Carrillo-Hermosilla, F., Garces, A., Otero, A., Lopez-Solera, I., Rodríguez, A.M., and Antinolo, A. (2010) *Organometallics*, **29**, 2789; (e) Lachs, J.R., Barrett, A.G.M., Crimmin, M.R., Kociok-Köhn, G., Hill, M.S., Mahon, M.F., and Procopiou, P.A. (2008) *Eur. J. Inorg. Chem.*, **2008**, 4173; (f) Ong, T.G., O'Brien, J.S., Korobkov, I., and Richeson, D.S. (2006) *Organometallics*, **25**, 4728.
22. Koller, J. and Bergman, R.G. (2010) *Organometallics*, **29**, 5946.
23. Schweizer, P.D., Wadepohl, H., and Gade, L.H. (2013) *Organometallics*, **32**, 3697.
24. (a) Gehrmann, T., Lloret Fillol, J., Scholl, S., Wadepohl, H., and Gade, L.H. (2011) *Angew. Chem. Int. Ed.*, **50**, 5757; (b) Gehrmann, T., Scholl, S.A., Lloret Fillol, J., Wadepohl, H., and Gade, L.H. (2012) *Chem. Eur. J.*, **18**, 3925.
25. Herrmann, H., Gehrmann, T., Wadepohl, H., and Gade, L.H. (2008) *Dalton Trans.*, 6231.

26. (a) van Wijngaarden, I., Hamminga, D., van Hes, R., Standaar, P.J., Tipker, J., Tulp, M.T.M., Mol, F., Olivier, B., and de Jonge, A. (1993) *J. Med. Chem.*, **36**, 3693; (b) Shi, Z., Zhang, C., Li, S., Pan, D., Ding, S., Cui, Y., and Jiao, N. (2009) *Angew. Chem. Int. Ed.*, **48**, 4572–4576.
27. Scholl, S., Wadepohl, H., and Gade, L.H. (2013) *Organometallics*, **32**, 937.

14
Sequential Catalysis for the Stereoselective Synthesis of Complex Polyketides

Thomas Debnar and Dirk Menche

14.1
Complex Polyketides

The exquisite and varied architectures of natural products provide a rich pallet for the discovery of synthetic chemicals. Irrespective of whether they are used to probe biological mechanisms or to provide the basis for pharmaceutical drug discovery, natural products continue to command attention. Synthetic chemists are drawn to these structures as testing grounds for new methodologies as well as inspirations for the design of novel synthetic ventures. Polyketides are structurally a very diverse family of natural products with an extremely broad range of biological activities and pharmacological properties. Polyketide antibiotics, antifungals, cytostatics, antiparasitics, and natural insecticides are in commercial use. In many cases, specific molecular targets are addressed at a molecular level, which adds to their attractiveness for further advancement [1]. Prominent examples include the manzacidins (**1**, Figure 14.1) [2], potent serotonin receptor antagonists; the halichondrins (**2**) [3], which show extraordinary antitumor activity both *in vitro* and *in vivo* rendering them promising leads for the development of new anticancer agents; rhizopodin (**3**) [4], a highly potent antiproliferative agent that interacts with the actin cytoskeleton; and leupyrrin A_1 (**4**) [5], a proteinkinase inhibitor.

From a structural perspective, polyketides are characterized by sequences of methyl- and hydroxyl-bearing stereogenic centers, enabling large numbers of possible stereochemical permutations. Biosynthetically, these are derived by iterative condensations of propionyl subunits and subsequent reduction of the derived β-keto-esters. Mimicking this biosynthetic pathway, the aldol reaction presents the most important method available for the stereocontrolled formation of propionates, and many variants for regio-, stereo-, and enantioselective carbon–carbon bond formation have been reported. Furthermore, also alternative strategies are of increasing importance. Well-established, more recently developed methods for the stereoselective assembly of polyketides range from aldol reactions, reductive couplings, crotylations, allenylations, selective radical processes, sequential substitutions, epoxide openings, rearrangements, and intramolecular approaches (Scheme 14.1) [6].

Molecular Catalysts: Structure and Functional Design, First Edition. Edited by Lutz H. Gade and Peter Hofmann.

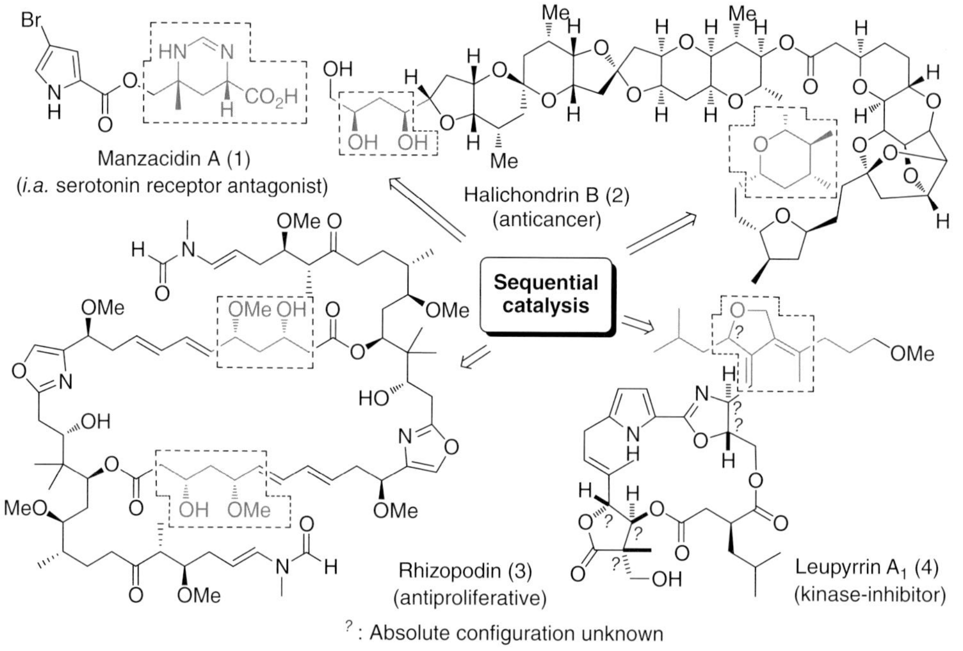

Figure 14.1 Complex polyketides: attractive targets for the development of sequential catalytic reactions.

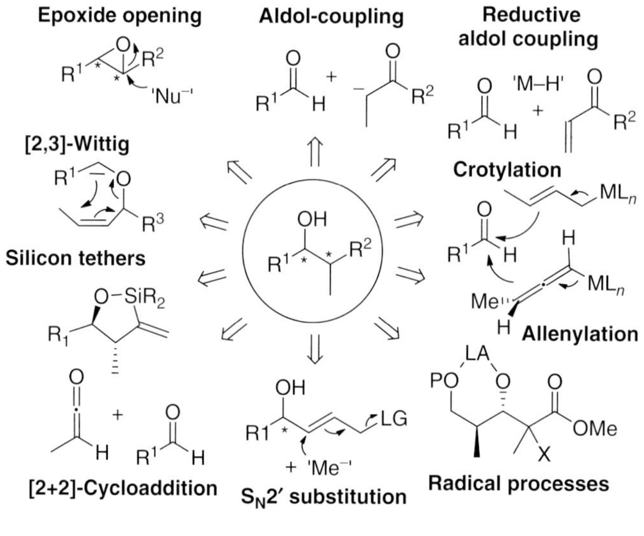

Scheme 14.1 Existing methods for complex polyketides synthesis: an overview.

In general, the synthesis of complex polyketides presents an important research goal, not only to enhance the often sparse natural supply of these compounds, support further biological applications, and enable structure–activity relationship (SAR) but also for unambiguous stereochemical assignment. Despite the impressive accomplishments and progress in the total synthesis of these structures in recent years, there remains a high need for the development of efficient synthetic methods to assemble these structures in a more direct and convergent manner. Domino sequences are particularly desirable along these lines because they combine several synthetic transformations in a one-pot process and thus enable the rapid assembly of complex architectures. Ideally, such methods should be based on catalytic procedures to allow high degrees of atom economy and synthetic efficiency, but also convergence and modularity. Accordingly, the development and design of relay catalytic processes for the rapid assembly of polyketide structures presents a highly desirable research goal. Metal complexes have been successfully applied to a broad range of organic transformations and occupy a central position in preparative organic chemistry. In recent years, the notion of combining several metal-mediated processes in relay-type domino sequences is gaining more and more attention [7]. However, despite these advancements, the development of more efficient methods that are truly applicable also in complex target synthesis remains an important research goal. This chapter details the design, development, and application of novel types of sequential processes to access key structural features of polyketides. These include a novel domino reaction relying on a combination of a nucleophilic addition and a Tsuji–Trost reaction as well as a sequence relying on an oxidative diyne cyclization and regioselective opening of the intermediate metallacycles. Importantly, the true applicability of these methods has enabled a highly concise and stereoselective synthesis of complex polyketides to be demonstrated.

14.2 Domino Nucleophilic Addition–Tsuji–Trost Reaction

The palladium-catalyzed allylic substitution reaction presents one of the most versatile and efficient methods for C–C and C–X bond formation, and consequently a broad range of procedures have been reported [8]. However, despite this importance, the implementation of such Tsuji–Trost type reactions in sequential processes do not appear to have been fully exploited, even though it has obvious potential in enabling a rapid increase of structural complexity from simple starting materials by combining several synthetic transformations in a one-pot manner.

14.2.1 Concise Synthesis of Tetrahydropyrans by a Tandem oxa-Michael–Tsuji–Trost Reaction

Substituted tetrahydropyrans (THPs) present prevalent constitutional chemotypes and underlying structural motifs in numerous natural products, registered drugs, and bioactive synthons. Various strategies for the construction of such systems have been reported [9], including cyclizations onto oxocarbenium ions and epoxides, hetero-Diels–Alder reactions, Prins cyclizations, intramolecular nucleophilic reactions, Michael reactions, reduction of cyclic hemi-acetals, cyclization on nonactivated double bonds, or one-pot procedures based on alkene–alkyne couplings followed by ether formation. Driven by the present targets in our group, in combination with certain limitations of these existing methods, we looked for a more direct and concise sequence for THP synthesis. As shown in Scheme 14.2, our synthetic concept capitalizes on a three-step sequential process involving an oxa-Michael addition and a Tsuji–Trost coupling. Following this sequence, readily available homoallylic alcohol **5** would first add to a suitably acceptor-substituted alkene **6**, giving enolate **7** (step 1). Second, a π-allyl complex **8** would be generated (step 2), which would finally be intramolecularly trapped by an allylic substitution reaction, generating the desired THP motif in a highly direct manner (step 3). Notably, three new stereogenic centers are assembled along this process, demonstrating a considerable increase in structural complexity from very simple starting materials.

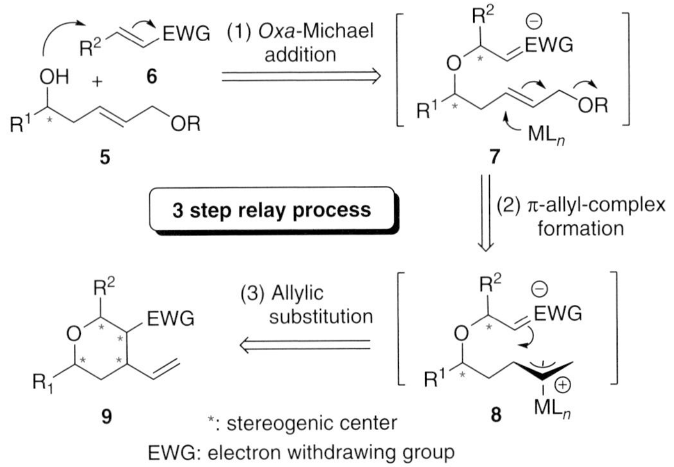

Scheme 14.2 Three-step tandem concept for tetrahydropyran synthesis.

Based on initial experiments with different Michael acceptors, nitro-olefins were shown to be most effective for the desired transformation. After evaluation of various reagents (bases, catalysts, ligands) and parameters (temperature, solvent), the synthetic strategy could be successfully implemented. The most effective conditions included catalytic amounts of $Pd_2(dba)_3$ (dba, dibenzylideneacetone) with PPh_3 in combination with Li*Ot*Bu as base. As shown in Scheme 14.3, the major products differ only in the configurations of the propyl- and the nitro-bearing centers, while both the phenyl and vinyl substituent reside in equatorial positions. As shown in Scheme 14.3a, various further THPs were readily obtained by this domino process. Furthermore, after modification of reaction conditions, this procedure could also be applied to THPs bearing a tetrasubstituted carbon center (Scheme 14.3b). Again, good selectivities and yields were obtained considering the stereochemical complexity of this process [10, 11]. Notably, the heterocyclic products bear two functional handles (NO_2, alkene), which may be readily further elaborated, adding to the synthetic usefulness of the process.

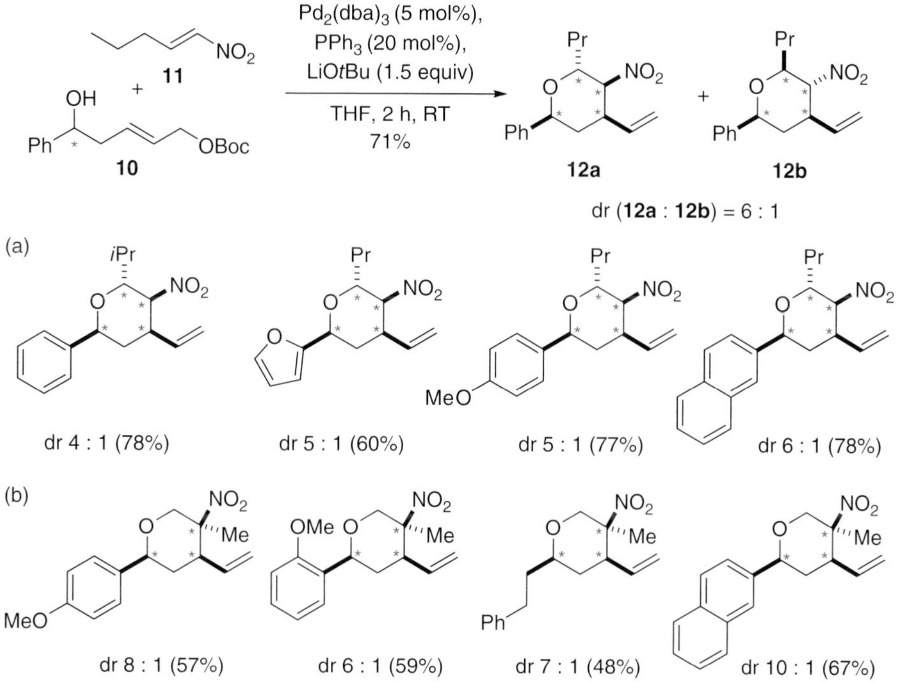

Scheme 14.3 Stereoselective tetrahydropyran synthesis by a domino oxa-Michael–Tsuji–Trost reaction. (a) Substrate scope. (b) Generation of tetrasubstituted carbon centers.

Mechanistically, the observed selectivities and yields of these domino reactions may be explained by a reversible oxa-Michael reaction. As shown in Scheme 14.4, this first addition may not be stereo-discriminating but the intermediate **14** may be reversibly transformed into the more favorable diastereomer **13**, leading to the major product **12a**, together with minor amounts of **12b**. The observed stereoselectivity may arise from a Zimmerman–Traxler type transition state bearing the C2 and C4 substituents in equatorial positions, in agreement with the observed stereochemical outcome. Generation of the axial configuration at C5, in turn, may be explained by chelation of the metal counterion to the ether oxygen and the nitro group, which would be more favorable with the nitro group being in an axial conformation, namely **13**. Alternatively, also minimization of dipole–dipole interactions of the nitronate with the π-allyl complex would be more favorable for **13** as compared to **14**. Chelate **13** may also rationalize the observed selectivity at C6, as this substituent would reside in a pseudo-equatorial conformation in such a six-membered chelate.

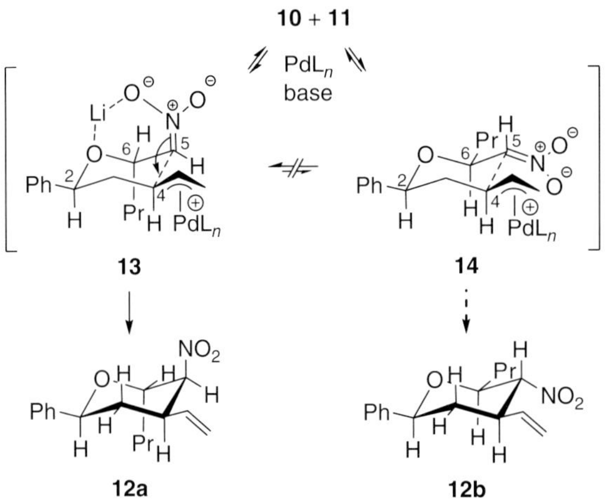

Scheme 14.4 Mechanistic rationale for the stereochemical outcome.

14.2.2
Concise Synthesis of Acetal-Protected 1,3-*syn*-Diols by a Tandem Hemiacetal/Tsuji–Trost Reaction

As exemplified by the macrolide antibiotic RK-397 (**15**, Scheme 14.4), proximal 1,3-arrays of hydroxyl-bearing stereogenic centers present the most prevalent structural phenotypes in a wide variety of polyketides [1]. Inspired by the cascade concept reported above, we desired a more direct synthesis to this structural synthon. As shown in Scheme 14.5, this involves the addition of a homo-allylic alcohol **5** to a suitable carbonyl compound **18**, giving the corresponding hemi-acetal alkoxide (step 1). Formation of an electrophilic π-allyl complex (step 2) then results in the generation of intermediate **17**, which finally undergoes an intramolecular allylic substitution reaction to the desired 1,3-allylic alcohols **16** in a suitably protected form (step 3). Notably, this sequence presents one of the first examples of hemiacetal alkoxides as nucleophiles in allylic substitution reactions.

As shown in Scheme 14.6, this novel domino sequence could indeed be realized [12]. It was found that optimized conditions for the selective generation of the 1,3-*syn*-dioxane products involved treatment of homoallylic alcohol **10** in acetaldehyde (**19**) as co-solvent with a slight excess of potassium hexamethyldisilazide (KHMDS) (1.5 equiv) and catalytic amounts (10 mol%) of $[Pd(allyl)Cl]_2$ and PPh_3 (30 mol%), and conducting the reaction in toluene at room temperature.

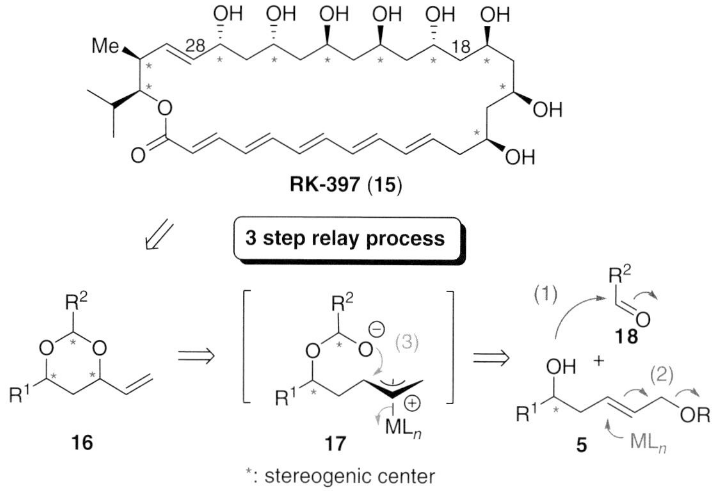

Scheme 14.5 Three-step tandem concept for 1,3-diol synthesis.

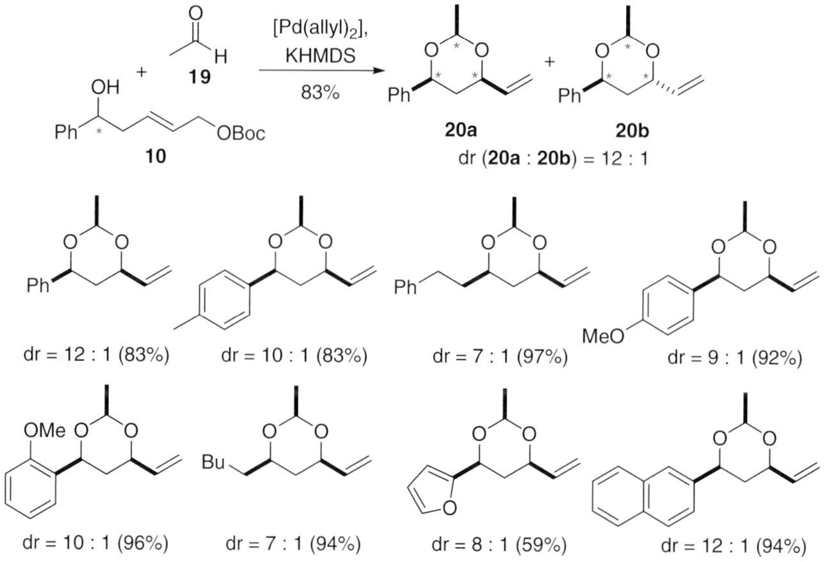

Scheme 14.6 Concise synthesis of acetal-protected 1,3-*syn*-diols by a tandem hemiacetal/Tsuji–Trost reaction.

This enabled an efficient access to a broad range of suitably protected 1,3-*syn*-diols with good stereoselectivities and excellent yields (Scheme 14.6). The substrate scope ranged from simple phenolic or benzylic to aliphatic compounds.

This novel concept may be readily applicable in convergent polyketide synthesis, as demonstrated in the concise synthesis of the C18–C28 fragment of macrolide

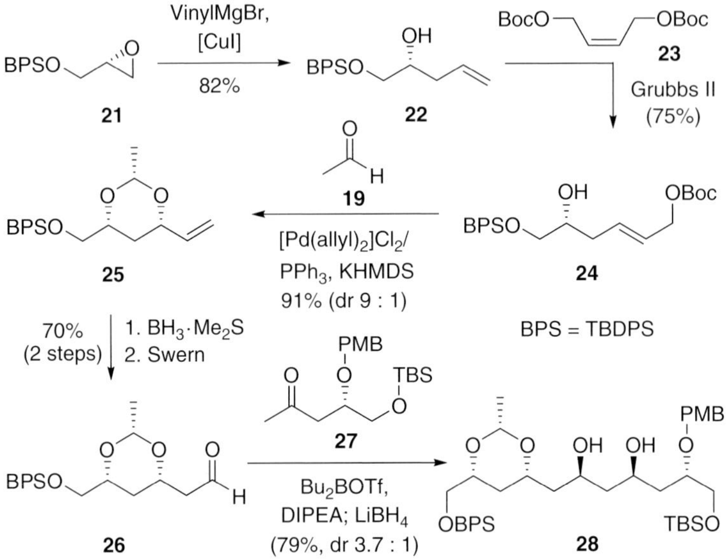

Scheme 14.7 Concise synthesis of the C18–C28 fragment **23** of RK-397.

RK-397 (Scheme 14.7). Importantly, the required chiral homoallylic alcohol substrates **24** are readily accessible, for example, by opening of the terminal epoxides (i.e., transformation of **21** to **22**) or asymmetric allylation of aldehydes (not shown) and subsequent cross metathesis of the derived terminal alkene (see conversion of **22** to **24** with **23**). The pivotal two-component coupling of **24** with acetaldehyde proceeded in high yields and selectivities, demonstrating again the general usefulness of this method. Importantly, the terminal alkene of the derived 1,3-*syn*-diol product **25** may be efficiently used as a functional handle for further elaboration, as shown here for the conversion of **24** to the corresponding aldehyde **26**. Further elaboration of **26** with methyl ketone **27** by a 1,5-*anti*-aldol coupling and subsequent *in situ* 1,3-*syn* reduction gives the desired polyol **28** in a concise and effective manner.

14.2.3
General Concept and Further Applications for Diamine and Aminoalcohol Synthesis

Based on these findings, a general concept for the tandem combination of a nucleophilic substitution reaction and a Tsuji–Trost coupling may be proposed. As shown in Scheme 14.8, this includes the coupling of various homoallylic nucleophiles, with X being oxygen, nitrogen, or sulfur, to different types of electrophiles. These may be Michael acceptors **6** (left part), heteroolefins **34** (e.g., imines or carbonyls) (middle part), or allene homologs **37** (e.g., cyanates or thiocyanates) (right part) to give the corresponding six-membered heterocycles **9**, **36**, or **38**, respectively, with X being oxygen, nitrogen, sulfur, or carbon. In all cases, these reactions should proceed by anionic relay.

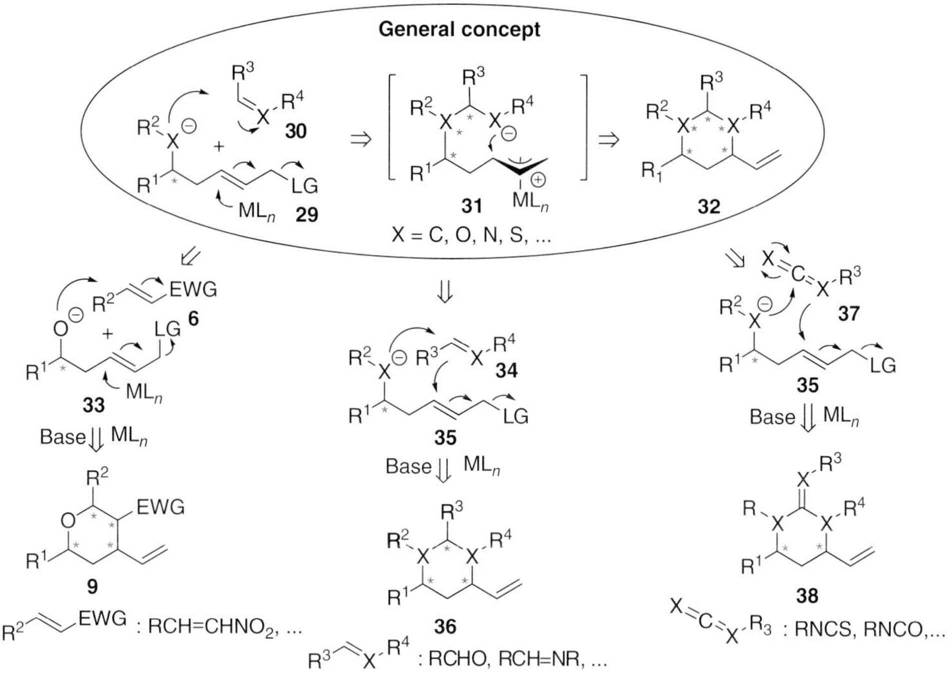

Scheme 14.8 General domino concept based on a nucleophilic addition and Tsuji–Trost coupling.

In an effort to further evaluate the generality of this concept, we were particularly attracted to evaluate novel methods for the synthesis of chiral 1,3-diamine motifs and of 1,3-aminoalcohol functionalities, both of which constitute particularly important structural elements in various bioactive natural products and medicinal compounds. Because of their prevalence, a number of strategies for the construction of such systems have been reported. Inspired by our domino concept in combination with certain limitations of existing methods, in particular, with respect to modularity and convergence, a more direct and flexible procedure for 1,3-diamine and 1,3-amino alcohol synthesis based on stereoselective intramolecular allylic substitution reaction seemed desirable.

As shown in Scheme 14.9a,b, an efficient protocol for the stereoselective synthesis of 1,3-*syn* and -*anti* tetrahydropyrimidinones (*syn*- and *anti*-**42**) could be developed [13]. The modular procedure is based on a stereodivergent cyclization of readily available urea-type substrates **41** by intramolecular allylic substitution. The cyclization proceeds with excellent yield (up to 99%) and selectivity (up to dr > 20 : 1), purely based on substrate control. The product pyrimidines **42** can be readily transformed into the corresponding free *syn*- and *anti*-amines **43**. Furthermore, also a novel domino sequence for the rapid assembly of 1,3-*syn*-substituted oxazines **46** could be realized [14]. Mechanistically, the one-pot procedure is based on a three-step sequential process involving a hemi-aminalization

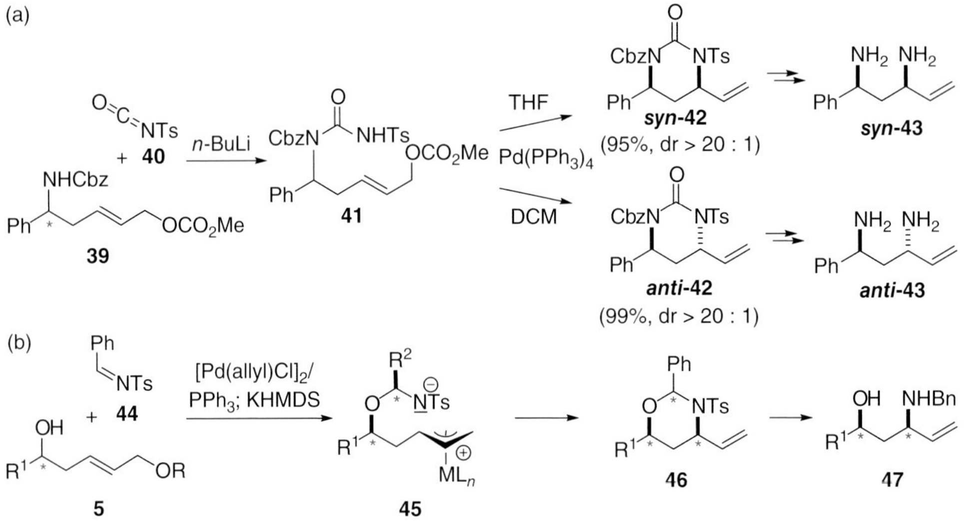

Scheme 14.9 Tandem processes for the stereoselective synthesis of (a) 1,3-diamines and (b) 1,3-amino-alcohols.

and Tsuji–Trost reaction. The process generates up to two new stereogenic centers in a concise and convergent manner from simple and readily available starting materials. The product oxazines **46** could be readily transformed to the 1,3-*syn*-aminoalcohols **47**.

14.3
Sequential Diyne Cyclization and Regioselective Opening of Zirconacyclopentadienes

Oxidative diyne cyclizations using low-valent zirconcene and titanocene species of type "$Cp_2M(II)$" are one of the most important methods for alkyne functionalization [15]. As shown in Scheme 14.10 for the cyclization of **48**, they may be efficiently employed for the directed synthesis of exocyclic dienes of type **51**. Mechanistically, they are based on intermediate metalla-cyclopentadienes **49**, which can be cleaved with various reagents. While symmetric cleavages by complete hydrolysis or dihalogenation are well established, a regioselective, stepwise opening via **50** is much less advanced. A first promising result was published by the group of Takahashi in 1998 [16], in which it was reported that a zirconacyclopentadiene with a mixed aliphatic–aromatic substitution pattern may be selectively monohalogenated, on the basis of the distinguishable basicity of both α-carbon atoms. In contrast, the regioselective opening of metallacycles with sterically similar aliphatic functions and the potential of using remote directing groups for such purposes has not been advanced, despite the importance of such procedures to significantly amplify the usefulness of this process and allow for a rapid and selective access to structural complexity from simple

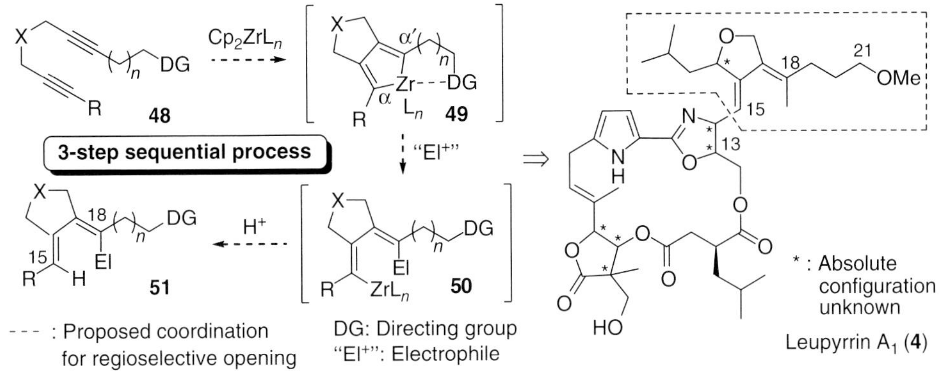

Scheme 14.10 Regioselective opening of zirconacyclopentadienes: proposed one-pot synthesis of the furan core of the leupyrrins.

starting materials. Such a sequence would also enable a highly concise entry to the tetrahydrofuran (THF) core of the leupyrrins (**4**) [5], which are structurally unique myxobacterial metabolites that exhibit potent biological activities against various fungi and eukaryotic cells. Notably, this unsymmetrically substituted THF with two *E*-configured exocyclic double bonds presents an unprecedented structural element, and consequently no synthetic routes have been reported.

Based on this synthetic design, the development, mechanistic background, and synthetic application of a novel one-pot process involving a zirconocene-mediated cyclization of unsymmetric aliphatic 1,6-diynes and a subsequent regioselective monohalogenation could be realized [17]. As shown in Scheme 14.11, this sequence involves a highly regioselective opening of aliphatic zirconacyclopentadienes of type **49** with *N*-bromosuccinimide (NBS), which are derived by a cyclization from 1,6-diyne **52**. This sequence enables a highly specific entry into monobrominated products **54** or **55**.

The selectivity is purely based on the side-chain functionality and may be inverted by suitable protective groups. The remarkable selectivity profile was analyzed by density functional theory (DFT) calculations using Gaussian09 at B3LYP level using LANL2DZ basis set employing effective core potentials (ECPs) for transition metals in combination with the polarizable continuum model (PCM) and THF as solvent. These studies revealed a pronounced tendency of the side-chain functionality to interact with the metal (Scheme 14.11). Importantly, the relative free enthalpy values of the two possible regioisomers (15-bromo **53a** or 18-bromo **53b**) are highly differentiating. The optimized geometry of 18-bromo-**53b** suggests that this stabilization is caused by a diminished distance of 2.5 Å of the OMe group to C18 (Scheme 14.11b). Consequently, the Cp rings are pushed closer to C15, which then shields this position for bromination. In contrast, for *para*-methoxybenzyl (PMB)-protected metallacycle **53a**, a stabilization of the 18-bromo isomer (40.5 kJ mol^{-1}) was found, presumably on the basis of an electrostatic interaction between the PMB-protected oxygen and the Zr metal (15-bromo-**53a**), as shown in Scheme 14.11a. This short distance

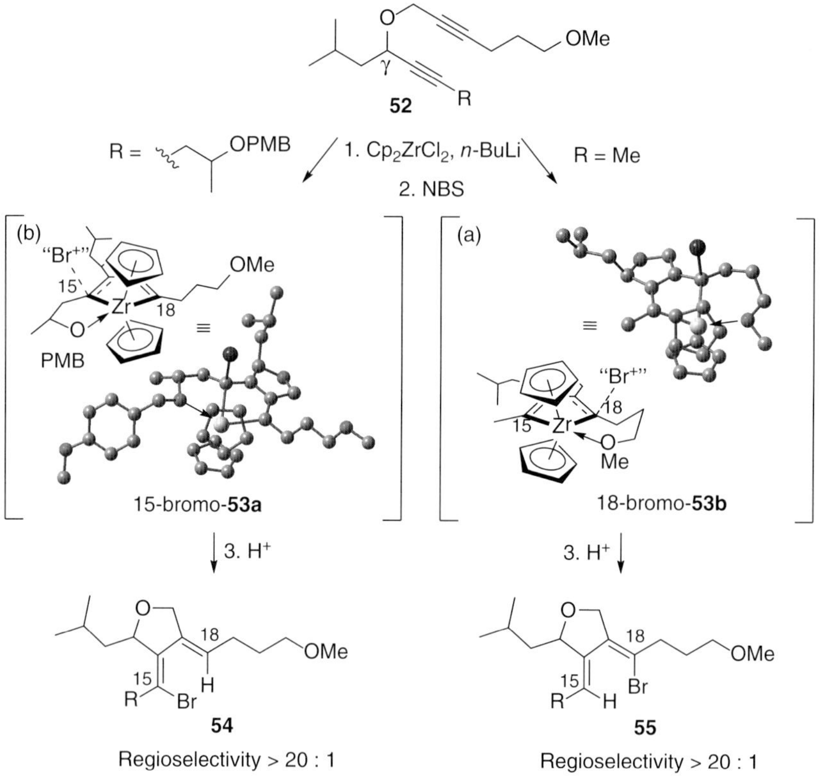

Scheme 14.11 Tandem zirconocene-mediated diyne-cyclization and regioselective opening of zirconacyclopentadienes.

(2.6 Å) again affects the position of the Cp rings, which are now pushed in the opposite direction, close to C18, and consequently this position is shielded for bromination. In agreement with these results, the PMB-bearing oxygen provides much more electron density in comparison with an OMe group.

With this strategy in mind, efforts were then directed to access the authentic furan core of the leupyrrins by introducing the methyl group at C18. Accordingly, the transformation of **60**, bearing already all the desired oxygen functionalities, namely an oxygen at C13 and C21 of the target function in suitably protected form, was studied. This compound may be readily prepared by the addition of the corresponding alkyne **57** to 3-methyl-1-butanal and subsequent etherification with **58**. Importantly, the zirconocene-mediated one-pot cyclization–regioselective opening of **59** proceeded again in high yields and selectivities, showing the general usefulness of our approach. Finally, the required methyl group may be installed by treatment of **60** with *t*-BuLi, affording a lithium species, which was trapped with Me_2SO_4 giving the desired methyl-alkylidene **61** in 80% yield (Scheme 14.12).

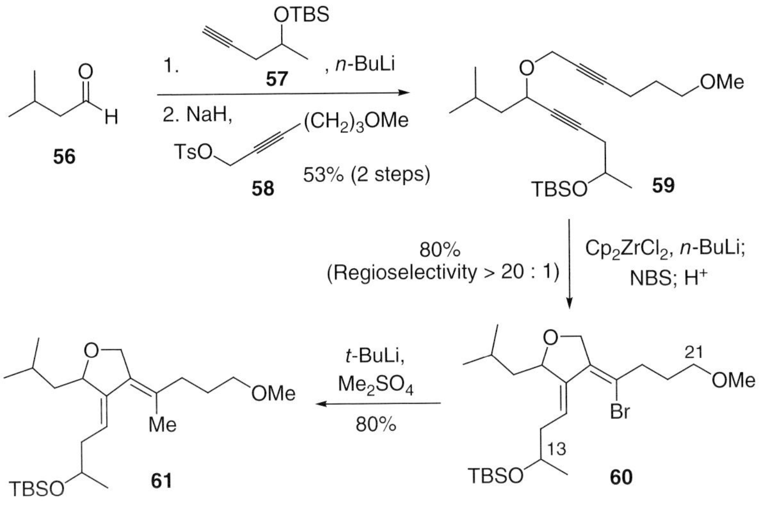

Scheme 14.12 Concise synthesis of the furan core of the leupyrrins.

14.4
Conclusion and Perspectives

In summary, based on the innovative design of novel catalytic domino concepts we have developed efficient one-pot strategies for the efficient synthesis of key structural elements of polyketides. Importantly, these sequential processes combine several synthetic transformations in a one-pot process and thus enable the rapid assembly of structural complexity. In detail, a novel domino concept based on a combination of a nucleophilic addition reaction and a Tsuji–Trost coupling has been devised and developed, which allows a highly concise entry into most diverse building blocks, including THPs, 1,3-amino alcohols, -diamines, or -diols. Importantly, these methods proceed with high stereoselectivity. Also, the high degrees of asymmetric induction are purely based on substrate control, adding to the general usefulness of these procedures. Furthermore, a novel sequential process involving a zirconocene-mediated diyne-cyclization and highly regioselective opening of the intermediate metalcyclopentadienes has been developed as an efficient entry into complex target synthesis. The true applicability of all these methods in complex target synthesis has been demonstrated. It is expected that these novel types of catalytic domino sequences will be further explored and developed and may also initiate the development of further sequential processes. We also believe that these methods will be further developed and applied in the synthesis of diverse polyketides and structurally related complex targets, which will be crucial for further advancing these promising bioactive agents.

References

1. (a) Hertweck, C. (2009) *Angew. Chem. Int. Ed.*, **48**, 4688–4716; (b) Weissman, K.J. and Müller, R. (2010) *Nat. Prod. Rep.*, **27**, 1276–1295.
2. (a) Kanda, F., Ishibashi, M.H., Shigemori, H., and Kobayashi, H.J. (1991) *J. Org. Chem.*, **56**, 4574–4576; (b) Namba, K., Shinada, T., Teramoto, T., and Ohfune, Y. (2000) *J. Am. Chem. Soc.*, **122**, 10708–10709; (c) Wehn, P.M. and Bois, J.D. (2002) *J. Am. Chem. Soc.*, **124**, 12950–12951.
3. For a leading reference, see: Namba, K., Jun, H.-S., and Kishi, Y. (2004) *J. Am. Chem. Soc.*, **126**, 7770–7771.
4. For a leading reference, see: Dieckmann, M., Kretschmer, M., Li, P., Rudolph, S., Herkommer, D., and Menche, D. (2012) *Angew. Chem. Int. Ed.*, **51**, 5667–5670.
5. Bode, H.B., Irschik, H., Wenzel, S.C., Reichenbach, H., Müller, R., and Höfle, G. (2003) *J. Nat. Prod.*, **66**, 1203–1206.
6. Li, J. and Menche, D. (2009) *Synthesis*, **41**, 2293–2315.
7. For a review on tandem reactions by binary catalytic systems, see: (a) Wasilke, J.-C., Obrey, S.J., Baker, R.T., and Bazan, G.C. (2005) *Chem. Rev.*, **105**, 1001–1020.
8. For a review on Pd-catalyzed AA, see: Crawley, M.L. and Trost, B.M. (2003) *Chem. Rev.*, **103**, 2921–2944.
9. For a review, see: Clarke, P.A. and Santos, S. (2006) *Eur. J. Org. Chem.*, **2006**, 2045–2053.
10. Wang, L., Li, P., and Menche, D. (2010) *Angew. Chem. Int. Ed.*, **49**, 9270–9273.
11. Wang, L. and Menche, D. (2012) *J. Org. Chem.*, **77**, 10811–10823.
12. Wang, L. and Menche, D. (2012) *Angew. Chem. Int. Ed.*, **51**, 9425–9427.
13. Morgen, M., Bretzke, S., Li, P., and Menche, D. (2010) *Org. Lett.*, **12**, 4494–4497.
14. Tang, B., Wang, L., and Menche, D. (2013) *Synlett*, **24**, 625–629.
15. For Ti-mediated examples, see: (a) Nugent, W.A. and Calabrese, J.C. (1984) *J. Am. Chem. Soc.*, **106**, 6422–6424; (b) Nugent, W.A., Thorn, D.L., and Harlow, R.L. (1987) *J. Am. Chem. Soc.*, **109**, 2788–2796; For Zr-mediated examples, see: (c) Negishi, E., Cederbaum, F.E., and Takahashi, T. (1986) *Tetrahedron Lett.*, **27**, 2829–2835; (d) Negishi, E., Holmes, S.J., Tour, J.M., Miller, J.A., Cederbaum, F.E., Swanson, D.R., and Takahashi, T. (1989) *J. Am. Chem. Soc.*, **111**, 3336–3346; (e) Kemp, M.I., Coote, S.J., and Whitby, R.J. (1998) *Synthesis*, (Suppl. 1), 557–568; (f) Chen, R.T. and Wong, K.T. (2002) *Tetrahedron Lett.*, **43**, 3313–3317.
16. Ubayama, H., Xi, Z., and Takahashi, T. (1998) *Chem. Lett.*, **27**, 517–518.
17. Debnar, T., Dreisigacker, S., and Menche, D. (2013) *Chem. Commun.*, **49**, 725–727.

15
Modular Assembly of Chiral Catalysts with Polydentate Stereodirecting Ligands

Lutz H. Gade

15.1
Introduction

The field of asymmetric catalysis has grown rapidly during the past decades and plays a key role in many areas of synthetic chemistry including the discovery of new drugs and the production of fine chemicals [1]. Rational ligand design and high-throughput screening methodologies have become vital tools in catalysis research because a broad scope of substrates requires rapid development of customized ligands. Although only a limited number of "privileged" chiral structures have become established [2], the development of synthetic pathways facilitating access to new classes of chiral ligands remains a key challenge [3].

In practical terms, the quest for successful strategies to solve a given problem is of particular interest. A few years ago, Hoveyda *et al.* [4] pointed out that efficient routes to novel catalysts may be based on the design of a *search pathway* rather than the preconception of an individual catalyst structure derived from a complete mechanistic analysis of a catalytic process. Such search strategies are conveniently based on highly modular catalyst systems, their basic structural motifs being derived from the available mechanistic insight into the catalytic process combined with the known ligand properties of its components [5].

In an optimized "search algorithm," the ancillary (stereodirecting) ligands are assembled in just a few steps from readily accessible building blocks. The ligand assembly is followed by a metallation step giving the (pre-)catalyst. The coordination of the ancillary ligand to the metal should be kinetically inert in order to give a relatively well-defined active system. Its structure should therefore combine at least one "anchor" unit and – in the case of asymmetric catalysis – a stereodirecting element.

This chapter will focus on four case histories that illustrate the strategies of catalyst development outlined above. These are taken from our own research activities in the field, which are related to those of others, and provide examples for efficient ligand design and straightforward catalyst optimization for a broad variety of stereoselective transformations. Rotational symmetry may greatly simplify this task, as well as the investigation of reaction mechanisms, thereby

Molecular Catalysts: Structure and Functional Design, First Edition. Edited by Lutz H. Gade and Peter Hofmann.

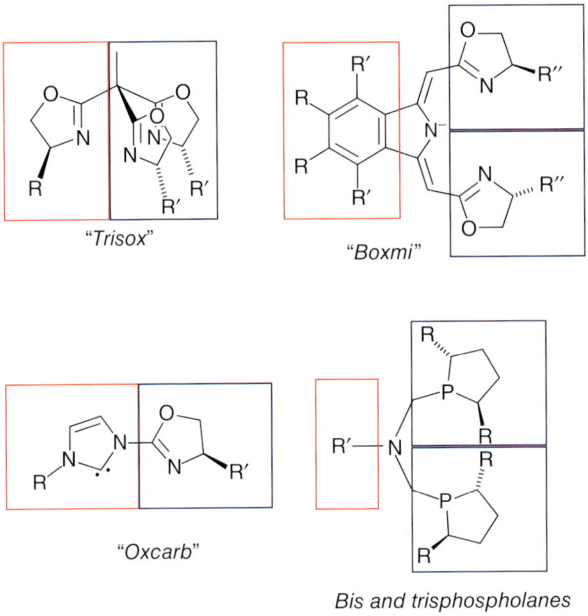

Figure 15.1 Four families of ligands discussed in this chapter representing case histories for modular ligand design in enantioselective catalysis. The colored boxes represent the molecular units that are assembled in the ligand synthesis.

reducing the number of diastereomeric intermediates and transition states to be considered [6, 7].

The *trisox*, *oxcarb*, as well as the bis and trisphospholane ligands – which we will focus on – may be assembled in one step from readily accessible building blocks (Figure 15.1). The key step in the three-step synthesis of the chiral pincer bis(oxazolinylmethylidene)isoindolines (*boxmi*) is a Wittig reaction involving the phthalimide starting material. Finally, the oxazoline–carbene (*oxcarb*) system provides an example of the combination of an "anchor" unit (N-heterocyclic carbene, NHC) and a stereodirecting element (chiral oxazoline).

15.2
A Modular Synthesis of C_3- and C_1-Chiral 1,1,1-Tris(oxazolyl)ethanes ("Trisox")

Oxazolines are nowadays essential ligands in asymmetric catalysis and also important synthons for stereoselective synthesis [8]. The success of the C_2-symmetric bis(oxazolines) ("BOX") and pyridine-bis(oxazolines) ("Pybox") discovered in the early 1990s has established them as a "privileged" class of ligands [9]. In contrast, the development and application of trisoxazolines lagged behind for a long time. Katsuki and collaborators [10] reported the first example of a chiral trisoxazoline in 1995 and their use in the allylic oxidation of alkenes (Kharasch–Sosnovsky reaction), as well as the enantioselective addition of diethylzinc to aldehydes.

The results obtained were promising, though, ultimately, not superior to those of the well-established C_2-symmetric BOX systems. However, a class of non-C_3-symmetrical trisoxazolines was developed by Zhou and Tang [11] and successfully employed in a range of highly enantioselective catalytic transformations.

1,1,1-Tris(oxazolyl)methane or -ethane ligands provide a geometry of the metal binding site which is most adapted to facial tripodal coordination of the metal center. This is thought to lead to a relatively rigid and well-defined coordination geometry. Such trisoxazoline ligands, first proposed in the 1990s, proved to be elusive for a long time [7]. Their attempted synthesis from methane-/ethanetris(carboxylate) or -trinitrile precursors, or from BOX carboxalates or nitriles led at best to BOX derivatives because of a kinetically favored decarboxylation step. As trisoxazolines of that type, generated via cyclization at a common carbon atom, were inaccessible, the obvious alternative was to cyclize first and then to assemble the three oxazoline rings in a final step. We therefore reverted to a strategy in which a readily accessible BOX derivative was coupled with a preformed, activated mono(oxazoline) ring. This was readily achieved by reaction of the lithium salt of 1,1-bis(oxazolinyl)ethane with 2-bromooxazoline [12] and allowed the high-yield access to chiral symmetrically substituted derivatives, such as 1,1,1-tris{2-[(*S*)-4-isopropyl]oxazolinyl}ethane (**A**) (Scheme 15.1).

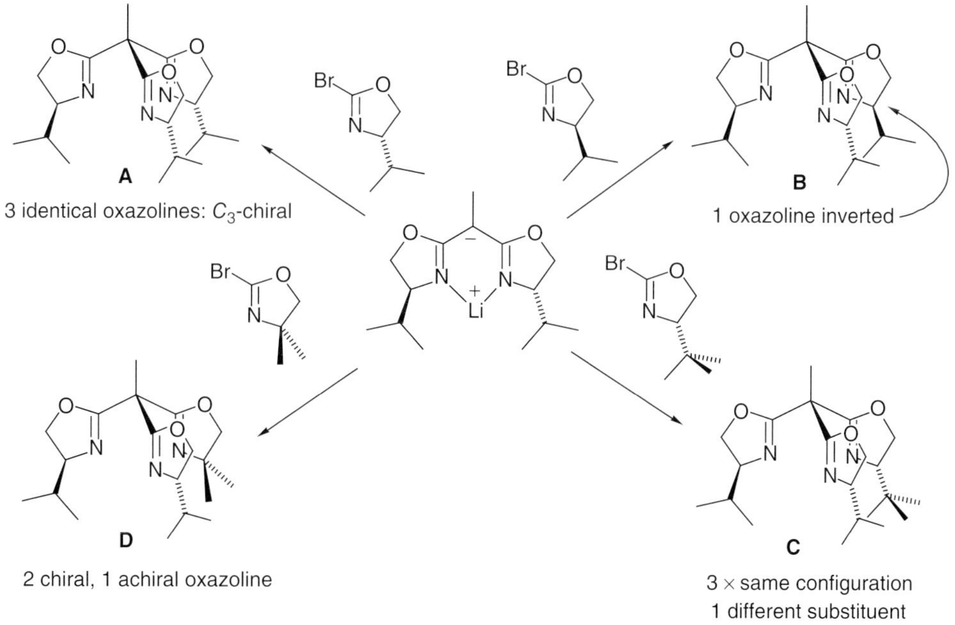

Scheme 15.1 Modular assembly of trisoxazoline ligands by reaction of the metallated bisoxazolines with 2-bromooxazoline derivatives. This allows the synthesis of C_3-chiral ligands (**A**), the inversion of one chiral center in an otherwise symmetrically substituted tripod (**B**), the combination of different chiral oxazoline rings (**C**), or the combination of chiral and achiral oxazoline rings (**D**).

Additionally, the synthesis of tripods with mixed substitution patterns (**B–D**, including the inversion of chiral centers as in **B**) was achieved in excellent yields [13]. This opened up the possibility to approach the synthesis of such polydentate oxazolines in a *modular* way, including tripods with mixed stereochemistry, that is, partially inverted chiral centers as well as achiral oxazoline rings.

The ability of trisox ligands to coordinate facially to transition metals has been established in a systematic study of its coordination chemistry. Provided the stereoelectronic properties of the metal do not strongly favor other coordination geometries, tridentate binding is observed [14].

Small molecular models mimicking hydrolase enzyme activity have been extensively studied during the past decade. The work on scorpionate and related tripod-Zn complexes has elucidated key mechanistic features of Zn enzyme activity [15]. The trisox ligands were viewed as models emulating *both* the tris(histidine) binding sites and the chiral environment of the protein skeletal structure. This provided the opportunity for developing model compounds for hydrolases or trans-esterases with the potential for *stereoselective* transformations normally observed only for the enzymatic system. We found that the zinc triflate complex $[Zn_2(\mu\text{-}O_3SCF_3)_3(i\text{Pr-trisox})_2](O_3SCF_3)$ displayed catalytic activity in the stereoselective transesterification of various phenyl esters (conducted as a kinetic chiral resolution) [16].

15.2.1
C_3-Chirality in Polymerization Catalysis with Rare-Earth Complexes

The efficient control of the tacticity and molecular weight distribution in polymers by ligand design provides the basis for the success of group 4 metallocene catalysis of α-olefins [17]. Prior to our work, there was no report of the use of C_3-chiral stereodirecting ligands in polymerization catalysis, although they confer an element of molecular helicity to a complex which is thought to be beneficial for face selection in the key migratory insertion step.

The trialkyl complex $[Sc(i\text{Pr-trisox})(CH_2SiMe_3)_3]$ was prepared by reaction of the trialkyl precursor $Sc(CH_2SiMe_3)_3(thf)_2$ (thf, tetrahydrofuran) with an equimolar amount of *i*Pr-trisox (Scheme 15.2) [18]. Early attempts to polymerize 1-hexene with *in situ* generated $[Sc(i\text{Pr-trisox})(CH_2SiMe_3)_2][B(C_6F_5)_4]$ suffered from low activity and somewhat variable reproducibility as far as tacticity control was concerned, as well as bimodal molecular mass distributions. The latter indicated the presence of at least two catalytically active species.

In view of Okuda's previous suggestion that in certain rare-earth polymerization catalysts a dicationic species may be involved [19], we investigated the reaction of the trialkyl complex with 2 equiv of $[Ph_3C][B(C_6F_5)_4]$, the product of which was tentatively formulated as the dicationic complex $[Sc(i\text{Pr-trisox})(CH_2SiMe_3)][B(C_6F_5)_4]_2$, although this species has defied isolation and complete characterization to date. This *in situ* generated cationic complex was found to be highly active for the polymerization of 1-hexene and displayed good

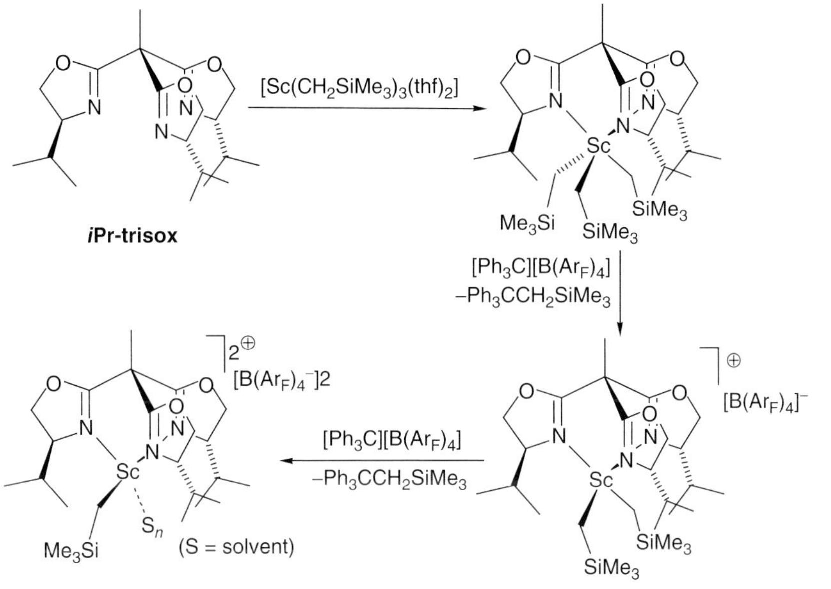

Scheme 15.2 Synthesis of the Sc complex of [Sc(*i*Pr-trisox)(CH_2SiMe_3)$_3$] and its conversion to the mono and (possibly) dicationic catalysts by alkyl abstraction [18a].

tacticity control. Polymerization studies were carried out at various temperatures, and the activities and polymer characteristics are provided in Table 15.1 [18].

The activity of 36 200 kg mol^{-1} h^{-1} (Table 15.1) is comparable to the extremely high activities reported for zirconium amine-bis(phenoxide) complexes [20]. Upon reducing the polymerization temperature to −30 °C, the activity dropped to 2030 kg mol^{-1} h^{-1}; however, at this temperature the poly(1-hexene) produced

Table 15.1 1-Hexene polymerization data for the highly active catalyst generated by abstraction of two alkyl groups from [Sc(*i*Pr-trisox)(CH_2SiMe_3)$_3$] with 2 M equiv of [Ph_3C][$B(C_6F_5)_4$] [18a].

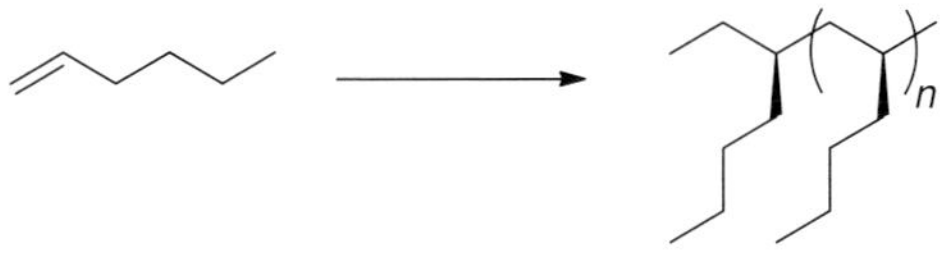

Temperature (°C)	Reaction time (min)	Yield (g)	Activity[a)]	M_w	M_w/M_n
−30	3	1.01	2 030	750 000	1.18
−20	1.5	1.90	7 600	552 000	1.87
0	1	2.18	13 080	354 000	2.36
21	0.5	3.02	36 230	227 000	2.22

a) Activity in kg mol^{-1} h^{-1}. Yields based on 3.4 g of hexene.

was highly isotactic (*mmmm* > 90% by $^{13}C\{^1H\}$ NMR spectroscopy). Gel permeation chromatography (GPC) analysis of the polymer obtained under these conditions established a very narrow monomodal molecular mass distribution with $M_w = 750\,000$ and a polydispersity index (PDI) of 1.18, indicating that the isoselective catalytic polymerization carried out at low temperature shows living-type behavior.

Given these early results obtained with the scandium complex [Sc(*i*Pr-trisox)(CH_2SiMe_3)$_3$], we continued our investigations with yttrium and the lanthanides in the order of increasing ionic radius. For the polymerization of *n*-hexene, *n*-heptene, and *n*-octene, M_w/M_n values of between 1.58 and 2.08 were obtained and isotacticities of 80–95% as determined by ^{13}C NMR spectroscopy [21]. However, none of the activities was comparable to those observed with the scandium congener [Sc(*i*Pr-trisox)(CH_2SiMe_3)$_3$] described above [22].

15.2.2
Trisox as a Bidentate Ligand: Chiral Trisoxazolines in Copper(II) Lewis Acid Catalysis and Palladium-Catalyzed Asymmetric Allylic Substitutions

The discussion of catalyst symmetry in connection with stereodirecting podands has mainly focused on static models of the complexes [7]. However, the high rotational symmetry of the chiral tripodal ancillary ligand may render reversible complex transformations into the active catalytic species equivalent. An example of the exploitation of molecular symmetry in this particular context is thought to be underlying the highly efficient applications of trisox derivatives in Cu^{II}-based Lewis acid catalysis, the principal domain of the well-established BOX ligands [8, 9]. A major practical disadvantage of the latter is the generally high catalyst loadings, which are due to the kinetic lability of copper(II). In this context, the facial coordination by a chiral tridentate ligand was thought to stabilize the *resting state* of the copper complexes.

The additional oxazoline ligation is expected to deactivate the complexes in their Lewis acidity, as was shown in a theoretical study on (BOX)Cu catalysts [23]. The transformation of the resting state into the active (17e Cu^{II}) species therefore necessitates the decoordination of an oxazoline unit and the opening up of the system (Scheme 15.3, top). The required "hemilability" is provided *stereoelectronically* by the strong dynamic Jahn–Teller effect of the d^9 Cu^{II} center. As a consequence of the threefold rotational symmetry of the system, all the possible dicoordinated catalytically active species (**A**–**C**) are equivalent (Scheme 15.3, bottom).

To test this concept, we applied [Cu^{II}(trisox)] complexes in the asymmetric Mannich reaction [24] of a β-ketoester with an activated *N*-tosyl-α-imino ester, a reaction that had been previously reported by Jørgensen *et al.* [25] using chiral copper(II)-BOX catalysts (10 mol%). After optimization of the reaction conditions, the reaction product was obtained with an excellent enantiomeric excess of 90% using 10 mol% of the catalyst (Table 15.2) [24a].

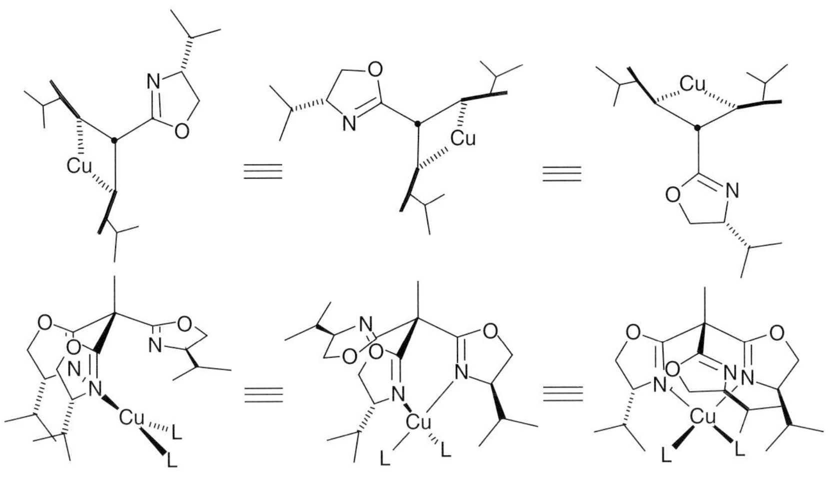

A B C

Scheme 15.3 Coordination/decoordination equilibrium between the proposed resting and active states of the trisox-Cu catalysts giving symmetry-equivalent active species [24].

Table 15.2 Enantioselective Mannich reaction of ethyl 2-methylacetoacetate with *N*-tosyl-α-imino methyl ester catalyzed by ^{i}Pr-trisox/$Cu(ClO_4)_2$ or ^{i}Pr-BOX/$Cu(ClO_4)_2$ [24a].

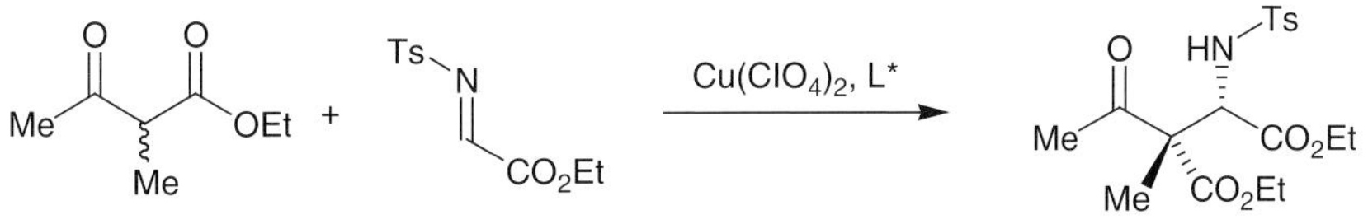

Entry	Catalyst loading (%)	Ligand [*i*Pr-BOX] Yield (%)	ee (%)	[*i*Pr-trisox] Yield (%)	ee (%)
1	10	84	84	84	90
2	1	70	84	75	89
3	0.1	56	80	59	91
4	0.01	35	66	36	90

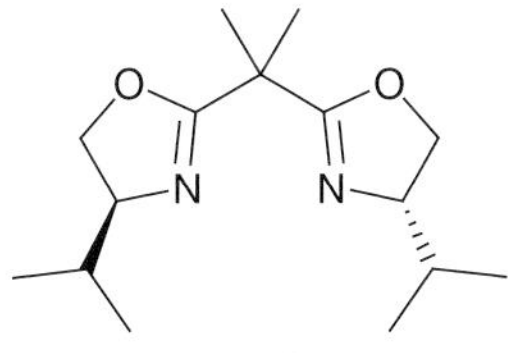

[*i*Pr-BOX]

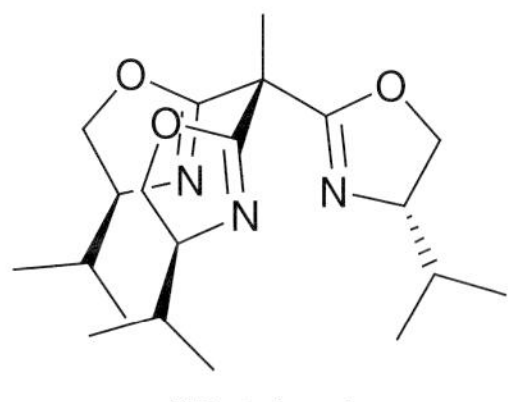

[*i*Pr-trisox]

Experimental conditions: acetone/diethyl ether 1/3, −28° C, and 36 h.

Upon stepwise reduction of the catalyst loading by a factor of 10^3, that is, in the presence of only 0.01 mol% of catalyst, the enantioselectivity remained unchanged (90% ee, the diastereoselectivity being throughout the dilution series at about syn/anti = 13/87). As a direct comparison, we also investigated the reaction in the presence of 10 mol% of [(*i*Pr-BOX)Cu]$(ClO_4)_2$ for which 84% ee (84% yield) was observed. Reducing the catalyst loading for these systems led to a decrease of the stereoselectivity, with enantiomeric excesses of 80% and 66% ee being observed at catalyst concentrations of 0.1 and 0.01 mol%, respectively [24a].

Direct evidence for our assumption of a partially decoordinated trisox ligand in the active state of the catalyst was obtained *inter alia* from an X-ray diffraction study of [Cu^{II}(*i*Pr-trisox)(κ^2-*O,O′*-MeCOCHCOOEt)]$^+$(BF_4^-) (Figure 15.2).

As another test reaction, we investigated the direct α-amination of α-substituted β-ketoesters with azodicarboxylates for which an efficient copper(II)-BOX-catalyzed version had previously been reported by Jørgensen and coworkers [26]. Here again, the trisox systems proved to be superior to the corresponding BOX-Cu^{II} catalysts.

It has been argued above that a threefold symmetrical chiral podand may simplify the stereochemistry of key catalytic intermediates for cases in which it only acts as bidentate ligand in the stereoselectivity-determining step: in other words, for metal complexes with a stereoelectronic preference for non-deltahedral coordination geometries. Palladium(II)-catalyzed allylic substitutions provide appropriate test reactions along these lines [27], and it was possible to study the dynamic exchange in model systems for both the Pd^{II} and Pd^0 intermediates of this catalytic reaction.

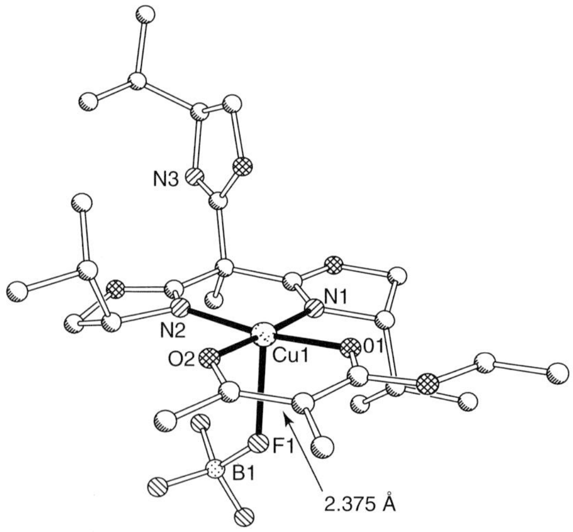

Figure 15.2 Molecular structure of the copper complex [Cu^{II}(*i*Pr-trisox)(κ^2-*O,O′*-MeCOCHCOOEt)]$^+$(BF_4^-) [24].

Table 15.3 Results of asymmetric allylic alkylations with various bisoxazolines and their trisox analogs [28].

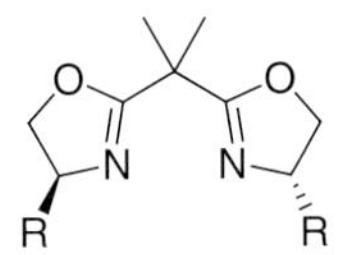

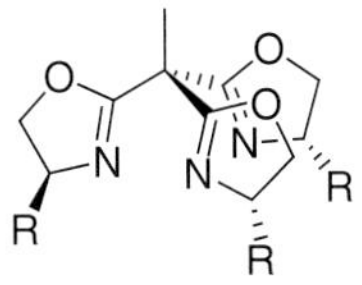

Entry	R	Yield (%)	ee (%)	Yield (%)	ee (%)
1	(*S*)-*i*Pr	89	89	90	95
2	(*R*)-Ph	7	−72	28	−88
3	(*S*)-Bn	88	83	92	88
4	(4*R*,5*S*)-Ind	13	−93	95	−98

Experimental conditions: catalytic precursor generated *in situ* from $[PdCl(C_3H_5)]_2$ and ligand (1 mol%, ratio Ligand/Pd = 1.1) in THF at 50 °C for 1.5 h; catalysis carried out at room temperature; yields and enantiomeric excess determined after 72 h.

Pfaltz and coworkers [29] had previously investigated the allylic alkylation of 1,3-diphenylprop-2-enyl acetate with dimethyl malonate as nucleophile (in the presence of *N*,*O*-bis(trimethylsilyl)acetamide, BSA) with the well-established BOX ligands as stereodirecting ligands, and this particular system therefore provided the point of reference for the catalytic study at hand (Table 15.3, entry 1) [28]. Under the reaction conditions displayed in Scheme 15.4, the trisoxazoline-based catalysts generally induce a better enantioselectivity compared to their BOX analogs, and this behavior appears to be independent of the substituent, as shown in Table 15.3.

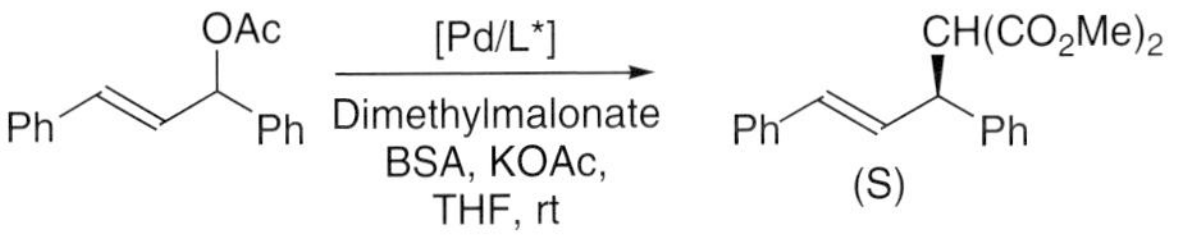

Scheme 15.4 Allylic alkylation of 1,3-diphenylprop-2-enyl acetate with dimethyl malonate (in the presence of *N*,*O*-bis(trimethylsilyl)acetamide – BSA).

Apart from the effect on the catalyst selectivity, the most notable observation is the rate acceleration with the tripods compared to the BOX ligands for all substitution patterns (Table 15.4). The rate of the reaction is strongly dependent on the substituent of the respective ligand, with the *i*Pr substituent yielding the most active BOX derivative and the indanyl substituent leading to the highest rate for the trisox-based catalysts [29]. With *i*Pr, Ph, and Bn-based ligands, the turnover frequencies (TOFs) differ by a factor of 4 in favor of the tripod, while a striking 64-fold acceleration was found for the indanyl derivative!

Table 15.4 Maximum turnover frequencies (TOF (h^{-1})) of asymmetric allylic alkylations with various bisoxazolines and their trisox analogs [28].

R	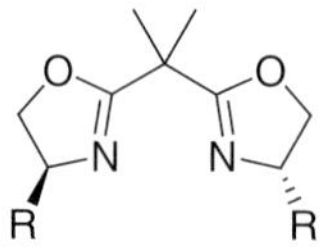	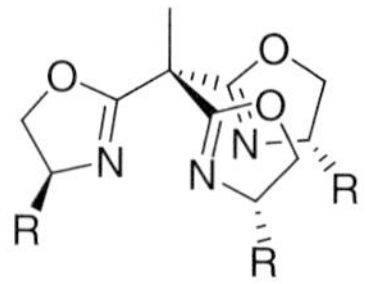	TOF ratio Trisox/BOX
(*S*)-*i*Pr	1.37	5.02	3.7
(*R*)-Ph	0.05	0.2	4
(*S*)-Bn	0.91	3.73	4.1
(4*R*,5*S*)-Ind	0.19	12.2	64.2

The adaptability of the trisox ligand to various types of metal centers, coordination geometries, and enantioselective catalytic transformations is now well established. Although originally designed to coordinate facially to a metal center, the potential of C_3-chiral trisox ligands in enantioselective catalytic transformations in which the intermediate in the selectivity-determining step does not adopt deltahedral coordination geometry and thus does not allow tripodal coordination of the stereodirecting ligand has been demonstrated. The symmetry-related simplified behavior with regard to potential catalyst equilibria in solution along with the stereochemical nonambiguity of the active catalytic species is to be considered in targeted catalyst-screening processes.

15.3 The *Boxmi* Pincer System: a Highly Efficient Modular Stereodirecting Ligand for a Broad Range of Catalytic Reactions

Meridionally coordinating chiral tridentate ligands, frequently referred to as *pincers* [30], provide the structural platform for the construction of efficient stereodirecting molecular environments. While many of the known chiral systems of the "pincer" type perform relatively poorly in enantioselective catalysis because of a certain lack of control of substrate orientation, the assembly from rigid heterocyclic units recently has given rise to several highly enantioselective catalysts [31]. These ligand systems, such as bis(oxazolinyl)phenyl (phebox) [32], bis(oxazolinyl)carbazole [33], and chiral bis(pyridylimino)isoindole (bpi) derivatives [22], have been proven to act as efficient stereodirecting ligands in a variety of applications in molecular catalysis.

We recently reported the synthesis of a new class of chiral NNN-pincer ligands, *boxmi*, which are readily accessible in the modular three-step synthesis (depicted in Scheme 15.5) starting from easily available phthalimides [34]. The backbones of the pincer ligands were prepared according to an improved Wittig procedure

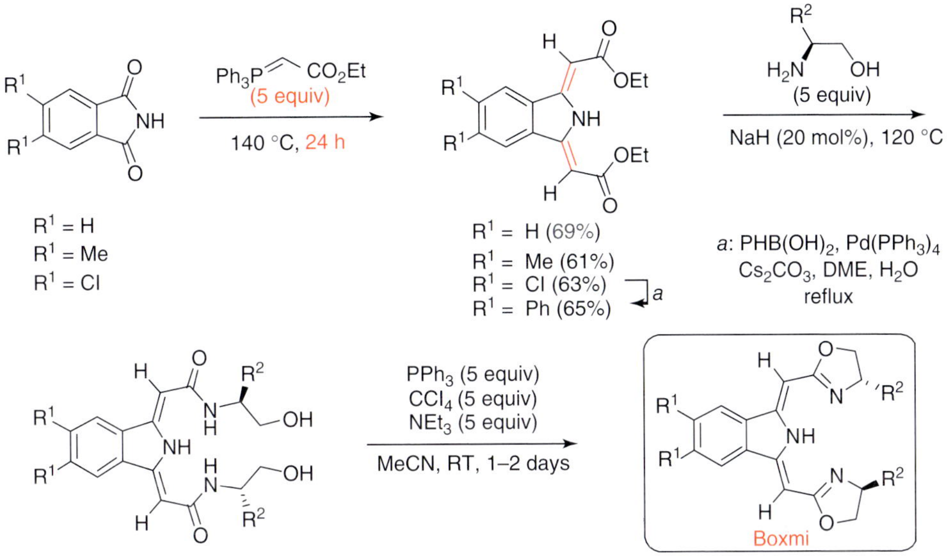

Scheme 15.5 Synthesis of the boxmi ligands: the key step is a Wittig coupling of the phthalimide [34].

(combined with a subsequent Suzuki coupling). The bisoxazolines were subsequently obtained in high yields by standard oxazoline cyclization methods.

These ligands were initially tested in the nickel(II)-catalyzed enantioselective fluorination of oxindoles and β-ketoesters, yielding the corresponding products with enantioselectivities of up to >99% ee and high yields. Application of the chiral pincer ligands in the chromium-catalyzed enantioselective Nozaki–Hiyama–Kishi reaction of aldehydes gave the corresponding alcohols with a maximum enantioselectivity of 93% [34].

To date, most applications of the boxmi system in enantioselective catalysis concern C–C and C–E (E = heteroelement) coupling reactions in which an electrophile reacts with an enolizable carbonyl derivative which coordinates to the Lewis acidic transition-metal center. In this context, structural data obtained by single-crystal X-ray diffraction have been instructive. In Figure 15.3, the molecular structures of the [Cu(Ph_2-boxmi)(OAc)] precatalyst employed in the enantioselective alkylations and trifluoromethylations (see below) and the [Fe(Ph_2-boxmi)OAc)(py)] precatalyst employed in the iron-catalyzed asymmetric electrophilic azidations are depicted. It is to be assumed that an ester enolate intermediate coordinates at the site occupied by the monoanionic co-ligand (acetate) in these structures and that one of the prochiral faces of the substrate is blocked by the phenyl group at the oxazolinyl rings in the Ph_2-boxmi ligand. The electrophilic coupling reagents will therefore preferentially approach from the *opposite* face of the substrate, which would explain the enantioselectivity of the transformation.

In a systematic study, the highly enantioselective alkylation of β-ketoesters was achieved using boxmi-copper(II) catalysts. In particular, benzylic and allylic

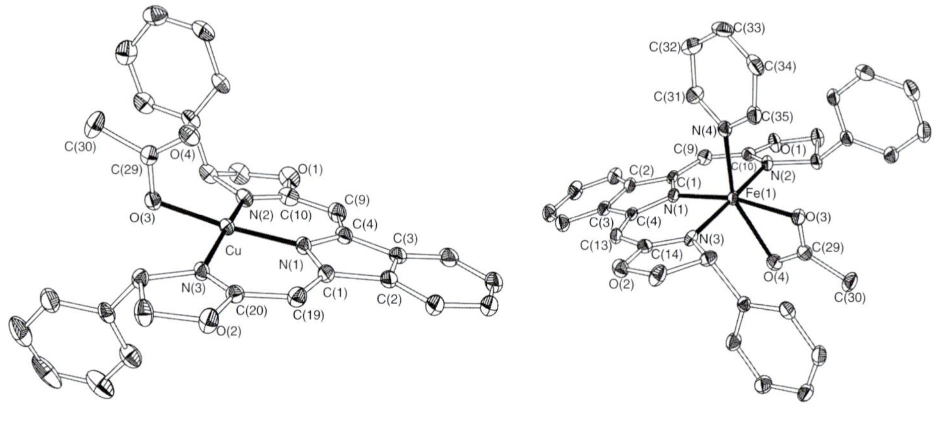

Figure 15.3 (a) Molecular structure of the [Cu(Ph$_2$-boxmi)(OAc)] precatalyst employed in the enantioselective alkylations and trifluoromethylations (see below) [35]. (b) Molecular structure of the [Fe(Ph$_2$-boxmi)OAc)(py)] precatalyst employed in the iron-catalyzed asymmetric electrophilic azidations of β-ketoesters and oxindoles (see below) [36].

alcohols were employed to prepare *in situ* the corresponding iodides as the alkylating reagents without further purification. The primary chiral alkylation products were cyclized subsequently in a one-pot procedure to generate spirolactones or bi-spirolactones by adding $BF_3{\cdot}Et_2O$ and in the presence of copper complex, as depicted in Scheme 15.6 [35].

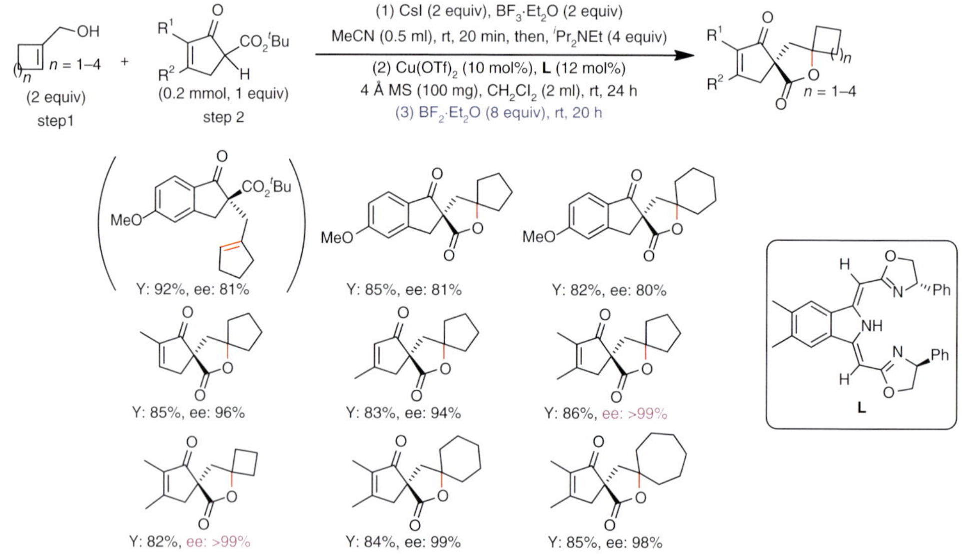

Scheme 15.6 Enantioselective synthesis of bi-spirolactones via Cu-catalyzed alkylation with allylic alcohols and subsequent acid-catalyzed cyclization [35].

Furthermore, β-ketoester-substituted allylsilanes were converted to spirolactones and bicyclic cyclopentanols with excellent enantioselectivities by subsequent treatment of the primary chiral allylation products (Scheme 15.7).

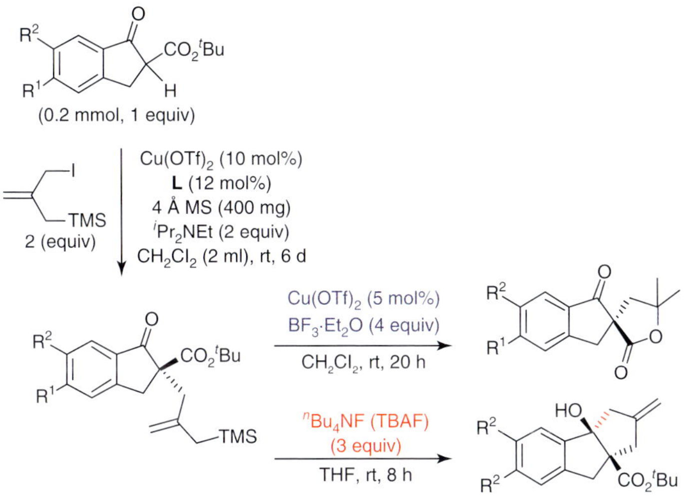

Scheme 15.7 Stereoselective synthesis of allylsilanes and their subsequent transformations (L = Ph-boxmi) [35].

It is well known that the physical, chemical, and biological properties of organic molecules may be significantly modified in the presence of trifluoromethyl groups [37]. While there is an ever-growing number of catalytic enantioselective nucleophilic trifluoromethylations [38], the corresponding electrophilic variants remain comparatively rare [39]. Employing the chiral Cu-pincer systems developed for enantioselective alkylations described above, we extended the concept to include the asymmetric catalytic trifluoromethylation of β-ketoesters under mild conditions by using Togni's electrophilic trifluoromethylating agent 3-dimethyl-1-(trifluoromethyl)-1,2-benziodoxole [40]. In this way, a broad range of cyclic five-membered ring β-ketoesters were trifluoromethylated with excellent enantioselectivity (Scheme 15.8) [41].

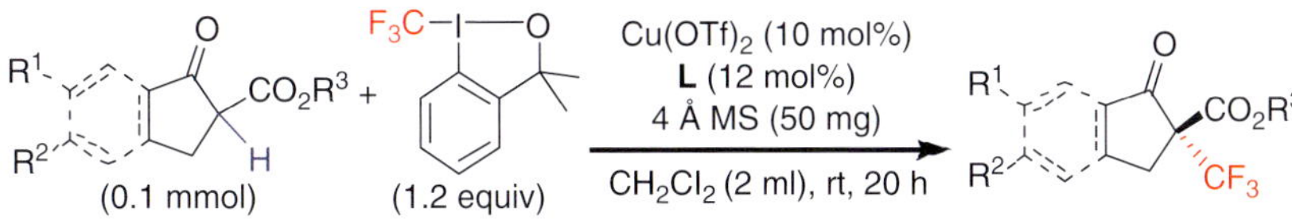

Scheme 15.8 Enantioselective trifluoromethylation of five-membered ring β-ketoesters (L = Ph-boxmi) [41].

In contrast to the type of substrates depicted in Scheme 15.8, the more easily enolizable cyclic six-membered ring β-ketoesters were trifluoromethylated with significant decrease in enantioselectivity under the conditions described above. However, upon replacing Togni's reagent by Umemoto's trifluoromethylsulfonium reagent, the corresponding six-membered ring systems were also trifluoromethylated with high enantioselectivity.

The usefulness of the enantioselective trifluoromethylation was demonstrated in further transformations of the trifluoromethylated products. As an example, the trifluoromethylated β-ketoesters reacted with Grignard reagents to give the corresponding α-CF_3 β-hydroxyesters in high yields with excellent diastereoselectivities (Scheme 15.9).

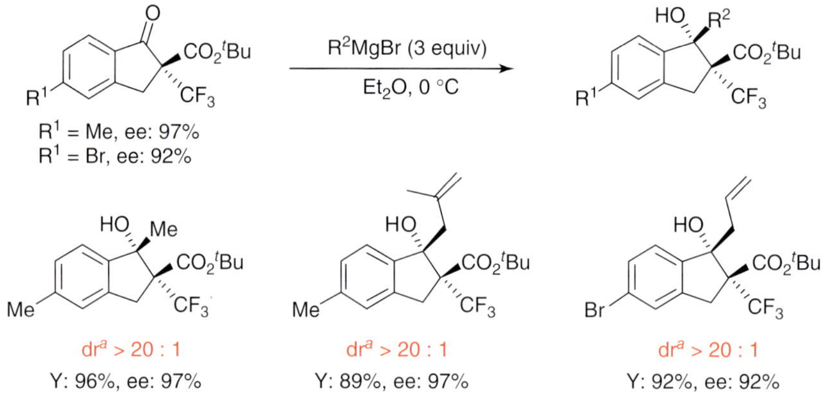

Scheme 15.9 Highly diastereoselective transformation of α-CF_3 β-ketoesters [41].

Organic azides are valuable intermediates in organic synthesis, in particular as powerful precursors for a wide range of nitrogen-containing synthetic targets. Despite extensive studies and significant advances in this field, the catalytic stereoselective introduction of azido group to organic compounds remains comparatively rare [42]. Exploiting a similar strategy as for the enantioselective trifluoromethylation described above, we developed a highly enantioselective Fe-catalyzed azidation of β-ketoesters and oxindoles by using a T-shaped iodine(III) compound as azido-transfer reagent [43]. In this way, an efficient protocol for enantioselective Fe-catalyzed azidation of cyclic β-ketoesters and 3-aryl oxindoles was achieved [36].

Cyclic β-ketoesters were converted to the corresponding products in high yields with up to 93% ee catalyzed by the combination of iron(II) chlorido complex and silver carboxylate. 3-Azido aryl oxindoles were obtained with up to 94% ee using the catalyst prepared by iron(II) propionate and the ligand *in situ* (Scheme 15.10).

To highlight the utility of the enantioselective azidation further, transformations of the resulting azides were carried out (Scheme 15.11). For example, an α-azido ester could be converted smoothly into α-amino ester by palladium-catalyzed hydrogenolysis, which may provide a useful method for the synthesis of highly substituted α-amino acid derivatives. On the other hand, the copper-catalyzed azide-alkyne 1,3-dipolar cycloaddition (CuAAC), as a "click" reaction, has been

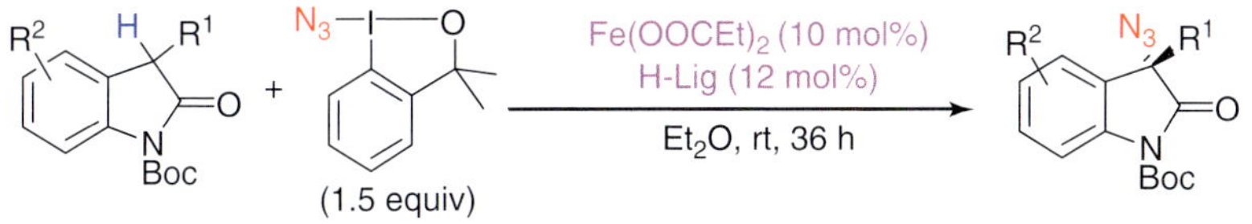

Scheme 15.10 Fe(boxmi)-catalyzed enantioselective azidation of cyclic β-ketoesters [36].

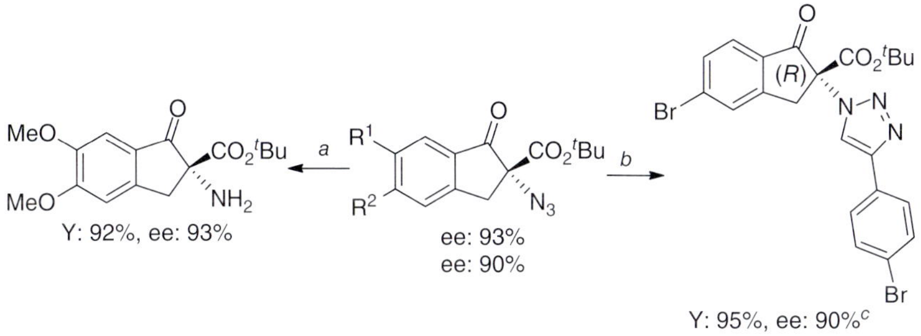

Scheme 15.11 Further transformation of the chiral azides. [a]H_2 (1 atm)/Pd/$BaSO_4$ (10 mol%), [b]1-bromo-4-ethynylbenzene (1.2 equiv), $CuSO_4$ (20 mol%), tris[(1-benzyl-1*H*-1,2,3-triazol-4-yl)methyl]amine. [c]The absolute configuration was determined by X-ray structure analysis [36].

established as a powerful coupling technology, and the α-azido ester **5c** was transformed into the corresponding triazole **11** in high yield [36].

15.4
Bidentate N-Heterocyclic Carbene Ligands Incorporating Oxazoline Units

As pointed out Section 15.1, the coordination of the ancillary ligand to the metal atom in a molecular catalyst should be kinetically inert in order to give a relatively well defined active system. NHCs [44, 45] are excellent "anchor" units for late transition metals which form strong metal–carbon bonds and have thus been widely used in homogeneous catalysis [46]. They are straightforward to synthesize, often chemically more stable than the extensively employed phosphines, and may be readily combined with other ligating units, in particular, by the appropriate functionalization of the N-atoms in their heterocyclic structures [47]. Using NHCs, therefore, allows the combination of at least one "anchor" unit with a stereodirecting element.

In 1998, Herrmann *et al.* reported the synthesis of the first chiral carbene containing an oxazoline unit. In this bidentate ligand, the oxazoline ring is linked in its 2-position to the imidazole ring via a methylene bridge [48]. The key step in the synthesis of the imidazolium precursor is the acid-catalyzed cyclization of the oxazoline by reaction of an iminoester, formed *in situ* from a nitrile function, and the amino alcohol (Scheme 15.12).

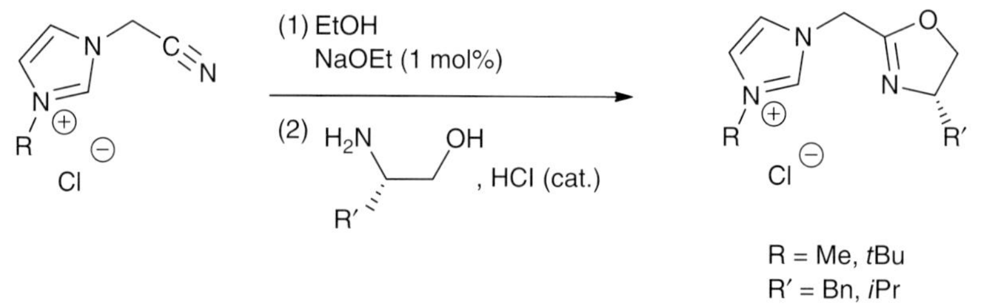

Scheme 15.12 Synthesis of the imidazolium precursor of Herrmann's oxazolinyl–carbene ligand.

The imidazolium salts depicted in Scheme 15.12 were subsequently coordinated as carbene–oxazoline ligands to rhodium(I) and palladium(II). The rhodium complex was tested in the hydrosilylation of ketones, giving the secondary alcohols with poor enantioselectivity [49].

A major step forward in the development of asymmetric catalysis with chiral NHC complexes was the work of Burgess *et al.* [50] on the asymmetric hydrogenation of alkenes using iridium(I) catalysts containing NHC-oxazolines. Their design was inspired by the chiral bidentate phosphine–oxazoline ligands (*Phox*) developed by Helmchen and Pfaltz, which had proved to be highly selective in the enantioselective hydrogenation of nonfunctionalized, trisubstituted alkenes [51]. Furthermore, Burgess and coworkers [52] had previously studied a novel family of P,N-ligands, dubbed *JM-Phos*, and were thus guided by the analogy between phosphanes and NHCs in the design of the new class of oxazoline–carbenes.

In the imidazolium salts, obtained by way of a nucleophilic substitution of the iodo-derivative by an imidazole, the oxazoline is linked by the carbon atom in 4-position. Coordination of the bidentate ligand to the $\{Ir(COD)\}^+$ (COD, cyclooctadiene) complex fragment was then achieved by *in situ* deprotonation (Scheme 15.13). This modular design allowed facile and rapid access to a large ligand library by variation of the substituents in 2-position of the oxazoline and at the "terminal" N-atom of the heterocyclic carbene.

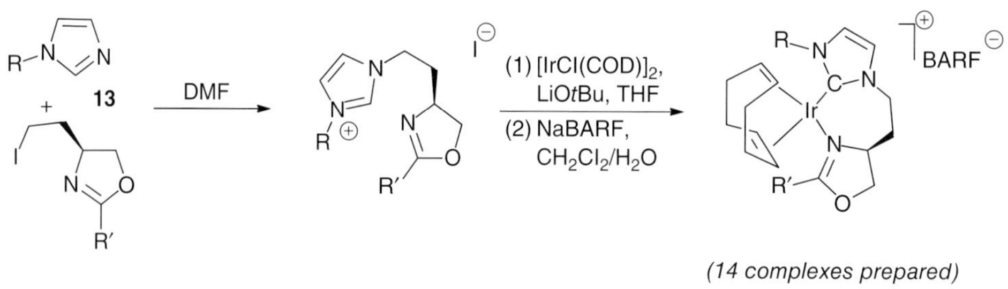

Scheme 15.13 Synthesis of an iridium(I) complex bearing Burgess's chiral oxazoline–imidazolylidene ligand.

Complexes of this type were tested *inter alia* in the asymmetric hydrogenation of *E*-1,2-diphenylpropene, and the derivative containing an adamantyl-substituted oxazoline ring proved to be the most active and selective for this reaction (Scheme 15.14).

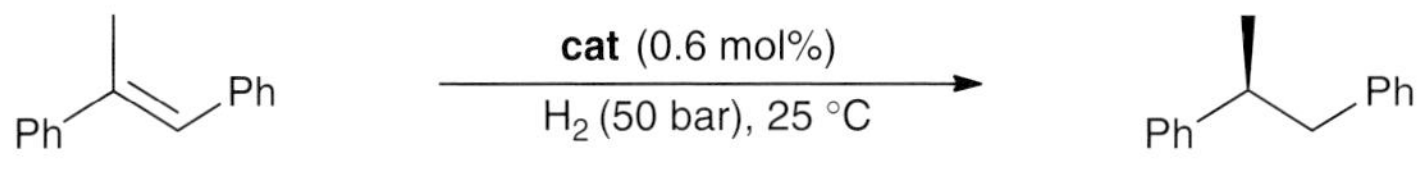

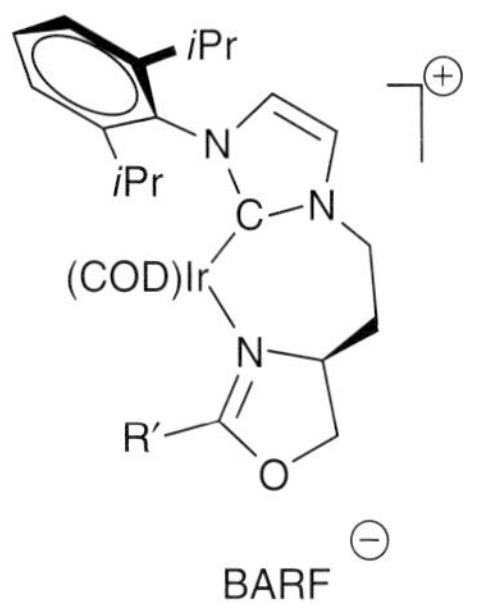

R	ee (%)	Yield (%)
Ph	13	25
$CHPh_2$	25	12
*t*Bu	50	81
1-Ad	98	99

Scheme 15.14 Catalytic hydrogenation of *E*-1,2-diphenylpropene with Burgess's iridium complexes.

The authors explained the high selectivity of this type of catalyst, pointing out the key structural features leading to an efficient chiral induction with this class of complexes: the bulky 2,6-$(i\mathrm{Pr})_2$-C_6H_3 group effectively blocks one of the quadrants of the active space in the catalyst, allowing good control of the geometry of the coordination sphere around the metal. A density functional theory (DFT) study of the Ir-catalyzed hydrogenation of trisubstituted alkenes revealed that an Ir(III)/Ir(IV) pathway is probably involved. Moreover, it revealed that the alkene is coordinated in trans disposition to the carbene, where it interacts directly with the oxazoline substituent [53]. Burgess's complexes were also successfully employed in the stereoselective hydrogenation of dienes, the mechanism of which was subsequently investigated in detail [54].

In view of these previous developments, we directly coupled various N-substituted imidazoles, which display nucleophilic reactivity, with 2-bromo-oxazolines to give the imidazolium precursors (Scheme 15.15) [55]. This direct condensation of an oxazoline and an imidazole salt provided a straightforward and modular route to the development of a new family of stereodirecting ligands. NHC–rhodium complexes could be obtained by reaction of the imidazolium salt with [{Rh(μ-O*t*Bu)(nbd)}$_2$] (nbd, norbornadiene) generated *in situ* [56].

Based on this strategy, a highly stereoselective Rh^{I} catalyst for the asymmetric hydrosilylation of ketones was developed [57]. An initial screening of catalyst structures identified the complex depicted in Figure 15.4 as the most selective catalyst.

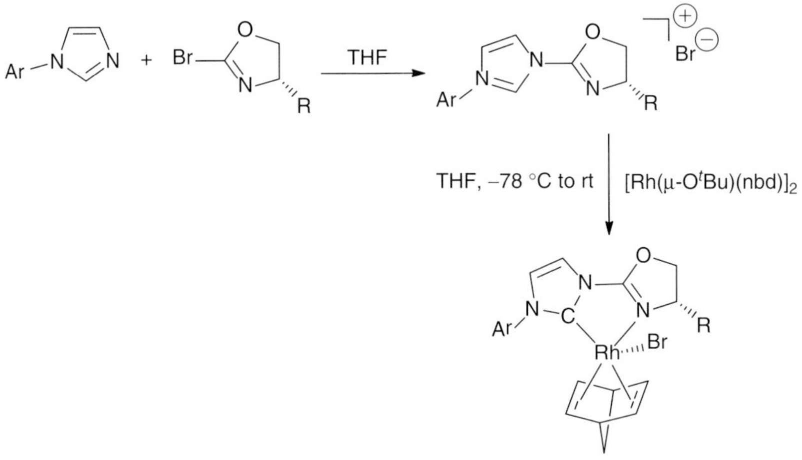

Scheme 15.15 Synthesis of the imidazolium–oxazoline ligand precursor and complexation with rhodium(I).

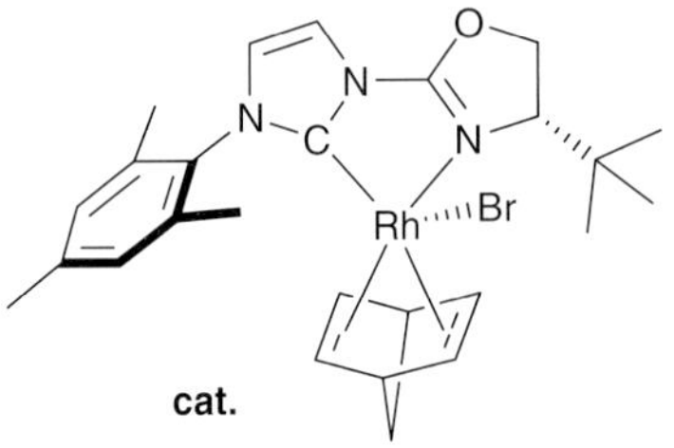

Figure 15.4 Molecular structure of the optimized hydrosilylation (pre-)catalyst.

Whereas, for example, the asymmetric hydrosilylation of 2-naphthyl methyl ketone with this catalyst was carried out with 99% yield and 91% ee, the enantioselectivities for most aryl alkyl ketones were found to be slightly below those of the most efficient phosphane-based systems. However, the system was found to be exceptionally selective in the hydrosilylation of unsymmetrical dialkyl ketones (Table 15.5), which are "difficult" substrates [58]. The selectivity for the reduction of prochiral dialkyl ketones was comparable or even superior to the best previously reported for prochiral nonaromatic ketones.

Whereas cyclopropyl methyl ketone was hydrosilylated with an enantioselectivity of 81% ee, the increase of the steric demand of one of the alkyl groups led to improved enantiomeric excesses, reaching 95% ee in the case of *tert*-butyl methyl ketone. Linear-chain *n*-alkyl methyl ketones, which are particularly challenging substrates, were reduced with good asymmetric induction, such as in the case of 2-octanone (79% ee) and even 2-butanone (65% ee) [58].

Table 15.5 Asymmetric hydrosilylation of ketones with the catalyst depicted in Figure 15.4.

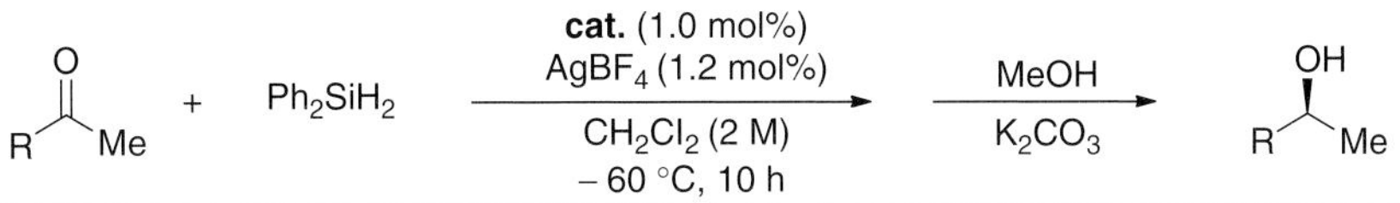

Entry	Ketone	ee (%)	Yield (%)
1	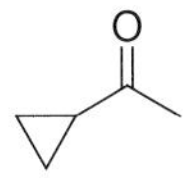	81	63[a]
2	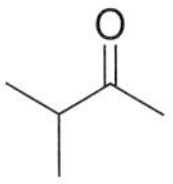	88	53[a]
3	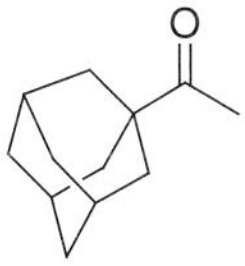	89	98
4	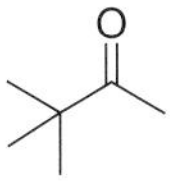	95	70[a]
5	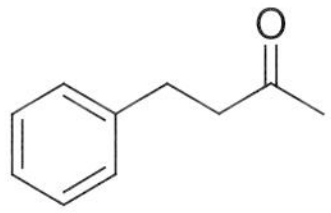	77	97
6	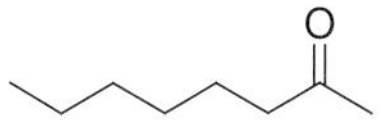	79	95
7	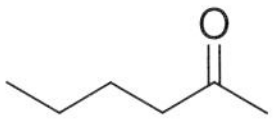	74	88[b]
8	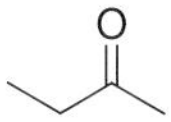	65	n. d.[b]

a) Moderate yield due to the volatility of the product.
b) Reaction carried out at −40 °C.

Unexpected nonlinearities in the optimization of this type of hydrosilylation catalyst led us to examine the mechanism of the Rh-catalyzed hydrosilylation of ketones more closely. Notable experimental results included the extreme rate enhancement upon using R_2SiH_2 compared to R_3SiH as silylating agents as well as a markedly different temperature dependence of enantiomeric excesses for these two reagents. Furthermore, the observation of an *inverse* kinetic H/D isotope effect of 0.8 for R_2SiH_2 compared to the absence of a kinetic isotope effect (KIE) for R_3SiH indicated that at least two mechanistic scenarios might apply depending on the nature of silane. A detailed computational study

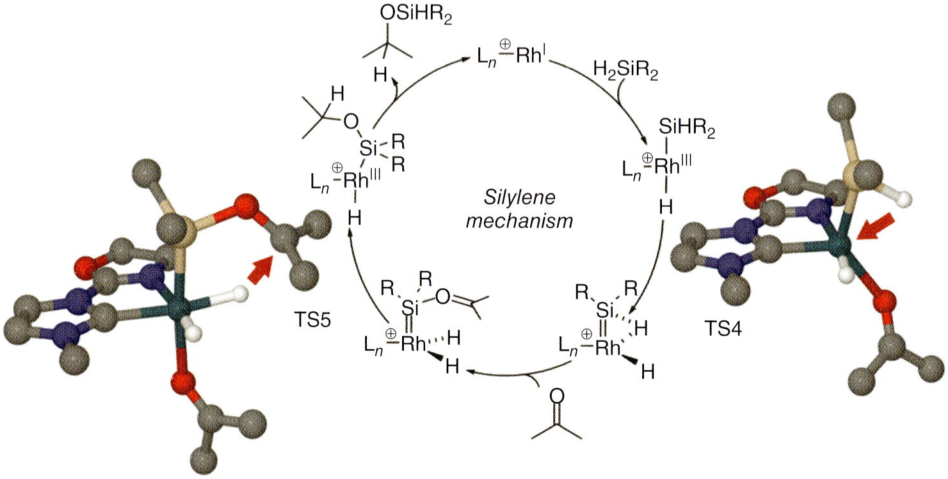

Figure 15.5 Silylene mechanism of the Rh-catalyzed hydrosilylation of ketones. The two modeled structures represent the key transition states. Right: The second oxidative addition of a Si–H bond. Left: hydride transfer from rhodium to the carbonyl C atom [59].

revealed a new mechanism for the hydrosilylation with secondary silanes, which involves rhodium-silylene intermediates as depicted in Figure 15.5 and which is consistent with all the experimental data obtained for that system [59].

15.5
New Modular Di- and Tridentate Phospholane Ligands

A well-known piece of phosphorus chemistry is the functionalization of a P–H bond by direct reaction with a twofold excess of formaldehyde in acidic conditions, which leads to the formation of an air- and moisture-stable phosphonium salt [60, 61]. Subsequent treatment of a phosphonium salt with a base was followed by the addition of a nucleophile, typically an amine (including ammonia, primary, or secondary amines) [62]. In this way, achiral mono- [63], di-[60, 64], and tridentate [65] ligands have been synthesized by combining two modules, namely the phosphine unit and the amine group.

We therefore focused on the development of a modular synthesis of bi- and tridentate chiral phospholane ligands and selected the 2,5-dimethyl (DMP) and 2,5-diphenyl (DPP) substituted phospholanes as stereodirecting groups, as these two privileged chiral units had shown remarkable reactivity and enantioselectivities with special impact in catalytic asymmetric hydrogenations [66].

The one-pot synthesis of a large family of bi- and tridentate phospholanes represented in Scheme 15.16 employed the air-stable chiral phospholanium chloride salts and primary amines or NH_4Cl as starting materials. These were transformed into the C_2-symmetric dimethylphospholane and diphenylphospholane ligands which reacted with $[Rh(COD)_2]BF_4$ to yield the rhodium complexes $[Rh(L)(COD)]BF_4$ (L = DMP or DPP bisphospholane ligands) [67].

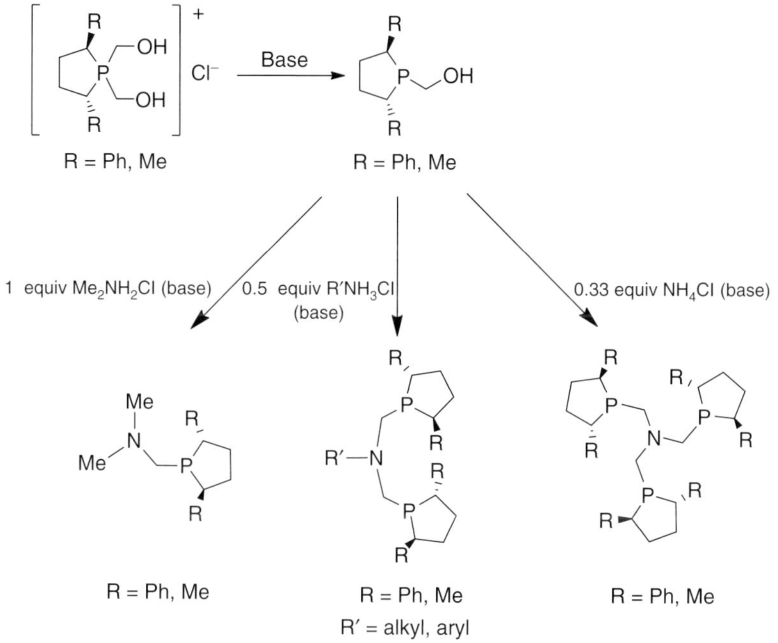

Scheme 15.16 Synthesis of mono-, di-, and tridentate phospholane ligands using amines (or NH_4Cl) and phospholanium salts as key intermediates [67].

The corresponding trisphospholane complexes were also obtained in high yields. In the derivative containing the tridentate 2,5-dimethylphospholane, the tripod ligand adopted a symmetrical tridentate binding form (Figure 15.6). A different coordination behavior was found in the case of the diphenyl-substituted (DPP) complex in which the coordination of only two of the three phospholane moieties to the metal center was observed.

The noncoordinated phospholane unit present in the DPP-rhodium complex allowed further modification through the coordination of Au(I)-X (X = Cl, C_6F_5, and ^{F}Mes) fragments to the pendant phosphine (Scheme 15.17).

To investigate the potential of the new ligands, the enantioselective hydrogenation of a series of prochiral olefins, serving as benchmark substrates using isolated Rh complexes as catalysts, were studied. The substrates included methyl esters of three dehydro-α-acetamido acids, namely methyl (*Z*)-α-acetamidocinnamate

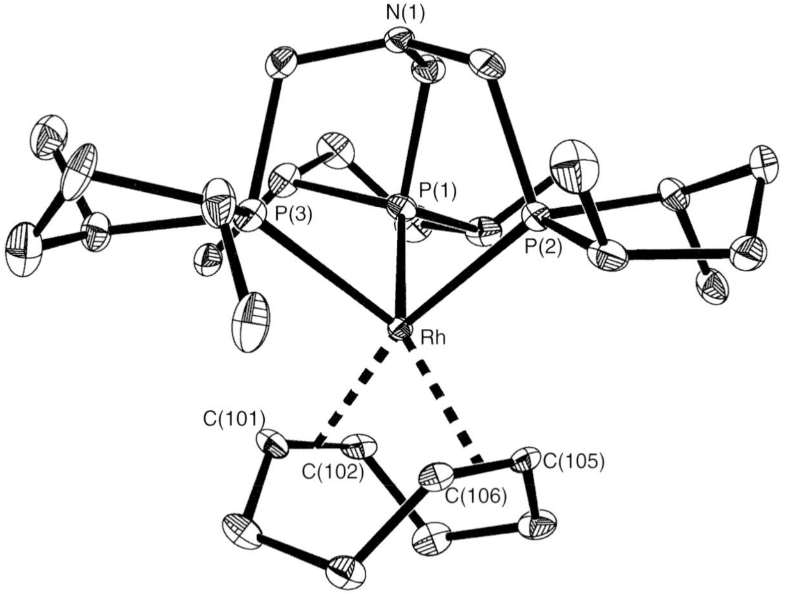

Figure 15.6 Molecular structure of a rhodium(I)–COD complex bearing the tripodal DME-derived trisphospholane [67].

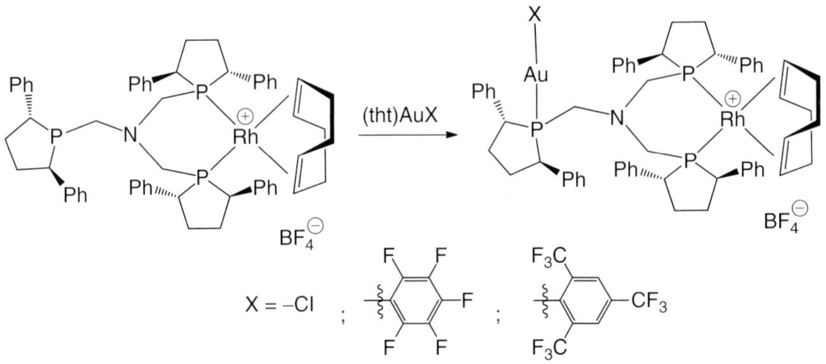

Scheme 15.17 Synthesis of heterodinuclear Rh–Au complexes [67].

(MAC), methyl 2-acetamidoacrylate (MAA), and methyl 2-acetamidobutenoate (MAB) (Figure 15.7) as well as two itaconic acid derivatives, namely dimethyl itaconate (DMI) and (*E*)-dimethyl-2-phenyl itaconate (DMPI). In general, good to excellent enantioselectivities (of up to >99% ee) were observed. Ligand backbone modification by coordination of bulky Au–X substituents to the free phospholane unit of the tripodal DPP ligand led to a remarkable enhancement of the catalyst

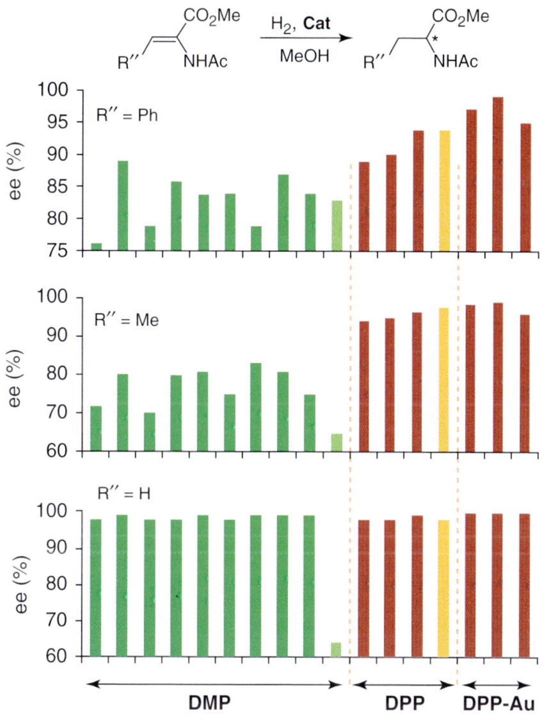

Figure 15.7 Summary of enantioselectivities of the different catalyts for (a) MAC, (b) MAB, (c) and MAA. A clear dependence on the substituents of phospholane and amine is observed [67].

performance, suggesting a clear correlation between the properties of the complex periphery and the enantioselectivity of the catalyst [67].

15.5.1
Cyclohydroaminations of γ-Allenyl Sulfonamides with Mono-, Bis-, and Trisphospholane Gold(I) Catalysts

A series of chiral mono, di-, and trinuclear gold(I) phospholane complexes were prepared, providing the first examples of chiral trinuclear gold(I) complexes employed in enantioselective catalysis (Scheme 15.18). This opened up the possibility to study a series of chiral catalysts with increasing nuclearity combining *symmetry-equivalent* sites [68].

The gold(I)-catalyzed cycloamination of γ-allenyl sulfonamides as reference reaction allowed us to study the influence of the nuclearity of the catalysts on their catalytic performance. Interestingly, the C_3-symmetric *trinuclear* complex was found to surpass its mono- and dinuclear congeners both in terms of activity and enantioselectivity (up to 95% ee) (Scheme 15.19).

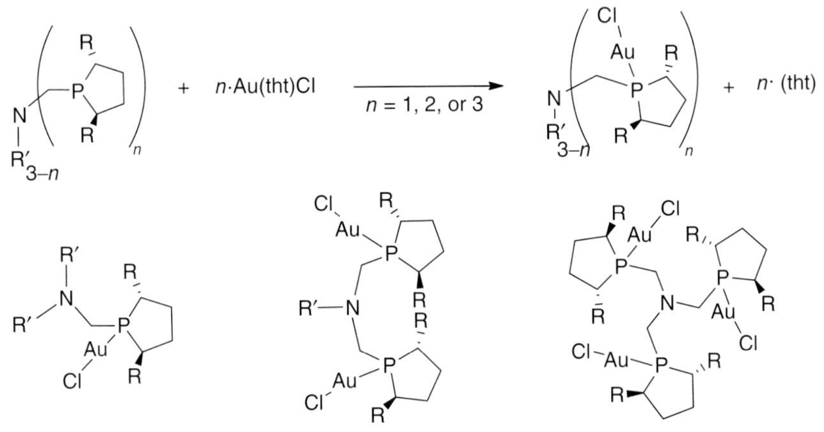

Scheme 15.18 General synthesis of chiral mono-, di-, and trinuclear phospholane gold(I) complexes [68].

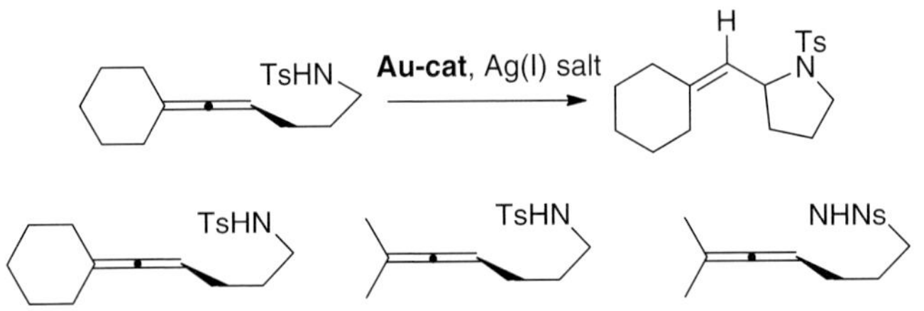

s1 s2 s3

Scheme 15.19 Top: The asymmetric cyclohydroamination reaction of allene **s1** employing the trinuclear gold complex as catalyst. Bottom: Allenes employed in this study.

While the C_3-symmetric complex provided high enantioselectivities (up to 95% ee), the corresponding dinuclear complex gave moderate enantiomeric excess values (22–40% ee). Surprisingly, the mononuclear complex produced racemic mixtures when substrate **s1** was studied but, unexpectedly, yielded the opposite enantiomer of the desired pyrrolidine derived from **s2** and **s3**. Figure 15.8 illustrates the trend observed in the enantioselectivity versus the DPP gold complex employed.

A similar trend was observed in a series of mono-, di-, and trinuclear 2,5-dimethylphospholane (DMP) gold(I) complexes. Aurophilic interactions were established in the solid-state structures of the trinuclear gold(I) complexes [68], raising the question whether these secondary forces are responsible for the different catalytic behavior observed.

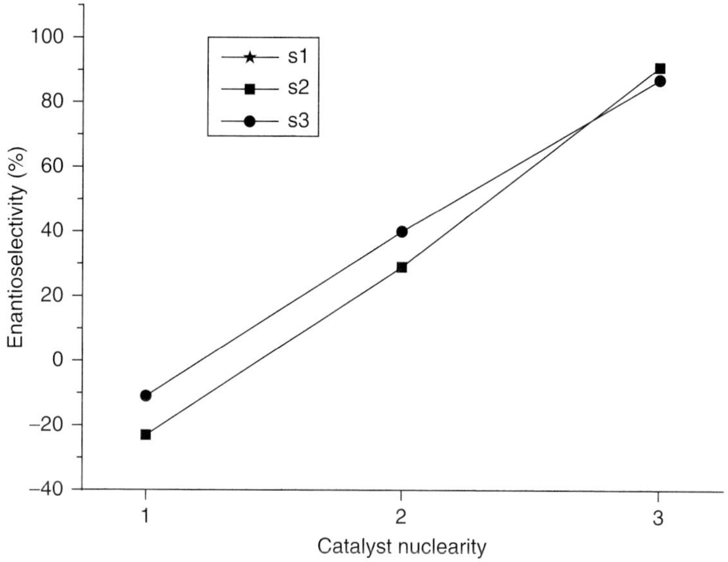

Figure 15.8 Enantioselectivity versus nuclearity of the gold complexes employed. Ag-*para*-nitrobenzoate was used as the halide abstractor [68].

References

1. (a) Jacobsen, E.N., Pfaltz, A., and Yamamoto, H. (eds) (2000) *Comprehensive Asymmetric Catalysis I-III*, Springer, Heidelberg; (b) Blaser, H.-U. and Federsel, H.J. (eds) (2010) *Asymmetric Catalysis on Industrial Scale*, 2nd edn, Wiley-VCH Verlag GmbH, Weinheim; (c) Ojima, I. (ed.) (2010) *Catalytic Asymmetric Synthesis*, 3rd edn, John Wiley & Sons, Inc., Hoboken, NJ; (d) Caprio, V. and Williams, J.M.J. (eds) (2009) *Catalysis in Asymmetric Synthesis*, John Wiley & Sons, Ltd, Chichester; (e) Mikami, K. and Lautens, M. (eds) (2007) *New Frontiers in Asymmetric Catalysis*, John Wiley & Sons, Ltd, Chichester; (f) Noyori, R. (1994) *Asymmetric Catalysis in Organic Synthesis*, John Wiley & Sons, Inc., New York.
2. Yoon, T.P. and Jacobsen, E.N. (2003) *Science*, **299**, 1691–1693.
3. (a) Gennari, C. and Piarulli, U. (2003) *Chem. Rev.*, **103**, 3071–3100; (b) Hoveyda, A. (2002) in *Handbook of Combinatorial Chemistry*, vol. 2 (eds K.C. Nicolaou, R. Hanko, and W. Hartwig), Wiley-VCH Verlag GmbH, Weinheim, pp. 991–1016; (c) Reetz, M.T. (2002) *Angew. Chem. Int. Ed.*, **41**, 1335–1338; (d) Carboni, S., Gennari, C., Pignataro, L., and Piarulli, U. (2011) *Dalton Trans.*, **40**, 4355–4373; (e) van Leeuwen, P.W.N.M. (ed.) (2008) *Supramolecular Catalysis*, Wiley-VCH Verlag GmbH, Weinheim; (f) Goudriaan, P.E., van Leeuwen, P.W.N.M., Birkholz, M.-N., and Reek, J.N.H. (2008) *Eur. J. Inorg. Chem.*, **2008**, 2939–2958; (g) Burk, M.J. (2000) *Acc. Chem. Res.*, **33**, 363–372.
4. Hoveyda, A.H., Hird, A.W., and Kaprzynski, M.A. (2004) *Chem. Commun.*, 1779.
5. See for example: (a) Mc Manus, H.A. and Guiry, P.J. (2004) *Chem. Rev.*, **104**, 4151; (b) Helmchen, G. and Pfaltz, A. (2000) *Acc. Chem. Res.*, **33**, 336.

6. The exploitation of C_2-symmetry in enantioselective catalysis was first reported (and explicitly formulated) by Kagan: (a) Kagan, H.B. and Dang, T.P. (1972) *J. Am. Chem. Soc.*, **94**, 6429; Reviews: (b) Whitesell, J.K. (1989) *Chem. Rev.*, **89**, 1581; (c) Brunner, H. (1988) *Top. Stereochem.*, **18**, 129.
7. For the exploitation of C_3-symmetry, see: (a) Moberg, C. (1998) *Angew. Chem. Int. Ed.*, **37**, 248; (b) Keyes, M.C. and Tolman, W.B. (1997) *Adv. Catal. Proc.*, **2**, 189.
8. (a) McManus, H.A. and Guiry, P.J. (2004) *Chem. Rev.*, **104**, 4151; (b) Desimoni, G., Faita, G., and Jørgensen, K.A. (2006) *Chem. Rev.*, **106**, 3561; (c) Desimoni, G., Faita, G., and Quadrelli, P. (2003) *Chem. Rev.*, **103**, 3119; (d) Desimoni, G., Faita, G., and Jørgensen, K.A. (2012) *Chem. Rev.*, **112**, PR284; (e) Meyers, A.I. (2005) *J. Org. Chem.*, **70**, 6137; (f) Rechavi, D. and Lemaire, M. (2002) *Chem. Rev.*, **102**, 3467; (g) Hargaden, G.C. and Guiry, P.J. (2009) *Chem. Rev.*, **109**, 2505.
9. (a) Lowenthal, R.E., Abiko, A., and Masamune, S. (1990) *Tetrahedron Lett.*, **31**, 6005; (b) Evans, D.A., Woerpel, K.A., Hinman, M.M., and Faul, M.M. (1991) *J. Am. Chem. Soc.*, **113**, 726; (c) Corey, E.J., Imai, N., and Zhang, Y.H. (1991) *J. Am. Chem. Soc.*, **113**, 728; (d) Müller, D., Umbricht, G., Weber, B., and Pfaltz, A. (1991) *Helv. Chim. Acta*, **74**, 232; (e) Nishiyama, H., Sakaguchi, H., Nakamura, T., Horihata, M., Kondo, M., and Itoh, K. (1989) *Organometallics*, **8**, 846.
10. (a) Kawasaki, K., Tsumura, S., and Katsuki, T. (1995) *Synlett*, 1245; (b) Kohmura, Y. and Katsuki, T. (2000) *Tetrahedron Lett.*, **41**, 3941.
11. Zhou, J. and Tang, Y. (2005) *Chem. Soc. Rev.*, **34**, 664.
12. Bellemin-Laponnaz, S. and Gade, L.H. (2002) *Chem. Commun.*, 1286.
13. (a) Bellemin-Laponnaz, S. and Gade, L.H. (2002) *Angew. Chem. Int. Ed.*, **41**, 3473; (b)Review:Gade, L.H. and Bellemin-Laponnaz, S. (2008) *Chem. Eur J.*, **14**, 4142.
14. Gade, L.H., Marconi, G., Dro, C., Ward, B.D., Poyatos, M., Bellemin-Laponnaz, S., Wadepohl, H., Caneschi, A., and Poneti, G. (2007) *Chem. Eur. J.*, **13**, 3058.
15. (a) Vahrenkamp, H. (1999) *Acc. Chem. Res.*, **32**, 589; (b) Parkin, G. (2000) *Chem. Commun.*, 1971; (c) Parkin, G. (2004) *Chem. Rev.*, **104**, 699.
16. Dro, C., Bellemin-Laponnaz, S., Welter, R., and Gade, L.H. (2004) *Angew. Chem. Int. Ed.*, **43**, 4479.
17. (a) Kaminsky, W. and Laban, A. (2001) *Appl. Catal. Gen.*, **222**, 47; (b) Bochmann, M. (1996) *J. Chem. Soc., Dalton Trans.*, 255; (c) Brintzinger, H.H., Fischer, D., Mülhaupt, R., Rieger, B., and Waymouth, R.M. (1995) *Angew. Chem., Int. Ed. Engl.*, **34**, 1143.
18. (a) Ward, B.D., Bellemin-Laponnaz, S., and Gade, L.H. (2005) *Angew. Chem. Int. Ed.*, **44**, 1668; (b) Ward, B.D., Lukesova, L., Wadepohl, H., Bellemin-Lapponaz, S., and Gade, L.H. (2009) *Eur. J. Inorg. Chem*, **2009**, 866.
19. Arndt, S., Spaniol, T.P., and Okuda, J. (2003) *Angew. Chem. Int. Ed.*, **42**, 5075.
20. Tshuva, E.Y., Goldberg, I., and Kol, M. (2000) *J. Am. Chem. Soc.*, **122**, 10706.
21. (a) Lukešová, L., Ward, B.D., Bellemin-Laponnaz, S., Wadepohl, H., and Gade, L.H. (2007) *Dalton Trans.*, 920; (b) Lukešová, L., Ward, B.D., Bellemin-Laponnaz, S., Wadepohl, H., and Gade, L.H. (2007) *Organometallics*, **26**, 4652; (c)Review:Ward, B.D. and Gade, L.H. (2012) *Chem. Commun.*, **48**, 10587.
22. Langlotz, B.K., Wadepohl, H., and Gade, L.H. (2008) *Angew. Chem. Int. Ed.*, **47**, 4670.
23. Thorhauge, J., Roberson, M., Hazell, R.G., and Jørgensen, K.A. (2002) *Chem. Eur. J.*, **8**, 1888.
24. (a) Foltz, C., Stecker, B., Marconi, G., Wadepohl, H., Bellemin-Laponnaz, S., and Gade, L.H. (2005) *Chem. Commun.*, 5115; (b) Foltz, C., Stecker, B., Marconi, G., Bellemin-Laponnaz, S., Wadepohl, H., and Gade, L.H. (2007) *Chem. Eur. J.*, **13**, 9912.
25. Marigo, M., Kjærsgaard, A., Juhl, K., Gathergood, N., and Jørgensen, K.A. (2003) *Chem. Eur. J.*, **9**, 2359.
26. Marigo, M., Juhl, K., and Jørgensen, K.A. (2003) *Angew. Chem. Int. Ed.*, **42**, 1367.

27. (a) Trost, B.M. and Van Vranken, D.L. (1996) *Chem. Rev.*, **96**, 395; (b) Helmchen, G. (2004) *Pure Appl. Chem.*, **76**, 495; (c) Graening, T. and Schmalz, H.–.G. (2003) *Angew. Chem. Int. Ed.*, **42**, 2580.
28. von Matt, P., Lloyd-Jones, G.C., Minidis, A.B.E., Pfaltz, A., Macko, L., Neuburger, M., Zehnder, M., Rüeger, H., and Pregosin, P.S. (1995) *Helv. Chim. Acta*, **78**, 265.
29. Foltz, C., Enders, M., Bellemin-Laponnaz, S., Wadepohl, H., and Gade, L.H. (2007) *Chem. Eur. J.*, **13**, 5994.
30. Reviews covering "pincer" complex chemistry: (a) Albrecht, M. and van Koten, G. (2001) *Angew. Chem. Int. Ed.*, **40**, 3750; (b) van der Boom, M. and Milstein, D. (2003) *Chem. Rev.*, **103**, 1759; Selected references for chiral pincer-based catalysts: (c) Williams, B.S., Dani, P., Lutz, M., Spek, A.L., and Van Koten, G. (2001) *Helv. Chim. Act.*, **84**, 3519; (d) Soro, B., Stoccoro, S., Minghetti, G., Zucca, A., Cinellu, M.A., Manassero, M., and Gladiali, S. (2006) *Inorg. Chim. Act.*, **359**, 1879; (e) Aydin, J., Kumar, K.S., Sayah, M.J., Wallner, O.A., and Szabo, K.J. (2007) *J. Org. Chem.*, **72**, 4689.
31. (a) Jacobsen, E.N., Pfaltz, A., and Yamamoto, H. (eds) (1999) *Comprehensive Asymmetric Catalysis*, Springer, Berlin; (b) Noyori, R. (1994) *Asymmetric Catalysis in Organic Synthesis*, John Wiley & Sons, Inc., New York; (c) Beller, M. and Bolm, C. (eds) (2004) *Transition Metals for Organic Synthesis*, 2nd edn, Wiley-VCH Verlag GmbH, Weinheim; (d) Mikami, K. and Lautens, M. (eds) (2007) *New Frontiers in Asymmetric Catalysis*, John Wiley & Sons, Inc., Hoboken, NJ; (e) Caprio, V. and Williams, J.M.J. (2009) *Catalysis in Asymmetric Synthesis*, 2nd edn, John Wiley & Sons, Inc., Hoboken, NJ; (f) Ojima, I. (ed.) (2009) *Catalytic Asymmetric Synthesis*, 3rd edn, John Wiley & Sons, Inc., Hoboken, NJ.
32. (a) Nishiyama, H. and Ito, J.-I. (2010) *Chem. Commun.*, **46**, 203; (b) Nishiyama, H. and Ito, J.-i. (2007) *Chem. Rec.*, **7**, 159; (c) Nishiyama, H. (2007) *Chem. Soc. Rev.*, **36**, 1133.
33. (a) Inoue, M., Suzuki, T., and Nakada, M. (2003) *J. Am. Chem. Soc.*, **125**, 1140; (b) Inoue, M. and Nakada, M. (2004) *Org. Lett.*, **6**, 2977; (c) Inoue, M. and Nakada, M. (2005) *Angew. Chem. Int. Ed.*, **45**, 252; (d) Inoue, M. and Nakada, M. (2007) *J. Am. Chem. Soc.*, **129**, 4164; (e) Inoue, M., Suzuki, T., Kinoshita, A., and Nakada, M. (2008) *Chem. Rec.*, **8**, 169.
34. Deng, Q.-H., Wadepohl, H., and Gade, L.H. (2011) *Chem. Eur. J.*, **17**, 14922.
35. Deng, Q.-H., Wadepohl, H., and Gade, L.H. (2012) *J. Am. Chem. Soc.*, **134**, 2946.
36. Deng, Q.-H., Bleith, T., Wadepohl, H., and Gade, L.H. (2013) *J. Am. Chem. Soc.*, **135**, 5356.
37. (a) Bonnet-Delpon, D. (2008) *Bioorganic and Medicinal Chemistry of Fluorine*, John Wiley & Sons, Inc., Hoboken, NJ; (b) Purser, S., Moore, P.R., Swallow, S., and Gouverneur, V. (2008) *Chem. Soc. Rev.*, **37**, 320; (c) Ojima, I. (2009) *Fluorine in Medicinal Chemistry and Chemical Biology*, Wiley-Blackwell, Chichester; (d) Lin, G.-Q., You, Q.-D., and Cheng, J.-F. (2011) *Chiral Drugs: Chemistry and Biological Action*, John Wiley & Sons, Inc., Hoboken, NJ.
38. For recent reviews on trifluoromethylation reactions, see: (a) Prakash, G.K.S. and Hu, J. (2007) *Acc. Chem. Res.*, **40**, 921; (b) Ma, J.-A. and Cahard, D. (2007) *J. Fluorine Chem.*, **128**, 975; (c) Uneyama, K., Katagiri, T., and Amii, H. (2008) *Acc. Chem. Res.*, **41**, 817; (d) Lundgren, R.J. and Stradiotto, M. (2010) *Angew. Chem. Int. Ed.*, **49**, 9322; (e) Tomashenko, O.A. and Grushin, V.V. (2011) *Chem. Rev.*, **111**, 4475; (f) Parsons, A.T. and Buchwald, S.L. (2011) *Nature*, **480**, 184; (g) Furuya, T., Kamlet, A.S., and Ritter, T. (2011) *Nature*, **473**, 470; (h) Besset, T., Schneider, C., and Cahard, D. (2012) *Angew. Chem. Int. Ed.*, **51**, 5048.
39. (a)For reviews on asymmetric trifluoromethylation reactions, see: Shibata, N., Mizuta, S., and Kawai, H. (2008) *Tetrahedron: Asymmetry*, **19**, 2633; (b) Ma, J.-A. and Cahard, D. (2008) *Chem. Rev.*,

108, PR1; (c) Zheng, Y. and Ma, J.-A. (2010) *Adv. Synth. Catal.*, **352**, 2745; (d) Valero, G., Companyó, X., and Rios, R. (2011) *Chem. Eur. J.*, **17**, 2018.

40. (a) Eisenberger, P., Gischig, S., and Togni, A. (2006) *Chem. Eur. J.*, **12**, 2579; (b) Kieltsch, I., Eisenberger, P., and Togni, A. (2007) *Angew. Chem. Int. Ed.*, **46**, 754.

41. Deng, Q.-H., Wadepohl, H., and Gade, L.H. (2012) *J. Am. Chem. Soc.*, **134**, 10769.

42. Bräse, S., Banert, K., and Organic Azides, K. (2009) *Syntheses and Applications*, John Wiley & Sons, Ltd, Chichester.

43. (a) Zhdankin, V.V., Krasutsky, A.P., Kuehl, C.J., Simonsen, A.J., Woodward, J.K., Mismash, B., and Bolz, J.T. (1996) *J. Am. Chem. Soc.*, **118**, 5192; (b) Akai, S., Okuno, T., Takada, T., Tohma, H., and Kita, Y. (1996) *Heterocycles*, **42**, 47.

44. Gade, L.H. and Bellemin-Laponnaz, S. (2007) *Coord. Chem. Rev.*, 718.

45. aReviews: (a) Arduengo, A.J. III, (1999) *Acc. Chem. Res.*, **32**, 913; (b) Bourissou, D., Guerret, O., Gabbaï, F., and Bertrand, G. (2000) *Chem. Rev.*, **100**, 39; (c) Herrmann, W.A. and Köcher, C. (1997) *Angew. Chem., Int. Ed. Engl.*, **36**, 2162; (d) Glorius, F. (2007) *Top. Organomet. Chem.*, **21**, 1.

46. Reviews: (a) Cavallo, L., Correa, A., Costabile, C., and Jacobsen, H. (2005) *J. Organomet. Chem.*, **690**, 5407; (b) Crudden, C.M. and Allen, D.P. (2004) *Coord. Chem. Rev.*, **248**, 2247; (c) Díez-González, S. and Nolan, S.P. (2007) *Coord. Chem. Rev.*, **251**, 874.

47. Reviews: (a) Herrmann, W.A. (2002) *Angew. Chem. Int. Ed.*, **41**, 1290; (b) Colacino, E., Martinez, J., and Lamaty, F. (2007) *Coord. Chem. Rev.*, **251**, 726; (c) Kantchev, E.A.B., O'Brien, C.J., and Organ, M.G. (2007) *Angew. Chem. Int. Ed.*, **46**, 2768; (d) Bellemin-Laponnaz, S. and Gade, L.H. (2007) *Top. Organomet. Chem.*, **21**, 117.

48. Herrmann, W.A., Goossen, L.J., and Spiegler, M. (1998) *Organometallics*, **17**, 2162.

49. Goossen, L.J. (1997) Doktorarbeit, Technische Universität München.

50. (a) Powell, M.T., Hou, D.-R., Perry, M.C., Cui, X., and Burgess, K. (2001) *J. Am. Chem. Soc.*, **123**, 8878; (b) Perry, M.C., Cui, X., Powell, M.T., Hou, D.-R., Reibenspies, J.H., and Burgess, K. (2003) *J. Am. Chem. Soc.*, **125**, 113.

51. Review: Helmchen, G. and Pfaltz, A. (2000) *Acc. Chem. Res.*, **33**, 336.

52. (a) Hou, D.-R. and Burgess, K. (1999) *Org. Lett.*, **1**, 1745; (b) Hou, D.-R., Reibenspies, J., and Burgess, K. (2001) *J. Org. Chem.*, **66**, 206.

53. Fan, Y., Cui, X., Burgess, K., and Hall, M.B. (2004) *J. Am. Chem. Soc.*, **126**, 16688.

54. (a) Cui, X. and Burgess, K. (2003) *J. Am. Chem. Soc.*, **125**, 14212; (b) Cui, X., Fan, Y., Hall, M.B., and Burgess, K. (2005) *Chem. Eur. J.*, **11**, 6859.

55. César, V., Bellemin-Laponnaz, S., and Gade, L.H. (2002) *Organometallics*, **21**, 5204.

56. César, V., Bellemin-Laponnaz, S., and Gade, L.H. (2004) *Eur J. Inorg. Chem.*, **2004**, 3436.

57. César, V., Bellemin-Laponnaz, S., and Gade, L.H. (2004) *Angew. Chem. Int. Ed.*, **43**, 1014.

58. César, V., Bellemin-Laponnaz, S., Wadepohl, H., and Gade, L.H. (2005) *Chem. Eur. J.*, **11**, 2862.

59. (a) Schneider, N., Finger, M., Haferkemper, C., Bellemin-Laponnaz, S., Hofmann, P., and Gade, L.H. (2009) *Angew. Chem. Int. Ed.*, **48**, 1609; (b) Schneider, N., Finger, M., Haferkemper, C., Bellemin-Laponnaz, S., Hofmann, P., and Gade, L.H. (2009) *Chem. Eur. J.*, **15**, 11515.

60. Fawcett, J., Hoye, P.A.T., Kemmitt, R.D.W., Law, D.J., and Russell, D.R. (1993) *J. Chem. Soc., Dalton Trans.*, 2563.

61. (a) Buckler, S.A. and Epstein, M. (1961) *J. Am. Chem. Soc.*, **27**, 1090; (b) Grim, S.O., Shah, D.P., and Matienzo, L.J. (1980) *Inorg. Chem.*, **19**, 2475.

62. Maggini, S. (2009) *Coord. Chem. Rev.*, **253**, 1793.

63. Clarke, D.A., Miller, P.W., Long, N.J., and White, A.J.P. (2007) *Dalton Trans.*, **40**, 4556.

64. Keles, M., Aydin, Z., and Serindag, O. (2007) *J. Organomet. Chem.*, **692**, 1951.

65. Miller, P.W. and White, A.J.P. (2010) *J. Organomet. Chem.*, **695**, 1138.
66. (a) Burk, M.J., Gross, M.F., and Martinez, J.P. (1995) *J. Am. Chem. Soc.*, **117**, 9375; (b) Burk, M.J., Wang, Y.M., and Lee, J.R. (1996) *J. Am. Chem. Soc.*, **118**, 5142; (c) Heller, D., Holz, J., Drexler, H.-J., Lang, J., Drauz, K., Krimmer, H.-P., and Börner, A. (2001) *J. Org. Chem.*, **66**, 6816; (d) Burk, M.J. (1991) *J. Am. Chem. Soc.*, **113**, 8518; (e) Boaz, N.W. (1998) *Tetrahedron Lett.*, **39**, 5505; (f) Pilkington, C.J. and Zanotti-Gerosa, A. (2003) *Org. Lett.*, **5**, 1273; (g) Fox, M.E., Jackson, M., Lennon, I.C., Klosin, J., and Abboud, K.A. (2008) *J. Org. Chem.*, **73**, 775.
67. Lloret Fillol, J., Kruckenberg, A., Scherl, P., Wadepohl, H., and Gade, L.H. (2011) *Chem. Eur. J.*, **17**, 14047.
68. Rodríguez, L.-I., Roth, T., Lloret Fillol, J., Wadepohl, H., and Gade, L.H. (2012) *Chem. Eur. J.*, **18**, 3721.

Part IV
Structures and Mechanisms in Biological Systems

Andres Jäschke

Biocatalysis is a central research area at the interface of chemistry and biology. All essential processes of life are accelerated by enzymes, and the high efficiency and enormous selectivity of enzymes have fascinated generations of scientists. Understanding how metabolically relevant enzymes work has therefore been at the core of biochemical research for decades, and scientists have established an ever-growing toolbox to study and tailor biocatalysts. Numerous Nobel Prizes, both in chemistry and in physiology or medicine, have been awarded for ground-breaking studies on the roles, structures, or mechanisms of various enzymes.

Furthermore, biocatalysis has become an important tool for industrial chemical synthesis and sustains significant growth. In the past half century, many biocatalytic processes have been implemented to produce a wide variety of products in various industries. While many of them use naturally occurring enzymes or microorganisms as catalysts, progress in genome and metagenome sequencing, protein and metabolic engineering, as well as directed evolution has provided rich and novel sources for biocatalysts with the desired selectivities, high turnover numbers and frequencies, and sufficient stability. Significant progress has also been made recently in the *de novo* design of proteins with enzymatic properties by computational methods so that it is conceivable that soon biocatalysts can be rationally designed to act specifically in any chemical reaction of interest, fulfilling the holy grail of catalysis: catalysis by design.

An increased understanding of the structure and function of biological catalysts has also inspired scientists to apply the underlying principles to other artificial systems, or to use biopolymers in ways or for purposes not known in Nature. This field of bio-inspired catalysis has blossomed in recent years, and one example is the development of hybrid catalysts that combine homogeneous transition-metal catalysis and biocatalysis in a novel manner. These developments hold great promise, as they may allow in the future new chemical transformations under mild conditions and a better integration of chemical and biochemical reaction steps in practical syntheses.

In view of the high scientific interest and industrial relevance of biocatalysis, the investigation of enzyme mechanisms and the development of techniques for this purpose are of utmost importance. In the scientific context of the Heidelberg

Molecular Catalysts: Structure and Functional Design, First Edition. Edited by Lutz H. Gade and Peter Hofmann.

SFB "Molecular Catalysts: Structure and Functional Design," biological and bio-inspired catalysis therefore formed one of the pillars of scientific activities.

The following three chapters describe very different systems and methods in the field of biological and bio-inspired catalysis. In Chapter 16, Schlichting reviews structural investigations on hemoproteins of the cytochrome P450 group, an important class of metabolic enzymes. Particular attention is paid to identify the reaction intermediates in the catalytic cycle of two different cytochrome enzymes, namely $P450_{cam}$, a cytochrome enzyme that catalyzes the enantio- and regiospecific hydroxylation of camphor, and chloroperoxidase, which halogenates a wide range of substrates and has catalase, monooxygenase, and dehydrogenase activity. The focus of this chapter is on the interaction of photoelectrons with the objects to be studied: on one hand side they allow – in X-ray crystallography – detailed insight into macromolecular structures; on the other, they induce severe damages to the objects to be studied. The chapter also discusses the exciting developments of X-ray free- electron lasers that may ultimately allow the solution of 3D structures of macromolecules from individual molecules, rather than from crystal lattices.

In Chapter 17, the tools used are different. Kiani and Fischer apply the recently established discipline of computational enzymology to an essential reaction in biology, namely the hydrolysis of the P–O bond by two different enzymes: the restriction endonuclease *Eco*RV, and the molecular motor protein myosin. The authors present the combined quantum mechanical/classical simulations of DNA cleavage in *Eco*RV and ATP hydrolysis in myosin, and show that, although the structure and function of these two enzymes are dissimilar, their fundamental catalytic strategy has striking similarities. Both enzymes cleave their respective substrates through a three-pronged approach: (i) P–O bond dissociation by protein stabilization of a trigonal-planar metaphosphate, (ii) water activation assisted by a surrounding basic protein residue to generate a hydroxyl group, and (iii) nucleophilic attack of the metaphosphate by the hydroxyl.

In Chapter 18, the field of bioinspired catalysis is the focus, and we report on the generation of hybrid catalysts based on nucleic acids. Hybrid catalysis combines homogeneous chemical catalysts with biopolymers to obtain selective catalysts for organic reactions. While proteins have been used as hosts for various transition-metal complexes, relatively few published examples are based on nucleic acids. We present different DNA-based systems for the catalysis of organic reactions in aqueous media. After a brief review of the field, different methods for the site-specific modification of DNA with transition metal–ligand complexes and other moieties are discussed. We use Ir(I)–diene chemistry to catalyze an allylic substitution in aqueous medium. Toward this end, we covalently attach transition-metal ligands, such as phosphinoxazoline and diene ligands, to specific positions of oligonucleotides. Our approach uses a modular design where an oligonucleotide carrying a transition-metal ligand is combined with different DNA or RNA counter-strands, thereby forming perfect and imperfect duplexes, triplexes, and quadruplexes with or without bulges of varying size, thereby providing changes in the environment of the metal center. The covalent attachment of the ligand guarantees its specific, reproducible positioning on nucleic acid structures. We demonstrate that catalysis occurs in the presence of DNA and its numerous functional groups, and that the structure of the DNA modulates the stereochemical outcome of the reaction.

16
Beating and Employing X-Ray-Induced Radiation Damage in Structural Studies of Hemoproteins

Ilme Schlichting

16.1
Introduction

Metalloproteins are involved in many diverse biological processes despite their often highly similar cofactors. Oxygen-binding hemoproteins can be classified according to their functions in enzymes: for example, proteins that use oxygen for chemical reactions, storage proteins that "simply" bind oxygen for storage or transportation, and sensors that change their conformation upon oxygen binding thereby (in)activating a downstream signaling partner. To a large extent, this difference is rooted in the nature of the protein ligand involved in the proximal coordination of the heme iron, the architecture of the distal coordination site, and the use of different oxidation states of the heme iron. Thus, the iron may serve as a source or sink of electrons during electron transfer or redox chemistry. In addition, also the porphyrin ring may serve as an electron source, for example, in peroxidase reactions. While a great deal of information on the interactions of the heme cofactor, the protein, and the ligand and their changes during the function can be obtained by various spectroscopic methods, important insight is provided by crystallographic analysis. Despite the notion that this approach is rather slow, providing temporal averages of the conformations of the molecules making up the crystal over the time used for data collection, X-ray crystallography can be used for structure determination of short-lived species such as those occurring during a chemical reaction. In the following, we will describe the approaches used and the challenges encountered in the structural characterization of activated oxygen species occurring during the reaction pathways of cytochrome P450 enzymes [1]. As described here for P450 enzymes, many investigations on metalloproteins are afflicted by their photoreduction by the X-ray beam [2] and other forms of radiation damage. Since radiation damage may be "outrun" by using X-ray free-electron laser (X-FEL) sources, we also summarize their emerging potential for structural biology [3, 4].

Molecular Catalysts: Structure and Functional Design, First Edition. Edited by Lutz H. Gade and Peter Hofmann.

16.2 Cytochrome P450 Enzymes

Cytochrome P450s are ubiquitous hemoproteins that are named after their absorption band at 450 nm when complexed with carbon monoxide. This unusual feature is caused by a protein thiolate serving as the proximal ligand of the heme iron. It endows P450 enzymes with unique catalytic capabilities, such as their ability to hydroxylate nonactivated hydrocarbons. For this reason, P450s have been likened to molecular blow torches [5]. P450 enzymes form a large superfamily of hemoproteins [6] that have been found in all branches of life. Their biological functions can be divided in two main roles: the degradation of xenobiotics, substances that are foreign to the organism; and the biosynthesis and metabolism of a number of essential compounds such as steroid hormones, fat soluble vitamins, and the conversion of polyunsaturated fatty acids into biologically active ones [7]. In line with these different tasks, a large number of catalytic activities have been identified [8, 9]. In general, P450 enzymes are best known for their monooxygenase activity. P450s activate molecular oxygen for the incorporation of one oxygen atom in the substrate, while the second oxygen atom is two-electron reduced and forms a water molecule. The stepwise reaction is shown in detail in Figure 16.1: (i) oxygen binds to the reduced heme iron (2) and an Fe^{2+}-OO or Fe^{3+}-OO^- complex (3) is formed. (ii) Upon one-electron reduction of this complex to Fe^{3+}-OO^{2-}, a ferric-peroxo (5a) complex is formed, which is easily protonated to form Fe^{3+}-OOH^-, a ferric-hydroperoxo (5b) state. (iii) Upon a second protonation of the Fe^{3+}-OOH^- complex at the distal oxygen atom to yield a transient unstable peroxo-water immediate (5c), heterolytic scission of O–O bond follows. (iv) A high-valent metal–oxo complex forms, often described as a oxo-iron(IV) porphyrin π-cation radical species ($Fe^{IV}(O)(Por^{+\bullet})$) (6), known as *compound I*. In an “oxygen-rebound mechanism” [10], compound I abstracts a hydrogen from the substrate by the terminal oxygen atom of compound I to form a caged alkyl radical and a [Fe^{IV}(OH)(Por)] intermediate, and then the hydroxide ligand is transferred to the alkyl radical, yielding the hydroxylated product (7). The details of the last steps and the nature of the species involved are still under debate [11, 12].

16.2.1 The Reaction Cycle of $P450_{cam}$ at High Structural Resolution

A great deal of information on P450 structure and mechanism has been obtained by studying the role model of P450s, namely the bacterial $P450_{cam}$ (CYP101). It catalyzes the stereo- and regiospecific hydroxylation of camphor to 5-*exo*-hydroxycamphor, thus allowing *Pseudomonas putida* to use camphor as a carbon source. For a long time, major questions regarding P450 mechanism concerned the nature of the active oxygen species and the source of the protons required for the oxygen activation described above. While the first question requires all kinds of spectroscopic analysis, great expectations were placed in crystallographic

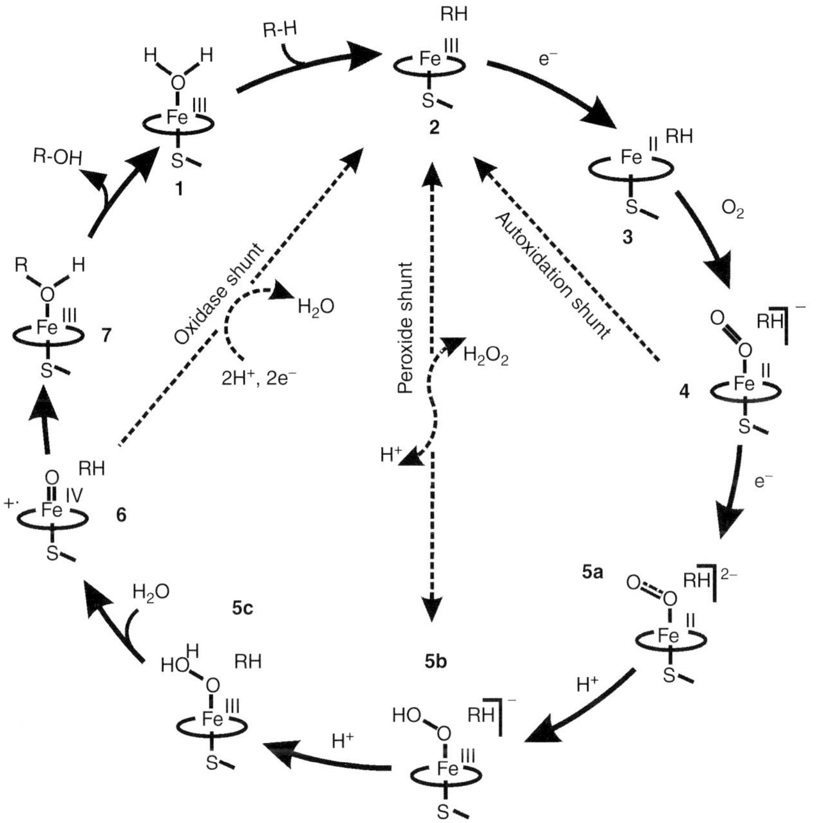

Figure 16.1 General reaction cycle of cytochrome P450 enzymes [1]. In general, the substrate binds to the low-spin ferric state (1), and the substrate complex (2) is reduced by redox partner proteins yielding the ferrous substrate complex (3) to which oxygen binds to afford the oxy complex (4). Upon a second one-electron reduction, a ferric peroxo complex forms (5a) that protonates easily to form a ferric hydroperoxo state. Upon a second protonation (5c) and heterolytic O–O bond scission, compound I, an oxo-iron(IV) porphyrin π-cation radical species is formed (6), which is believed to hydroxylate the substrate via an oxygen rebound mechanism.

analysis for addressing the latter. The carbonmonoxy complex of $P450_{cam}$ was considered a good analog for the oxygen complex, but its crystal structure did not provide any insight into the source of protons required for oxygen activation [13]. The structure determination of the oxy complex is complicated by the fact of its relatively short lifetime, about 4 min at 4 °C for $P450_{cam}$. First attempts to capture the structure of the $P450_{cam}$ oxy complex were made using the polychromatic Laue method which allows fast, snapshot-like diffraction data collection [14, 15]. $P450_{cam}$ crystals were mounted in a flow cell, reduced by dithionate, and then washed with oxygenated solution. Unexpectedly, however, the electron density did not show a terminal oxygen ligand to the heme but was more consistent with a hydroxylated camphor ligand and a five-coordinated heme. Ultimately, this was attributed to photoreduction by the X-ray beam, providing the second electron to complete the catalytic cycle to yield the product 5-hydroxy camphor. Therefore, a different approach was taken to determine the structure of the short-lived active oxy complex that does not rely on collecting the diffraction data faster than the oxy complex decays but by extending the lifetime of the oxy complex long enough for conventional monochromatic X-ray data collection approaches [16]. To this end, the oxy complex of $P450_{cam}$ was generated by briefly exposing chemically reduced crystals of $P450_{cam}$ to oxygen in a high-pressure cell before cryocooling the crystals in liquid nitrogen, thereby trapping the oxy complex. Monochromatic X-rays were used for data collection of a $P450_{cam}$ oxy crystal kept at cryogenic temperature for data collection. The electron density revealed an end-on-bound oxygen molecule and two new water molecules positioned within a hydrogen-bonding network of the bound oxygen, enabled by a structural change of the protein. Making use of the energy dependence of the photoelectric effect (inversely proportional to the cube of the photon energy), the number of photoelectrons was varied during data collection. First, diffraction data were collected at a short wavelength (high photon energy, low cross sections for electron production) to capture the structure of the $P450_{cam}$ oxy complex, and then the crystals were exposed to long wavelength X-rays (high cross sections for electron production) before another dataset was collected of a photoreduced $P450_{cam}$ state. Then the crystal was briefly thawed before a third dataset was collected. This last dataset clearly showed electron density for the product 5-*exo*-hydroxycamphor, while the electron density obtained from the second dataset was consistent with a monoatomic sixth ligand to the heme iron. Obviously, it was tempting and, given the enzymatic activity of the crystal, suggestive to tentatively assign this complex to an oxy ferryl intermediate. While the three structures (oxy, intermediate, product state) provided important insight into the source and pathway of the protons used for oxygen activation, they could not give any insight into the nature of the active oxygen species. Breakthrough mechanistic insight was obtained by analyzing a P450 enzyme derived from the thermophilic organism *Sulfolobus acidocaldarius*, CYP119 [17]. In this case, the ephemeral compound I species can be prepared with 75% yield by incubating CYP119 with *m*-chloroperbenzoic acid, and it lasts for microseconds. For the first time, this allowed high-yield accumulation by freeze–quench approaches, and thus direct analysis by UV–vis [18], Mössbauer, and electron paramagnetic resonance (EPR) spectroscopy, providing also the C–H bond activation kinetics [19].

16.2.2 Chloroperoxidase Compound

Chloroperoxidase (CPO) is one of the most versatile heme enzymes. In addition to its biological function to catalyze the hydrogen peroxide-dependent chlorination of cyclopentanedione during the biosynthesis of the antibiotic caldarioymcin [20, 21], CPO halogenates a wide range of substrates and has catalase, monoxygenase, and dehydrogenation activity. Structurally, CPO is unique, but it shares features with both peroxidases and P450 enzymes [22]. In particular, CPO and P450 enzymes share the proximal thiolate ligand to the heme iron but differ in the distal pocket which is much more polar in CPO. However, unlike other peroxidases, the normally conserved distal arginine is lacking in CPO and the catalytic acid base of CPO is a glutamic acid and not a histidine [23]. It is generally assumed that the active-site glutamate abstracts a proton from the hydrogen peroxide bound to heme, yielding the short-lived compound 0 [24], see Figure 16.2 [25]. In the next step, the glutamic acid is expected to protonate the terminal oxygen to facilitate the heterolytic cleavage of the O–O bond. Subsequently, water is released and compound I, the oxoferryl(IV) porphyrin π-cation radical, is formed.

Compound I can oxidize a chloride anion and has been proposed to form an Fe(III)-hypohalite intermediate [27]. However, it has not yet been entirely resolved whether this hypohalite intermediate directly chlorinates the organic substrate [28], or whether HOCl is released upon protonation of the intermediate and free HOCl or Cl_2 performs the chlorination reactions [29]. Interestingly, it has been shown by extended X-ray absorption fine structure (EXAFS) spectroscopy that the ferryl form is consistent with a protonated oxo species, which agrees well with density functional calculations on an Fe(IV)-OH species [30].

CPO has served as an important model system for the study of reaction intermediates of P450 enzymes because of their significantly longer lifetimes, greatly facilitating their study. This is particularly true for the high-valent intermediate compound I and its one-electron-reduced form compound II. Another common short-lived reaction intermediate is the ferric hydroperoxo intermediate

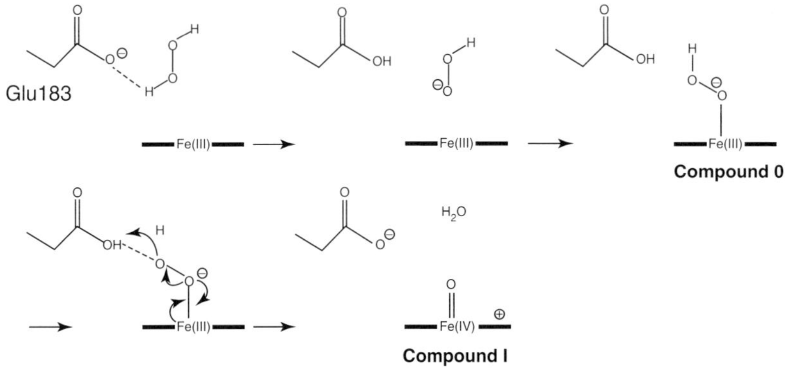

Figure 16.2 Proposed mechanism for compound I formation in chloroperoxidase [25]. The deprotonated active site Glu-183 first abstracts a proton from the incoming hydrogen peroxide. The generated hydroperoxo anion then binds to the heme yielding the ferric-hydroperoxo species, compound 0. In the next step, Glu-183 protonates the distal oxygen of compound 0, and the O–O bond is cleaved, water is released, and compound I forms, similar to the Poulos–Kraut mechanism for peroxidases [26].

Fe-OOH$^-$ (5b in Figure 16.1), also called *compound 0*, as mentioned above. This intermediate can be prepared via the reduction of oxy-ferrous heme proteins by cryogenic radiolysis [31–34] using, for example, γ-radiation from ^{60}Co sources. Thus, as in the case of P450$_{cam}$ described above [7], the CPO compound 0 complex structure determination included deliberate X-ray photoreduction of the oxy complex as a means to generate the hydroperoxo intermediate. As photoreduction happens rapidly and accumulates continuously during data collection [4] (see Figure 16.3), it is important to judge its extent by online (*in situ*) microspectrophotometric analysis of the cryocooled crystals while they are exposed to X-rays (Figure 16.3a). Partial datasets of several exposed oxy CPO crystals were merged such that composite datasets were obtained that had absorbed low, medium, or high dose, corresponding to different mixtures of the oxy complex, the hydroperoxo complex, and the subsequently formed species [35] (Figure 16.3b).

All datasets showed elongated density at the sixth ligand position of the heme iron, which was weaker in the high-dose dataset. The spectroscopic data showed that the medium-dose dataset contained the highest fraction of the hydroperoxo complex, and the electron density was modeled best with a hydroperoxo species that had a slightly longer bond length than the oxy complex. Quantum chemical calculations showed that the hydroperoxy species was more stable than the peroxo species and suggested a path for its formation [35]. Thus, a combination of spectroscopic analysis during acquisition of high-resolution diffraction data with quantum chemical calculations allowed the identification of a short-lived intermediate and determination of its crystal structure.

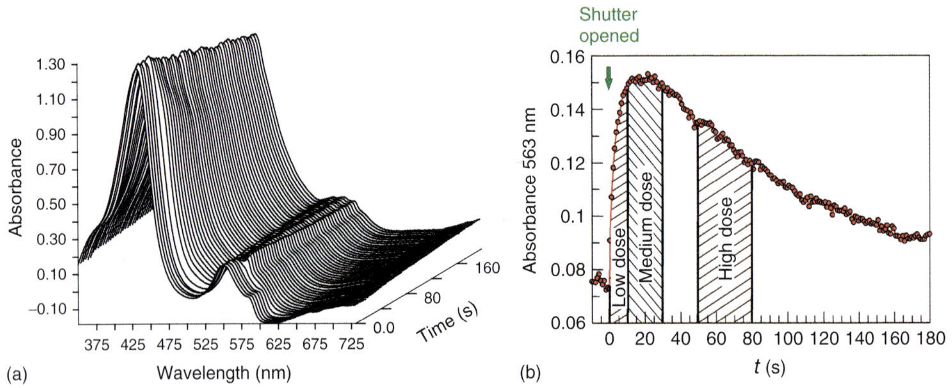

Figure 16.3 (a) Temporal evolution of the optical absorption spectra of a crystal of the oxy complex of CPO exposed to X-rays. Within seconds of opening the X-ray shutter, a new absorption band appears at 563 nm which belongs to the hydroperoxo or compound 0 complex. As shown in (b), this species also decays upon further X-ray exposure. Therefore, for the structure determination of the hydroperoxo complex, data of several crystals were merged that had absorbed a medium dose.

16.3
Photoelectrons – Friend and Foe

Both studies described above made deliberate use of the X-ray-induced photoelectrons to generate reaction intermediates by photoreduction followed by further chemical steps. In general, however, photoreduction is not a welcome feature of diffraction experiments but rather an undesired side effect that complicates the collection of chemically meaningful data. Photoreduction has been observed in many redox-sensitive systems. Although means have been suggested to slow the process, it is in general unavoidable because of the ratio of the cross sections for elastic Thomson scattering, Compton scattering, and photoabsorption/photoelectric effect (~10%, ~10%, and 80%, respectively, at 12 keV (~1 Å)). The latter dominates in the energy range up to about 20 keV. For the biologically relevant light elements (C, N, O, S, P), the ejection of a photoelectron from the inner shell is accompanied primarily by an Auger process emitting an outer-shell electron; subsequently, both electrons give rise to secondary ionization cascades (see e.g., [36]). These result in localized electronic changes, the formation of hydrated electrons, and long-timescale radical reactions and chemical changes [37–41] which may also decrease the overall scattering power of the sample. In particular, when studying redox-sensitive systems, it is important to analyze whether they are being photoreduced during data collection and, therefore, whether the data collected do in fact correspond to the desired functional state (see e.g., Figure 16.3a). Proteins containing heme [39–42] or flavin [43–45] cofactors are well known for their sensitivity toward photoreduction. Both online UV–vis and Raman [46] spectroscopy of crystals have been used to follow their oxidation state [4, 40–42] and to devise appropriate strategies for data collection. Photoreduction of chromophores in protein crystals has become a well-recognized issue and several synchrotron beamlines offer ready access to microspectroscopic analysis during data collection. However, this was not always the case, and photoreduction was often recognized only if results obtained by X-ray absorption spectroscopy and X-ray crystallography disagreed. A particularly prominent example is the analysis of the so-called oxygen-evolving complex in the large membrane protein complex photosystem II (PS II), the catalytic manganese cluster. EXAFS analysis [47] of a frozen solutions of PS II, as well as the recently published high resolution-structure, shows that the cluster readily photoreduces when exposed to X-rays and strongly indicates that the PS II crystal structures [48, 49] contain heavily damaged manganese clusters [49].

Cooling crystals to liquid helium or more typically liquid nitrogen temperature during data collection slows the long-timescale radiation damage due to strongly reduced diffusion rates of the generated radicals, leading to a tolerated dose of ~30 MGy [50]. In contrast, the generated photoelectrons move freely throughout the sample [36], which is the reason why photoreduction of metal centers happens rather quickly, typically on a seconds to minute timescale when using typical X-ray fluxes (10^{11} photons s^{-1}, 100 μm × 100 μm). Photoreduction can be slowed down somewhat by the addition scavengers such as ascorbate [51], but it cannot be eliminated. Thus, in the case of fast reduction rates, as is the case for the aforementioned PS II, composite data collection strategies become impractical because they rely on stitching together many partial low-dose datasets collected from fresh crystals. It has been predicted that the acceptable dose [50, 52, 53] can be increased significantly if the diffraction data is collected using highly brilliant femtosecond X-ray pulses that are short enough to pass through the sample before significant electronic rearrangement and atomic displacements occur ("diffraction before destruction" [54]). Such X-ray pulses are afforded by free-electron lasers (FELs).

16.4 X-ray Free-Electron Lasers

X-FELs are linear-accelerator-based X-ray sources which make use of the fact that electrons emit electromagnetic radiation when accelerated. Critical elements are very long undulators, which are periodic magnetic arrays of alternating poles that force the electrons on a "wiggly" sinusoidal path resulting in the spontaneous emission of photons. Depending of the length and design of the undulator, this "shot noise" can be amplified in a self-amplified stimulated emission (SASE) process in which the relativistic electrons and the emitted electromagnetic waves exchange energy, resulting in a bunching of the electrons. Then radiation is emitted in phase and amplified, resulting in coherent X-rays of very high brilliance. The radiated power builds up exponentially along the undulator.

Solem [55] suggested in 1986 that very intense X-ray pulses might be used to capture patterns of diffraction-limited resolution before the obliteration of the sample by radiation damage, thereby significantly extending the acceptable dose. With the emergence of X-FELs, this concept was pursued further by Hajdu and coworkers [54], who calculated the pulse length dependence of the ionization-induced atomic displacements and their effect on data quality using molecular dynamics simulations, suggesting that usable diffraction data can be obtained before explosion of the sample for very short FEL pulse lengths ("diffraction before destruction"). A proof-of-principle experiment validated this concept for biological materials when the first hard X-ray laser became operational, the Linac Coherent Light Source (LCLS) at SLAC in Stanford. It was shown that single mimivirus particles give interpretable two-dimensional (2D) diffraction patterns [56] and that tiny crystals of the large membrane protein complex photosystem I (PSI) exposed to femtosecond X-ray pulses yield diffraction data that are useful for structure analysis [57]. Since protein crystals contain a large fraction of solvent (80% in case of PSI), a gas-focused liquid microjet [58] was used to inject the randomly oriented micro- and nanocrystals into the FEL interaction region. The liquid jet not only allows keeping the crystals in their native mother liquor at ambient temperature but also provides a very convenient, fast, and gentle means to replenish the particles once they have interacted with a single femtosecond FEL pulse, upon which they are destroyed by Coulomb explosion. In reference to both the femtosecond pulses and the serial acquisition of the diffraction patterns, this new FEL-based approach of data collection was dubbed serial femtosecond crystallography (SFX) [57]. Since the first pioneering experiments on PSI which were limited to ~8 Å resolution by the available wavelength of 6.2 Å and the geometry of the experimental setup [57], it has been shown that high-resolution SFX data agree well with low-dose data collected at a synchrotron, using chicken egg white lysozyme as a well-characterized model system [59]. This established SFX as a viable and valuable complement to existing macromolecular crystallography techniques for the analysis of tiny crystals. Indeed, new structural information was obtained in a follow-up study on *in vivo* grown crystals of a trypanosomal protease [60], uncovering the mechanism of auto-inhibition [61].

SFX is a unique approach for crystallographic structure analysis beyond the limitations of conventional approaches. It allows the study of not only very small crystals but also of other systems that are highly sensitive to radiation damage. It was shown recently that the oxidation state of the oxygen-evolving manganese complex of PS II microcrystals exposed to femtosecond FEL pulses does not change [62]. This opens exciting, novel possibilities for the analysis of such systems, including time-resolved studies. In particular, the chemical timescale of femtoseconds is now accessible for diffraction and other studies using X-rays.

References

1. Denisov, I.G., Makris, T.M., Sligar, S.G., and Schlichting, I. (2005) *Chem. Rev.*, **105**, 2253–2277.
2. Beitlich, T., Kuhnel, K., Schulze-Briese, C., Shoeman, R.L., and Schlichting, I. (2007) *J. Synchrotron Radiat.*, **14**, 11–23.
3. Schlichting, I. and Miao, J. (2012) *Curr. Opin. Struct. Biol.*, **22**, 613–626.
4. Neutze, R. and Moffat, K. (2012) *Curr. Opin. Struct. Biol.*, **22**, 651–659.
5. Schlichting, I., Berendzen, J., Chu, K., Stock, A.M., Maves, S.A., Benson, D.E., Sweet, R.M., Ringe, D., Petsko, G.A., and Sligar, S.G. (2000) *Science*, **287**, 1615–1622.
6. Nelson, D.R. (2009) *Hum. Genomics*, **4**, 59–65.
7. Ortiz de Montellano, P.R. (1995) *Cytochrome P450. Structure, Mechanism, and Biochemistry*, 2nd edn, Springer, p. 652.
8. Guengerich, F.P. (2001) *Chem. Res. Toxicol.*, **14**, 611–650.
9. Isin, E.M. and Guengerich, F.P. (2007) *Biochim. Biophys. Acta*, **1770**, 314–329.
10. Groves, J.T., McClusky, G.A., White, R.E., and Coon, M.J. (1978) *Biochem. Biophys. Res. Commun.*, **81**, 154–160.
11. Costas, M. (2011) *Coord. Chem. Rev.*, **255**, 2912–2932.
12. Koppenol, W.H. (2007) *J. Am. Chem. Soc.*, **129**, 9686–9690.
13. Raag, R. and Poulos, T.L. (1989) *Biochemistry*, **28**, 7586–7592.
14. Moffat, K. (1998) *Acta Crystallogr. A*, **54**, 833–841.
15. Moffat, K. (2001) *Chem. Rev.*, **101**, 1569–1581.
16. Schlichting, I. (2000) *Acc. Chem. Res.*, **33**, 532–538.
17. Kellner, D.G., Hung, S.C., Weiss, K.E., and Sligar, S.G. (2002) *J. Biol. Chem.*, **277**, 9641–9644.
18. Rittle, J., Younker, J.M., and Green, M.T. (2010) *Inorg. Chem.*, **49**, 3610–3617.
19. Rittle, J. and Green, M.T. (2010) *Science*, **330**, 933–937.
20. Beckwith, J.R. and Hager, L.P. (1963) *J. Biol. Chem.*, **238**, 3091–3094.

21. Beckwith, J.R., Clark, R., and Hager, L.P. (1963) *J. Biol. Chem.*, **238**, 3086–3090.
22. Sundaramoorthy, M., Terner, J., and Poulos, T.L. (1995) *Structure*, **3**, 1367–1377.
23. Yi, X., Conesa, A., Punt, P.J., and Hager, L.P. (2003) *J. Biol. Chem.*, **278**, 13855–13859.
24. Wagenknecht, H.A. and Woggon, W.D. (1997) *Chem. Biol.*, **4**, 367–372.
25. Kuhnel, K., Blankenfeldt, W., Terner, J., and Schlichting, I. (2006) *J. Biol. Chem.*, **281**, 23990–23998.
26. Poulos, T.L. and Kraut, J. (1980) *J. Biol. Chem.*, **255**, 8199–8205.
27. Libby, R.D., Thomas, J.A., Kaiser, L.W., and Hager, L.P. (1982) *J. Biol. Chem.*, **257**, 5030–5037.
28. Libby, R.D., Beachy, T.M., and Phipps, A.K. (1996) *J. Biol. Chem.*, **271**, 21820–21827.
29. Griffin, B.W. (1983) *Biochem. Biophys. Res. Commun.*, **116**, 873–879.
30. Green, M.T., Dawson, J.H., and Gray, H.B. (2004) *Science*, **304**, 1653–1656.
31. Davydov, R., Macdonald, I.D.G., Makris, T.M., Sligar, S.G., and Hoffman, B.M. (1999) *J. Am. Chem. Soc.*, **121**, 10654–10655.
32. Davydov, R.M., Yoshida, T., Ikeda-Saito, M., and Hoffman, B.M. (1999) *J. Am. Chem. Soc.*, **121**, 10656–10657.
33. Denisov, I.G., Makris, T.M., and Sligar, S.G. (2001) *J. Biol. Chem.*, **276**, 11648–11652.
34. Denisov, I.G., Makris, T.M., and Sligar, S.G. (2002) *J. Biol. Chem.*, **277**, 42706–42710.
35. Kuhnel, K., Derat, E., Terner, J., Shaik, S., and Schlichting, I. (2007) *Proc. Natl. Acad. Sci. U.S.A.*, **104**, 99–104.
36. Nave, C. and Hill, M.A. (2005) *J. Synchrotron Radiat.*, **12**, 299–303.
37. Burmeister, W.P. (2000) *Acta Crystallogr., Sect. D: Biol. Crystallogr.*, **56**, 328–341.
38. Ravelli, R.B.G. and McSweeney, S.M. (2000) *Structure*, **8**, 315–328.
39. Schlichting, I., Berendzen, J., Chu, K., Stock, A.M., Maves, S.A., Benson, D.E., Sweet, R.M., Ringe, D., Petsko, G.A., and Sligar, S.G. (2000) *Science*, **287**, 1615–1622.
40. Berglund, G.I., Carlsson, G.H., Smith, A.T., Szoke, H., Henriksen, A., and Hajdu, J. (2002) *Nature*, **417**, 463–468.
41. Hersleth, H.P., Hsiao, Y.W., Ryde, U., Gorbitz, C.H., and Andersson, K.K. (2008) *Chem. Biodivers.*, **5**, 2067–2089.
42. Gumiero, A., Metcalfe, C.L., Pearson, A.R., Raven, E.L., and Moody, P.C. (2011) *J. Biol. Chem.*, **286**, 1260–1268.
43. Fedorov, R., Schlichting, I., Hartmann, E., Domratcheva, T., Fuhrmann, M., and Hegemann, P. (2003) *Biophys. J.*, **84**, 2474–2482.
44. Mees, A., Klar, T., Gnau, P., Hennecke, U., Eker, A.P., Carell, T., and Essen, L.O. (2004) *Science*, **306**, 1789–1793.
45. Rohr, A.K., Hersleth, H.P., and Andersson, K.K. (2010) *Angew. Chem. Int. Ed.*, **49**, 2324–2327.
46. Bourgeois, D., Katona, G., de Rosny, E., and Carpentier, P. (2009) *Methods Mol. Biol.*, **544**, 253–267.
47. Yano, J., Kern, J., Irrgang, K.D., Latimer, M.J., Bergmann, U., Glatzel, P., Pushkar, Y., Biesiadka, J., Loll, B., Sauer, K., Messinger, J., Zouni, A., and Yachandra, V.K. (2005) *Proc. Natl. Acad. Sci. U.S.A.*, **102**, 12047–12052.
48. Loll, B., Kern, J., Saenger, W., Zouni, A., and Biesiadka, J. (2005) *Nature*, **438**, 1040–1044.
49. Umena, Y., Kawakami, K., Shen, J.R., and Kamiya, N. (2011) *Nature*, **473**, 55–60.
50. Owen, R.L., Rudino-Pinera, E., and Garman, E.F. (2006) *Proc. Natl. Acad. Sci. U.S.A.*, **103**, 4912–4917.
51. Murray, J. and Garman, E. (2002) *J. Synchrotron Radiat.*, **9**, 347–354.
52. Henderson, R. (1995) *Q. Rev. Biophys.*, **28**, 171–193.
53. Meents, A., Gutmann, S., Wagner, A., and Schulze-Briese, C. (2010) *Proc. Natl. Acad. Sci. U.S.A.*, **107**, 1094–1099.
54. Neutze, R., Wouts, R., van der Spoel, D., Weckert, E., and Hajdu, J. (2000) *Nature*, **406**, 752–757.
55. Solem, J.C. (1986) *J. Opt. Soc. Am. B*, **3**, 1551–1565.
56. Seibert, M.M., Ekeberg, T., Maia, F.R., Svenda, M., Andreasson, J., Jonsson, O., Odic, D., Iwan, B., Rocker, A., Westphal, D., Hantke, M., DePonte, D.P., Barty, A., Schulz, J., Gumprecht, L., Coppola, N., Aquila, A., Liang, M., White, T.A.,

Martin, A., Caleman, C., Stern, S., Abergel, C., Seltzer, V., Claverie, J.M., Bostedt, C., Bozek, J.D., Boutet, S., Miahnahri, A.A., Messerschmidt, M., Krzywinski, J., Williams, G., Hodgson, K.O., Bogan, M.J., Hampton, C.Y., Sierra, R.G., Starodub, D., Andersson, I., Bajt, S., Barthelmess, M., Spence, J.C., Fromme, P., Weierstall, U., Kirian, R., Hunter, M., Doak, R.B., Marchesini, S., Hau-Riege, S.P., Frank, M., Shoeman, R.L., Lomb, L., Epp, S.W., Hartmann, R., Rolles, D., Rudenko, A., Schmidt, C., Foucar, L., Kimmel, N., Holl, P., Rudek, B., Erk, B., Homke, A., Reich, C., Pietschner, D., Weidenspointner, G., Struder, L., Hauser, G., Gorke, H., Ullrich, J., Schlichting, I., Herrmann, S., Schaller, G., Schopper, F., Soltau, H., Kuhnel, K.U., Andritschke, R., Schroter, C.D., Krasniqi, F., Bott, M., Schorb, S., Rupp, D., Adolph, M., Gorkhover, T., Hirsemann, H., Potdevin, G., Graafsma, H., Nilsson, B., Chapman, H.N., and Hajdu, J. (2011) *Nature*, **470**, 78–81.

57. Chapman, H.N., Fromme, P., Barty, A., White, T.A., Kirian, R.A., Aquila, A., Hunter, M.S., Schulz, J., DePonte, D.P., Weierstall, U., Doak, R.B., Maia, F.R.N.C., Martin, A.V., Schlichting, I., Lomb, L., Coppola, N., Shoeman, R.L., Epp, S.W., Hartmann, R., Rolles, D., Rudenko, A., Foucar, L., Kimmel, N., Weidenspointner, G., Holl, P., Liang, M., Barthelmess, M., Caleman, C., Boutet, S., Bogan, M.J., Krzywinski, J., Bostedt, C., Bajt, S., Gumprecht, L., Rudek, B., Erk, B., Schmidt, C., Homke, A., Reich, C., Pietschner, D., Struder, L., Hauser, G., Gorke, H., Ullrich, J., Herrmann, S., Schaller, G., Schopper, F., Soltau, H., Kuhnel, K.U., Messerschmidt, M., Bozek, J.D., Hau-Riege, S.P., Frank, M., Hampton, C.Y., Sierra, R.G., Starodub, D., Williams, G.J., Hajdu, J., Timneanu, N., Seibert, M.M., Andreasson, J., Rocker, A., Jonsson, O., Svenda, M., Stern, S., Nass, K., Andritschke, R., Schroter, C.D., Krasniqi, F., Bott, M., Schmidt, K.E., Wang, X., Grotjohann, I., Holton, J.M., Barends, T.R.M., Neutze, R., Marchesini, S., Fromme, R., Schorb, S., Rupp, D., Adolph, M., Gorkhover, T., Andersson, I., Hirsemann, H., Potdevin, G., Graafsma, H., Nilsson, B., and Spence, J.C.H. (2011) *Nature*, **470**, 73–77.

58. Weierstall, U., Spence, J.C., and Doak, R.B. (2012) *Rev. Sci. Instrum.*, **83**, 035108.

59. Boutet, S., Lomb, L., Williams, G.J., Barends, T.R.M., Aquila, A., Doak, R.B., Weierstall, U., Deponte, D., Steinbrener, J., Shoeman, R.L., Messerschmidt, M., Barty, A., White, T.A., Kassemeyer, S., Kirian, R.A., Seibert, M.M., Montanez, P.A., Kenney, C., Herbst, R., Hart, P., Pines, J., Haller, G., Gruner, S.M., Philipp, H.T., Tate, M.W., Hromalik, M., Koerner, L.J., van Bakel, N., Morse, J., Ghonsalves, W., Arnlund, D., Bogan, M.J., Caleman, C., Fromme, R., Hampton, C.Y., Hunter, M.S., Johansson, L., Katona, G., Kupitz, C., Liang, M., Martin, A.V., Nass, K., Redecke, L., Stellato, F., Timneanu, N., Wang, D., Zatsepin, N.A., Schafer, D., Defever, J., Neutze, R., Fromme, P., Spence, J.C.H., Chapman, H.N., and Schlichting, I. (2012) *Science*, **337**, 362–364.

60. Koopmann, R., Cupelli, K., Redecke, L., Nass, K., DePonte, D.P., White, T.A., Stellato, F., Rehders, D., Liang, M., Andreasson, J., Aquila, A., Bajt, S., Barthelmess, M., Barty, A., Bogan, M.J., Bostedt, C., Boutet, S., Bozek, J.D., Caleman, C., Coppola, N., Davidsson, J., Doak, R.B., Ekeberg, T., Epp, S.W., Erk, B., Fleckenstein, H., Foucar, L., Graafsma, H., Gumprecht, L., Hajdu, J., Hampton, C.Y., Hartmann, A., Hartmann, R., Hauser, G., Hirsemann, H., Holl, P., Hunter, M.S., Kassemeyer, S., Kirian, R.A., Lomb, L., Maia, F.R., Kimmel, N., Martin, A.V., Messerschmidt, M., Reich, C., Rolles, D., Rudek, B., Rudenko, A., Schlichting, I., Schulz, J., Seibert, M.M., Shoeman, R.L., Sierra, R.G., Soltau, H., Stern, S., Struder, L., Timneanu, N., Ullrich, J., Wang, X., Weidenspointner, G., Weierstall, U., Williams, G.J., Wunderer, C.B., Fromme, P., Spence, J.C., Stehle, T., Chapman, H.N., Betzel, C., and Duszenko, M. (2012) *Nat. Methods*, **9**, 259–262.

61. Redecke, L., Nass, K., DePonte, D.P., White, T.A., Rehders, D., Barty,

A., Stellato, F., Liang, M., Barends, T.R., Boutet, S., Williams, G.J., Messerschmidt, M., Seibert, M.M., Aquila, A., Arnlund, D., Bajt, S., Barth, T., Bogan, M.J., Caleman, C., Chao, T.C., Doak, R.B., Fleckenstein, H., Frank, M., Fromme, R., Galli, L., Grotjohann, I., Hunter, M.S., Johansson, L.C., Kassemeyer, S., Katona, G., Kirian, R.A., Koopmann, R., Kupitz, C., Lomb, L., Martin, A.V., Mogk, S., Neutze, R., Shoeman, R.L., Steinbrener, J., Timneanu, N., Wang, D., Weierstall, U., Zatsepin, N.A., Spence, J.C., Fromme, P., Schlichting, I., Duszenko, M., Betzel, C., and Chapman, H.N. (2013) *Science*, **339**, 227–230.

62. Kern, J., Alonso-Mori, R., Tran, R., Hattne, J., Gildea, R.J., Echols, N., Glockner, C., Hellmich, J., Laksmono, H., Sierra, R.G., Lassalle-Kaiser, B., Koroidov, S., Lampe, A., Han, G., Gul, S., Difiore, D., Milathianaki, D., Fry, A.R., Miahnahri, A., Schafer, D.W., Messerschmidt, M., Seibert, M.M., Koglin, J.E., Sokaras, D., Weng, T.C., Sellberg, J., Latimer, M.J., Grosse-Kunstleve, R.W., Zwart, P.H., White, W.E., Glatzel, P., Adams, P.D., Bogan, M.J., Williams, G.J., Boutet, S., Messinger, J., Zouni, A., Sauter, N.K., Yachandra, V.K., Bergmann, U., and Yano, J. (2013) *Science*, **340**, 491–495.

17
The Catalytic Strategy of P–O Bond-Cleaving Enzymes: Comparing *Eco*RV and Myosin

Farooq Ahmad Kiani and Stefan Fischer

17.1
Introduction

Enzymes are proteins capable of highly selective catalysis. They catalyze biochemical reactions in all living cells [1] with remarkable catalytic efficiency [2, 3]. They are of interest for chemists and biochemists because of their efficient catalysis *in vitro* and in physiological conditions. They can achieve very high substrate specificity and enantioselectivity and product enantiospecificity [4]. They are challenging to study because of their complex three-dimensional structure and multiple conformational sub-states.

17.1.1
How Do Enzymes Achieve Catalysis?

Enzymes achieve their marvelous catalytic properties by combining several approaches (reviewed here) [5, 6]:

1) First, an enzyme uses a "key-and-lock" mechanism to achieve the binding of a specific substrate (the key) into the binding site of the protein (the lock) [7]. This has two advantages: (i) it achieves a high substrate specificity, that is, it allows discriminating between substrates that have similar chemical structures but do not fit into the "lock," and (ii) it enriches the apparent substrate "concentration" in the vicinity of the actual catalyst (those protein residues in the binding pocket that are responsible for the lowering of the enthalpy barrier of activation).
2) Some of the free energy of binding then can be used by the enzyme to achieve ground-state destabilization. Ground-state destabilization is a modified form of key and lock, in which the substrate (key) and the enzyme (lock) do not perfectly fit into each other. Rather, the enzyme exerts a strain on the substrate to deform it in such a way that even without thermal activation the substrate already moves somewhat along the reaction coordinate. In other words, the

Molecular Catalysts: Structure and Functional Design, First Edition. Edited by Lutz H. Gade and Peter Hofmann.

equilibrium conformation of the bound substrate starts to resemble its transition state (TS) conformation. As a result, the remaining activation barrier is lowered [8].

3) Finally, as first proposed by Pauling [9, 10], the structural complementarity between the binding pocket and the activated complex of the reaction results in the stabilization of the TS, thus further lowering the energy barrier to thermal activation.

The sequence of amino acids in the polypeptide chain of the protein, which, once folded up, leads to the specific binding site of the enzyme, is evolved in such a way that it can bind both the reactant state as well as the TS. However, the binding of the enzyme to the TS is preferred to that of the reactant state. In fact, enzymatic catalysis is often simplified into a two-step process [11, 12]:

$$\mathrm{E + S} \overset{K_m}{\rightleftharpoons} \mathrm{ES} \xrightarrow{k_{cat}} \mathrm{EP} \rightleftharpoons \mathrm{E + P}$$

In the first step, the enzyme (E) binds with the substrate (S) to form a complex (ES). The equilibrium constant of this binding under standard state conditions is called K_m, where $K_m = [E]\,[S]/[ES]$. The catalytic reaction rate for crossing the TS between the reactant (ES) and product (EP) complexes is called k_{cat}. Strong reactant binding (i.e., a lower value of K_m) leads to higher specificity, but this can increase the activation barrier relative to ES (i.e., it can lower k_{cat}). Conversely, strong TS binding increases k_{cat}, but may result in a decreased initial binding of the substrate. Therefore, the overall catalytic efficiency of an enzyme is measured as k_{cat}/K_m, which embodies a compromise between achieving substrate enrichment and lowering the thermal activation barrier.

To achieve an even higher degree of substrate and reaction specificity or to couple the reaction with another biochemical process, enzymes often undergo an induced fit:

$$\mathrm{E + S} \rightleftharpoons \mathrm{ES} \rightleftharpoons \mathrm{E'S'} \rightleftharpoons \mathrm{E''S''} \xrightarrow{k_{cat}} \cdots\cdots \rightleftharpoons \mathrm{E + P}$$

A conformational change of the protein (E′, E″) is induced by the initial weak binding of the substrate. This results in a modified conformation of the substrate (S′, S″) and to a catalytically more active conformation of the binding site (E″) in which the protein residues are optimally positioned to carry out catalysis.

Both enzymes discussed below (*Eco*RV and myosin) undergo such an induced fit (see for instance Figure 17.1 for *Eco*RV). The catalytic strategy discussed here for these enzymes is primarily concerned with understanding k_{cat}, not K_m.

Many other tricks are used by some enzymes to improve K_m or k_{cat}. For example, in reactions where water decreases the rate of the reaction compared to the gas phase, one role of the enzyme binding site may be to desolvate the substrate.

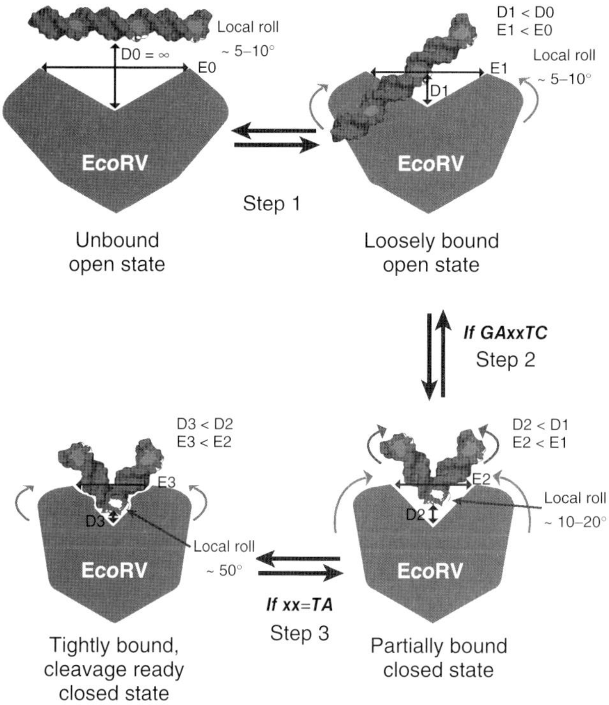

Figure 17.1 Sequence recognition by induced fit of DNA into the *Eco*RV enzyme. In the first step, a loosely bound complex is formed between *Eco*RV and the DNA. Once the DNA base pairs are recognized, a partially bound closed state is formed. DNA cleavage occurs in the fully bound closed state in which the DNA is highly bent. (Figure redrawn from Ref. [13] with permission of Elsevier.)

17.1.2
Computational Investigation of Enzymatic Mechanisms

By allowing the single-molecule study of the true substrate (as opposed to an analogous inhibitor) bound to the wild-type protein (as opposed to single-point mutants) under physiological solvation conditions (as opposed to a crystal or in vacuum), theoretical simulations can give a more meaningful and accurate understanding of the essential determinants of enzymatic strategy than experimental methods. Moreover, the simulation yields the 3D structure of the true TS at atomic resolution. The TS structure is by definition unstable and exists only for a very short time in nature, making it extremely difficult to describe it experimentally. With the TS structure at hand and comparing with the reactant, it is possible to analyze how each interaction between the substrate and protein moieties contribute to lowering the barrier of activation. This allows the full dissection of the mode of action of the enzyme and thus the ultimate understanding of its catalytic strategy. Computational approaches such as virtual screening have been effectively used to produce better therapeutic targets for drug discovery

[14]. Classical molecular mechanics (MM) and molecular dynamics (MD) simulations are employed to understand the structure and internal dynamics of enzyme/substrate complexes [15]. However, when bonds are broken or made during the catalytic reaction, quantum mechanical (QM) methods are needed [16]. Computational methods have been helpful not only in the elucidation of reaction mechanisms in naturally occurring [17] and synthetic enzymes [18] but also in the development of artificially designed enzymes [19, 20]. Increased computational power allows the employment of computationally expensive but very accurate electronic structure methods to study enzyme-catalyzed reactions. Enzymatic catalysis can be studied (i) with stand-alone quantum chemical methods (QM) in which only a truncated region of the active site is studied (the cluster approach) [21], or (ii) with combined quantum/classical mechanics (QM/MM) methods [22], in which parts of the catalytically important residues of the binding site are treated quantum mechanically whereas the remaining protein atoms surrounding the QM region are treated classically. The latter accounts for the flexible deformation of the active site in response to the reaction better and for potential long-range electrostatic effects on the energetics of the reaction. Today's method of choice in both stand-alone as well as combined QM/MM methods is density functional theory (DFT) using a hybrid B3-lyp functional [23], generally for up to 250 atoms in the QM region [21]. More advanced *ab initio* methods can be used when the catalytic site involves much less number of atoms [24]. For reaction mechanisms in which a larger number of atoms are directly involved, semiempirical QM methods can also be used [25].

17.1.3
Enzymes that Catalyze Reactions Involving Phosphate

Esters, diesters, and anhydrides of phosphoric acids are common examples of phosphates in biochemistry that act as intermediary metabolites as well as linking groups for the nucleic acids DNA and RNA [26]. They contain a phosphorus–oxygen bond which has remarkable kinetic stability in aqueous solution. The structure and function of enzymes that deal with phosphate groups in living cells are quite diverse. For example, protein kinases catalyze phosphorylation reactions, that is, they modify the structures of other proteins by transferring the terminal phosphate group of adenosine triphosphate (ATP) or guanosine triphosphate (GTP) to a polar protein side chain [27]. Phosphatases catalyze the dephosphorylation reaction in which they hydrolyze a water molecule to remove a phosphate group from the substrates [28]. Phosphorylases catalyze the phosphorolysis reaction in which they add a phosphate to the substrate [29]. These reactions are important for many cellular processes, for example, signal transduction, energy flow, and genetic inheritance [30]. The enzymes that catalyze the hydrolysis of a phosphorus–oxygen bond belong to the hydrolase class of enzymes, which has been given the Enzyme Commission Number 3 (EC 3). Hydrolase enzymes split one water molecule into its constituent ions H^+ and OH^- and add these to the substrate. Two such hydrolase enzymes are

*Eco*RV (EC 3.1.24) and myosin (EC 3.6.4.1), which catalyze the hydrolysis of phosphodiester and ATP, respectively.

17.1.4
Endonuclease Enzymes

*Eco*RV is a hydrolase enzyme that belongs to the nuclease enzymes (EC 3.1) and is responsible for cleaving the P–O bond of phosphodiester in DNA and RNA. Exonuclease enzymes hydrolyze the P–O phosphodiester bond of the terminal mononucleotide and are not sequence specific [31]. Endonuclease enzymes include both examples, that is, enzymes that are not sequence specific and those that are sequence specific (e.g., type II restriction endonucleases) [32]. Type II restriction endonucleases in bacteria, archaea, and algae use their sequence specificity to bind and cleave the foreign DNA of invading viruses. These enzymes are especially interesting, as they combine their remarkable sequence specificity with high catalytic rates of phosphodiester hydrolysis.

17.1.5
NTPase Enzymes

Myosin is an NTPase hydrolase enzyme. Nucleoside triphosphatases (NTPases) are one of the most important enzymes in living cells. A very large number of NTPase enzymes associated with a wide range of biological functions are known. They hydrolyze the nucleoside triphosphate (NTP), which is commonly referred to as *metabolic energy currency*, to obtain the chemical energy required to regulate and control vital cellular processes [33, 34]. Given that a wide range of catalytic efficiencies of hydrolysis are displayed by NTPases, it is likely that the catalytic efficiency of hydrolysis of each NTPase is specifically evolved according to the requirement of the particular biological process with which they are associated.

17.1.6
Hydrolysis Mechanism

Mechanistic details of how the hydrolysis mechanism proceeds in enzymes are largely controversial. Hydrolysis of esters, diesters, and anhydrides of phosphoric acids can occur through an associative or dissociative mechanism depending upon the relative changes in the P–O bond lengths of the P–O–P moiety during the hydrolysis reaction. If a P–O bond breaks before the nucleophilic attack, then the S_N1-like reaction mechanism is referred to as the *dissociative mechanism*. If the P–O bond breaks after the nucleophilic attack (i.e., involves the formation of a pentaphosphorane intermediate through an S_N2-like mechanism), then the reaction mechanism is referred to as the *associative mechanism*. Most phosphate-containing substrates can potentially undergo both associative and dissociative mechanisms. There is emerging consensus that the hydrolysis reaction in several enzymes is catalyzed via a dissociative mechanism.

In this chapter, we discuss the catalytic strategy of the P–O bond hydrolysis adopted by two enzymes: the *Eco*RV endonuclease and the myosin ATPase. The former cleaves the P–O–C phosphodiester linkage in the DNA backbone. Myosin cleaves the P–O–P anhydride linkage in ATP, which is the common energy currency in biology. Both these substrates have a kinetically very stable P–O bond. The phosphodiester P–O bond is exceptionally difficult to break [35–37] as indicated by its slow cleavage rate constant ($k_{sol} = 10^{-15}\,s^{-1}$) in aqueous solution [38], which translates to a free-energy barrier of 38 kcal mol^{-1}. Catalysis by *Eco*RV accelerates this reaction by a factor of $k_{cat}/k_{sol} = 10^{15}$ [39]. The resistance of ATP to spontaneous hydrolysis in water is also significant: At pH = 8.4 and 4–5, the reaction rate constants k_{sol} of $3.2 \times 10^{-5}\,s^{-1}$ and $17.5 \times 10^{-5}\,s^{-1}$, respectively, were determined [40, 41]. These correspond to an activation free-energy barrier in the 23–24 kcal mol^{-1} range [42]. In vacuum, we computed the hydrolysis barrier of ATP^{4-} complexed with $[Mg(H_2O)_4]^{2+}$ to be 42 kcal mol^{-1} high.[1] Myosin accelerates the reaction by a factor 10^7.

While there have been many studies of these two enzymes [43, 44] and catalytic reaction pathways have been proposed [45, 46], the strategies used by the protein to lower the activation barrier has not been quantitatively understood. Moreover, there have been no studies comparing these strategies. Using theoretical and simulation methods, we were able to provide a quantitative description of the strategy employed by these enzymes to weaken the P–O bond.

*Eco*RV [47, 48] is a type II restriction endonuclease enzyme from *Escherichia coli* that protects the host cell from the invading viral DNA. It specifically cleaves the DNA at the 5′-GAT↓ATC-3′ sequence (where ↓ marks the cleavage site) [49]. Upon recognizing the sequence of the invading DNA, the protein undergoes an induced fit, progressively binding the DNA tighter (Figure 17.1) and leading to the catalytically competent state in which the enzyme cleaves the poly(ribose-phosphate) backbone of the DNA. It does so by catalyzing the hydrolysis of the backbone phosphodiester bond:

$$5'-T-(CH_2-PO_2^{\,-}-O-CH_2)-A-3' + H_2O \rightarrow 5'-T-CH_2-OH + HPO_3^{\,-}-O-CH_2-A-3'$$

This is the rate-limiting step, with a turnover rate of $k_{cat} = 0.6\,s^{-1}$, which corresponds to a free-energy barrier of 17.8 kcal mol^{-1}. In absence of such catalysis, the DNA backbone is extremely stable ($t_{1/2} = 3 \times 10^7$ years) under physiological conditions of pH and temperature [35]. Figure 17.2a shows the active site in the closed, catalytically competent conformation of *Eco*RV.

Myosin II is a molecular motor responsible for muscle contraction and intracellular cargo transport [44]. It uses the hydrolysis of ATP to drive the Lymn–Taylor cycle, which describes the cyclic interactions between myosin and the actin filament that produce motion (Figure 17.3). Induced fit during steps I → II → III

1) Using B3LYP/6-31+G** level of density functional theory.

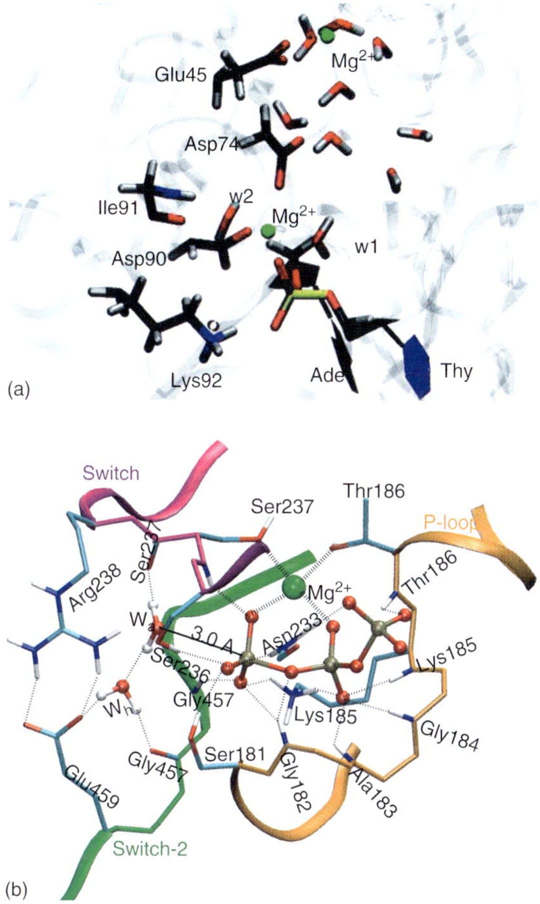

Figure 17.2 (a) Structure of the energy-optimized reactant state of *Eco*RV. Atoms that were treated quantum mechanically are shown in licorice. The dimethyl phosphate (phosphorus atom in yellow) is the part of the phosphodiester backbone strand that is cleaved. Two Mg^{2+} ions are in green. (Figure taken from Ref. [50], with permission of American Chemical Society.) (b) Structure of the reactant state III (see Figure 17.3) of myosin, with ATP bound. The triphosphate moiety (phosphorus atom colored in gold) is bound between three protein loops: switch-1 (in purple), switch-2 (in green), and the P-loop (in brown). The hydrogen-bonding pattern is shown in dotted lines.

(closing of switch-I and switch-II loops, see Figure 17.2b) leads to the catalytically competent active site in which myosin can catalyze the hydrolysis of the γ-phosphate:

$$ADP^{2-} - P_{\gamma}O_3{}^{2-} + H_2O \rightarrow ADP^{3-} + H_2PO_4{}^{-}$$

at a rate of $k_{cat} = 10^2\ s^{-1}$, which relates to a free-energy barrier of 14.4 kcal mol^{-1} [51]. The active site of state III is shown in Figure 17.2b.

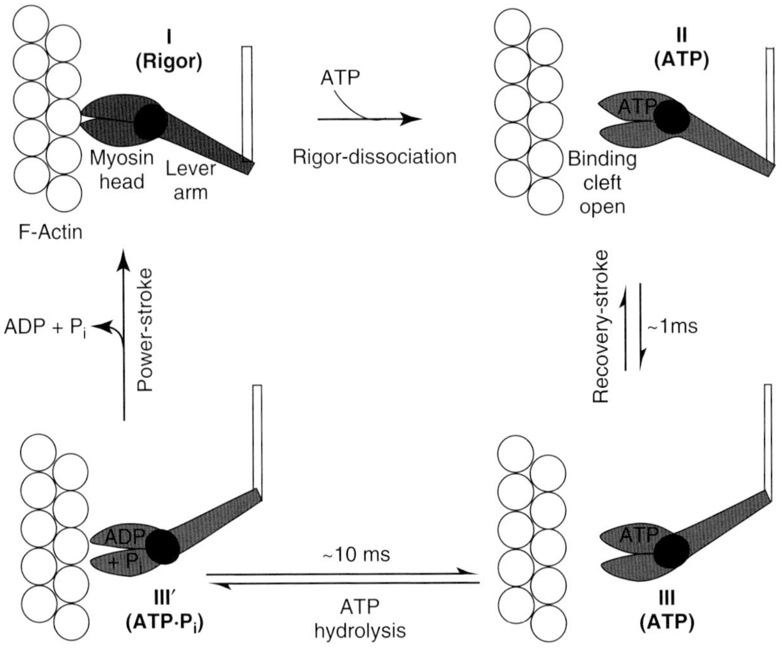

Figure 17.3 Structural events during the Lymn–Taylor muscle contraction cycle. In the rigor state (state I), myosin is strongly bound to actin in the absence of ATP. During rigor dissociation, that is, conformational transition from the rigor state to the prerecovery state (state II), myosin binds to ATP and dissociates from actin. As a result of the recovery stroke transition, myosin attains the pre-hydrolysis conformation (state III), which is the state competent for catalysis of ATP hydrolysis. The resulting state III′, in which myosin still binds to the hydrolysis products (ADP·P_i), can rebind to actin and perform the force generating the "power stroke," returning to state I while releasing the hydrolysis products.

The different mechanisms of phosphate hydrolysis are often distinguished on the basis of the order of the two main events, namely the breaking of the P–O bond and the attack of the hydrolyzed water on the phosphorus atom. In case of an associative mechanism, P–O bond cleavage takes place after attack by the hydroxyl has occurred, leading to a trigonal-bipyramidal pentaphosphorane TS ($-PO_5-$). The mechanism is called *dissociative* when this order is reversed, that is, the P–O bond breaks before the hydroxyl attack, proceeding through a trigonal-planar metaphosphate (PO_3). In each enzyme, we investigated many variants of both the associative and the dissociative mechanisms.

Here we describe how a simulation approach using combined QM/classical MM methodology enabled us to identify the energetically most favorable mechanism. We found that in both EcoRV and myosin the energetically most favorable pathway clearly proceeds through a dissociative mechanism. Based on the similarities between the EcoRV and myosin mechanisms for catalysis of phosphate hydrolysis, we can answer the following questions: (i) Why is the dissociative mechanism energetically more favorable than the associative mechanism? (ii) Is there a common catalytic strategy shared by these very dissimilar enzymes? (iii) Could this enzymatic strategy be valid for phosphate-hydrolyzing enzymes in general?

17.2 Results

Although the structure and function of *Eco*RV and myosin are very dissimilar, we find that their fundamental catalytic strategy has striking similarities (Figure 17.4). Both enzymes catalyze their respective substrates through a threefold approach: (a) favoring dissociation, that is, initial P–O bond cleavage to form a trigonal-planar metaphosphate; (b) assisting the hydrolysis of a water molecule by a basic

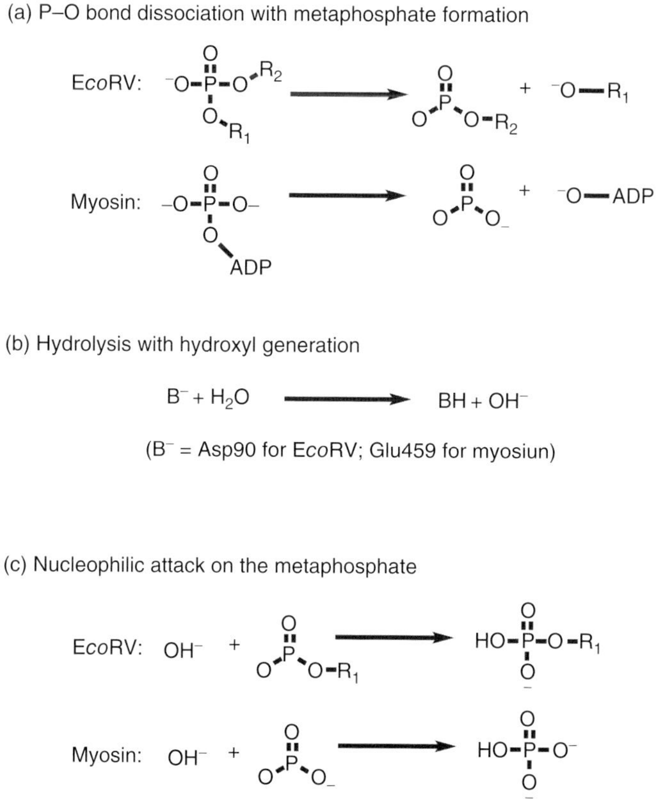

Figure 17.4 Similarity in the catalytic strategies of *Eco*RV and myosin. (a) Initial P–O bond dissociation and stabilization of a planar-trigonal PO_3 metaphosphate species. (b) General base-assisted generation of a hydroxyl. In *Eco*RV, the base is the carboxyl group of Asp90; in myosin the base is Glu459. (c) Nucleophilic attack of the hydroxyl group on the phosphorus of the metaphosphate produced in (a).

protein residue to generate an OH^- hydroxyl; and (c) the nucleophilic attack of the metaphosphate by the hydroxyl group.

The details of this catalytic strategy in *Eco*RV are illustrated in Figure 17.5. The first step (from panel R to a0 in Figure 17.5) is the rearrangement of the coordination sphere of Mg^{2+} from six-coordinated to four-coordinated, allowing for a translation of water 2 between the Asp90 and the phosphate in such a way that it can later act as the phosphorus-attacking water. This involves the crossing of a small barrier of 5 kcal mol^{-1} relative to the reactant state. Next (panels a0 to d1), P–O bond cleavage generates a trigonal-planar metaphosphate. This corresponds to step (a) in Figure 17.4. This transition is facilitated by protonation of the leaving methoxy R_1-O^- (where R_1 = ribose 3′) with the proton given by the Mg^{2+}-coordinated water w1. This involves the crossing of a 20 kcal mol^{-1} TS. The resulting metaphosphate state (panel d1 in Figure 17.5) has an energy of only 13 kcal mol^{-1} (always taken here relative to the reactant). The next two catalytic

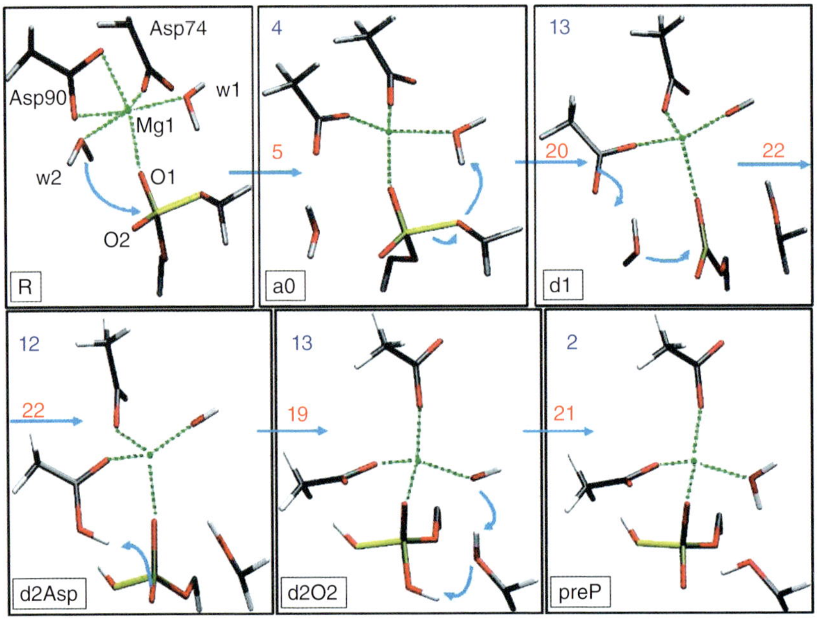

Figure 17.5 Energetically most favorable pathway for the DNA cleavage in *Eco*RV. (R) Reactant state, with the intact phospho-ribose backbone. The hexacoordinate Mg^{2+} is shown in green dotted lines. The blue arrow indicates the motion of water w2. (a0) water w1, activated by the Mg^{2+}, donates a proton to the oxygen of the P–O–C ester bond (blue arrow), thus breaking the P–O bond. (d1) Trigonal planar metaphosphate. Asp90 abstracts a proton from the water w2, thus generating an OH^- that can attack the metaphosphate (blue arrows). (d2Asp) and (d2O2) A proton is shuffled in two steps from Asp90 to water w1. (preP) Pre-product state. Numbers in the upper-left corner of each panel give the energy of the structure relative to the reactant (panel R). Numbers in red give the energy of the transition state between two successive panels. (Figure taken from Ref. [50], with permission of American Chemical Society.)

steps ((b) and (c) in Figure 17.4) are concerted in *Eco*RV (panel d1 to d2Asp in Figure 17.5). Water w2 gives a proton to the Asp90 side chain (which acts as a general base) and the resulting OH^- attacks the metaphosphate. With a barrier of 22 kcal mol^{-1}, this transition is the overall rate-limiting step of the catalytic reaction and is close to the experimentally measured barrier. Finally, two steps of proton transfer restore Asp90 and water w1 to their original protonation states (panel preP in Figure 17.5).

The details of the catalytic strategy of myosin are shown in Figure 17.6. The first step (R to m1 in Figure 17.6) is the breaking the $P_\gamma-O-P_\beta$ anhydride bond to generate adenosine diphosphate (ADP) and the planar metaphosphate PO_3^-, in accordance with Figure 17.4a. This structure is very stable, with only 2.7 kcal mol^{-1} higher in energy than the reactant state. The energy barrier for the transition is also low, at 8.9 kcal mol^{-1}. In going from the state m1 to m2, the helping water W_h reorients to make a hydrogen bond with the Ser181 side chain. This forms in m2 a hydrogen-bonding network which allows a concerted proton transfer during the m2–g1 transition: from the attacking water to the helping water to the Ser181 side chain to the γ-phosphate. This activates water W_a into a hydroxyl OH^- ion (as shown in Figure 17.4b), which can attack the γ-phosphorus, forming an $H_2PO_4^-$ molecule in state g1. The corresponding TS (panel m2–g1 in Figure 17.6) has an energy of 10.3 kcal mol^{-1}, making this the overall rate-limiting step with a value comparable to the experimental energy barrier. Finally, protons of Ser181 and the γ-phosphate reorient (panels g1–g2) and the helping water moves back to its original position, thus yielding the product state (panels g2–P).

17.3 Conclusions

Our QM/MM calculations of the minimum-energy catalytic pathways indicate that the catalytic strategy common to both *Eco*RV endonuclease and myosin ATPase can be formally seen as three-pronged approach.

1) *P–O bond dissociation with metaphosphate formation*: In both enzymes, that is, *Eco*RV and myosin, hydrolysis occurs through a dissociative mechanism in which the initial event in the enzyme is the P–O bond cleavage, prior to nucleophilic attack of the phosphorus atom. This yields a planar metaphosphate group. In *Eco*RV, cleavage takes place at the P–O–C ester bond, whereas in myosin the mechanism proceeds through cleavage of the P–O–P anhydride bond. In both studies, this metaphosphate structure is found to be a stable intermediate, that is, a local energy minimum on the potential energy surface. This is remarkable given that metaphosphate is very unstable in absence of the proteins. For instance, in myosin, when the surrounding protein atoms are removed, the potential energy difference between the reactant ($ATP^{4-}/Mg^{2+}/4H_2O$) and the metaphosphate ($ADP^{3-}/PO^{3-}/Mg^{2+}/4H_2O$) structure was 35.7 kcal mol^{-1} compared to 2.9 kcal mol^{-1} in presence of the

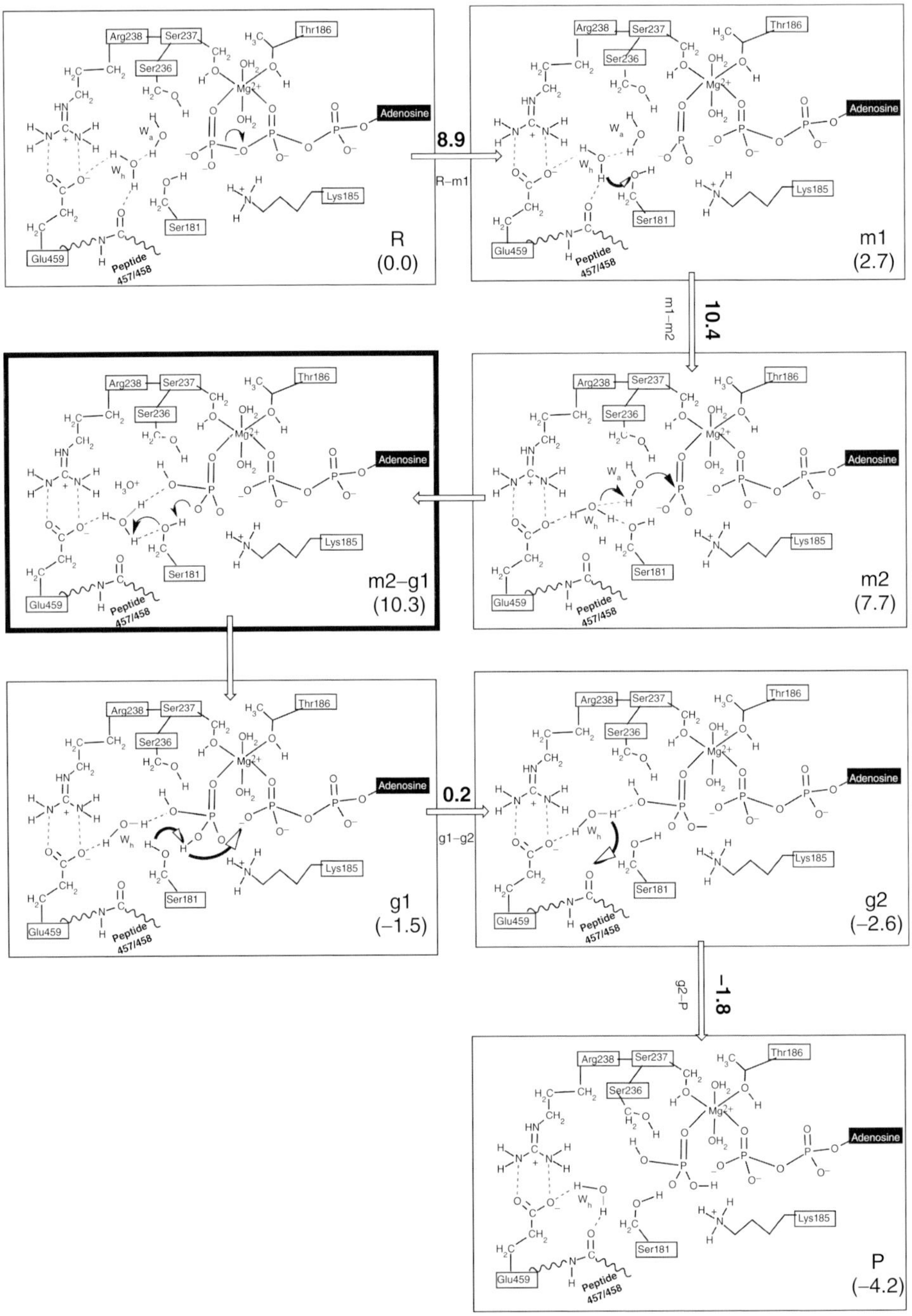
R
(0.0)
8.9
R–m1
m1
(2.7)
m1–m2
10.4
m2
(7.7)
m2–g1
(10.3)
g1
(−1.5)
0.2
g1–g2
g2
(−2.6)
g2–P
−1.8
P
(−4.2)
Arg238
Ser237
Ser236
Thr186
Ser181
Lys185
Glu459
Adenosine
Peptide 457/458

protein (as mentioned in Section 17.2). This shows that the protein is geared toward strongly stabilizing the metaphosphate state. The difference in charge distribution between the reactant and metaphosphate states (for example, in myosin, γ-phosphate going from −2 in ATP to −1 in metaphosphate and β-phosphate going from −1 in ATP to −2 in metaphosphate) is used for this stabilization of the metaphosphate state. The protein does so by placing a large number of H-bond donors in positions where they can interact with phosphates α and β. Note that, even though the three oxygen atoms bonded to the phosphorus atom in the metaphosphate are nearly coplanar with the phosphorus atom, this is not a bipyramidal structure as would be formed, for example, in an associative TS, because the attacking and leaving oxygen atoms are not equally distant.

2) *Hydrolysis with hydroxyl generation*: Both *Eco*RV and myosin place a general base to assist the water activation. The attacking water molecule is activated by a proximate general base, which accepts a proton from the water, thereby generating the hydroxyl nucleophile. This strategy is complementary to the metaphosphate generation, and together they are instrumental in lowering the catalyzed hydrolysis barrier. In the case of *Eco*RV, Asp90 directly abstracts a proton from the attacking water molecule, whereas in myosin Glu459 and a helping water together act as the general base.
3) *Nucleophilic attack on the metaphosphate*: After generation of both the metaphosphate and the attacking hydroxyl nucleophile in the binding pocket, the eventual hydrolysis in both enzymes is accomplished by the nucleophilic attack of the hydroxyl ion to the planar metaphosphate. The geometry of the planar metaphosphate group is much better for the nucleophilic attack by the OH^- hydroxyl group than the tetrahedral geometry of orthophosphate. In myosin, the γ-phosphate of the metaphosphate state contains only one formal negative charge, one less than the doubly charged $-P_\gamma O_4^{2-}$ of the ATP state (Figure 17.7). Similarly in *Eco*RV, a negative charge on reactant state phosphate ester shifts to the leaving $R{-}O^-$ group (see Figure 17.4) so that the phosphate group has one less negative charge in the metaphosphate state than in the reactant state. This charge shift makes the metaphosphate a much better nucleophile acceptor for the hydroxyl attack.

Figure 17.6 Energetically most favorable pathway for the hydrolysis of ATP in myosin. (R) The reactant state. The thin arrow indicates the breaking of the P–O bond. (m1) This results in the formation of a metaphosphate (PO_3^-), (m2) in which water W_h reorients to form a H-bond with Ser181 (thick arrow). (m2–g1) Transition state between panels m2 and g1. Protons transfer concertedly from the attacking water W_a, to the helping water W_h, to the Ser181 side chain, and to the γ-phosphate oxygen, while the resulting hydroxyl W_a^- attacks the γ-phosphate oxygen. (g1) Ser181 and the γ-phosphate protons reorient to switch the H-bond network. (g2) Helping water W_h rearranges to make the same H-bonds as in the reactant R. (P) Product state. The structures corresponding to panels R, m1, m2, g1, g2, and P are the local energy minima. Energies of the structures (relative to the reactant R) are given in parenthesis. Thin arrows indicate electronic rearrangements, and thick arrows with a white head indicate motions of the nuclei.

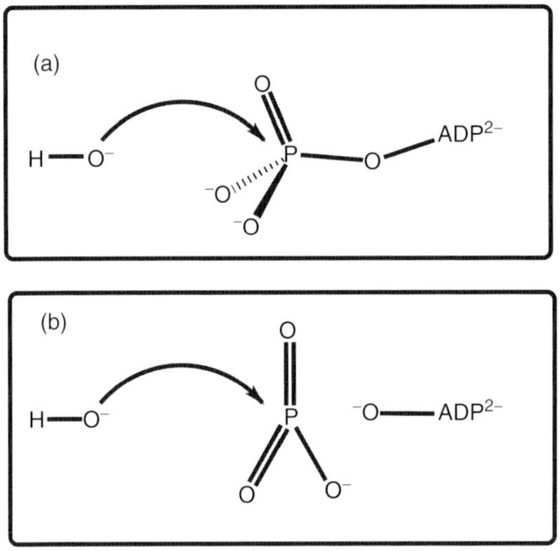

Figure 17.7 Formal negative charges and geometry of the phosphate groups in (a) associative and (b) dissociative mechanisms. The γ-phosphate in the associative mechanism is formally double negatively charged, whereas the metaphosphate in the dissociative mechanism is formally single negatively charged. The nucleophilic attack of the OH^- ion is easier on the single negatively charged metaphosphate moiety. Moreover, the tetrahedral geometry of the phosphate group in the associative mechanism hinders the nucleophilic attack as compared to the trigonal planar geometry of the metaphosphate in the dissociative mechanism.

Although *Eco*RV and myosin are two distinct proteins with different substrates, they adopt a strikingly similar catalytic strategy. In both proteins, P–O bond cleavage precedes the generation of the nucleophile hydroxyl and its nucleophilic attack on the phosphorus, that is, a dissociative mechanism is preferred over an associative mechanism. Both enzymes use a general base in order to abstract a proton from the water molecule. This can be rationalized in terms of the energy barriers associated with (i) the P–O bond dissociation and (ii) the water O–H bond dissociation upon hydroxyl formation. In the case of an associative mechanism, both steps (i) and (ii) are concerted, so that the energetic cost of (i) adds to the cost of (ii), raising the overall barrier. In a dissociative mechanism, the individual barriers of (i) and (ii) are crossed sequentially, effectively lowering the rate-limiting barrier. Moreover, this allows the enzyme to provide separate catalytic groups better aimed at specifically lowering the barrier of each step. As shown in Figures 17.4a and 17.7, the initial P–O bond cleavage transfers one formal negative charge from the phosphate group to the leaving oxygen. This leaves a less negatively charged phosphate group, thus making the nucleophilic attack of the hydroxyl group (OH^-) easier. Finally, the OH^- attack is easier with the geometry of a planar metaphosphate than it would be for the geometry of a tetrahedral orthophosphate. All these aspects come together in bringing down the overall barrier for the phosphate hydrolysis from a high value of 30–40 to the

10–15 kcal mol^{-1} range. This large reduction of the barrier is probably not easily achieved by other means at physiological conditions of pH and temperature. Therefore, it is very likely that other enzymes that have evolved to catalyze the hydrolysis of P–O–C and P–O–P bonds may use the same catalytic strategy.

17.4 Methods

The E*co*RV starting structure for the QM/MM minimization and subsequent minimum-energy path calculations was obtained from the three crystallographically resolved structures with protein data bank (PDB) codes 1SX8, 1SUZ [52], and 1RVB [53]. A thorough description of the classical treatment of the E*co*RV structure before it was subjected to the QM/MM minimization has been described elsewhere [50]. The energies mentioned here for E*co*RV were obtained with the AM1-d method for the QM region (shown in Figure 17.2a) as implemented in the MNDO96 program [54] interfaced with CHARMM [55].

For myosin, QM/MM minimization of the crystallographically resolved 1VOM structure [56] was carried out using CHARMM interfaced to the Turbomole5.9 [57]. The QM region (shown in Figure 17.2b) was optimized using the hybrid B3LYP functional [23, 58] with a 6-31G* basis set.

In both enzymes, four regions were treated differently: (i) a quantum mechanically treated region; (ii) a classically treated region around (i), with atoms in regions (i) and (ii) free to move; (iii) a region surrounding the region (ii) in which the atoms are positionally constrained using a harmonic restoring force; and (iv) a fixed region which consists of the rest of the protein. Nonuniform charge scaling (NUCS) implicit solvent method [59] was used. This method rescales the electrostatic charges on the classically treated atoms to reproduce the electrostatic interaction energies obtained from a solution of a Poisson–Boltzmann analysis. QM/MM boundaries were treated using the link-atom approach [60].

Minimum energy paths and their TSs were calculated using the conjugate peak refinement (CPR) method [61] as implemented in the TREK module of CHARMM.

References

1. (a) Buchholz, K., Kasche, V., and Bornscheuer, U.T. (2012) *Biocatalysts and Enzyme Technology*, 2nd edn, Wiley-Blackwell, Weinheim. (b) Frey, P.A. and Hegeman, A.D. (2007) *Enzymatic Reaction Mechanisms*, Oxford University Press, New York.
2. Szefczyk, B., Mulholland, A.J., Ranaghan, K.E., and Sokalski, W.A. (2004) *J. Am. Chem. Soc.*, **126**, 16148–16159.
3. Bruice, T.C. (2002) *Acc. Chem. Res.*, **35**, 139–148.
4. Bugg, T.D.H. (2012) *Introduction to Enzyme and Coenzyme Chemistry*, John Wiley & Sons, Ltd, Chichester.
5. Benkovic, S.J. and Hammes-Schiffer, S. (2003) *Science*, **301**, 1196–1202.
6. Garcia-Viloca, M., Gao, J., Karplus, M., and Truhlar, D.G. (2004) *Science*, **303**, 186–195.

7. Fischer, E. (1894) *Ber. Dtsch. Chem. Ges.*, **27**, 3189–3232.
8. Haldane, J.B.S. (1930) in *Monographs on Biochemistry* (eds R.H.A. Plimmer and F.G. Hopkins), Longmans, Green & Co., London.
9. Pauling, L. (1946) *Chem. Eng. News*, **24**, 1375–1377.
10. Pauling, L. (1948) *Nature*, **161**, 707–709.
11. Michaelis, L. and Menten, M.L. (1913) *Biochem. Z.*, **49**, 333–369.
12. Johnson, K.A. and Goody, R.S. (2011) *Biochemistry*, **50**, 8264–8269.
13. Zahran, M., Daidone, I., Smith, J.C., and Imhof, P. (2010) *J. Mol. Biol.*, **401**, 415–432.
14. Kitchen, D.B., Decornez, H., Furr, J.R., and Bajorath, J. (2004) *Nat. Rev. Drug Discov.*, **3**, 935–949.
15. Karplus, M. and McCammon, J.A. (2002) *Nat. Struct. Biol.*, **9**, 646–652.
16. Ramos, M.J. and Fernandes, P.A. (2008) *Acc. Chem. Res.*, **41**, 689–698.
17. See for example, (a) Claeyssens, F., Harvey, J.N., Manby, F.R., Mata, R.A., Mulholland, A.J., Ranaghan, K.E., Schütz, M., Thiel, S., Thiel, W., and Werner, H.-J. (2006) *Angew. Chem. Int. Ed.*, **118**, 7010–7013. (b) Himo, F. (2006) *Theor. Chem. Acc.*, **116**, 232–240. (c) Gao, J., Ma, S., Major, D.T., Nam, K., Pu, J., and Truhlar, D.G. (2006) *Chem. Rev.*, **106**, 3188–3209. (d) Nemukhin, A.V., Grigorenko, B.L., Lushchekina, S.V., and Varfolomeev, S.D. (2012) *Russ. Chem. Rev.*, **81**, 1011–1025.
18. Ke, Z., Abe, S., Ueno, T., and Morokuma, K. (2012) *J. Am. Chem. Soc.*, **134**, 15418–15429.
19. Nanda, V. and Koder, R.L. (2010) *Nat. Chem.*, **2**, 15–24.
20. Röthlisberger, D., Khersonsky, O., Wollacott, A.M., Jiang, L., DeChancie, J., Betker, J., Gallaher, J.L., Althoff, E.A., Zanghellini, A., Dym, O., Albeck, S., Houk, K.N., Tawfik, D.S., and Baker, D. (2008) *Nature*, **453**, 190–195.
21. Lind, M.E.S. and Himo, F. (2013) *Angew. Chem. Int. Ed.*, **125**, 4661–4665.
22. (a) Warshel, A. and Levitt, M. (1976) *J. Mol. Biol.*, **103**, 227–249. (b) Ranaghan, K.E. and Mulholland, A.J. (2010) *Int. Rev. Phys. Chem.*, **29**, 65–133. (c) Senn, H.M. and Thiel, W. (2007) *Top. Curr. Chem.*, **268**, 173–290. (d) Siegbahn, P.E.M. and Borowski, T. (2006) *Acc. Chem. Res.*, **39**, 729–738.
23. (a) Becke, A.D. (1993) *J. Chem. Phys.*, **98**, 5648–5652. (b) Lee, C., Yang, W., and Parr, R.G. (1988) *Phys. Rev.*, **B 37**, 785–789.
24. Harvey, J.N. (2011) *J. Biol. Inorg. Chem.*, **16**, 831–839.
25. Riccardi, D., Schaefer, P., Yang, Y., Yu, H., Ghosh, N., Prat-Resina, X., Koenig, P., Li, G., Xu, D., Guo, H., Elstner, M., and Cui, Q. (2006) *J. Phys. Chem. B*, **110**, 6458–6469.
26. Westheimer, F.H. (1987) *Science*, **235**, 1173–1178.
27. Adams, J.A. (2001) *Chem. Rev.*, **101**, 2271–2290.
28. Jackson, M.D. and Denu, J.M. (2001) *Chem. Rev.*, **101**, 2313–2340.
29. Nakai, H., Kitaoka, M., Svensson, B., and Ohtsubo, K. (2013) *Curr. Opin. Chem. Biol.*, **17**, 301–309.
30. (a) Vetter, I.R. and Wittinghofer, A. (1999) *Q. Rev. Biophys.*, **32**, 1–56. (b) Cleland, W.W. and Hengge, A.C. (2006) *Chem. Rev.*, **106**, 3252–3278. (c) Benkovic, S.J. and Schray, K.J. (1973) in *The Enzymes*, vol. VIII (ed. P.D. Boyer), Academic Press, New York, pp. 201–238. (d) Kirby, J.A. and Warren, S.G. (1967) *The Organic Chemistry of Phosphorous*, Elsevier, Amsterdam. (e) Westheimer, F.H. (1981) *Chem. Rev.*, **81**, 313–326. (f) Cox, J.R. and Ramsay, O.B. (1964) *Chem. Rev.*, **64** (4), 317–352.
31. Dupureur, C.M. (2010) *Metallomics*, **2**, 609–620.
32. Pingoud, A., Fuxreiter, M., Pingoud, V., and Wende, W. (2005) *Cell. Mol. Life Sci.*, **62**, 685–707.
33. Teru, O. and Wilkinson, A.J. (2001) *Genes Cells*, **6**, 575–597.
34. Etienne-Manneville, S. and Hall, A. (2002) *Nature*, **420**, 629–635.
35. Bunton, C.A., Mhala, M.M., Oldham, K.G., and Vernon, C.A. (1960) *J. Chem. Soc.*, **1960**, 3293–3301.
36. Kumamoto, J., Cox, J.R., and Westheimer, F.H. (1956) *J. Am. Chem. Soc.*, **78**, 4858–4860.
37. Kirby, A.J. and Younas, M. (1970) *J. Chem. Soc. B*, 510–513.

38. Schroeder, G.K., Lad, C., Wyman, P., Williams, N.H., and Wolfenden, R. (2006) *Proc. Natl. Acad. Sci. U.S.A.*, **103**, 4052–4055.
39. (a) Horton, N.C., Otey, C., Lusetti, S., Sam, M.D., Kohn, J., Martin, A.M., Ananthnarayan, V., and Perona, J.J. (2002) *Biochemistry*, **41**, 10754–10763. (b) Martin, A.M., Horton, N.C., Lusetti, S., Reich, N.O., and Perona, J.J. (1999) *Biochemistry*, **38**, 8430–8439.
40. Trentham, D.R., Eccleston, J.F., and Bagshaw, C.R. (1976) *Q. Rev. Biophys.*, **9**, 217–281.
41. Miller, D.L. and Westheimer, F.H. (1966) *J. Am. Chem. Soc.*, **88**, 1507–1511.
42. Schwarzl, S.M. (2006) Understanding the ATP hydrolysis mechanism in myosin using computer simulation techniques. PhD thesis. Ruprecht-Karls University of Heidelberg.
43. See, for example, for *EcoRV*, (a) Kovall, R.A. and Matthews, B.W. (1999) *Curr. Opin. Chem. Biol.*, **3**, 578–583, and references therein. (b) Stanford, N.P., Halford, S.E., and Baldwin, G.S. (1999) *J. Mol. Biol.*, **288**, 105–116.
44. Sweeney, H.L. and Houdusse, A. (2010) *Annu. Rev. Biophys.*, **39**, 539–557, and references therein.
45. (a) Groll, D.H., Jeltsch, A., Selent, U., and Pingoud, A. (1997) *Biochemistry*, **36**, 11389–11401. (b) Halford, S.E. and Goodall, A.J. (1988) *Biochemistry*, **27**, 1771–1777. (c) Sam, M.D. and Perona, J.J. (1999) *Biochemistry*, **38**, 6576–6586.
46. (a) Grigorenko, B.L., Rogov, A.V., Topol, I.A., Burt, S.K., Martinez, H.M., and Nemukhin, A.V. (2007) *Proc. Natl. Acad. Sci. U.S.A.*, **104**, 7057–7061. (b) Grigorenko, B.L., Kaliman, I.A., and Nemukhin, A.V. (2011) *J. Mol. Graphics. Modell.*, **31**, 1–4.
47. (a) Williams, N.H., Takasaki, B., Wall, M., and Chin, J. (1999) *Acc. Chem. Res.*, **32**, 485–493. (b) Baldwin, G.S., Sessions, R.B., Erskine, S.G., and Halford, S.E. (1999) *J. Mol. Biol.*, **288**, 87–103.
48. Perona, J.J. (2002) *Methods*, **28**, 353–364.
49. Schildkraut, I., Banner, C., Rhodes, C.S., and Parekh, S. (1984) *Gene*, **27**, 327–329.
50. Imhof, P., Fischer, S., and Smith, J.C. (2009) *Biochemistry*, **48**, 9061–9075.
51. Schwarzl, S.M., Smith, J.C., and Fischer, S. (2006) *Biochemistry*, **45**, 5830–5847.
52. Horton, N.C. and Perona, J.J. (2004) *Biochemistry*, **43**, 6841–6857.
53. Winkler, F.K., Banner, D.W., Oefner, C., Tsernoglou, D., Brown, R.S., Heatman, S.P., Bryan, R.K., Martin, P.D., Petratos, K., and Wilson, K.S. (1993) *EMBO J.*, **12**, 1781–1795.
54. Thiel, W. (2004) MNDO, Version 6.1, Max-Planck-Institut fuer Kohlenforschung, Mülheim an der Ruhr.
55. Brooks, B.R., Bruccoleri, R.E., Olafson, B.D., States, D.J., Swaminathan, S., and Karplus, M. (1983) *J. Comput. Chem.*, **4**, 187–217.
56. Smith, C.A. and Rayment, I. (1996) *Biochemistry*, **35**, 5404–5417.
57. Ahlrichs, R., Bär, M., Häser, M., Horn, H., and Kölmel, C. (1989) *Chem. Phys. Lett.*, **162**, 165–169.
58. (a) Vosko, S.H., Wilk, L., and Nusair, M. (1980) *Can. J. Phys.*, **58**, 1200–1211. (b) Stephens, P.J., Devlin, F.J., Chabalowski, C.F., and Frisch, M.J. (1994) *J. Phys. Chem.*, **98**, 11623–11627.
59. Schwarzl, S.M., Huang, D., Smith, J.C., and Fischer, S. (2005) *J. Comput. Chem.*, **26**, 1359–1371.
60. (a) Waszkowycz, B., Hillier, I.H., Gensmantel, N., and Payling, D.W. (1991) *J. Chem. Soc., Perkin Trans. 2*, 225–231. (b) Waszkowycz, B., Hillier, I.H., Gensmantel, N., and Payling, D.W. (1991) *J. Chem. Soc., Perkin Trans. 2*, 1819–1832. (c) Waszkowycz, B., Hillier, I.H., Gensmantel, N., and Payling, D.W. (1991) *J. Chem. Soc., Perkin Trans. 2*, 2025–2032.
61. Fischer, S. and Karplus, M. (1992) *Chem. Phys. Lett.*, **194**, 252–261.

18
Selective Hybrid Catalysts Based on Nucleic Acids

Andres Jäschke

18.1
Introduction

Hybrid catalysis combines homogeneous chemical catalysts with biopolymers to obtain selective catalysts for organic reactions. While proteins have been used as hosts for various transition-metal complexes, relatively few published examples are based on nucleic acids. In these reports, high stereoselectivities were obtained in Diels–Alder reactions (DARs), Michael additions, and other reactions, with DNA as sole source of chirality, but most of these systems relied on Lewis acid catalysis by Cu(II) ions. Our goal is the application of DNA-conjugated transition-metal complexes in organometallic catalysis, as this kind of catalysis is widespread among synthetically useful reactions. Furthermore, we focus on site-specific covalent attachment of the ligand–metal complexes to DNA, as this will allow better understanding and control of catalytic behavior.

Here we present different DNA-based systems for the catalysis of organic reactions in aqueous media. After a brief review of the field, different methods for the site-specific modification of DNA with transition metal–ligand complexes and other moieties are presented. We use Ir(I)–diene chemistry to catalyze an allylic substitution in aqueous medium. Toward this end, we covalently attach transition-metal ligands, such as phosphinoxazoline (PHOX) and diene ligands, to specific positions of oligonucleotides. Our approach uses a modular design where an oligonucleotide carrying a transition-metal ligand is combined with different DNA or RNA counter-strands, thereby forming perfect and imperfect duplexes, triplexes, quadruplexes with or without bulges of varying size, thereby providing changes in the environment of the metal center. The covalent attachment of the ligand guarantees its specific, reproducible positioning on nucleic acid structures. We demonstrate that catalysis occurs in the presence of DNA and its numerous functional groups, and that the structure of the DNA modulates the stereochemical outcome of the reaction.

Molecular Catalysts: Structure and Functional Design, First Edition. Edited by Lutz H. Gade and Peter Hofmann.

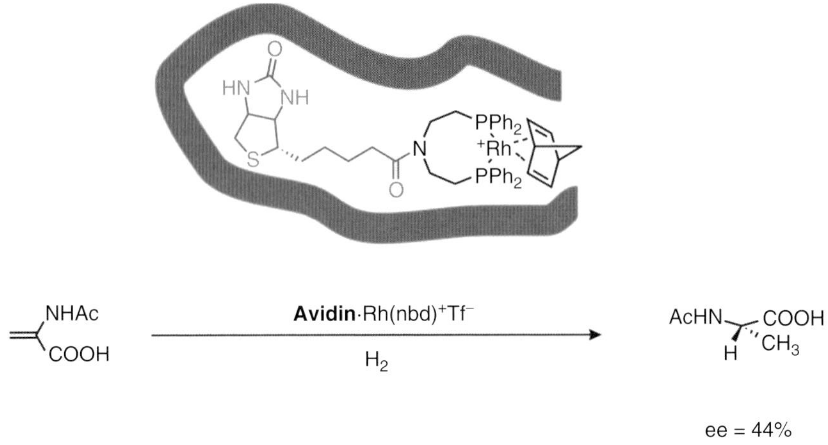

Figure 18.1 Asymmetric hydrogenation with avidin containing a biotinylated phosphine–rhodium (I) complex [1].

18.2 Hybrid Catalysis

Hybrid catalysis is a concept that combines biocatalysis with homogeneous chemical catalysis. Typically, a chemical catalyst (most often a transition-metal complex) is anchored inside a pocket of a biomolecule, and, according to the original concept, the metal complex should provide the reactivity while the chiral environment of the biopolymer determines and fine-tunes the selectivity. Meanwhile, several systems have been developed where such a clear-cut distribution of work is not possible anymore and the biomolecule contributes significantly to rate acceleration.

The earliest example of hybrid catalysis originates from Wilson and Whitesides [1], who implanted a biotinylated rhodium(I) catalyst into avidin (Figure 18.1). The selectivity remained modest, and practical applications were lacking at that time. Twenty-five years later, Ward, Reetz, and others [2] developed approaches of directed evolution, combining either rational or combinatorial mutagenesis with screening in an iterative process. This work ultimately led to remarkable catalysts with high turnover frequencies and stereoselectivities for numerous chemical reactions. The field of protein-based hybrid catalysts has been extensively reviewed, and the reader is directed to these instructive reviews for further reference [3].

18.3 DNA-Based Hybrid Catalysis

In Nature, DNA serves primarily as a storage medium for genetic information. Composed of four major monomers, it predominantly occurs in the form of a right-handed double helix obeying the Watson–Crick base-pairing rules. The helical chirality and the chirality of the sugar 2′-deoxy-D-ribose make DNA a potential host that could modulate the stereoselectivity of unselective catalysts, but the limited structural diversity of double-stranded DNA does not offer much variety.

In addition to the major double-helical form, DNA also exists as triplexes and quadruplexes and forms complex three-way or four-way junctions. Single-stranded DNA has been shown *in vitro* to fold into complex, stable three-dimensional folds and to bind to small molecules (such DNA molecules are named *aptamers*) or to catalyze reactions in a way similar to proteins (so-called deoxyribozymes or DNAzymes) [4]. This increased structural diversity makes it a more attractive host molecule for hybrid catalysis [5]. Furthermore, directed evolution approaches are easier to conduct with nucleic acids than with proteins, as nucleic acids can be directly replicated and amplified.

The first published reports on DNA-based hybrid catalysis are from Roelfes, Feringa, and coworkers [6]. They could demonstrate that DNA isolated from salmon sperm (an inexpensive commodity chemical) could dramatically influence the stereochemistry of the Cu^{2+}-catalyzed DAR in water between an aza-chalcone and cyclopentadiene when used in combination with different aminoacridine–aminomethylpyridine conjugates (Figure 18.2). In these experiments, Cu^{2+} ions act as Lewis acids and the aminomethylpyridine is a bidentate Cu^{2+} chelator, while the acridine moiety intercalates into DNA and thereby brings the copper catalyst into the chiral environment. In their first report, the authors could achieve moderate enantioselectivities (up to 49% ee for the endo

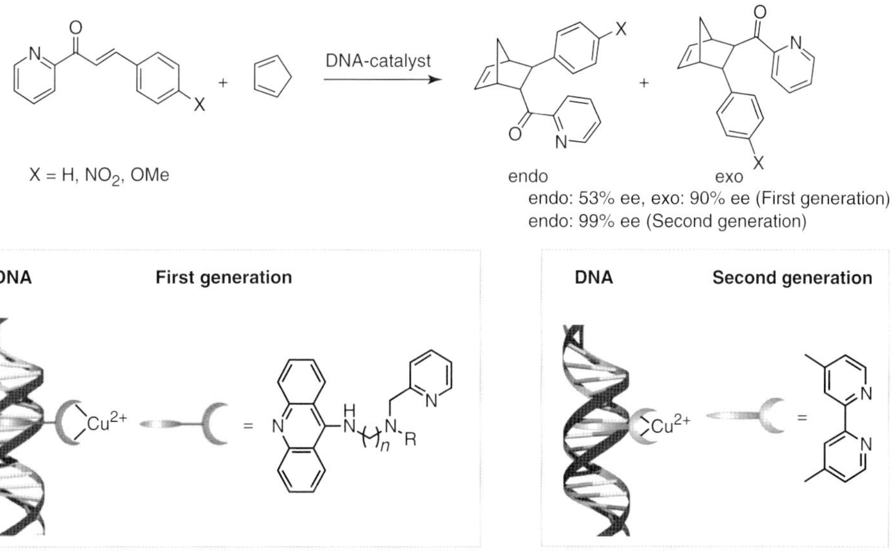

Figure 18.2 Diels–Alder reaction of cyclopentadiene with aza-chalcone catalyzed by copper complexes in the presence of DNA [6c,d].

isomer) [6d]. Furthermore, the stereochemical outcome was found to depend on the structure linking the chelator and the intercalator. Consequently, the authors eliminated the linker entirely in their second generation of ligands and chose simple ligands that can intercalate and chelate at the same time, such as 4,4′-dimethyl-2,2′-bipyridine (dmbipy) [7]. Impressive diastereo- and enantioselectivities (>99 de, >99% ee) were reported using 33 mol% of catalyst, and high selectivities were also observed at low catalyst loading (5 mol%). The enormous stereo-discrimination is remarkable considering that salmon sperm DNA is a heterogeneous mixture of many different DNA sequences and the intercalation-based approach does not provide much specificity with respect to the site of interaction between DNA and the acridine-based ligand. Roelfes and coworkers could demonstrate that salmon sperm DNA actually not just provides a chiral environment for the DAR but also provides ~60-fold rate acceleration compared to Cu^{2+}–dmbipy alone. Employing synthetic oligonucleotides of defined composition, both rate acceleration and stereoselectivity were found to be strongly sequence-dependent. Interestingly, the sequences that gave rise to the highest enantiomeric excess values also caused the largest rate accelerations [6a].

The concept was soon expanded to other reactions, keeping salmon sperm DNA, Cu^{2+}, and intercalating ligands (mostly dmbipy). Using these systems, asymmetric fluorinations, Michael and oxa-Michael additions, Friedel–Crafts reactions were reported, as well as a syn-hydration (in water), and an intramolecular cyclopropanation [6c, 8].

Rather than using double-stranded DNA, other authors utilized quadruplex-forming DNA oligonucleotides and observed pronounced effects of the modification position and the secondary structure of the oligonucleotide as well as the nature and concentration of cations and polymeric additives on the activity and selectivity of the catalyst [9]. G-quadruplex structures are topologically more complex, and their structural diversity is greater than double-stranded DNA. Using G-quadruplexes in combination with Cu(II) ions, asymmetric Diels–Alder and Friedel–Crafts reactions were reported. Two different, well-characterized, G-quadruplex-forming sequences were used as chiral source for these reactions.

A detailed understanding of structure–function relations could not, however, be established so far in DNA-based hybrid catalysis. Likewise, applications in practical (preparative) synthesis are still lacking [3a, 5b].

18.4 Organometallic Chemistry with Nucleic Acids

In contrast to the work described above, which involved mainly Lewis-acid catalysis, our laboratory had focused primarily on organometallic reactions in which the (transient) formation of a carbon–metal bond is instrumental to the process. Furthermore, we decided not to investigate intercalation-based systems, as they do not allow positional control of the DNA–metal interaction, thereby precluding a structure-based understanding. Methodologically, we developed two different approaches: a rational one, in which systematic variations of structural parameters were carried out, and a combinatorial one, in which catalytically active (and hopefully selective) DNA or RNA ligands should be isolated (selected) on the basis of their catalytic properties from huge combinatorial libraries. Reactions to be studied were iridium-catalyzed allylic aminations and rhodium-catalyzed conjugate additions.

These approaches required the development of several new methods, which will be described in the following paragraphs.

18.5
Combinatorial Selections of Catalysts from Nucleic Acid Libraries

Since the early 1990s, numerous catalysts have been identified from combinatorial RNA libraries by iterative deconvolution using the systematic evolution of ligands by exponential enrichment (SELEX) approach [10]. In these experiments, a population of different RNAs (typically $\sim 10^{14}$ sequences, i.e., different chemical compounds) is challenged for a specific task (like binding to a target molecule or reacting with a partner molecule in a catalytic manner), and the selection process is designed such that few active sequences are retained and then enzymatically amplified. To observe a significant enrichment of active sequences over background, 8–15 iterative rounds are usually conducted, at which point these species dominate the enriched library.

This combinatorial work has led to fundamentally new insight into the catalytic properties of RNA [11]. However, it only allowed the selection of RNA-modifying reactions, and most published "artificial ribozymes" turned out to catalyze only a single turnover. The restriction to RNA-modifying reactions could, however, be overcome by the introduction of tethered reactants, which – theoretically – gave access to all kinds of bimolecular bond-forming reactions [12]. A Diels-Alderase ribozyme identified in our lab using this methodology arguably constitutes the most characterized artificial ribozyme known to date [12a, 13].

For the combinatorial selection of RNA (or DNA)–transition-metal catalysts, further elements have to be developed and integrated into the scheme (Figure 18.3). In addition to a tethered substrate, a site-specifically attached transition metal ligand needs to be present in each molecule of the library. After loading with the metal, it should allow formation of the catalytically active species, preferably with the reactant tethered to the same RNA molecule. The other reactant carries a purification tag, allowing the selective isolation of only those species in which a reaction had taken place. A further nontrivial requirement is that the attachment of the metal–ligand complex to DNA or RNA does not interfere with the enzymatic copying steps (transcription, reverse transcription (RT), polymerase chain reaction (PCR)).

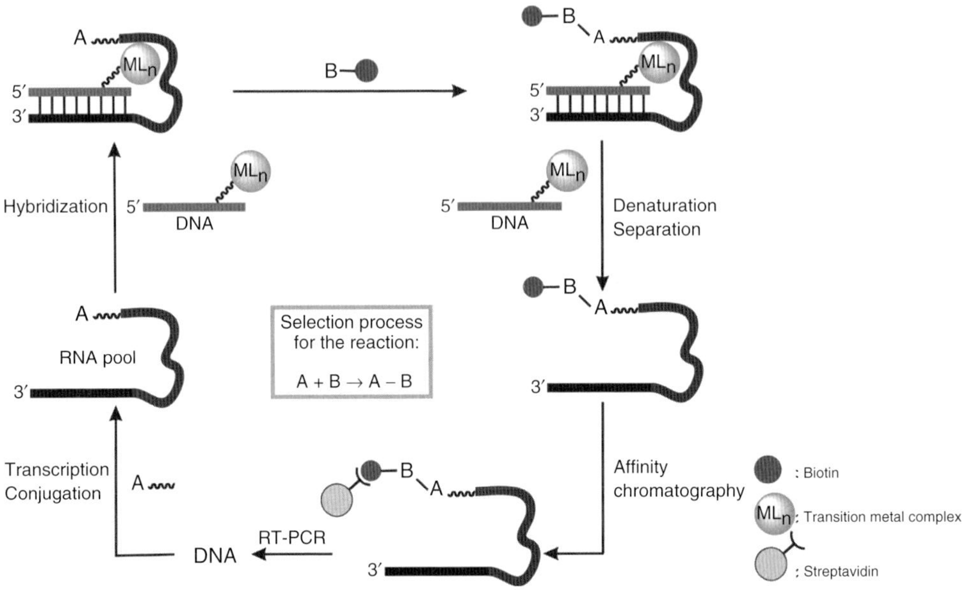

Figure 18.3 *In vitro* selection of RNA-based hybrid catalysts with DNA-appended transition-metal complexes.

18.6
Site-Specific Internal Functionalization of Nucleic Acids with Transition-Metal Ligands and Other Moieties

Both the rational and the combinatorial approaches require the site-specific functionalization of nucleic acids with transition-metal ligands. As many of these ligands are not stable under the conditions of phosphoramidite solid-phase synthesis, mild postsynthetic approaches are preferred. We first explored Verdine's convertible nucleoside approach [14] in which a uridine triazolide is incorporated during the solid-phase synthesis, and the triazole moiety is later substituted by the desired ligand which – for this purpose – has to be presented as an amine derivative. This approach was utilized for the incorporation of various structurally highly diverse ligands into DNA [15]. Representative examples are shown in Scheme 18.1. While the conditions for derivatization are indeed rather mild, the convertible nucleoside approach suffers from two disadvantages: (i) For attachment, one of the regular Watson–Crick positions is utilized, so that the attachment of the ligand may disturb base pairing and higher order structure formation of the modified oligonucleotide. (ii) Efficient derivatization typically requires a huge excess (~1000-fold) of the ligand to be coupled, which renders the method synthetically inefficient. Therefore, alternative approaches for the site-specific internal modification are desired.

Diels–Alder cycloaddition reactions with normal electron demand were first used for nucleic acid derivatization and conjugation in 1999 [17], but did not

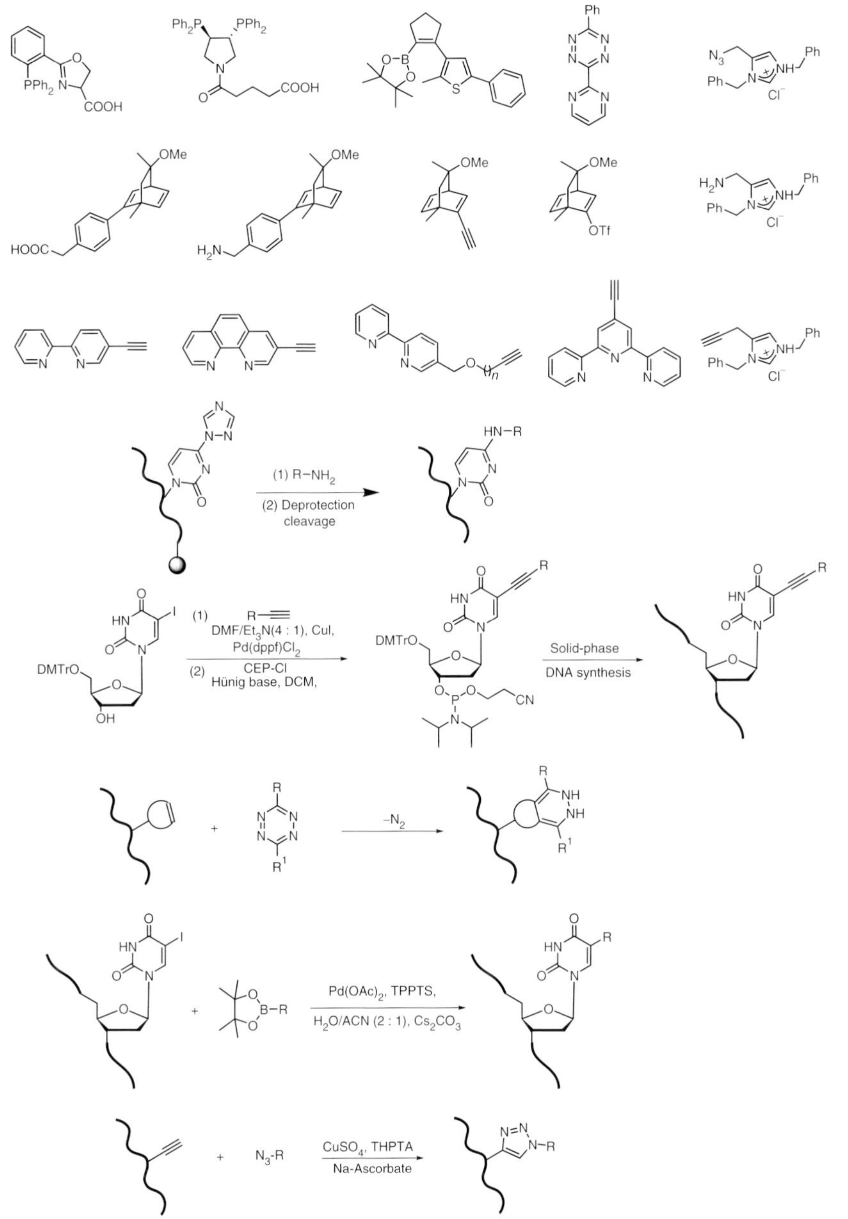

Scheme 18.1 Collection of different ligands and other moieties attached to oligonucleotides (top) using different chemical approaches. From top: convertible nucleoside approach, Sonogashira coupling, inverse-electron-demand Diels–Alder reaction, Suzuki–Miyaura coupling, and CuAAC [15, 16].

become standard tools in nucleic acid chemistry, mainly because of their low reaction rate. On the other hand, DAR with inverse electron demand, and in particular those that employ tetrazine dienes, became popular in recent years because of their high reactivity and mild reaction conditions [18]. Until 2009, however, they had not found application in nucleic acid chemistry. We developed phosphoramides and derivatized solid supports that allow the site-specific functionalization of oligonucleotides with a dienophile at virtually any position: 3′-terminal, 5′-terminal, and internal [16b, 19]. The subsequent reaction with a tetrazine-based reactant proceeds quantitatively in water at room temperature, rendering this chemistry very attractive for the attachment of sensitive functionalities. While the first norbornene-based generation of building blocks yielded rate constants of $\sim 20\,M^{-1}\,s^{-1}$ and therefore required reaction times of 1 h and more [16b], the next generation (*trans*-cyclooctenes) was orders of magnitude more reactive (k up to $400\,000\,M^{-1}\,s^{-1}$), guaranteeing completion of the reaction within seconds even at low micromolar concentration of both reactants, as typical in biomolecular conjugations [19c]. Recently, this approach could also be extended to deoxyribonucleoside triphosphates, allowing the enzymatic incorporation of dienophiles into DNA [19b]. These modified nucleotides were efficiently incorporated by an engineered DNA polymerases, giving access to highly functionalized replicable polymers with a wide variety of potential applications. The inverse-electron-demand DAR furthermore turned out to be fully orthogonal to the copper-catalyzed alkyne–azide cycloaddition (CuAAC), thereby allowing site-specific double modification with different moieties [19c].

Currently, copper-catalyzed and copper-free alkyne–azide cycloadditions are the most widely applied cycloaddition reactions for the modification of biomolecules. We developed an approach based on the combination of CuAAC or strain-promoted copper-free variants with enzymatic tailing (and optional enzymatic ligation), which allows the convenient postsynthetic introduction of an azide into existing unmodified RNA strands (chemically or enzymatically synthesized or even isolated from biological sources) at a desired position, and the subsequent functionalization with a wide variety of commercially available or self-synthesized functional alkynes [16c].

Besides cycloadditions, cross-coupling reactions were applied to the site-specific modification of DNA. In addition to Sonogashira couplings, we developed a selective variant of the Suzuki–Miyaura coupling reaction that allows the one-step postsynthetic modification of commercially available iodo-modified DNA with a variety of boronic acids to form functional DNA derivatives (in this case, photoswitchable derivatives) [16a]. This methodology shortened the synthetic pathway to such modified DNA by 10 synthetic steps. Recent work identified Stille–Migita coupling as a much milder and potentially very useful reaction (Krause and Jäschke, to be published).

While these methods were so far mostly shown to allow the incorporation of dyes, affinity handles, and convertible groups, they also enrich the arsenal of methods for the introduction of transition-metal ligands. Cycloaddition and cross-coupling reactions were successfully utilized to attach a variety of diene, bipyridine, terpyridine, phenanthroline, and N-heterocyclic carbene (NHC) precursor ligands to DNA under mild conditions (Scheme 18.1). CuAAC turned out to cause problems, as contaminating copper ions interfered with metallation of the ligands at later stages.

18.7 Metallation of DNA–Ligand Conjugates

The characterization of DNA–ligand–metal complexes is generally difficult and ambitious, as these complexes are typically labile, dissociate during purification, and the amounts prepared are not sufficient for NMR characterization. The iridium(I) complexes of the PHOX ligands, however, were exceptionally stable and could be preparatively purified by reversed-phase HPLC (Figure 18.4). The chromatogram indicates full conversion. Furthermore, the DNA–metal complexes could be characterized by electrospray ionization Fourier transform ion cyclotron resonance (ESI-FT-ICR) mass spectrometry. The spectrometric information confirmed the direct interaction of the iridium ion with the PHOX ligand and ruled out unspecific DNA–metal interactions.

18.8 Site-Specific Terminal Functionalization of Nucleic Acids with Substrates

The selection of catalysts from large combinatorial RNA libraries based on their ability to form a covalent bond between the two substrates requires one substrate to be covalently tethered to all members of the library while the other substrate to carry a purification tag that allows the subsequent distinction of reacted from unreacted RNA species (Figure 18.3). Different one- or two-step protocols were developed for the selective modification of the 5′-end enzymatically synthesized RNA species with potential substrates and other groups [17, 21]. These include allylic substrates, dienes (both normal and inverse electron demand), and also potent nucleophiles. One very efficient strategy involves the incorporation of aldehyde groups (protected as acetals) by T7 RNA polymerase using suitably protected guanosine monophosphates [22]. This initiator nucleotide was efficiently incorporated into RNA, and the modified RNAs were quantitatively coupled to reaction partners displaying a hydrazine moiety at one end, such as peptide derivatives. As the conjugation reaction occurs postsynthetically under mild conditions, rather complex conjugates could be assembled, allowing the introduction of additional features, such as an enzyme cleavage site [22a]. In an extension of this work, initiator nucleotides were synthesized that consisted of a guanosine monophosphate, a polydisperse polyethylene glycol spacer, and a protected aldehyde, and introduced into RNA with 52% yield. Postsynthetic conversion with substrate hydrazides allowed easy and mild access to complex RNA–tether–substrate conjugates [22b].

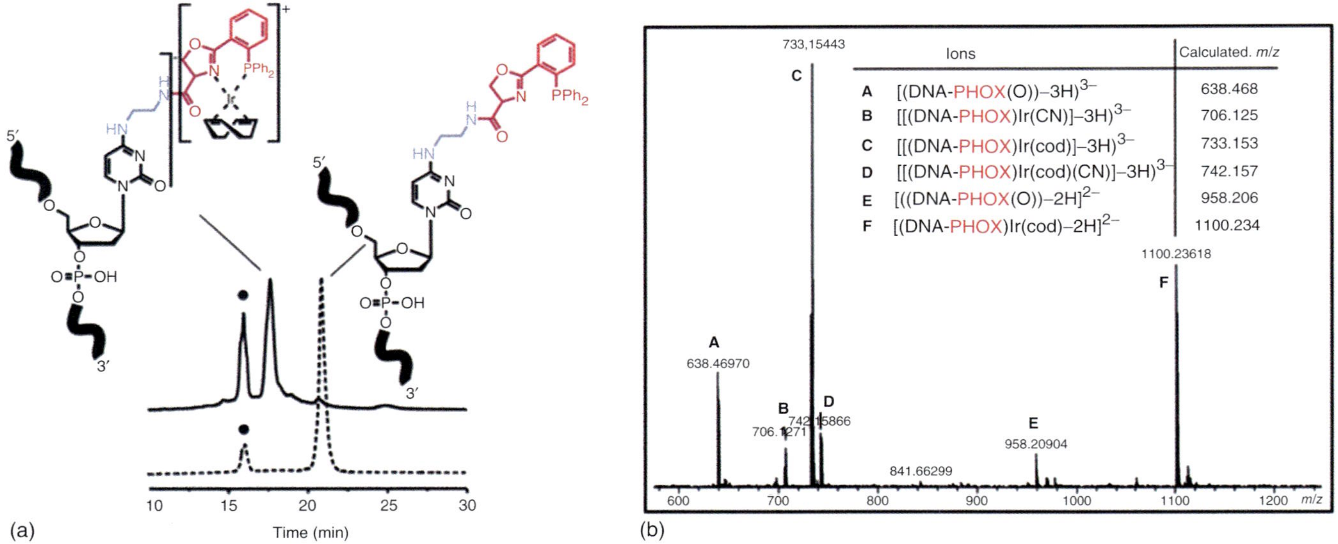

Figure 18.4 Metallation reaction between a PHOX–DNA pentanucleotide and $[Ir(cod)Cl]_2$. (a) HPLC analysis; dashed line: starting material; solid line: reaction mixture; peak with dot: PHOX oxidation product. (b) ESI-FT-ICR mass spectrum [20].

The inverse-electron-demand DAR described above can also be applied to the synthesis of terminally modified RNA transcripts. A norbornene-derivatized initiator nucleotide was synthesized, efficiently incorporated into RNA, and used for the mild preparation of RNA–peptide conjugates [19a]. A selection of tethered substrates is shown in Figure 18.5.

18.9
Allylic Aminations by DNA-Based Hybrid Catalysts

In this project, a rational design approach was used in which a 19mer oligodeoxynucleotide (ODN) carrying a diene ligand was combined with different DNA (cDNA) or RNA (cRNA) complementary strands, thereby creating diversity [15b]. The attachment of the ligand allows its specific, reproducible positioning on rationally predictable nucleic acid duplexes.

Although diene ligands have recently drawn considerable interest, they had not been used as anchoring ligands in hybrid catalysis previously. We selected the bicyclo[2.2.2]octadiene scaffold since it had shown high activity in Ir(I)-catalyzed allylic substitutions and provides convenient positions for derivatization. Both enantiomers of the diene ligand **L2** were synthesized, allowing us to vary the DNA–metal distance and the absolute ligand configuration and thereby to modulate the interaction between the carrier ODN and metal complex. The resulting ligands were tested in the Ir-catalyzed allylic substitution of phenyl allyl acetate with morpholine in aqueous medium (Figure 18.6).

The DNA-based dienes formed highly active catalysts, and no difference in activity was observed between single- and double-stranded hybrid catalysts. Thus, DNA and its numerous functional groups do not interfere with the metal ion coordination. The stability of the catalyst under these conditions is remarkable: up to 4600 turnovers were observed.

For evaluating the stereoselectivity of the hybrid catalysts in the kinetic resolution of phenyl allyl acetate, the diene-carrying DNA oligonucleotides were allowed to hybridize with various DNA and RNA complementary strands in order to modulate the shape of the environment surrounding the iridium complex. While the obtained enantiomeric excesses remained modest, a strong chirality transfer from the nucleic acid to the iridium complex could be observed. In the absence of a complementary strand, the hybrid catalysts displayed similar enantiomeric excess values as the corresponding free dienes, and enantiomeric dienes showed enantiomeric excess values opposite of each other: they behaved as if they felt hardly any influence of the DNA they were bound to. The enantiomeric excess values, however, changed if a complementary sequence was added, the most striking example being the addition of the perfectly complementary RNA strand to the DNA-carrying ODN, which triggered a reversal of the stereoselectivity, with the enantiomeric excess for the product going from +23 to −27. Interestingly, the effect was much less pronounced when the complementary sequence was a DNA strand. The introduction of bulges had an unpredictable

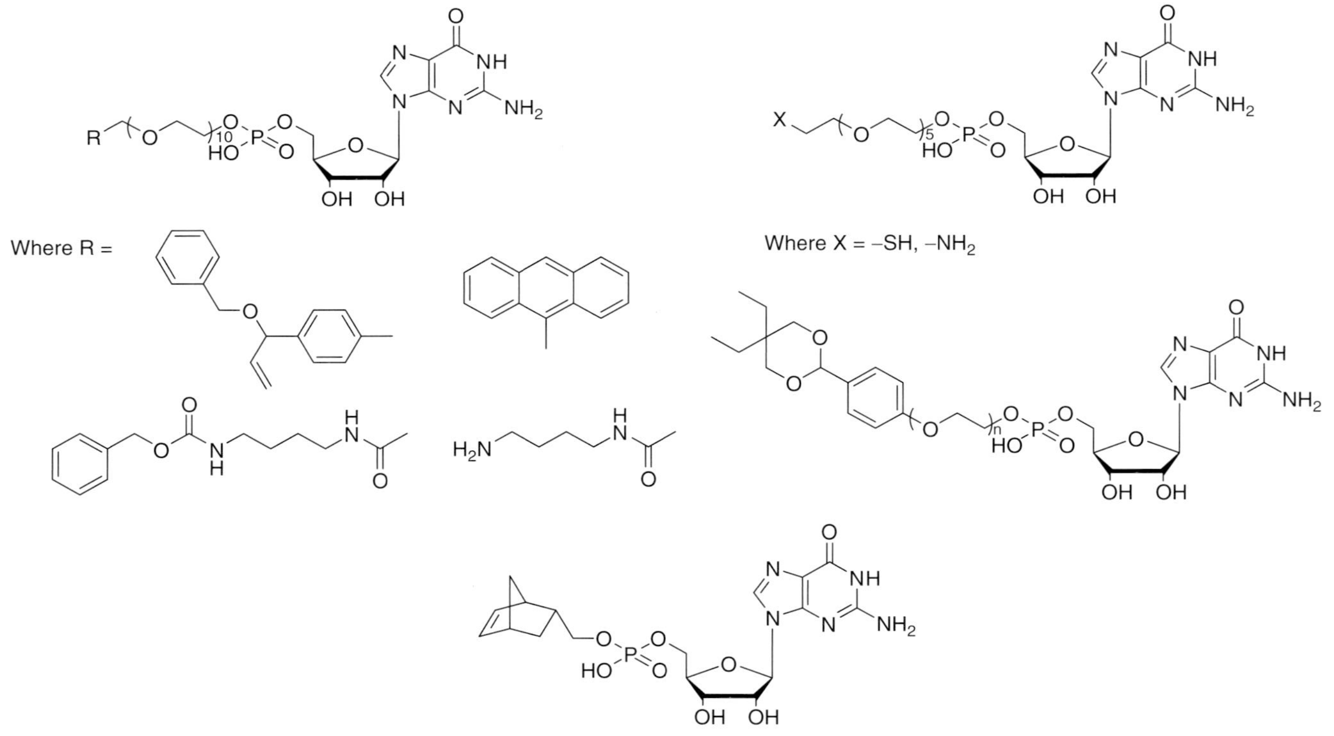

Figure 18.5 Different initiator nucleotides synthesized for one-step or two-step incorporation into enzymatically synthesized RNA [19a, 21, 22].

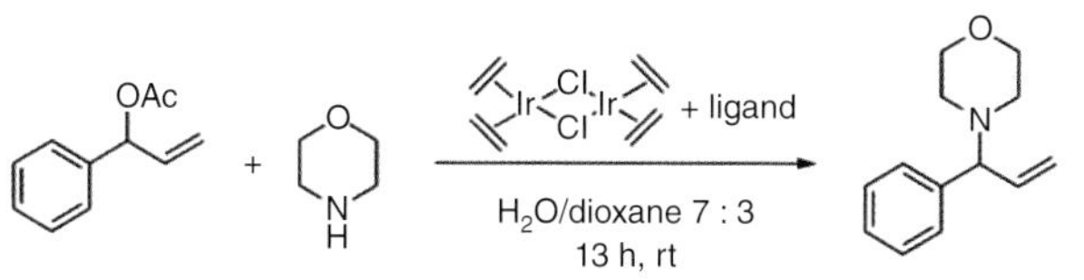

Entry	Ligand	Yield (%)	ee (%)
1	–	3	
2	**L2**	81	
3	**L2** + DNA	80	
4	L	88	
5	**L2**	48	28
6	L	49	23
7	L DNA	45	9
8	L RNA	48	–27

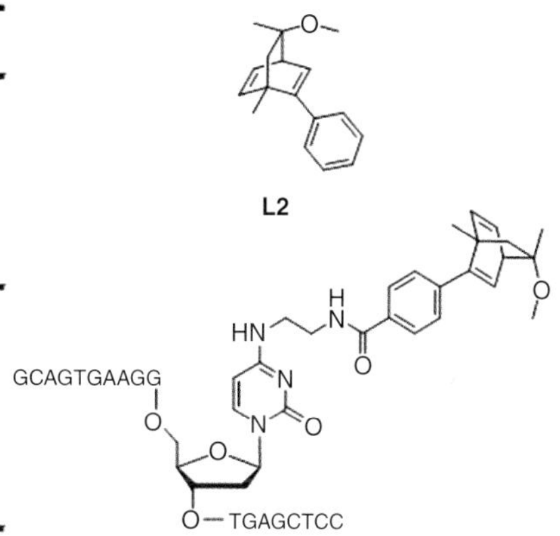

Figure 18.6 Stereoselective allylic amination by DNA–diene conjugates [15b].

influence on the stereoselectivity, demonstrating that small structural variations may have strong effects, and thereby highlighting the potential of screening large oligonucleotide libraries.

PHOX ligands, on the other hand, were also found to be active in allylic aminations. Hybrid catalysts were generated *in situ* from $[Ir(cod)Cl]_2$ (cod, cyclooctadiene) and DNA–PHOX conjugates. The activity was found to be strongly sequence-dependent; guanosine residues in the direct vicinity of the tethered PHOX ligands led to a reduction in activity. As guanosine is known to interact with transition metals via its N7 ring nitrogen, this position was assumed to be responsible for this effect. This assumption was supported by the observation that 7-deazaguanosine-containing oligonucleotides were fully active. In contrast to the Ir–diene system described above, enantioselectivity was very low, and the effects of added perfect and imperfect duplexes were small [20] (Figure 18.7).

18.10
Summary and Outlook

The approaches and results summarized here illustrate the potential and advantages of DNA-based hybrid catalysts. While practical applications in preparative organic synthesis have still not been reported, it is now clear that DNA can efficiently modulate the stereochemical outcome of various Lewis acid-catalyzed and organometallic reactions, and can also provide rate acceleration. Both,

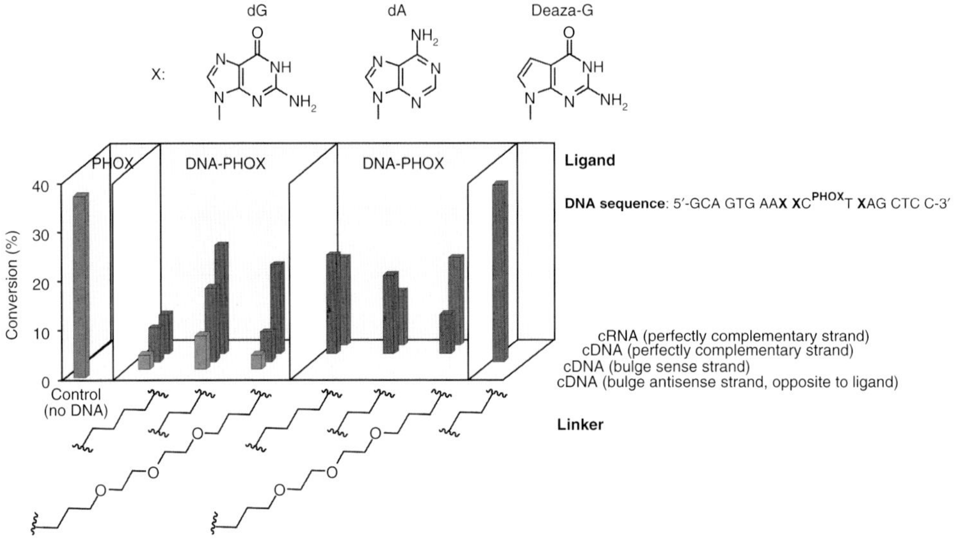

Figure 18.7 Allylic amination by DNA–PHOX–Ir(I) conjugates. Same reaction as in Figure 18.6.

noncovalent and covalent modification approaches have led to novel catalysts, with the former performing better and the latter providing a better basis for a rational understanding. While theoretically nucleic acids have many advantages over proteins for the direct evolution of catalysts, this superiority has still to be practically demonstrated.

References

1. Wilson, M.E. and Whitesides, G.M. (1978) *J. Am. Chem. Soc.*, **100**, 306.
2. (a) Reetz, M.T., Peyralans, J.J., Maichele, A., Fu, Y., and Maywald, M. (2006) *Chem. Commun. (Camb.)*, 4318; (b) Skander, M., Humbert, N., Collot, J., Gradinaru, J., Klein, G., Loosli, A., Sauser, J., Zocchi, A., Gilardoni, F., and Ward, T.R. (2004) *J. Am. Chem. Soc.*, **126**, 14411.
3. (a) Deuss, P.J., den Heeten, R., Laan, W., and Kamer, P.C. (2011) *Chem. Eur. J.*, **17**, 4680; (b) Ringenberg, M.R. and Ward, T.R. (2011) *Chem. Commun. (Camb.)*, **47**, 8470; (c) Ward, T.R. (2011) *Acc. Chem. Res.*, **44**, 47.
4. Höbartner, C. and Silverman, S.K. (2007) *Biopolymers*, **87**, 279.
5. (a) Park, S. and Sugiyama, H. (2012) *Molecules*, **17**, 12792; (b) Silverman, S.K. (2010) *Angew. Chem. Int. Ed.*, **49**, 7180.
6. (a) Boersma, A.J., Klijn, J.E., Feringa, B.L., and Roelfes, G. (2008) *J. Am. Chem. Soc.*, **130**, 11783; (b) Roelfes, G. (2007) *Mol. Biosyst.*, **3**, 126; (c) Coquiere, D., Feringa, B.L., and Roelfes, G. (2007) *Angew. Chem. Int. Ed.*, **46**, 9308; (d) Roelfes, G. and Feringa, B.L. (2005) *Angew. Chem. Int. Ed.*, **44**, 3230.
7. Roelfes, G., Boersma, A.J., and Feringa, B.L. (2006) *Chem. Commun. (Camb.)*, 635.
8. (a) Boersma, A.J., Coquiere, D., Geerdink, D., Rosati, F., Feringa, B.L., and Roelfes, G. (2010) *Nat. Chem.*, **2**, 991; (b) Boersma, A.J., Feringa, B.L., and Roelfes, G. (2009) *Angew. Chem. Int. Ed.*,

48, 3346; (c) Megens, R.P. and Roelfes, G. (2012) *Chem. Commun. (Camb.)*, **48**, 6232; (d) Oelerich, J. and Roelfes, G. (2013) *Chem. Sci.*, **4**, 2013; (e) Shibata, N., Yasui, H., Nakamura, S., and Toru, T. (2007) *Synlett*, 1153.
9. (a) Roe, S., Ritson, D.J., Garner, T., Searle, M., and Moses, J.E. (2010) *Chem. Commun. (Camb.)*, **46**, 4309; (b) Wang, C., Jia, G., Li, Y., Zhang, S., and Li, C. (2013) *Chem. Commun. (Camb.)*; (c) Wang, C., Jia, G., Zhou, J., Li, Y., Liu, Y., Lu, S., and Li, C. (2012) *Angew. Chem. Int. Ed.*, **51**, 9352; (d) Wang, C., Li, Y., Jia, G., Liu, Y., Lu, S., and Li, C. (2012) *Chem. Commun. (Camb.)*, **48**, 6232.
10. (a) Ellington, A.D. and Szostak, J.W. (1990) *Nature*, **346**, 818; (b) Tuerk, C. and Gold, L. (1990) *Science*, **249**, 505.
11. Wilson, D.S. and Szostak, J.W. (1999) *Annu. Rev. Biochem.*, **68**, 611.
12. (a) Seelig, B. and Jäschke, A. (1999) *Chem. Biol.*, **6**, 167; (b) Tarasow, T.M., Tarasow, S.L., and Eaton, B.E. (1997) *Nature*, **389**, 54.
13. (a) Kobitski, A., Nierth, A., Helm, M., Jäschke, A., and Nienhaus, G. (2007) *Nucleic Acids Res.*, **35**, 2047; (b) Kraut, S., Bebenroth, D., Nierth, A., Kobitski, A.Y., Nienhaus, G.U., and Jäschke, A. (2012) *Nucleic Acids Res.*, **40**, 1318; (c) Seelig, B., Keiper, S., Stuhlmann, F., and Jäschke, A. (2000) *Angew. Chem., Int. Ed.*, **39**, 4576; (d) Serganov, A., Keiper, S., Malinina, L., Tereshko, V., Skripkin, E., Höbartner, C., Polonskaia, A., Phan, A., Wombacher, R., Micura, R., Dauter, Z., Jäschke, A., and Patel, D. (2005) *Nat. Struct. Mol. Biol.*, **12**, 218; (e) Stuhlmann, F. and Jäschke, A. (2002) *J. Am. Chem. Soc.*, **124**, 3238.
14. MacMillan, A.M. and Verdine, G.L. (1990) *J. Org. Chem.*, **55**, 5931.
15. (a) Caprioara, M., Fiammengo, R., Engeser, M., and Jäschke, A. (2007) *Chem. Eur. J.*, **13**, 2089; (b) Fournier, P., Fiammengo, R., and Jäschke, A. (2009) *Angew. Chem. Int. Ed.*, **48**, 4426.
16. (a) Cahova, H. and Jäschke, A. (2013) *Angew. Chem. Int. Ed.*, **52**, 3186; (b) Schoch, J., Wiessler, M., and Jäschke, A. (2010) *J. Am. Chem. Soc.*, **132**, 8846; (c) Winz, M.L., Samanta, A., Benzinger, D., and Jäschke, A. (2012) *Nucleic Acids Res.*, **40**, e78.
17. Seelig, B. and Jäschke, A. (1997) *Tetrahedron Lett.*, **38**, 7729.
18. Devaraj, N.K. and Weissleder, R. (2011) *Acc. Chem. Res.*, **44**, 816.
19. (a) Schoch, J., Ameta, S., and Jäschke, A. (2011) *Chem. Commun. (Camb.)*, **47**, 12536; (b) Schoch, J. and Jäschke, A. (2013) *RSC Adv.*, **3**, 4181; (c) Schoch, J., Staudt, M., Samanta, A., Wiessler, M., and Jäschke, A. (2012) *Bioconjug. Chem.*, **23**, 1382.
20. Caprioara, M. (2007) DNA-Based Ligands for Use in Asymmetric Catalysis and Development of Metallo-(deoxy) Ribozymes, PhD thesis. Ruprecht-Karls-Universität, Heidelberg.
21. (a) Fiammengo, R., Musilek, K., and Jäschke, A. (2005) *J. Am. Chem. Soc.*, **127**, 9271; (b) Schlatterer, J.C. and Jäschke, A. (2006) *Biochem. Biophys. Res. Commun.*, **344**, 887.
22. (a) Pfander, S., Fiammengo, R., Kirin, S.I., Metzler-Nolte, N., and Jäschke, A. (2007) *Nucleic Acids Res.*, **35**, e25; (b) Wahl, F. and Jäschke, A. (2012) *Biochem. Biophys. Res. Commun.*, **417**, 1224.

Part V
Studies of Immobilized Catalysts – Introduction

Oliver Trapp and Johannes Troendlin

V.1
Introduction

Catalysis of organic reactions represents one of the most sustainable technologies that allow us to convert raw materials into valuable chemicals in an efficient and economical way [1]. Despite their high activity and selectivity, the application of (asymmetric) homogeneous catalysts in industrial processes is still rare partly because of the difficulties of separation and recycling of the sometimes rather expensive catalysts and ligands. To overcome this problem, many strategies were tested for almost a century. In recent years, immobilizations of homogeneous catalysts on suitable supports are of growing academic and industrial interest, as this approach combines the advantages of homogeneous and heterogeneous catalysts [2–9]. These resulting heterogeneous catalysts are characterized by high activity and selectivity and can be recycled and reused many times because separation from the reaction mixtures is facilitated. Additional advantages are minimization of metal traces in the product, improved handling and process control, and the possibility to screen various reactions quickly by performing high-throughput (ht) measurements [10] when immobilized ligands attached to an inorganic or organic support are used to generate large libraries of catalysts with different metals and coordinated ligands [11]. The application of the "one-bead-one-compound" concept used in combinatorial chemistry can be envisioned. Furthermore, in some cases immobilized catalysts are even more selective or active than their nonimmobilized counterparts.

For immobilized catalysts, a (chiral) ligand or a preassembled metal complex, which has previously proven to be efficient in nonsupported transformations, is used as starting material. Covalent attachment of catalysts, either via copolymerization or via covalent attachment (anchoring), is the most often used immobilization method, which can also be applied to synthesize soluble immobilized catalysts.

Noncovalent attachment of the catalytic systems can be realized by the adsorption of the catalyst to the support, by electrostatic interactions between the ligand and the support, or by entrapment in porous solid supports. Alternatively, the

Molecular Catalysts: Structure and Functional Design, First Edition. Edited by Lutz H. Gade and Peter Hofmann.

ligand can be synthesized on the support. This part of the book focuses mainly on covalently attached metal–ligand complexes.

V.2 Covalent Immobilization of Catalysts

V.2.1 Selection of Appropriate Spacers

The linkers are the crucial, according to our experience.

Independent of the procedure, covalent binding always requires modification of the ligand structure by introducing a linker and a spacer between the catalyst and the chemical support.

The linker provides a functional group that allows the formation of a chemical bond between the ligand and the support material, whereas the spacer serves to ensure accessibility of the bonded catalyst to possible substrates. It therefore represents a crucial part in any immobilization, as it greatly influences the activity, selectivity, and lifetime of any given immobilized catalyst.

There are two principal strategies for the selection of an appropriate linker for a given set of catalyst and support material. One is to choose a long linker to avoid interference of the catalyst with the support material; the other is to use a short linker in order to generate additional interactions with the solid support material.

The choose the optimal linker length for a specific task, many factors have to be considered, such as (i) the size of the catalyst, (ii) the desired catalyst loading, (iii) the accessibility of the active site, and (iv) the solvent. Larger catalysts usually perform better with longer spacers, as do systems with high catalyst loading. Typically, a very long linker will decrease the accessibility, and a very short linker will cause increased interactions with the support material. Therefore, in most cases a spacer length of 7–11 carbon atoms was found to be optimal, although many exceptions with very short and very long spacer lengths are described in the literature.

V.2.2 Immobilization Strategies

There are many factors that determine the immobilization strategy: (i) the nature of the catalyst, (ii) the position of the active site, (iii) the remaining functional groups that can be used for immobilization, (iv) the kind of functional groups that should not be influenced or changed by the immobilization, and, finally, (v) the desired solid support material.

Taking all these factors into account, an immobilization strategy can be chosen. An overview of the numerous possibilities is given in Figure 1.

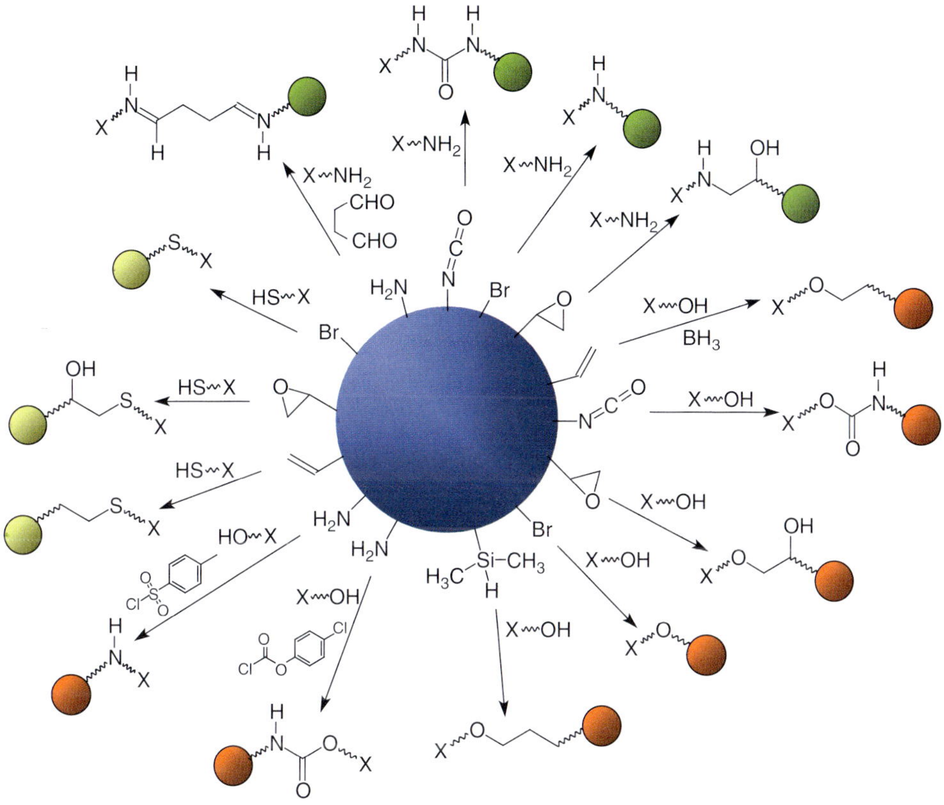

Figure 1 Methods for the covalent binding of catalysts to a solid support. X: selector and O: support material.

V.3
Support Materials

V.3.1
General Aspects

In contrast to classical homogeneous synthesis, reactions can also be carried out on or in insoluble support materials. These support materials might be inorganic silica particles or organic polymers. The physical properties of the solid support materials and the possible applications vary according to the material from which the carrier is constructed.

Still, there are some general aspects that apply to all types of solid support: They

- must be physically stable under the applied conditions;
- must be inert to all the reagents and solvents used;
- must allow the attachment of a catalyst.

V.3.2
Silica

Silica is one of the first materials that were used in column chromatography. Silica gel is insensitive to most solvents and stable even under high pressure. Various particle sizes are used for different separation purposes, ranging from a few micrometers to about 100 μm. As it is made from pure sodium silicate, the hydroxyl groups on the surface lead to a high polarity, which results in longer retention times for polar compounds. Functionalizing the hydroxyl groups, for example, by chemically bonding C18 groups, affords a drastic change in the stationary phase parameters. Now the nonpolar compounds are retained most. This type of stationary phases is called *reversed phase* (RP).

Other functionalizations with more complex selectors or catalysts are very common. These materials are used as chromatographic stationary phases or insoluble reagents, for example, in heterogeneous catalysis.

V.3.3
Organic Resins

Since the development of solid-phase peptide synthesis by Merrifield in the 1960s [12], the approach of covalently bonded (chiral) ligands and catalysts has gained increased attention.

There are several organic resins in use today. All of them have in common that the reaction sites are not only attached to the surface of the material, as the term *solid support* would suggest, but the reaction sites are distributed within the support material itself. This requires a certain amount of swelling to take place, in order for the reactant molecules to be able to penetrate the support material and reach the reaction sites.

Examples of the organic solid support materials used today are as follows:

- *Gel-type supports*: Highly solvated polymers with an equal distribution of functional groups, for example, polystyrene, polyacrylamide (hydrophilic alternative to polystyrene), and poly(ethylene glycol) (PEG).
- *PEG-based supports*: Composed of a PEG–polypropylene glycol network or PEG with polyamide or polystyrene.
- *Surface-type supports*: Materials developed for surface functionalization, for example, cellulose fibers, and highly cross-linked polystyrene.
- *Composites*: Gel-type polymers supported by rigid matrices.

V.3.4
Polysiloxanes

Poly(dimethyl siloxane)s are optically clear, inert, nontoxic, and nonflammable, and possess excellent thermal, oxidative, chemical, and biological stability. Siloxanes do not swell in water, but most organic solvents are able to penetrate into the material and cause it to swell. Linear polysiloxanes, which are routinely used as stationary phases for gas chromatography [13] or in microfluidic devices, have only been rarely used as supporting material for catalyst immobilization [14, 15]. This is surprising because polysiloxanes can show positive effects in catalysis [16].

Immobilization of homogeneous catalysts onto a polysiloxane or silica matrix can be realized by a hydrosilylation reaction of the polymeric Si–H groups with a suitable ligand system. This strategy, developed by Schurig *et al.* [17], offers many advantages, and can not only be applied to modify the polysiloxane coating of microcapillaries but also be used to modify the capillary wall of fused silica capillaries. Modified capillaries, produced with or without coating, can then be used in different modes as capillary format on column chromatographic reactors. The capillaries coated with catalyst-modified polysiloxanes are characterized by an extraordinarily large specific interfacial area per volume. Because the modified polymer simultaneously serves as a catalyst and a chromatographic separation material, catalysis and separation can in such systems be combined in a single capillary [18–20]. This permits performing ht measurements of large substrate libraries. Such reactors are not limited to gas chromatographic systems but can be transferred to a broad range of stationary phases, for example, for gas–liquid, gas–liquid–solid, or liquid–liquid systems [21].

V.4
Examples of Immobilized Catalyst Systems

V.4.1
Immobilized Phosphine Ligands

Blümel *et al.* [22–32] presented a strategy where previously successful homogeneous molecular catalysts are covalently bound to oxidic supports. Subsequent analysis and investigation was performed by conventional and suspension solid-state NMR spectroscopy.

Silica was used as support material for the immobilization of phosphine ligands. As the surface of the silica is able to react with the phosphine ligands, an improved spacer/linker system had to be applied in order to shield the substituents and to prolong the lifetime of the catalyst. The spacer/linker system is therefore designed to prevent entirely the contact of the catalysts with the aggressive surface. Two new linker systems were tested to achieve this goal.

Linkers of the tetraphenylmethane type, such as depicted in Figure 2a, shield the phosphine groups and their coordinated metal complexes from the surface by the rigidity of the backbone. Such ligand systems allow the investigation of cooperative effects, even with heterobimetallic systems. These materials are also well suited to study the mobility of the surface species.

The second approach utilizes several alkyl chains as substituents at the coordinated chelate phosphines. These are able to efficiently shield the catalysts from the reactive surface of the support material (cf. Figures 2b,c). The alkyl chains also prevent dimerization or even oligomerizations, even in cases with high catalyst

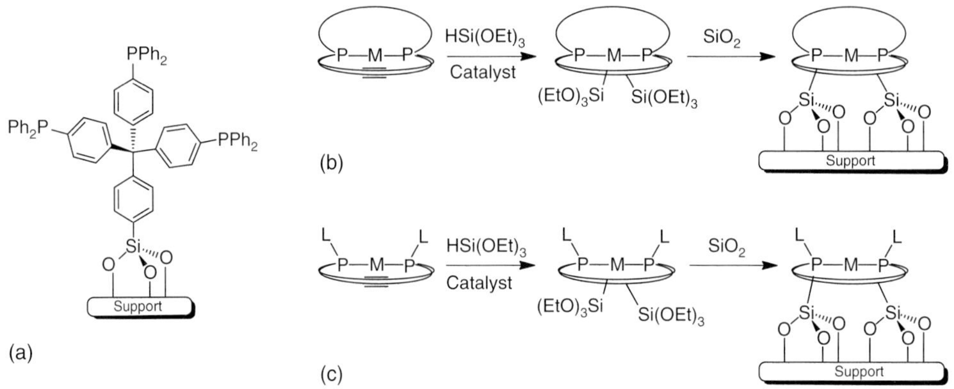

Figure 2 (a–c) Immobilization of phosphine ligands to a silica support.

loading. Varying the length and the number of the alkyl chains allows tuning the product selectivity.

V.4.2
Immobilized NHC Ligands

Blechert *et al.* described the covalent bonding of Grubbs and Hoveyda–Blechert type metathesis catalysts to a support material via the N-heterocyclic carbene (NHC) ligands [33–35]. They used a Merrifield-type resin as the polymeric backbone of the immobilized catalysts (cf. Figure 3a,b)).

The supported catalyst in Figure 3a by Fürstner *et al.* [36] was characterized by the excellent recyclability for up to three cycles with comparable activity to the homogeneous soluble analog.

A very elegant approach was developed by Hoveyda *et al.* [37], where NHC-based catalysts have been immobilized by sol–gel chemistry. The thus-obtained

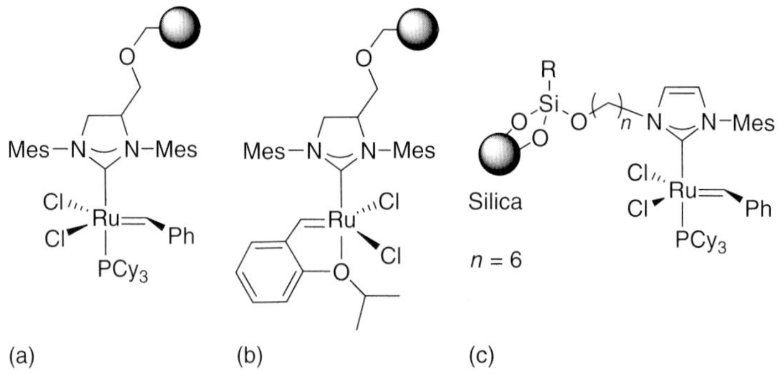

Figure 3 (a–c) Examples of Grubbs-type metathesis catalysts immobilized via the NHC ligand.

beads may be removed from the reaction solution and reused up to 10 times in reactions.

In 2011, Trapp *et al.* [38] presented a strategy for bonding 2-(pentafluorophenyl) imidazolidine to a polysiloxane and permanently coating this modified polymeric matrix on the inner surface of fused-silica capillaries (id 250 μm) to obtain thin films of defined thicknesses (50 – 1000 nm).

The immobilized NHC preligands obtained by this strategy were then converted to free carbene ligands by flushing these capillaries with suitable reagents. The use of different organometallic precursors (Ru, Au) allowed the production of catalytically active and stable separation columns. For the synthesis of the NHC preligand system, *N*,*N*′-dimesityl-2,3-diamino-propane-1-ol served as starting material. The conversion of the hydroxy group of the alcohol into the allylether was simply achieved by a Williamson's ether synthesis. This alkene chain was reacted in a final step with hydridomethyldimethylpolysiloxanes (HMPSs) via a hydrosilylation reaction to yield a polymer-bound 1,3-bismesityl-2-imidazolidinylidene ligand (cf. Figure 4). It is important to note that the 1,3-bismesityl-2-imidazolidinylidene ligand can also be converted into ligand precursors with various spacer lengths. However, a total spacer length of about 5 – 10 Å has proven to be ideal because intramolecular interactions of the ligand with the polymer backbone caused by gauche conformations [39] are diminished while sufficient flexibility of the ligand is retained.

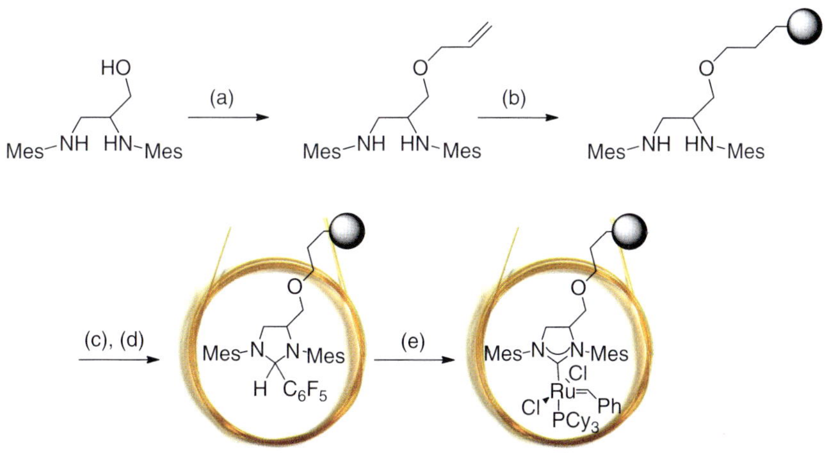

Figure 4 Synthesis of permanently bonded polymeric Grubbs second-generation catalyst in a fused-silica capillary. (a) 3-Bromoprop-1-ene, NaH, tetrahydrofuran (THF), 64 °C, 24 h. (b) HMPS, Karstedt's catalyst, THF ultrasonication, 3 h. (c) Static coating of modified polysiloxane onto the inner surface of fused-silica capillary, 180 °C, N_2. (d) pentafluorobenzaldehyde, AcOH, 24 h. (e) $(Cy_3P)_2RuCl_2CHPh$, *n*-pentane, 60 °C.

To generate permanently bonded polymeric NHC ligand-based catalysts, the absence of a free base, for example, K*t*BuO, is necessary to avoid decomposition of the catalyst caused by the base and the remaining Si–H groups of the polymeric support.

As the immobilization requires high temperatures ($T = 180\,°C$), which would cause catalyst decomposition, the preligand system was permanently bonded onto the glass surface of the fused-silica capillary via Si–O–Si bonds before the *in situ* generation of the catalytically active system. Therefore, it was necessary to convert the permanently bonded diamine to the permanently bonded polymeric 2-(pentafluorophenyl)imidazolidine by using pentafluorobenzaldehyde [40].

Slight heating ($T = 60\,°C$) of the surface-modified capillaries produces the immobilized free carbene. Together with the appropriate organometallic precursors, the immobilized carbene yields the desired NHC ligand-based catalyst. Using this strategy, the reaction of the free carbene with Grubbs first-generation catalyst leads to the permanently bonded Grubbs second-generation catalyst by a simple ligand exchange against the tricyclohexylphosphine ligand, which can be monitored by ^{31}P-NMR spectroscopy.

Compared to Grubbs second-generation catalyst dissolved in poly(dimethyl siloxane) and coated onto the inner surface of fused-silica capillaries, permanently bonded polymeric Grubbs second-generation catalyst capillaries are moisture- and air-insensitive. Such immobilized catalyst capillaries can be installed several times over many weeks without detecting any loss of activity or leaching. In addition, no decomposition of the catalyst is observed even at elevated temperatures of 105 °C over several hours.

This modified catalyst capillary was applied in on-column reaction gas chromatography. As this efficient technique combines catalysis and analysis in one single step, *in situ* reaction monitoring is possible.

On-column catalysis was performed by coupling the Grubbs second-generation catalyst capillary between a pre-separation capillary for thermal equilibration and a separation column connected to electroionization gas chromatography/mass spectrometry (EI GC/MS).

Injection of *N*,*N*-diallyltrifluoroacetamide into the described experimental setup leads to the formation of *N*-trifluoroacetamide-3-pyrroline in 61% yield considering a response factor of 1.66 : 1 in the signal intensities of the reactant to the product (Figure 5).

The reaction temperature was set to 50 °C, and helium was used as the inert carrier gas (inlet pressure 50 kPa). From the obtained chromatographic data, the reaction rate constants were directly accessible by assuming a pseudo-first-order reaction law for the ring-closing metathesis (RCM). The contact time Δt of the reactant on the catalytically active column was determined to be only 1.75 s, which corresponds to a reaction rate constant of $0.54\,s^{-1}$ and an activation barrier $\Delta G^{\ddagger}_{(323\,K)}$ of $81\,kJ\,mol^{-1}$. The high activity of the permanently bonded polymeric Grubbs second-generation catalyst is corroborated by the activation barrier determined for the formation of *N*-trifluoroacetamide-3-pyrroline.

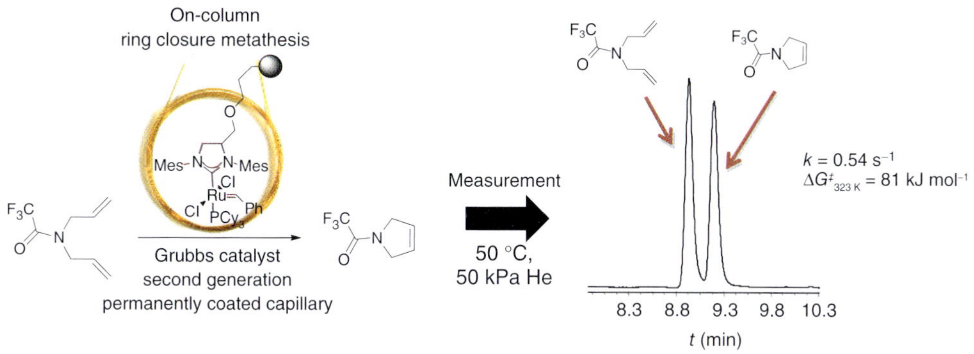

Figure 5 On-column closing metathesis over permanently bonded polymeric Grubbs second-generation catalyst of *N,N*-diallyltrifluoroacetamide at 50 °C.

V.4.3
Immobilized Salen Ligands

Salen ligands can be bound to different materials. The synthesis of polymer-bound chiral Mn(III)–salen complexes was independently reported by Sivaram *et al.* [41, 42] and Minutolo *et al.* [43]. Here, the monomeric units of the Jacobsen catalyst [44] were functionalized with two vinyl groups, which were copolymerized with styrene and divinylbenzene.

A similar strategy was developed by Kureshy *et al.* [45]. In 1998, Sherrington *et al.* [46] reported the application of polymer-bound Mn(III)–salen catalyst in the epoxidation of 1-phenylcyclohex-1-ene, resulting in 49% yield and 91% ee, which are comparable to results obtained with the nonimmobilized Jacobsen catalyst. The synthesis of these catalysts started from polymer-bound salicylaldehyde derivatives, which were first treated with 1,2-diamino cyclohexane and then with a second salicylaldehyde derivative. In the last step, the manganese complex was formed. Polymethacrylate was used as polymeric backbone. A similar approach was employed by Peukert and Jacobsen [47] to immobilize this catalyst to polystyrene.

In 2006, Weck *et al.* [48, 49] succeeded in synthesizing a highly efficient catalyst for the kinetic resolution of epichlorohydrin by radical homo- and copolymerization of an asymmetric monostyryl-substituted salen ligand with styrene.

The employment of Co(II)–salen occluded in the zeolite as a "ship-in-the-bottle" catalyst and its use in enantioselective transfer hydrogenations of acetophenone were published by Hölderich *et al.* [50]. Whereas the reactivity was high, the observed enantioselectivities were low (only in the range of 2–22%).

In a further approach, Trapp *et al.* [51] immobilized covalently bound chiral vanadium(IV)–salen complexes with a tunable spacer length on polysiloxanes. These polymer-modified salen-type ligands were used as catalytically active stationary phase in on-column reaction gas chromatography, on-column reaction electrophoresis, and flow-through reactors, and as reactive coating for laboratory glassware.

Starting from 2-*tert*-butyl-hydroquinone, which was converted to a modified salicylaldehyde bearing an ether moiety with terminal double bond, the synthesis of salen ligands anchored to polysiloxanes was realized in a linear four-step reaction sequence, in which a one-pot condensation of (*R*,*R*)-diaminocyclohexane with salicylaldehyde derivatives represented a key step leading to the modified salen ligands [52]. Consecutive immobilization of the ligands to HMPS [53, 54] was achieved by hydrosilylation of the terminal alkene chain using the Karstedt's catalyst under ultrasonication (Figure 6). The vanadium catalysts themselves were prepared *in situ* by stirring the immobilized salen ligands with $VO(acac)_2$ (acac, acetylacetonate) in dichloromethane for 1 h.

The freshly prepared catalysts were applied to study the influence of the alkyl spacer length on the catalytic activity of the immobilized salen-type ligands in the presence of the nonpolar polysiloxane backbone.

Enantioselective sulfoxidation of benzylphenylsulfide using hydrogen peroxide as oxidant was chosen as the model reaction. Because the enantioselectivity of this reaction is relatively low, this kind of transformation represents an ideal indicator to tune the reaction conditions. Experiments of the chiral sulfoxidation revealed a strong influence of the alkyl spacer length on the initial reaction rates, conversions, and enantioselectivities.

This reaction was investigated using the same salen-type vanadium catalyst but varying the spacer length ($n = 1, 4, 6$). The highest activity and enantioselectivity was achieved with a C_8 linker ($n = 6$), resulting in 76% yield and 38% ee. Reduction of the spacer length led to an apparent decrease in enantioselectivity (C_6 linker: 35% ee; C_3 linker: 14% ee), possibly caused by the greater steric and interaction effects of the polysiloxane backbone with decreased linker length (cf. Table 1).

Additionally, the determined values for the initial reaction rate constants (k_{init}) and the Gibbs free activation energies indicated that the accessibility of the catalysts is improved by reduced intramolecular interactions and the greater flexibility of the spacer.

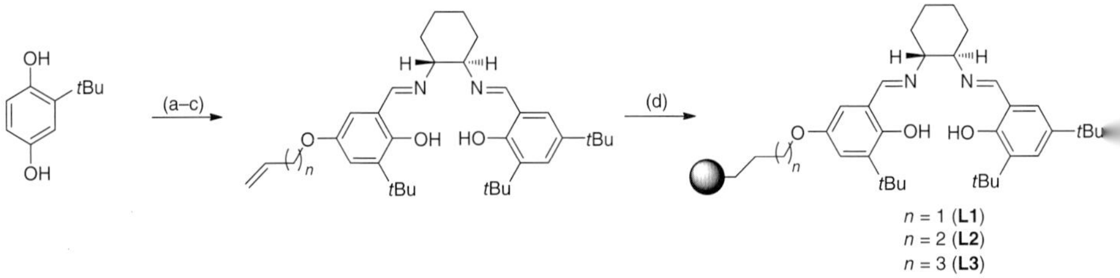

Figure 6 Salen-type ligand synthesis. (a) Bromo 1-alkene, K_2CO_3, KI, MeCN, reflux, 3 h. (b) 2,6-Lutidine, $SnCl_4$, paraformaldehyde, toluene, 90 °C, 6 h. (c) (*R*,*R*)-Diaminocyclohexane, salicylaldehyde derivative, EtOH, MeOH, NEt_3, dichloromethane (DCM), rt, 8 h. (d) HMPS, Karstedt's catalyst, toluene, ultrasonication, 3 h.

Table 1 Enantioselective sulfoxidation of benzylphenylsulfide.

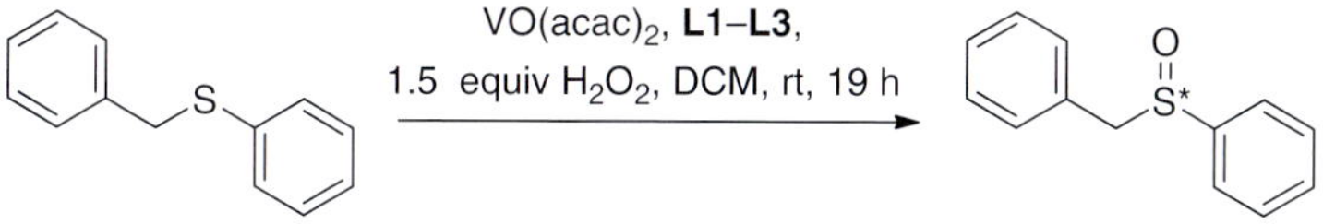

Entry	Ligand	Metal	Conversion (%)[a)]	Enantiomeric excess (%)[b)]
1	**L1**	$VO(acac)_2$	76 ± 2	14 ± 1
2	**L2**	$VO(acac)_2$	71 ± 9	35 ± 3
5	**L3**	$VO(acac)_2$	76 ± 3	38 ± 1

a) GC/MS measurements (150 °C, 90 kPa, and He).
b) EKC (150 mg ml^{-1} sulfated β-cyclodextrin, 10 mM Na_2HPO_4 (pH 8.3), 50 m bar, 3 s, +10 kV). Reactions were repeated three times each.

Because of the low concentration of the remaining polymeric Si–H groups, the application of polysiloxane as supporting material offered the possibility of permanently bonding the used salen ligands on silica and glass surfaces via condensation reaction with surface silanol groups (Figure 7).

These materials also opened the possibility to coat glassware or even stir bars with the immobilized vanadium salen catalyst. This is another approach toward a recyclable system, because such stir bars can be used in several reactions.

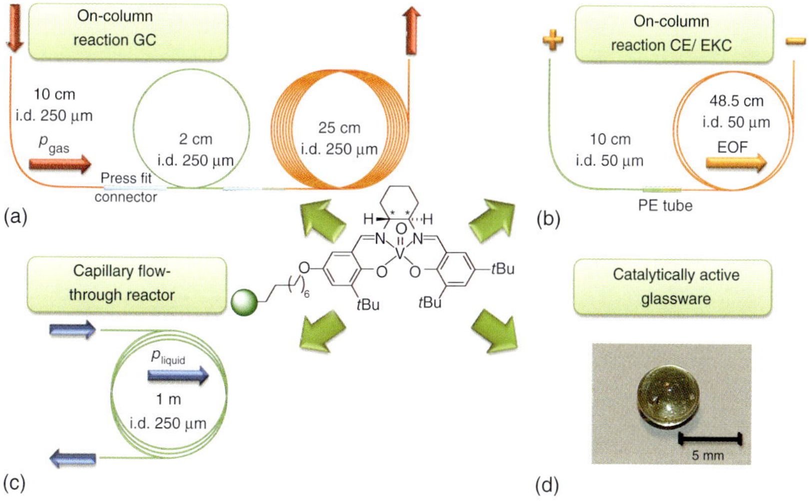

Figure 7 Overview of the strategy of immobilized salen-type catalysts by (a) on-column reaction GC, (b) on-column reaction capillary electrophoresis (CE)/electrokinetic chromatography (EKS), (c) flow-through microreactor, and (d) coated glassware.

References

1. Blaser, H.-U. (2010) *Asymmetric Catalysis on Industrial Scale*, Wiley-VCH Verlag GmbH, Weinheim.
2. DeVos, D.E., Vankelcom, I.F.J., and Jacobs, P.A. (2000) *Chiral Catalyst Immobilization and Recycling*, Wiley-VCH Verlag GmbH, Weinheim.
3. Leznoff, C.C. (1974) *Chem. Soc. Rev.*, **3**, 65–85.
4. Hartley, F.R. and Vezey, P.N. (1977) in *Advances in Organometallic Chemistry*, vol. 15 (eds F.G.A. Stone and W. Robert), Academic Press, pp. 189–234.
5. McMorn, P. and Hutchings, G.J. (2004) *Chem. Soc. Rev.*, **33**, 108–122.
6. Cozzi, F. (2006) *Adv. Synth. Catal.*, **348**, 1367–1390.
7. Heitbaum, M., Glorius, F., and Escher, I. (2006) *Angew. Chem.*, **118**, 4850–4881.
8. Clapham, B., Reger, T.S., and Janda, K.D. (2001) *Tetrahedron*, **57**, 4637–4662.
9. Fan, Q.-H., Li, Y.-M., and Chan, A.S.C. (2002) *Chem. Rev.*, **102**, 3385–3466.
10. Gennari, C. and Piarulli, U. (2003) *Chem. Rev.*, **103**, 3071–3100.
11. Dahmen, S. and Bräse, S. (2001) *Synthesis*, 1431–1449.
12. Merrifield, R.B. (1963) *J. Am. Chem. Soc.*, **85**, 2149–2154.
13. Owen, M.J. (2000) in *Silicon-Containing Polymers* (eds R.G. Jones, W. Ando, and J. Chojnowski), Kluwer Publishers, Dordrecht, pp. 213–231.
14. Keller, F., Weinmann, H., and Schurig, V. (1997) *Chem. Ber.*, **130**, 879–885.
15. Grunlan, M.A., Regan, K.R., and Bergbreiter, D.E. (2006) *Chem. Commun.*, 1715–1717.
16. DeClue, M.S. and Siegel, J.S. (2004) *Org. Biomol. Chem.*, **2**, 2287–2298.
17. Schurig, V., Schmalzing, D., and Schleimer, M. (1991) *Angew. Chem. Int. Ed.*, **30**, 987–989.
18. Trapp, O., Weber, S.K., Bauch, S., and Hofstadt, W. (2007) *Angew. Chem. Int. Ed.*, **46**, 7307–7310.
19. Trapp, O., Weber, S.K., Bauch, S., Bäcker, T., Hofstadt, W., and Spliethoff, B. (2008) *Chem. – Eur. J.*, **14**, 4657–4666.
20. Weber, S.K., Bremer, S., and Trapp, O. (2010) *Chem. Eng. Sci.*, **65**, 2410–2416.
21. Kobayashi, J., Mori, Y., Okamoto, K., Akiyama, R., Ueno, M., Kitamori, T., and Kobayashi, S. (2004) *Science*, **304**, 1305–1308.
22. Reinhard, S., Behringer, K.D., and Blümel, J. (2003) *New J. Chem.*, **27**, 776–778.
23. Reinhard, S. and Blümel, J. (2003) *Magn. Reson. Chem.*, **41**, 406–416.
24. Reinhard, S., Behringer, K.D., and Blümel, J. (2003) *New J. Chem.*, **27**, 776–778.
25. Reinhard, S. and Blümel, J. (2003) *Magn. Reson. Chem.*, **41**, 406–416.
26. Merckle, C. and Blümel, J. (2003) *Adv. Synth. Catal.*, **345**, 584–588.
27. Sommer, J., Yang, Y., Rambow, D., and Blümel, J. (2004) *Inorg. Chem.*, **43**, 7561–7563.
28. Posset, T., Rominger, F., and Blümel, J. (2005) *Chem. Mater.*, **17**, 586–595.
29. Bogza, M., Oeser, T., and Blümel, J. (2005) *J. Organometal. Chem.*, **690**, 3383–3389.
30. Piestert, F., Fetouaki, R., Bogza, M., Oeser, T., and Blümel, J. (2005) *Chem. Commun.*, **41**, 1481–1483.
31. Merckle, C. and Blümel, J. (2005) *Top. Catal.*, **34**, 5–15.
32. Guenther, J., Reibenspies, J., and Blümel, J. (2011) *Adv. Synth. Catal.*, **353**, 443–460.
33. Schürer, S.C., Gessler, S., Buschmann, N., and Blechert, S. (2000) *Angew. Chem. Int. Ed.*, **39**, 3898–3901.
34. Randl, S., Buschmann, N., Connon, S.J., and Blechert, S. (2001) *Synlett*, 1547–1550.
35. Buchmeiser, M.R. (2008) *Chem. Rev.*, **108**, 303–321.
36. Prühs, S., Lehmann, C.W., and Fürstner, A. (2003) *Organometallics*, **23**, 280–287.
37. Kingsbury, J.S., Garber, S.B., Giftos, J.M., Gray, B.L., Okamoto, M.M., Farrer, R.A., Fourkas, J.T., and Hoveyda, A.H. (2001) *Angew. Chem. Int. Ed.*, **40**, 4251–4256.
38. Lang, C., Gartner, U., and Trapp, O. (2011) *Chem. Commun.*, **47**, 391–393.
39. Raitza, M., Wegmann, J., Bachmann, S., and Albert, K. (2000) *Angew. Chem. Int. Ed.*, **39**, 3486–3489.

40. Nyce, G.W., Csihony, S., Waymouth, R.M., and Hedrick, J.L. (2004) *Chem. Eur. J.*, **10**, 4073–4079.
41. De, B.B., Lohray, B.B., Sivaram, S., and Dhal, P.K. (1994) *Macromolecules*, **27**, 1291–1296.
42. De, B.B., Lohray, B.B., Sivaram, S., and Dhal, P.K. (1997) *J. Polym. Sci. A: Polym. Chem.*, **35**, 1809–1818.
43. Minutolo, F., Pini, D., and Salvadori, P. (1996) *Tetrahedron Lett.*, **37**, 3375–3378.
44. Zhang, W., Loebach, J.L., Wilson, S.R., and Jacobsen, E.N. (1990) *J. Am. Chem. Soc.*, **112**, 2801–2803.
45. Kureshy, R.I., Khan, N.H., Abdi, S.H.R., and Iyer, P. (1997) *React. Funct. Polym.*, **34**, 153–160.
46. Canali, L., Cowan, E., Deleuze, H., Gibson, C.L., and Sherrington, D.C. (1998) *Chem. Commun.*, 2561–2562.
47. Peukert, S. and Jacobsen, E.N. (1999) *Org. Lett.*, **1**, 1245–1248.
48. Holbach, M., Zheng, X., Burd, C., Jones, C.W., and Weck, M. (2006) *J. Org. Chem.*, **71**, 2903–2906.
49. Zheng, X., Jones, C.W., and Weck, M. (2006) *Chem. Eur. J.*, **12**, 576–583.
50. Kahlen, W., Wagner, H.H., and Hölderich, W.F. (1998) *Catal. Lett.*, **54**, 85–89.
51. Sandel, S., Weber, S.K., and Trapp, O. (2012) *Chem. Eng. Sci.*, **83**, 171–179.
52. Holbach, M., Zheng, X., Burd, C., Jones, C.W., and Weck, M. (2006) *J. Org. Chem.*, **71**, 2903–2906.
53. Figge, H., Deege, A., Köhler, J., and Schumburg, G. (1986) *J. Chromatogr. A*, **351**, 393–408.
54. Cousin, H., Trapp, O., Peulon-Agasse, V., Pannecoucke, X., Banspach, L., Trapp, G., Jiang, Z., Combret, J.C., and Schurig, V. (2003) *Eur. J. Org. Chem.*, **2003**, 3273–3287.

19
Dendrimers as Platforms for Stereoselective Catalysis

Lutz H. Gade

19.1
Introduction

Homogeneous catalysts, which are the focus of this monograph, are generally well-defined molecular systems frequently displaying high activity and selectivity in catalytic reactions as well as good reproducibility of the catalyst performance [1]. In contrast, most of the catalysts used in industry belong to the class of heterogeneous catalysts because they can be easily separated from reactants and products and thus reused (or employed in continuous processes!) [2]. To transfer this feature to molecular catalysts, supported molecular catalysts have been developed using polymers as soluble – but separable – supporting materials. Dendrimers are highly branched macromolecules that combine a high level of structural regularity, molecular monodispersity, and a wide variety of modes of attachment of well-defined catalytic sites [3].

In the assessment of multisite macromolecular catalysts such as dendrimer catalysts [4, 5], it is essential to establish whether the immobilized catalyst units retain their identity and are not altered by the nature of the dendrimer backbone. The linker and spacer units employed in the fixation of the catalysts may be crucial in this respect as well as the functional groups present in the dendrimer. Regarding the dendrimer core structure itself, the length and conformational rigidity of the branches and spacers are important factors when evaluating a dendritic catalyst. For immobilized asymmetric catalysts, even subtle conformational changes may significantly influence their stereoselectivity. The interplay of all these factors will generally determine the detrimental or beneficial "dendrimer effects" on catalyst performance [6].

19.2
Fixation of Chiral Catalysts on Dendrimers and Hyperbranched Polymers

Chiral dendrimer catalysts have been generated either by attachment of achiral complexes to chiral dendrimer structures or by immobilization of chiral catalysts

Molecular Catalysts: Structure and Functional Design, First Edition. Edited by Lutz H. Gade and Peter Hofmann.

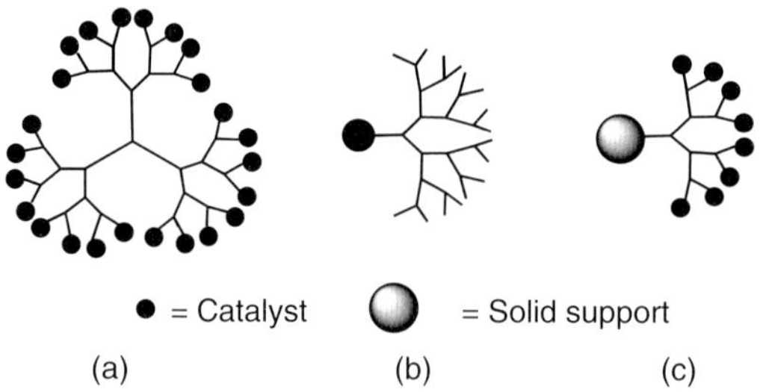

Figure 19.1 Fixation of catalytic metal centers (represented by the black spheres) in (a) exodendral and (b) endodendral positions of dendrimers or on (c) dendronized polymers.

to nonchiral dendrimers. As monodispersed macromolecular supports, they provide ideal model systems for less regularly structured but commercially more viable supports such as hyperbranched polymers [7], and have been successfully employed in continuous-flow membrane reactors.

Catalyst recycling has been achieved using membrane reactors, biphasic solvent systems, as well as catalyst precipitation and subsequent filtration, although frequently with deteriorating catalyst performance over time. This utilitarian aspect of dendrimer catalysis has provided the motivation for much of the work on chiral dendrimer catalysts.

Dendrimer fixation may be achieved by attachment of catalysts at the periphery of dendrimers (Figure 19.1a) in the way first established by van Koten, van Leeuwen, and coworkers [8] in their pioneering work on the Karasch reaction. The second possibility is the attachment of one or more dendritic wedges to the catalysts, which are then located at the core of the resulting functionalized dendrimers (Figure 19.1b). The first example of such catalysts was reported by Brunner and Altmann [9] ("dendrizymes"), who studied the influence of a chiral dendritic periphery on the performance of asymmetric cyclopropanation catalysts. Finally, a third type of chiral dendritic catalysts has been developed that are based on an insoluble polymer support loaded with dendritic wedges. These are functionalized by the attachment of molecular catalysts to their periphery (Figure 19.1c). The structures obtained following this strategy combine the well-defined dendritic architecture near the active sites of the catalytic phase with the ease of reuse characteristic of common heterogeneous catalysts.

19.3 Case Histories

The field of dendrimer catalysis has been comprehensively reviewed [5, 6]. In particular, there are several reviews covering the development of asymmetric dendrimer catalysis during the past 15 years [10, 11]. The following sections will therefore focus on several case histories from our own activities in the field. These illustrate the key challenges and the way they may be addressed in practice. They also serve to show the limitations of this approach.

19.3.1 "Dendritic Effects" Observed for Immobilized Pyrphos-Based Hydrogenation Catalysts

The first example of asymmetric rhodium-catalyzed hydrogenation of prochiral olefins in dendrimer catalysis was reported in 1998 by Togni *et al.* [12] who immobilized the chiral ferrocenyl diphosphine "Josiphos" at the end groups of dendrimers, thus obtaining systems of up to 24 chiral metal centers in the periphery. The fact that the catalytic properties of the dendrimer catalysts were almost identical to those of the mononuclear catalysts was interpreted as manifestation of the independence of the individual catalytic sites in the macromolecular systems. In a comprehensive study carried out by the same group, the asymmetric hydrogenation of diethyl itaconate, asymmetric allylic substitutions, and asymmetric hydroboration reactions catalyzed by the multi-Josiphos rhododendrimers were investigated [13]. The stereoselectivities obtained with the monodisperse dendrimer catalysts with up to 16 metal centers in the periphery were again found to be very similar to those of the mononuclear reference systems. The authors concluded that the absence of negative dendrimer effects is probably the best possible result when cooperativity effects between single catalyst units do not play any relevant role. The small losses in selectivity observed upon going to higher dendrimer generations were thought be due to local concentration effects which become important in particular for cationic catalyst species.

Inspired by Togni's approach, we synthesized a series of chiral phosphine-functionalized polypropyleneimine (PPI) and poly(amido)amine (PAMAM) dendrimers by reaction of carboxyl-linked C_2-chiral pyrphos ligands (pyrphos, 3,4-bis(diphenylphosphino)pyrrolidine) with zeroth- to fifth-generation PPI (zeroth- to fifth-generation PAMAM) using ethyl-*N*,*N*-dimethylaminopropyl-carbodiimide (EDC)/1-hydroxybenzotriazole (HOBT) as a coupling reagent (Scheme 19.1). The functionalized dendrimers were characterized by NMR spectroscopy and elemental analysis, as well as by fast atom bombardment (FAB) and MALDI-TOF mass spectrometry, establishing their molecular masses of up to 20 700 amu for the PPI derivatives [14, 15].

Metallation of the multisite phosphines with $[Rh(COD)_2]BF_4$ (COD, cyclooctadiene) or $[PdCl_2(NCCH_3)_2]$ cleanly yielded the cationic rhododendrimers or palladodendrimers respectively, containing up to 64 metal centers (Scheme 19.2). The relationship between the size/generation of the rhododendrimers and their catalytic properties was established, *inter alia*, in the asymmetric hydrogenation of *Z*-methyl-α-acetamidocinnamate and dimethyl itaconate. Generally, a decrease in activity of the dendrimer catalysts was observed on going to the higher generations for the PPI- and PAMAM-based dendrimers whereas the enantioselectivity in the hydrogenation of *Z*-methyl-acetamidocinnamate remained mostly unaffected (Figure 19.2) [14, 16].

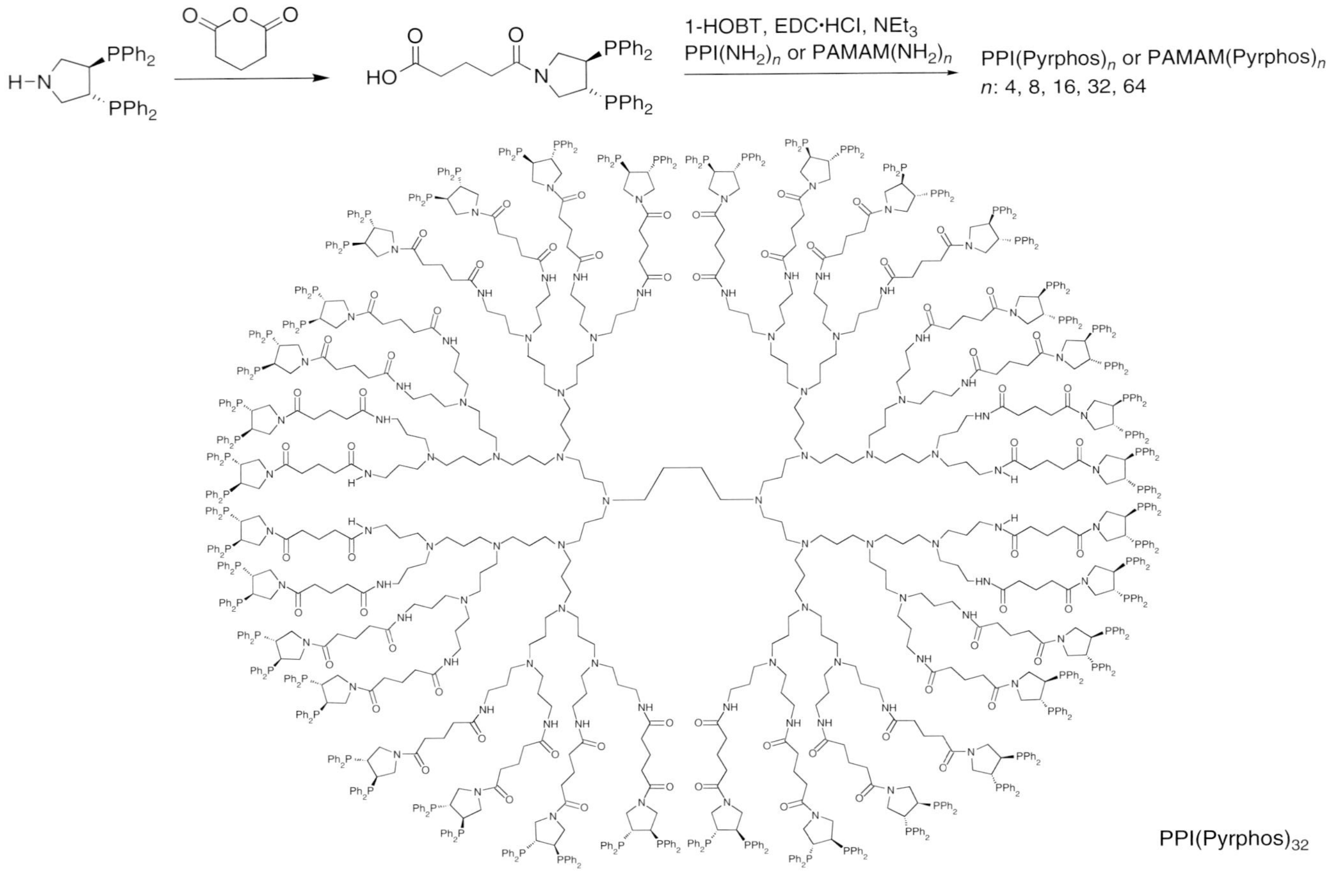

Scheme 19.1 Synthesis of pyrphos-functionalized PPI and PAMAM dendrimers by means of peptide coupling methods.

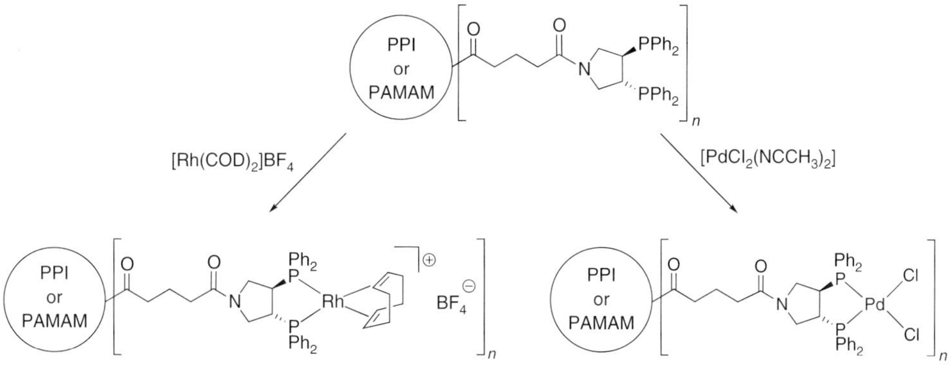

Scheme 19.2 Metallation of the pyrphos-functionalized PPI and PAMAM dendrimers.

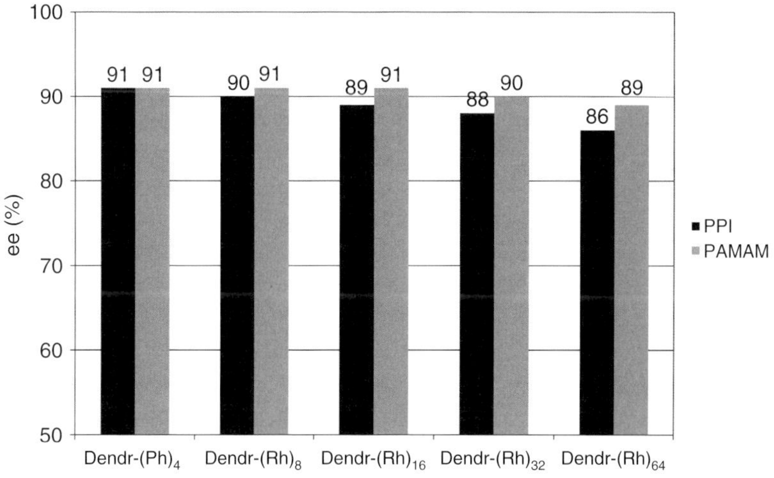

Figure 19.2 Comparison of the enantioselectivities obtained with the PPI- and PAMAM-bound pyrphos–Rh(NBD) (NBD, norbornadiene) complexes [16].

In addition to dendrimers, hyperbranched polymers have been used by several groups as soluble supports for catalysts [7, 17]. These supports are polydisperse and randomly branched, and, since they are prepared in a single reaction step, are generally much cheaper materials. Nevertheless, it has been shown that catalysts immobilized on hyperbranched polymers may possess similar properties as dendritic systems [18]. Therefore, dendritic catalysts serve as ideal model systems for catalysts attached to hyperbranched polymers. We functionalized several hyperbranched polyethyleneimines (PEIs) employing the peptide coupling protocol in reactions with the pyrphos linker system. The pyrphos–rhodium complexes bound to the hyperbranched polymers were also found to be active catalysts for the hydrogenation of *Z*-methyl-α-acetamidocinnamate [16]. As observed for the regular dendrimers, the activity of these systems decreased with increasing size of the supporting material (Table 19.1).

It was our goal to work out conditions for catalyst recycling in a biphasic solvent system employing a suitable ionic liquid (IL) to trap the polycationic rhodium catalysts. A biphasic reaction medium consisting of the IL [BMIM][BF_4] (BMIM, 1-butyl-3-methylimidazolium) and isopropanol [19] was thought to be suitable for such catalysts, the IL dissolving the catalyst and the isopropanol the substrate *Z*-methyl-α-acetamido-cinnamate, as well as the corresponding amino acid as reaction product. This was thought to allow a facile separation of the phases after complete conversion, with the possibility to reuse the ionic catalyst phase by simple addition of new substrate in *i*PrOH.

Each catalyst was tested in three successive catalytic runs of 2 h at a reaction temperature of 55 °C in order to take advantage of the *thermomorphic* behavior of the solvent mixture. This allowed the actual catalytic conversion to take place in a homogeneous reaction medium under these conditions, with the phases separating again upon cooling to ambient temperature at the end of the reaction. The catalysts tethered to the hyperbranched PEIs were also tested under similar reaction conditions, and the results obtained are represented in Figure 19.3.

Table 19.1 Turnover frequencies (TOFs) and enantioselectivities obtained with the hyperbranched catalysts [16].

Catalyst	TOF (min^{-1})	ee (%)
$PEI_{0.8}-(Rh)_n$	37	89
$PEI_2-(Rh)_n$	31	87
PEI_{25}-$(Rh)_n$	13	84

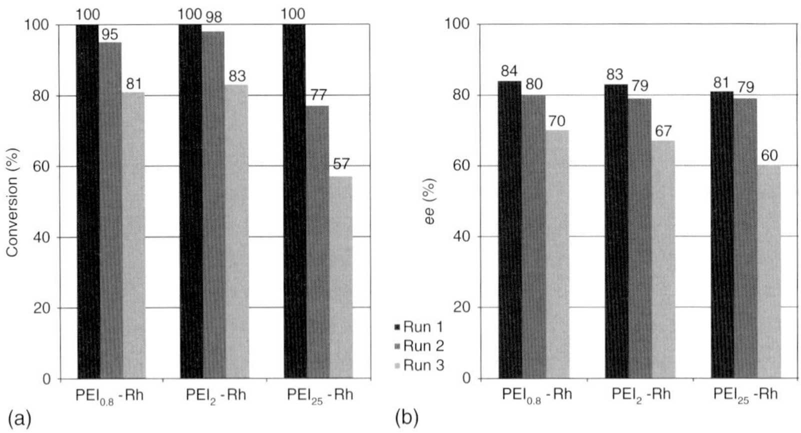

Figure 19.3 (a,b) Conversions and enantioselectivities of three successive runs with hyperbranched systems as catalysts [16].

During the effort to extend the use of the pyrphos-derived dendrimers to asymmetric Pd-catalyzed coupling reactions, strongly positive selectivity effects were observed upon going to very large multisite chiral dendrimer catalysts. This enhancement of the catalyst selectivity was observed in palladium-catalyzed allylic substitutions, such as displayed in Scheme 19.3, which are known to be particularly sensitive to small changes in the chemical environment of the active catalyst sites [15].

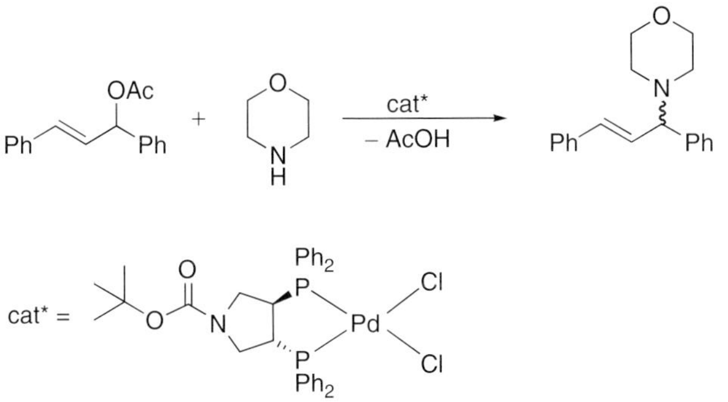

PPI$(PyrphosPdCl_2)_4$ – PPI$(PyrphosPdCl_2)_{64}$
PAMAM$(PyrphosPdCl_2)_4$ – PAMAM$(PyrphosPdCl_2)_{64}$

Scheme 19.3 Asymmetric allylic amination of 1,3-diphenyl-1-propene-3-acetate catalyzed by pyrphos–palladium complexes.

The mononuclear catalyst [(Boc-pyrphos)$PdCl_2$], which was found to be very unselective for this transformation (9% ee), provided the point of reference for the subsequent studies with the dendrimer catalysts. This system and the metalladendrimers PPI$(PyrphosPdCl_2)_4$–PPI$(PyrphosPdCl_2)_{64}$ and PAMAM(Pyrphos $PdCl_2)_4$–PAMAM$(PyrphosPdCl_2)_{64}$ in 0.3 mol% catalyst concentration were studied in the catalytic amination of 1,3-diphenyl-1-acetatopropene (Scheme 19.3), which gave the results displayed in Figure 19.4.

A remarkable increase in catalyst selectivity was observed as a function of the dendrimer generation. This steady increase in enantiomeric excess values for the allylic amination was less pronounced for the PPI-derived catalysts (40% ee for PPI$(PyrphosPdCl_2)_{64}$) than for the palladium-PAMAM dendrimer catalysts for which an increase in selectivity from 9% ee for the mononuclear complex to 69% ee for PAMAM$(PyrphosPdCl_2)_{64}$ was found. The same general trend was observed in the asymmetric allylic alkylation of 1,3-diphenyl-1-acetatopropene with sodium dimethylmalonate, which indicates that the results of the amination reaction may be typical for allylic substitutions in general.

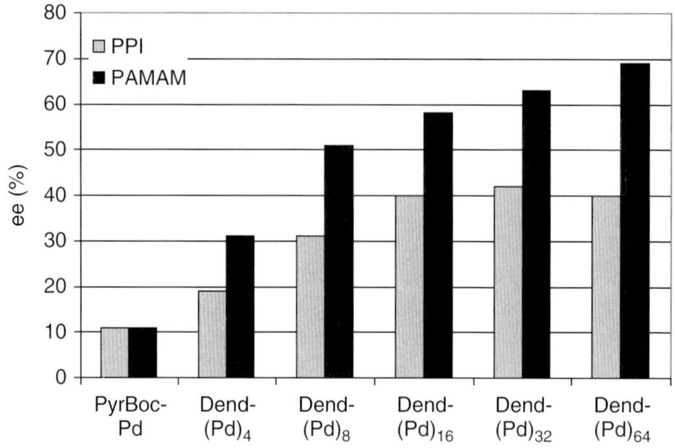

Figure 19.4 Dependence of the enantiomeric excesses found for the allylic amination in Scheme 19.3 on the dendrimer generation for both precatalyst series $PPI(PyrphosPdCl_2)_4$ $-PPI(PyrphosPdCl_2)_{64}$ and $PAMAM(PyrphosPdCl_2)_4-PAMAM(PyrphosPdCl_2)_{64}$ [15].

19.3.2 BINAP–Copper(I) Hydrosilylation with Functionalized PPI and PAMAM Dendrimers as well as Hyperbranched Polymers

The immobilization of the chiral diphosphine BINAP (2,2′-bis(diphenylphosphino)-1,1′-binaphthalene) at the periphery of a series of PPI and PAMAM dendrimers was carried out in order to employ these polyfunctional phosphines in the Cu(I)-catalyzed hydrosilylation of acetophenone as reference reaction [20]. The synthesis of the functionalized dendrimers required a quantitative coupling reaction of glutaroyl-AMINAP with the amino end groups using ethyl-*N*,*N*-dimethylcarbodiimide (EDC)/1-hydroxybenzotriazole as coupling reagents. Using this method, AMINAP-functionalized PPI and PAMAM dendrimers bearing 4–64 diphosphine ligands in their periphery were obtained in good yields (Scheme 19.4) [21].

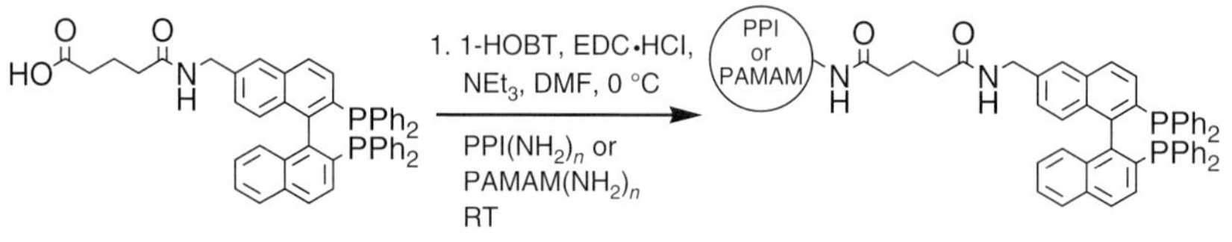

Scheme 19.4 Coupling of glutaroyl-AMINAP with the amino end groups of the PPI and PAMAM dendrimers [21].

In Figure 19.5, the results of the catalytic tests carried out with the functionalized PPI and PAMAM dendrimers are shown. On the whole, it was found that the selectivity remained almost unchanged at around 93% ee, a value that is slightly

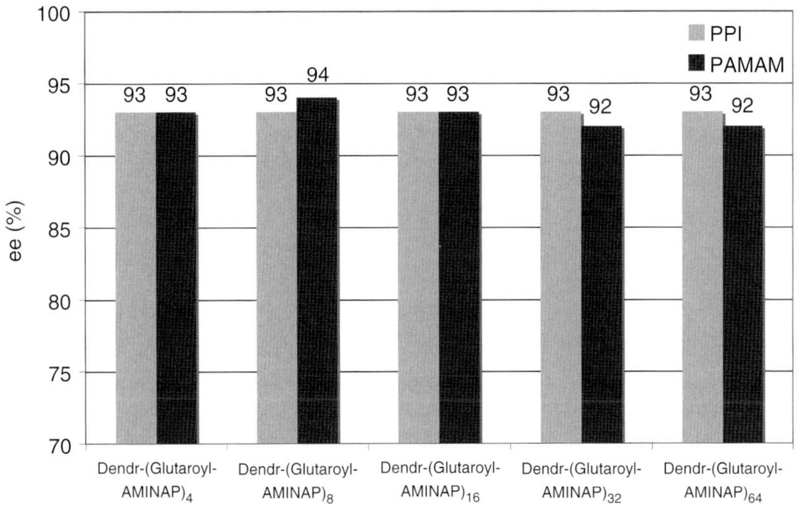

Figure 19.5 Comparison of the selectivity of the asymmetric hydrosilylation of acetophenone using functionalized PPI or PAMAM dendrimers as ligands [21].

higher than that of the unfunctionalized BINAP ligand. This absence of a "dendritic" effect indicated that the catalytic conversion was entirely controlled by the first coordination sphere of the copper. This is in part due to the marked coordination preference of Cu(I) for the soft diphosphine compared to the chemically hard functionalities in the dendrimer backbone.

In contrast to the PPI and PAMAM dendrimers that essentially possess only primary amino groups (due to the almost perfect branching of these molecules), there are also secondary amino groups in the hyperbranched PEIs that result from incomplete conversion at some branching points in their one-step synthesis. It was therefore important to functionalize both primary and secondary amino groups in the PEIs because free NH functionalities were thought to act as potential (undesired) ligands in the catalyst assembly [22]. The coupling reagents used for dendrimer functionalization HOBT and EDC were found to be appropriate for both types of amino functionalities. In this way, we were able to synthesize carbo-BINAP- and glutaroyl-AMINAP-functionalized hyperbranched PEIs depicted in Figure 19.6.

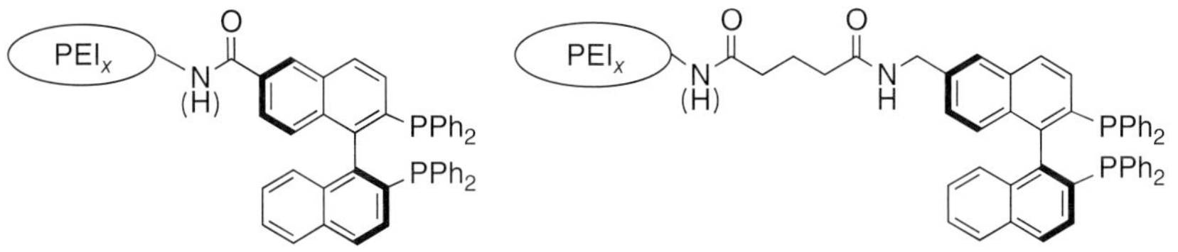

Figure 19.6 (a) Carbo-BINAP- and (b) glutaroyl-AMINAP-functionalized hyperbranched PEIs ($x = 0.8$, 2, 25) [22].

Table 19.2 Enantioselectivity of the asymmetric hydrosilylation of acetophenone for the different hyperbranched Carbo-BINAP and AMINAP ligands (in tetrahydrofuran (THF)/toluene).

Ligand	Run	Yield (%)	ee (%)
$PEI_{0.8}$ – (Carbo-BINAP)	—	96	94
PEI_{2} – (Carbo-BINAP)	—	97	93
PEI_{25} – (Carbo-BINAP)	1	100	93
PEI_{25} – (Carbo-BINAP)	2	100	93
PEI_{25} – (Carbo-BINAP)	3	99	93
$PEI_{0.8}$ – (Glutaroyl-AMINAP)	—	98	92
PEI_{2} – (Glutaroyl-AMINAP)	—	98	93
PEI_{25} – (Glutaroyl-AMINAP)	1	95	93
PEI_{25} – (Glutaroyl-AMINAP)	2	95	92
PEI_{25} – (Glutaroyl-AMINAP)	3	94	92

The catalysts were recyclable by simple precipitation by addition of diethyl ether [22].

For the carbo-BINAP ligands attached to the hyperbranched PEIs, uniform high enantioselectivities of 93–94% ee and good activities were observed (Table 19.2) and thus the hyperbranched systems display the "ideal" behavior of the mononuclear reference ligands.

19.3.3 "Catalysis in a Tea Bag" with Dendrimer-Immobilized Bis- and Trisoxazoline Copper Catalysts

Ultra and nanofiltration techniques based on dialysis in the reaction engineering of catalytic processes were originally developed for biotechnological applications, whereas their application to organometallic homogeneous catalysis has been a more recent development. Key contributions to the development of continuously operating membrane reactors for this type of catalytic systems are due to Kragl and others [23], while van Koten and van Leeuwen and their coworkers [24, 25] first reported the application of this technique to dendrimer catalysis which led to its establishment in dendrimer chemistry.

The membranes employed in such reactors need to be adapted to the reaction conditions to achieve good catalytic performance [26]. This includes the choice of the appropriate pore size as well as mechanical, thermal, and chemical stability in the required solvent and at the required temperature. Driven by a gradient, the substrates/reactants are transported through the membrane, whereas the catalyst is retained because of its size. The driving force may be a pressure, concentration, or temperature gradient, or a difference in electrical potential.

The simplest approach is based on catalyst trapping within a nanoporous membrane "bag." It provides adequate dispersion of the catalyst and guarantees minimal interaction between the catalyst and the polymer, thus allowing the use of relatively labile catalyst systems such as the copper(II)-based Lewis acid catalysts employed in our study. These were attached to dendrimer supports (Scheme 19.5) and were held within a dialysis membrane "tea bag" [27] which was "dipped" into a reactant solution and recycled several times [28].

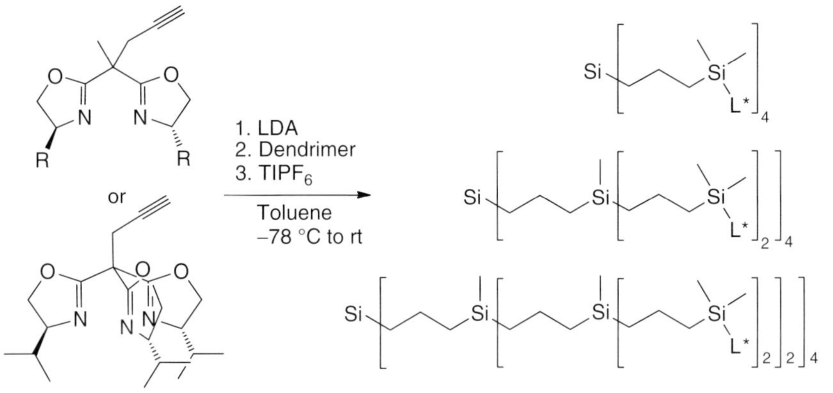

Parent dendrimers: {G0}-$(SiMe_2Cl)_4$ for **G0**, {G1}-$(SiMe_2Cl)_8$ for **G1**, {G2}-$(SiMe_2Cl)_{16}$ for **G2**.

Scheme 19.5 Synthesis of the BOX- and trisox-functionalized carbosilane dendrimers **G0-L***, **G1-L***, and **G2-L*** [28].

As test reactions, the α-hydrazination of ethyl 2-methylacetoacetate and the Henry reaction of nitromethane and 2-nitrobenzaldehyde were chosen. In both cases, a significant difference between the bis-oxazoline (BOX) and trisox-based catalysts was notable. The performance of all the trisox-based catalysts proved to be inferior for both reactions, both in terms of the activities and enantioselectivities. While the attachment of a linker at the ligand backbone and the immobilization only moderately affected the BOX-based catalysts for both reactions, the effect on the trisox-based catalysts led to a dramatic decrease in activity.

This observation was explained on the basis of an equilibrium between κ^3- and κ^2-trisox coordinated complexes. The coordination of the third ligand arm stabilizes the resting state but leads to a deactivation of the copper complexes by reduction of the Lewis acidity [29]. The transformation of the stabilized but inactive resting state into the active 17 electron Cu(II) species therefore required the decoordination of an oxazoline unit which then adopts a remote position from the center, as is depicted in Scheme 19.6. The latter is sterically hindered for the trisox systems, which explains the reduced activity.

As stated previously, the dendrimer catalysts had been developed with the purpose of catalyst recycling based on dialysis, using membrane bags fabricated from a commercially available dialysis membrane (Sigma-Aldrich: benzoylated dialysis tubing, MWCO 2000). The general principle and its practical implementation are depicted in Figure 19.7.

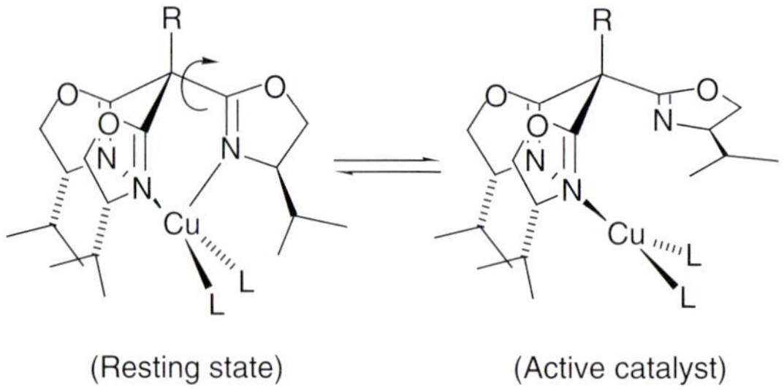

Scheme 19.6 Equilibrium between κ^3- and κ^2-trisox coordinated complexes. In the κ^2-coordinated species, the third oxazoline adopts a remote position from the metal center.

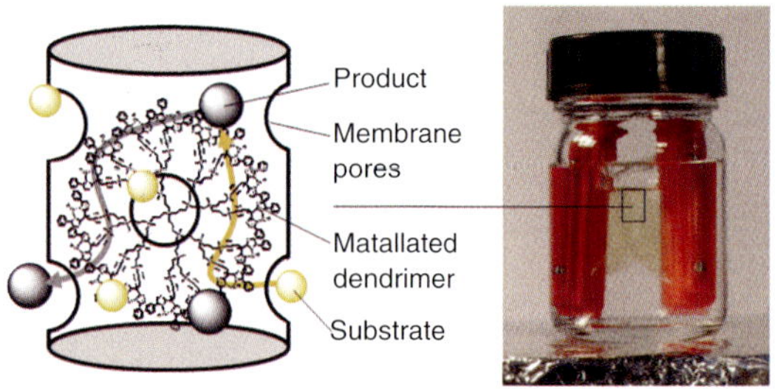

Figure 19.7 General setup for the recycling using the "catalyst in a tea bag" principle. An enlarged schematic view clarifies the operational principle [28].

The transport of the substrates and the product in and out of the membrane bag occurred by diffusion, which was accelerated by operating at an elevated temperature of 40 °C, consequently resulting in somewhat lower enantioselectivity. The substrates migrated into the membrane bag, where they interacted with the catalytically active terminal groups of the dendrimer and were converted to the product. Consequently, the latter was enriched in the interior and then passed through the membrane to the exterior part of the reactor following the concentration gradient.

The practical handling of such a catalyst "tea bag" is illustrated in Figure 19.8. Initially, the dendrimer-filled membrane bag was placed into the vial containing the yellow solution of the substrates, the color being due to the diazodicarboxylate (Figure 19.8a). After their complete conversion to the product, which was accompanied by a decoloration of the solution (Figure 19.8b), the bag was transferred into another vial containing fresh substrates (Figure 19.8c). The results obtained for seven successive runs are summarized in Figure 19.9.

The reason for the different behaviors of the two dendrimer catalysts was thought to be primarily the markedly lower activity of the trisoxazoline derivative compared to BOX-derived system, the probable origin of which has been discussed above. This necessitated increased reaction times for each cycle, leading to higher levels of metal ion leaching as well as the modification of the membrane structure due to its exposure to trifluoroethanol at 40 °C.

Figure 19.8 Series showing the course of one recycling: (a) dendrimer-filled membrane bag in the yellow solution of the substrates; (b) colorless solution after complete conversion; (c) transfer of the bag into another vial; and (d) subsequent reaction cycle [28].

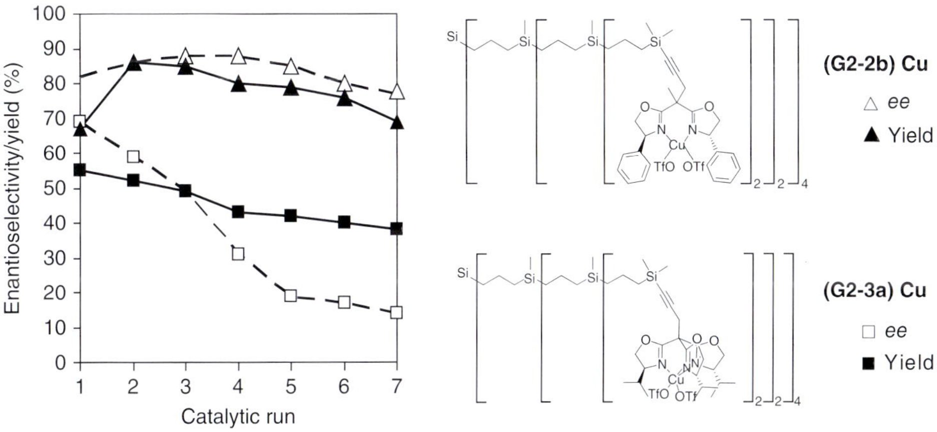

Figure 19.9 Results of the catalytic performance obtained in the recycling study. Seven successive runs with second-generation Box–Cu and Trisox–Cu [28].

19.4 Conclusion and Outlook

The kinetically controlled stereoselection depends on very small increments of free activation enthalpy. It is therefore an excellent sensitive probe for "dendrimer effects." As monodispersed macromolecules, chiral dendrimer catalysts provide ideal model systems for less regularly structured but commercially more viable supports such as hyperbranched polymers. However, the results obtained by a considerable number of research groups in the field have also established that the structural characteristics of the established dendrimer systems, such as the absence of a well-defined secondary structure, have limited the development of efficient abiotic enzyme mimics based on dendrimers. To achieve this ambitious goal, more effort in dendrimer synthesis will be necessary. The use of dendritic catalysts in biphasic solvent systems has only just begun and appears to be a particularly fruitful field for further developments.

References

1. (a) Cornils, B. and Herrmann, W.A. (eds) (2002) *Applied Homogeneous Catalysis with Organometallic Compounds: A Comprehensive Handbook*, 2nd edn, Wiley-VCH Verlag GmbH, Weinheim; (b) Behr, A. (2008) *Angewandte Homogene Katalyse*, Wiley-VCH Verlag GmbH, Weinheim.
2. (a) Ertl, G., Knözinger, H., and Weitkamp, J. (eds) (1997) *Handbook of Heterogeneous Catalysis*, 1st edn, Wiley-VCH Verlag GmbH, Weinheim; (b) Thomas, J.M. and Thomas, W.J. (1996) *Principles and Practice of Heterogeneous Catalysis*, Wiley-VCH Verlag GmbH, Weinheim.
3. (a) Vögtle, F., Richardt, G., and Werner, N. (2007) *Dendritische Moleküle – Konzepte, Synthesen, Eigenschaften, Anwendungen*, 1st edn, B. G. Teubner, Wiesbaden; (b) Newkome, G.R., Moorefield, C.N., and Vögtle, F. (1998) *Dendritic Molecules: Concepts, Syntheses, Perspectives*, 2nd edn, Wiley-VCH Verlag GmbH, Weinheim; (c) Fischer, M. and Vögtle, F. (1999) *Angew. Chem.*, **111**, 934; *Angew. Chem. Int. Ed.*, (1999), **38**, 884; (d) Bosman, A.W., Janssen, H.M., and Meijer, E.W. (1999) *Chem. Rev.*, **99**, 1665.
4. (a) Oosterom, G.E., Reek, J.N.H., Kramer, P.C.J., and van Leeuwen, P.W.N.M. ; *Angew. Chem.*, (2001), **113**, 1878; *Angew. Chem. Int. Ed.*, (2001), **40**, 1828; (b) van Heerbeek, R., Kamer, P.C.J., van Leeuwen, P.W.N.M., and Reek, J.N.H. (2002) *Chem. Rev.*, **102**, 3717; (c) Dijkstra, H.P., van Klink, G.P.M., and van Koten, G. (2002) *Acc. Chem. Res.*, **35**, 798.
5. Gade, L.H. (ed) (2006) *Dendrimer Catalysis*, Topics in Organometallic *Chemistry*, vol. **20**, Springer, Heidelberg, New York.
6. Caminade, A.-M., Servin, P., and Majoral, J.-P. (2008) *Chem. Soc. Rev.*, **37**, 56.
7. (a) Schlenk, C., Kleij, A.W., Frey, H., and van Koten, G. (2000) *Angew. Chem.*, **112**, 3587; *Angew. Chem. Int. Ed.*, (2000), **39**, 3445; (b) Slagt, M.Q., Stiriba, S.-E., Kautz, H., Klein Gebbink, R.J.M., Frey, H., and van Koten, G. (2004) *Organometallics*, **23**, 1525; (c) Hajji, C., Roller, S., Beigi, M., Liese, A., and Haag, R. (2006) *Adv. Synth. Catal.*, **348**, 1760; (d) Beigi, M., Haag, R., and Liese, A. (2008) *Adv. Synth. Catal.*, **350**, 919.
8. (a) Knapen, J.W.J., van der Made, A.W., de Wilde, J.C., van Leeuwen, P.W.N.M., Wijkens, P., Grove, D.M., and van Koten, G. (1994) *Nature*, **372**, 659; overview:(b) Gossage, R.A., van de Kuil, L.A., and van Koten, G. (1998) *Acc. Chem. Res.*, **31**, 423.
9. (a) Brunner, H. and Altmann, S. (1994) *Chem. Ber.*, **127**, 2285; (b) Brunner, H. (1995) *J. Organomet. Chem.*, **500**, 39.
10. Ribourdouille, Y., Engel, G.D., and Gade, L.H. (2003) *C.R. Chim.*, **6**, 1087.
11. Kassube, J. and Gade, L.H. (2006) *Top. Organomet. Chem.*, **20**, 61.
12. (a) Köllner, C., Pugin, B., and Togni, A. (1998) *J. Am. Chem. Soc.*, **120**, 10274; (b) Schneider, R., Köllner, C., Weber, I., and Togni, A. (1999) *Chem. Commun.*, 2415; (c) Togni, A., Dorta, R., Köllner, C., and Pioda, G. (1998) *Pure Appl. Chem.*, **70**, 1477.
13. Köllner, C. and Togni, A. (2001) *Can. J. Chem.*, **79**, 1762.
14. Engel, G.D. and Gade, L.H. (2002) *Chem. Eur. J.*, **8**, 4319.
15. Ribourdouille, Y., Engel, G.D., Richard-Plouet, M., and Gade, L.H. (2003) *Chem. Commun.*, 1228.
16. Kassube, J.K. and Gade, L.H. (2009) *Adv. Synth. Catal.*, **351**, 739.
17. Kirkorian, K., Ellis, A., and Twyman, L.J. (2012) *Chem. Soc. Rev.*, **41**, 6138.
18. (a) Krämer, M., Pérignon, N., Haag, R., Marty, J.-D., Thomann, R., Lauth-de Viguerie, N., and Mingotaud, C. (2005) *Macromolecules*, **38**, 8308; (b)

Beigi, M., Roller, S., Haag, R., and Liese, A. (2008) *Eur. J. Org. Chem.*, 2135.
19. (a) Monteiro, A.L., Zinn, F.K., de Souza, R.F., and Dupont, J. (1997) *Tetrahedron: Asymmetry*, **2**, 177; (b) Fráter, T., Gubicza, L., Szöllösy, Á., and Bakos, J. (2006) *Inorg. Chim. Acta*, **359**, 2756.
20. Lipshutz, B.H., Noson, K., and Chrisman, W. (2001) *J. Am. Chem. Soc.*, **123**, 12917.
21. Kassube, J., Wadepohl, H., and Gade, L.H. (2008) *Adv. Synth. Catal.*, **350**, 1155.
22. Kassube, J.K., Wadepohl, H., and Gade, L.H. (2009) *Adv. Synth. Catal.*, **351**, 607.
23. Kragl, U., Vasic-Racki, D., and Wandrey, C. (1992) *Chem. Ing. Tech.*, **64**, 499.
24. See for example: (a) Hovestad, N.J., Eggeling, E.B., Heidbüchel, H.J., Jastrzebski, J.T.B.H., Kragl, U., Keim, W., Vogt, D., and van Koten, G. (1999) *Angew. Chem.*, **111**, 1763; *Angew. Chem. Int. Ed.*, (1999), **38**, 1655; (b) Kleij, A.W., Gossage, R.A., Jastrzebski, J.T.B.H., Boersma, J., and van Koten, G. (2000) *Angew. Chem.*, **112**, 179; *Angew. Chem. Int. Ed.*, (2000), **39**, 176; (c) Kleij, A.W., Gossage, R.A., Klein Gebbink, R.J.M., Brinkmann, N., Reijerse, E.J., Kragl, U., Lutz, M., Spek, A.L., and van Koten, G. (2000) *J. Am. Chem. Soc.*, **122**, 12112; (d) Dijkstra, H.P., Ronde, N., van Klink, G.P.M., Vogt, D., and van Koten, G. (2003) *Adv. Synth. Catal.*, **345**, 364.
25. (a) van Heerbeek, R., Kamer, P.C.J., van Leeuwen, P.W.N.M., and Reek, J.N.H. (2002) *Chem. Rev.*, **102**, 3717; (b) Fan, Q.-H., Li, Y.M., and Chan, A.S.C. (2002) *Chem. Rev.*, **102**, 3385.
26. Vankelecom, I.F.J. (2002) *Chem. Rev.*, **102**, 3779.
27. (a) The term "teabag catalyst" has been first coined by J. M. Thomas for certain types of heterogeneous catalyst systems: Thomas, J.M. (1990) *Philos. Trans.: Phys. Sci. Eng.*, **333**, 173; The concept of "catalyst in a teabag" has been discussed in: (b) Berger, A., Klein Gebbink, R.J.M., and van Koten, G. (2006) *Top. Organomet. Chem.*, **20**, 1.
28. Gaab, M., Bellemin-Laponnaz, S., and Gade, L.H. (2009) *Chem. Eur. J.*, **15**, 5450.
29. (a) Foltz, C., Stecker, B., Marconi, G., Wadepohl, H., Bellemin-Laponnaz, S., and Gade, L.H. (2005) *Chem. Commun.*, 5115; (b) Foltz, C., Stecker, B., Marconi, G., Bellemin-Laponnaz, S., Wadepohl, H., and Gade, L.H. (2007) *Chem. Eur. J.*, **13**, 9912.

20
Solid Phases as Protective Environments for Biomimetic Catalysts

Katja Heinze

20.1
Introduction

One of the most fascinating and inspiring approaches to catalyst design is to mimic natural catalytic systems, namely enzymes and – from the perspective of (bio)inorganic and coordination chemistry – especially metalloenzymes [1]. Insights into the mechanism of natural enzymes can be gained with biomimetic model systems, and, vice versa, enzymes inspire chemists to design new artificial single-site catalysts (bioinspired single-site catalysts) [2–5]. On one hand, biomimetic active site models of metalloenzymes are typically employed to structurally and spectroscopically mimic the active site with regard to the type and oxidation state of the central metal, type and number of ligating atoms, coordination geometry, ligand field strength, redox potentials, and spectroscopic properties such as IR/Raman, UV–vis, Mößssbauer, or electron paramagnetic resonance (EPR) spectroscopic signatures. On the other hand, biomimetic models can aim at functional mimicry, that is, modeling elementary reaction steps pertaining to the enzymatic reactivity (ligand exchange, electron transfer, proton transfer, oxidative addition, migratory insertion, atom transfer, etc.) or even the complete catalytic cycle of an enzymatic reaction, without necessarily featuring perfect structural or spectroscopic similarity.

One step beyond pure active site modeling is the emulation of the microenvironment, that is, the protein surrounding the active site, with its specific tasks such as active site protection (e.g., against dioxygen, dimerization, etc.), substrate/product selectivity (size, regio, stereo, reaction-type selectivity), and reagent/substrate delivery at optimal rates (e.g., electrons or protons, and the substrate itself). In this regard, polymeric matrixes of all flavors (e.g., organic/inorganic, shape-persistent/swellable, conducting/nonconducting, with/without functional groups for cooperative binding, cooperative hybrid organic/inorganic interfaces [4]) have found considerable interest as active and synergistic participants in the reactions. Especially, the so-called site isolation effect, that is, the restricted mobility of active units, prevents the active moieties from self-deactivation, for example, by dimerization, oligomerization, or detrimental

Molecular Catalysts: Structure and Functional Design, First Edition. Edited by Lutz H. Gade and Peter Hofmann.

redox chemistry. Using molecularly imprinted polymers (MIPs) [6], even aspects of substrate selectivity can be addressed. With electrically conducting matrixes (e.g., metals, metal oxides, carbon-based, or silicon-based materials) serving both as electrodes and as hosts of biomimetic, catalytically active species, the timely delivery of electrons (or holes) to the active site can be realized.

The aim of this chapter is to highlight specific aspects of the utilization of polymeric matrixes and their added value in biomimetic catalysis apart from the obvious extra bonus of potential recyclability and ease of product separation [7]. Rather than being a comprehensive review, instructive and inspiring examples are discussed and future developments toward artificial enzymes are envisioned.

As the site isolation effect plays an important role in biomimetic catalysis using immobilized complexes, this aspect will be illuminated in the first part. The potentially differing chemistry in homogeneous solutions will be discussed exemplarily by comparing the structure, spectroscopic signatures, and reactivity of homogeneous complexes with those of their immobilized congeners.

In the second part, selected immobilized structural and spectroscopic active site models will be discussed and aspects of characterization and analytics of immobilized transition metal complexes will be exemplarily disclosed. Typical techniques include spectroscopic methods addressing the immobilized biomimetic species and determination of metal ion leaching and active site integrity, for example, by selective extraction of the intact biomimetic metal complex – the prosthetic group – from the matrix – the apoenzyme (prosthetic group extraction). The third section gives a short overview of the elementary reaction steps in the catalytic processes and their observation on solid matrixes. Selected immobilized biomimetic functional active site models will be discussed in detail in the last section.

20.2
Site Isolation Experienced by Matrix-Bound Transition-Metal Complexes

The effective density of active sites on a typically loaded polymer (~0.1 – 1 mmol g^{-1}) is quite high when translated into concentration ($mol\,l^{-1}$). However, the restricted translational and rotational mobility often prevents effective bimolecular collisions between active species, for example, redox reactions (disproportionation or comproportionation), dimerization/oligomerization, or ligand scrambling. This spatial isolation of reactive entities is termed the *site isolation effect*. Several consequences can be envisioned resulting from this site isolation.

Unusual oxidation states of mononuclear metal complexes can be realized when the complexes are confined in the pores of zeolites or silicoaluminophosphates. The matrix mechanically prevents the thermodynamically favored and kinetically facile disproportionation to higher and lower oxidation states. Using this concept of site isolation, mononuclear $[Au^{II}(en)_2]^{2+}$ complexes (en = 1,2-ethylenediamine) and Zn^{I} species have been prepared within the pores of the host material and characterized by EPR spectroscopy [8, 9].

Electronic interaction of chromophores prone to stacking and aggregation, for example, *meso*-tetraphenyl porphyrins (TPPs), can also be suppressed by mechanically incorporating and site-isolating the chromophores in a polysiloxane matrix prepared by atmospheric-pressure plasma polymerization of the metalloporphyrins and hexamethyldisiloxane [10, 11].

The thermodynamically downhill comproportionation of $[Mo^{VI}O_2(N\cap N')_2]$ and $[Mo^{IV}O(N\cap N')_2]$ complexes **1** and **2** to μ-oxido Mo^{V} complexes **3** has been successfully retarded by *double attachment* of the complexes to polymeric matrixes via two N∩N′ chelate ligands (Figure 20.1a) [12]. The well-known oxidation of Fe^{II}(TPP) complexes **4** by dioxygen or hydrogen peroxide to stable μ-oxido Fe^{III} complexes **5** [13] has been suppressed by immobilization onto solid supports (Figure 20.1b) [14]. To prevent μ-oxido dimer formation in homogeneous solutions, highly sophisticated picket-fence porphyrins, as introduced by Collman [15], are typically needed as sterically demanding and protecting ligands. Hence, active site protection can be achieved either by carefully designed shielding ligands with the drawback of diminished substrate accessibility, or by immobilization onto a support, preferably by double attachment, or even by multipoint fixation.

Immobilization can also help to realize metal : ligand stoichiometries which are difficult to obtain in homogeneous solutions. Burzlaff *et al.* have demonstrated this feature with several $Cu^{II}(NNO)_n$ complexes **6–9** (NNO = facially coordinating tridentate NNO ligand based on imidazole or pyrazole, $n = 1, 2$). UV–vis spectroscopy showed that the 1 : 1 complexes **8** and **9** formed on the matrix feature bathochromically shifted absorption bands as compared to the 1 : 2 complexes **6** and **7** formed in solution (Figure 20.2a) [16–19]. Similarly, Heinze *et al.* could show by EPR spectroscopy that immobilized dipyrromethene ligands (N∩N) form planar mixed-ligand Cu^{II}(acac)(N∩N) (acac, acetylacetonate) complexes **11** with $Cu^{II}(acac)_2$ instead of the favored $Cu^{II}(N\cap N)_2$ complexes **10** with a twisted geometry in solution (Figure 20.2b). The mixed-ligand Cu^{II} complex **11** features a larger hyperfine splitting constant (hfc, hyperfine coupling) to the copper nuclei (**11**: $A_{iso}(^{63/65}Cu) = 77$ G; **10**: $A_{iso}(^{63/65}Cu) = 63$ G)) because of the planar geometry of **11** [20].

A *template approach* has been invoked by Stack *et al.* to immobilize a metastable $[Fe^{II}(OTf)_2(phen^R)_2]$ ($phen^R$, substituted 9,10-phenanthroline) complex **12** in a preformed binding pocket shaped by a thermodynamically stable $[Cu(phen^R)_2]^+$ template. After replacing Cu^+ by Fe^{2+}, the iron(II)-containing polymer **12** was obtained. A $[Fe^{II}(OTf)_2(phen^R)_2]$ intermediate is unstable in a homogeneous solution and rapidly leads to the tris(chelate) $[Fe(phen^R)_3]^{2+}$ complex and Fe^{2+} ions [21]. Thus, the matrix prevents ligand scrambling and again kinetically stabilizes a metastable species by site isolation.

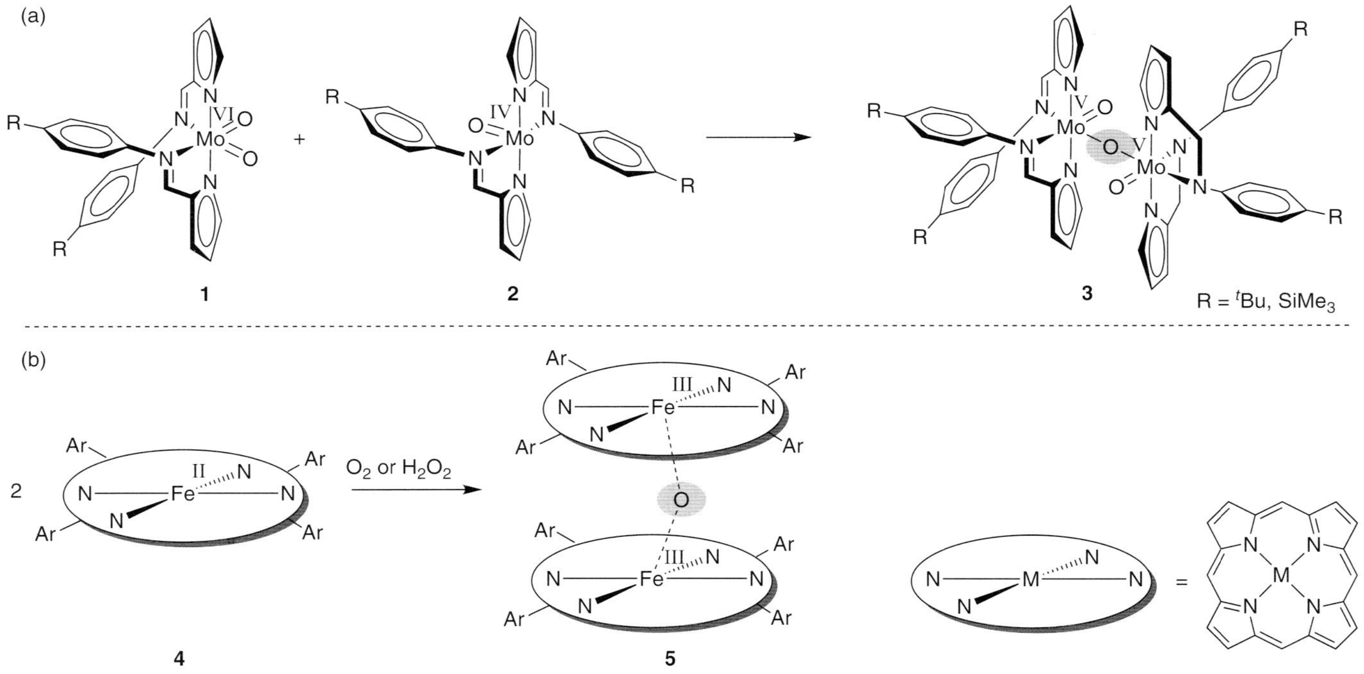

Figure 20.1 (a) Comproportionation of Mo^{IV} and Mo^{VI} oxido complexes **1** and **2** to **3** and (b) oxidation of Fe^{II}(TPP) complexes **4** to μ-oxido Fe^{III} dimers **5**.

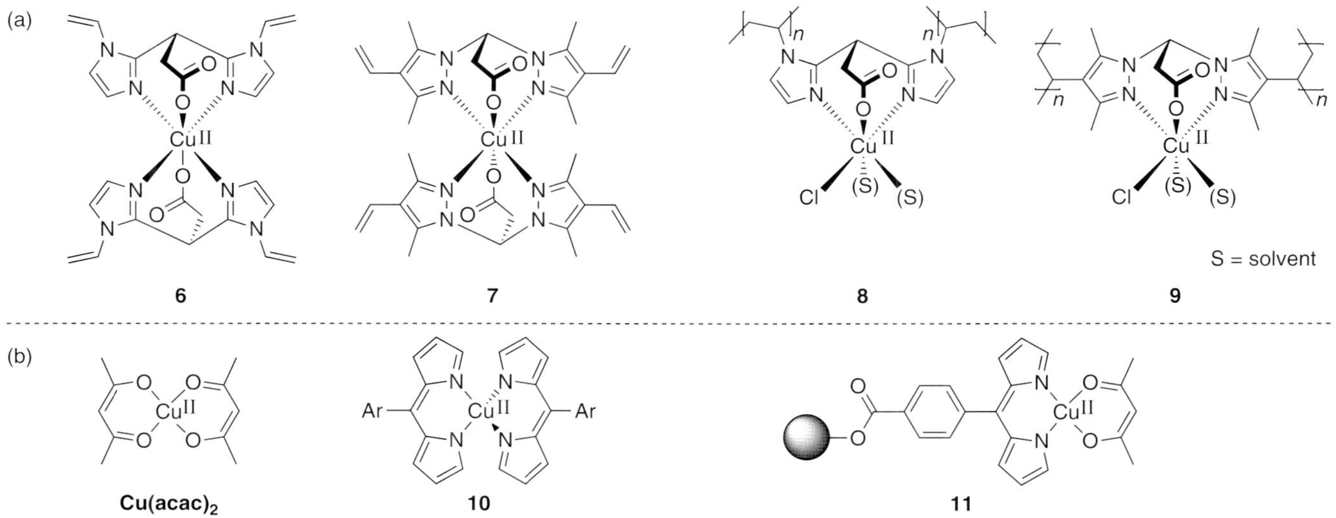

Figure 20.2 (a) 2 : 1 and 1 : 1 complexes of NNO ligands **6–9** and (b) 2 : 1 complexes Cu(acac)$_2$ and **10** and immobilized mixed-ligand Cu^{II}(N∩N)(acac) complex **11**.

20.3 Immobilized Structural and Spectroscopic Active Site Models

Selective analytic and spectroscopic methods for analyzing biomimetic model complexes bound to the support are rather limited, and a detailed characterization of such hybrid materials is much more challenging than of soluble molecular complexes. If local spectroscopic probes are available, IR spectroscopy, gel-phase or solid-state NMR spectroscopy, EPR spectroscopy, or (diffuse reflectance) UV–vis spectroscopy is applicable. Useful IR spectroscopic labels include CO, C≡N, E–H, M≡E, and M=E stretching vibrations as long as they are either very intense or are separated from the matrix vibrations. For EPR spectroscopic analysis, metal isotopes with nuclear spin $I \neq 0$ are suitable. Biologically interesting metal ions include ^{51}V ($I = 7/2$, 99.76%), ^{55}Mn ($I = 5/2$, 100%), ^{59}Co ($I = 7/2$, 100%), $^{63/65}Cu$ ($I = 3/2$, 69.09%; $I = 3/2$, 30.91%), and $^{95/97}Mo$ ($I = 5/2$, 15.72%; $I = 5/2$, 9.46%). For NMR characterization, some heteroatoms of coordinated ligands are suitable, for example, ^{19}F ($I = 1/2$, 100%), ^{31}P ($I = 1/2$, 100%) or ^{13}C ($I = 1/2$, 1.108%), and ^{15}N ($I = 1/2$, 0.37%) especially after enrichment. The local environment of the metal center can be probed by extended X-ray absorption fine structure (EXAFS) measurements. The incorporated metal ions can be quantitatively measured by atomic absorption spectroscopy (AAS), inductively coupled plasma mass spectrometry (ICP-MS), or neutron activation analysis.

To corroborate the structural identity of matrix-bound complexes, a release protocol compatible with the stabilities of several transition-metal complex types has been developed which is based on the mild cleavage of Si–O bonds by fluoride ions (Figure 20.3a) [22, 23]. For biomimetic model complexes, this method can be termed *prosthetic group extraction.* This way, the identity and integrity of the resin-bound complexes ("prosthetic group") can be easily probed by standard spectroscopic methods in solution (e.g., by solution NMR, UV–vis or IR spectroscopy, and mass spectrometry). The remaining Si–F-functionalized support ("apoenzyme") can be conveniently analyzed by mass spectrometry. The technique has been successfully applied to a wide range of complex types (Figure 20.3b), namely $M^0(CO)_3L(N\cap N')$ complexes **13** (M = Cr, Mo, W; L = CO, PPh_3, C≡N-R) [22, 24, 25], $Mo^{II}(allyl)(Cl)(CO)_2(N\cap N')$ and $Mo^{II}(Cl)(SnCl_2Ph)(CO)_3(N\cap N')$ complexes **14** and **15** [26], $Pt^{II}Cl_2(N\cap N')$ and $Pt^{II}(cat)$ $(N\cap N')$ complexes **16** and **17** [27, 28] ($N\cap N'$ = neutral bidentate Schiff base ligand, cat = substituted catecholato ligand), and, furthermore, biomimetic $Mo^{VI}O_2(N\cap N')_2$ complexes **18** [12] ($N\cap N'$ = anionic bidentate Schiff base ligand). Immobilized complexes attached to polymers by other linker types [29], which are in principle compatible with the stability of the complexes, for example, the Rink linker [29] often employed by Darensbourg [30–32], might also be released under mild conditions but this has not been demonstrated yet.

Selected examples of matrix-confined structural and spectroscopic models of mono- and binuclear metalloenzymes will be highlighted in the following paragraphs, in addition to aspects of the synthesis, characterization, and special properties.

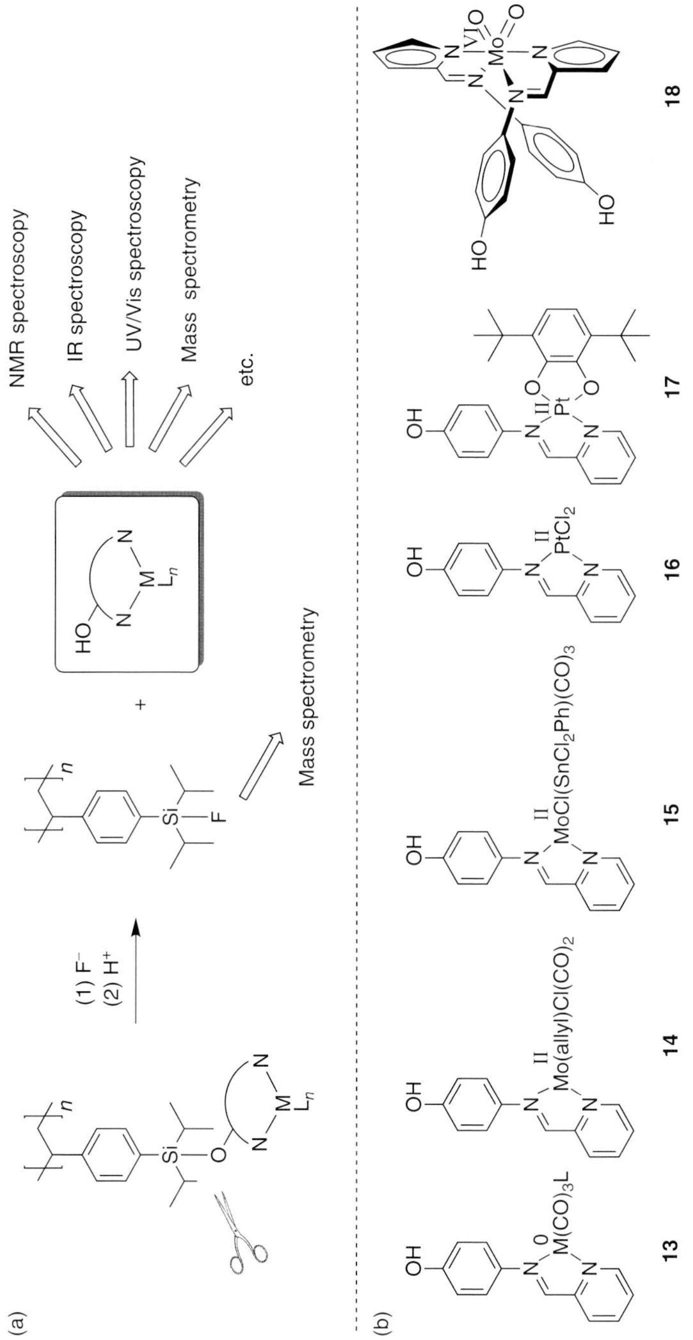

M = Cr, Mo, W
L = CO, PPh_3, CN-R

Figure 20.3 (a) Concept of prosthetic group extraction using Si–O linker [22, 23] and (b) complex types **13–18** successfully released.

Burzlaff *et al.* [16–19] have designed NNO heteroscorpionato ligands based on imidazole or pyrazole nitrogen donors and carboxylato oxygen donors with additional functional groups amenable to polymerization or to grafting onto preformed polymers. These ligands are therefore ideally suited to emulate the 2-His-1-carboxylate facial triad [33] observed in 2-oxoglutarate-dependent iron oxygenases containing mononuclear iron centers [34]. However, the high binding affinity of the NNO ligands typically leads to rather unreactive and non-biomimetic 2 : 1 bisligand complexes of the type **6**/**7** under homogeneous conditions (Figure 20.2a). A further challenge is to prevent the formation of thermodynamically stable μ-oxido dimers. Incorporation of the NNO ligand into polymeric materials either by grafting to preformed polymers or by copolymerization with methyl methacrylate or ethylene glycol dimethacrylate (EGDMA) has been devised as a successful method to suppress the formation of 2 : 1 complexes and μ-oxido species [16–19]. Burzlaff *et al.* used $Mn(CO)_3$ and $Re(CO)_3$ fragments to elegantly confirm the facial tridentate coordination mode of the resin-bound ligands by IR spectroscopy. The resulting immobilized organometallic complexes *fac*-$M(CO)_3$(NNO) **19** and **20** (M = Mn, Re) [35] feature the characteristic IR pattern with one A′ and two close A″ and A′ absorptions of the $M(CO)_3$ moiety coordinated to a tripodal NNO ligand (Figure 20.4a) [16–19]. Complexation of the NNO heteroscorpionato ligands with copper(II) salts leads to 2 : 1 complexes in solution but indeed yields the desired 1 : 1 complexes in the heterogenized systems as shown by UV–vis spectroscopy. Iron(II) complexes are rather challenging to obtain and characterize because of their high spin character and oxygen sensitivity. However, Burzlaff *et al.* [17] succeeded in the preparation of diamagnetic ruthenium(II) complexes such as Ru(NNO)(κ*O*,κ*O*-benzoylformato)(PPh_3) with a 2-oxocarboxylato ligand as a close structural mimic of 2-oxoglutarate-dependent iron oxygenases. These ruthenium complexes could be successfully imprinted by copolymerization with EGDMA, giving **21** and maintaining the facial coordination mode of the NNO ligand (Figure 20.4b). The generation of iron(II)-based imprinted polymers as

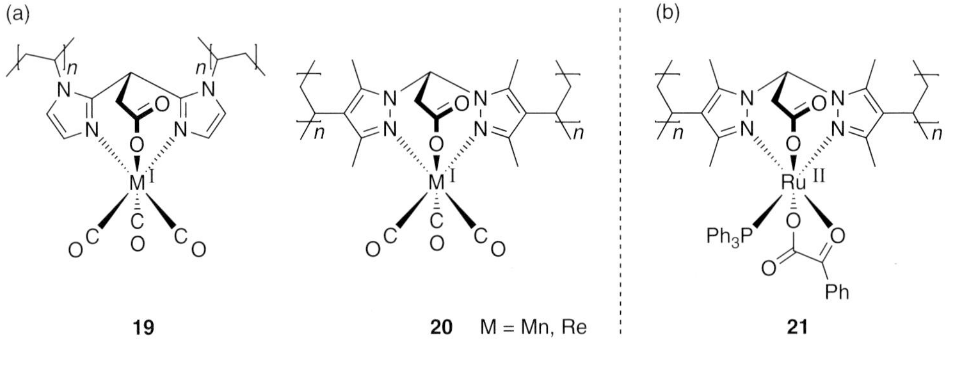

19 **20** **21**

Figure 20.4 (a) IR-spectroscopic probes $M(CO)_3$(NNO) **19** and **20** and (b) substrate-coordinated structural model **21** of 2-oxoglutarate-dependent iron oxygenase.

improved active site models by replacing the ruthenium(II) ion by iron(II) centers is the next logical step, but it has not yet been reported.

An enzyme-like MIP has been reported by Li *et al.* [36], featuring an iron(III) porphyrin (hemine) active site **22** including the bound substrate homovanillic acid as template (Figure 20.5a). As this system even displays catalytic activity mimicking natural peroxidases such as horseradish peroxidase (HRP) [37], it will be discussed in Section 20.5 in more detail (see below). Phosphinato-(η^6-arene) ruthenium(II)-based MIPs **23**, again including a bound transition-state analog (phosphinato ligands) for the targeted catalysis (transfer hydrogenation), were reported by Severin *et al.* Substitution of the phosphinato ligands by labile chlorido ligands generated a form-selective cavity in >80% yield according to ICP-MS analysis. It was convincingly demonstrated that double fixation of the precatalyst to the polymer promotes a higher activity as compared to singly attached precatalysts. Obviously, in the doubly bound case, the precatalyst is more precisely and rigidly positioned to the substrate binding site [38].

Metal(II) complexes with $M^{2+} = Fe^{2+}$, Cu^{2+}, or Zn^{2+} featuring bidentate or tridentate N-donor co-ligands often display considerable hydrolase activity owing to their high ligand substitution rates and Lewis acidity. In this respect, these complexes can be viewed as esterases or phosphor(di)esterases [39]. Severin *et al.* used a molecular imprinting approach and polymerized the stable metal complex template $[Mo^{II}(methallyl)(CO)_2(NNN)]^+$ (NNN = tris[2-(1-vinylimidazolyl)]phosphine) with three vinyl groups in the tripodal NNN ligand to the template polymer **24** (Figure 20.5b). This ensured a favorable preorganization of the tridentate ligand NNN as His_3 motif by three-point fixation within the resulting polymer [40]. The successful incorporation of the molybdenum(II) template was proven by IR spectroscopy (absorption bands of the CO stretching vibrations at 1939 and 1847 cm^{-1}). Removal of the molybdenum(II) template was performed by a consecutive HNO_3/NaOH treatment. IR spectroscopy confirmed the absence of CO ligands, and the ^{31}P NMR spectrum proved the stability of the chelate ligand under these harsh conditions. However, ICP-MS analysis suggested the presence of some residual molybdenum in the MIP. Re-metallation of the empty binding sites by $CuCl_2$ gave a green polymer **25**, likely incorporating the 1 : 1 copper : NNN ligand complex (Figure 20.5b). This polymer gave an EPR spectrum ($g = 2.10$) with partially resolved hfc to copper, which suggests a weak spin–spin interaction between "neighboring" sites within the MIP. Interaction between adjacent active sites can be beneficial for hydrolysis applications because of cooperative effects with respect to substrate activation [39]; however, too small a distance will lead to inactive μ-hydroxo complexes $[L_nCu^{II}(\mu\text{-}OH)_2Cu^{II}L_n]$. In an aqueous buffered suspension (pH 9), the Cu^{II} MIP is more stable toward metal ion leacing (as CuO) compared to a conventionally polymerized NNN ligand lacking the preorganized binding site. The stability has been ascribed to the favorable tridentate NNN coordination mode in the Cu^{II} MIP in contrast to some bidentate coordination sites in the conventionally prepared Cu^{II} polymers which are prone to Cu^{2+} loss. In catalytic test reactions, bis(*p*-nitrophenyl)phosphate was more rapidly hydrolyzed by the Cu^{II} MIP than by the conventional Cu^{II}

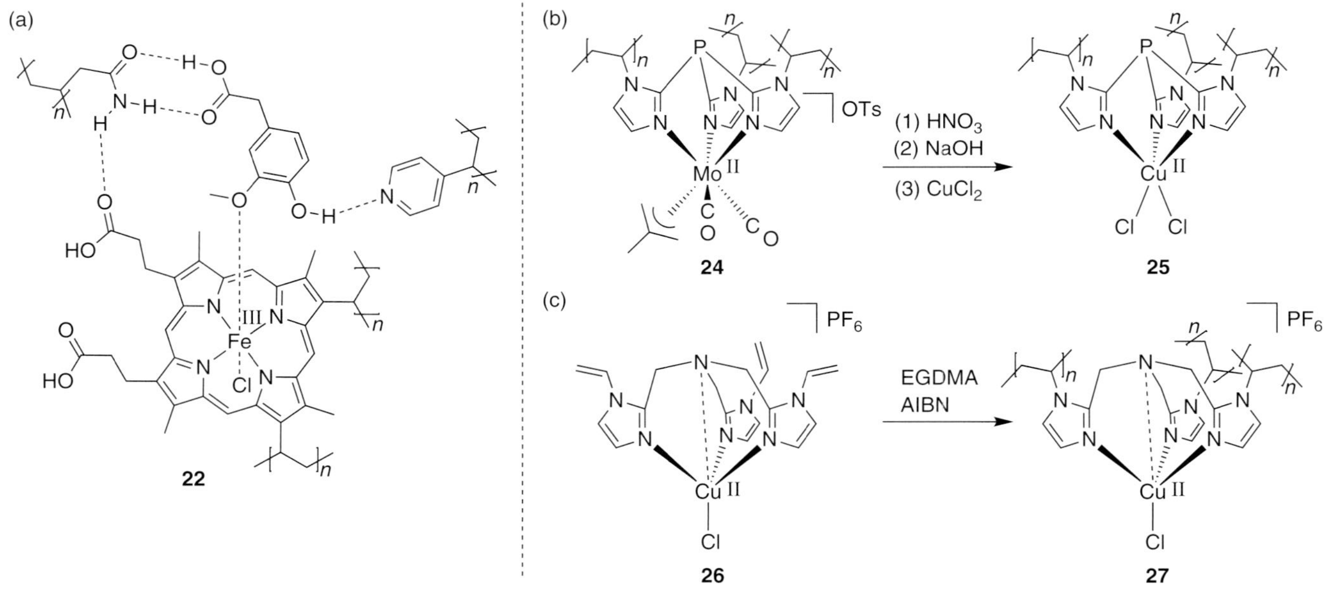

Figure 20.5 (a) Fe^{III}(porphyrin) MIP **22** with substrate-bound active site as peroxidase mimic, (b) Mo^{II} template **24** for the Cu^{II} MIP **25** as artificial phosphoresterase, and (c) imprinting of the tris-vinyl substituted complex **26** to MIP **27**.

polymer with the same total copper amount, while the reverse was observed for the (*p*-nitrophenyl)phosphate substrate. This might be due to some cooperative or inhibition effects or to the use of different solvent systems, although diffusion itself has been shown to play only a minor role [40]. This example nicely illustrates that, even after a thorough characterization of the immobilized/imprinted active site, some questions remain unanswered.

A similar ligand tris[(1-vinylimidazol-2-yl)methyl]amine NNN forms the ionic copper(II) complex $[CuCl(NNN)]PF_6$ **26** with $CuCl_2$ and NH_4PF_6 (Figure 20.5c). After copolymerization of the copper complex with EGDMA, the supported complex **27** was used in the hydrolysis of bis(*p*-nitrophenyl)phosphate at pH 9.5. The highly cross-linked Cu^{II} copolymer **27** was shown to be 56 times more active than its homogeneous congener **26**, which has been attributed to a favorable partitioning effect that increases the local concentration of the substrate in the polymer. In this case, the matrix plays an active role by increasing the substrate concentration and thus accelerating the bimolecular rate-limiting reaction step and ensuring a timely supply of the substrate [41].

Acetyl CoA synthetase (ACS) is capable of forming acetyl coenzyme A (AcSCoA) from carbon monoxide, CH_3^+, and HSCoA. The so-called A cluster of ACS features a tetradentate tripeptide motif Cys-Gly-Cys accommodating two nickel centers [42] for oxidative addition of CH_3^+, migratory insertion of CO, and, finally, reductive elimination of CoA-$SCOCH_3$. Interestingly, this mechanism strongly resembles the Monsanto acetic acid process using *cis*-$[Rh^{I}I_2(CO)_2]^-$ as catalyst. Thus, Darensbourg *et al.* prepared an N-acetylated Cys-His-Cys tripepetide **28** on TentaGel® resin by solid-phase peptide synthesis (SPPS) [43] and coordinated a Ni^{2+} ion by using $Ni^{II}(acac)_2$ and KOH to deprotonate the amides and thiols to give the dianionic metalloligand **29** [30] (Figure 20.6a). A rhodium(I) dicarbonyl unit was attached to **29** as second metal center to give the heterometallic monoanionic system **30** by treating **29** with $Rh_2Cl_2(CO)_4$. Notably, the CO stretching vibrations of **30** (2059 and 1988 cm^{-1}) are very similar to those of the Monsanto catalyst [*cis*-$RhI_2(CO)_2]^-$ (2067 and 1990 cm^{-1}) [44], attesting the immobilized metalloligand **29** a high electron donating power. However, unspecific binding of carbonyl-free rhodium species to the TentaGel polymer has also been reported. The loading has been assessed by 9-fluorenylmethoxycarbonyl Fmoc cleavage (peptide loading) and neutron activation analysis (metal content) [31]. The immobilized bimetallic complex **30** appears to be more stable toward air than its soluble counterpart [30]. No further reactivity, for example, toward alkylation agents and thiolates [45], has been reported so far, so **30** represents an elegant partial structural and spectroscopic biomimetic but not yet a functional ACS model (Figure 20.6a).

Hydrogenases (H_2ases) catalyze the reduction of protons to dihydrogen and the reverse reaction [46]. Some of these enzymes feature a bimetallic active site – with two iron centers ([FeFe]-H_2ase) or with one iron and one nickel center ([FeNi]-H_2ase) – bridged by thiolates and coordinated by terminal CO and CN^- ligands. Darensbourg *et al.* [32, 47] have prepared resin-bound models such as **31** of the [FeFe]H_2ase active site (Figure 20.6b), characterized them by IR spectroscopy (CO label), and investigated CO substitution reactions by

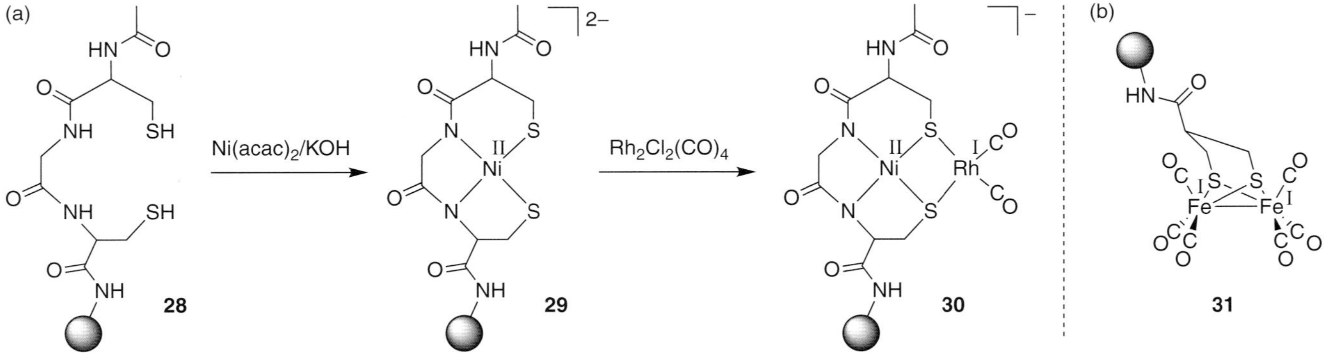

Figure 20.6 (a) Immobilized ACS model **30** prepared from immobilized CysGlyCys tripeptide **28** and (b) [FeFe]-H_2ase model **31**.

phosphanes and cyanides to give substituted derivatives. Furthermore, the polyethylene glycol (PEG) chains of the TentaGel support seem to facilitate photolytic CO loss and subsequent degradation of the immobilized complex. In this specific case, the PEG-containing polymer imparts a lower stability to the immobilized complex. Attempts to prepare hydrido complexes by protonation failed [32]. Attachment of [FeFe]H_2ase model complexes to amino-functionalized highly ordered pyrolytic graphite (HOPG) electrodes as conductive support was successful. Unfortunately, no electrocatalytic proton reduction has been observed with this structural and spectroscopic biomimetic model upon electrolysis at sufficiently negative potentials [47]. Electrocatalytic proton reduction was, however, achieved by a functionalized polypyrrole polymer attached to a glassy carbon electrode [48]. A preformed dinuclear thiolate-bridged iron carbon monoxide complex has been employed as functional [FeFe]-H_2ase mimic in the polymer-modified electrode **32**. The construct **32** is capable of proton reduction upon application of a sufficiently negative potential in the presence of a soluble proton source [48]. Obviously, the polypyrrole sustains charge propagation and delivers the required electrons. This system crudely but efficiently emulates the elaborated electron supply chain of iron–sulfur clusters found, for example, in natural [FeNi]-H_2ases and [FeFe]-H_2ases [46].

Bimetallic active sites are also present in soluble methane mono-oxygenase (sMMO) [49], which features two iron centers bridged by carboxylates (acetate and glutamate) and a hydroxide, the rest of the ligand sphere being composed of two histidine and three glutamate residues. The His and most Glu residues are present as His-X-X-Glu peptide sequences and thus Pirngruber *et al.* [50] attached a His-Gly-Gly-Glu peptide onto a silica support and allowed the peptides to self-assemble around iron(II) **33** and copper(II) **34** (Figure 20.7a). Detailed EXAFS K-edge analyses provided insight into the first and second coordination sphere of the immobilized copper complex **34**. The data suggested N(His), N(amide), and O(amide) coordination, but no indication for bimetallic complexes was obtained [50].

Bis(copper) centers are found in catechol oxidases, with a His_3 motif ligating each metal center and shaping an O_2 and substrate binding pocket [51]. A tripodal mononuclear copper(II) complex **35** was hence immobilized onto graphite felt with the hope of forming dinuclear species (Figure 20.7b). However, the proximity of two grafted copper complexes could not be demonstrated, and only mononuclear species seemed to be present. Furthermore, during the catalytic oxidation of 3,5-di-*tert*-butyl-catechol in a flow cell, four-electron reduction of O_2 to water was not observed but only the two-electron reduction to H_2O_2 unlike the natural enzymatic reaction [52]. Although not explicitly stated by the authors, these findings might be ascribed to the site isolation effect. The last two examples suggest that bimetallic active site models should be better self-assembled before immobilization onto supports in order to ensure a defined single-site bimetallic unit.

A last example of considerable historic relevance comprises the immobilization of metalloporphyrins onto surfaces and matrices for potential (biocatalytic)

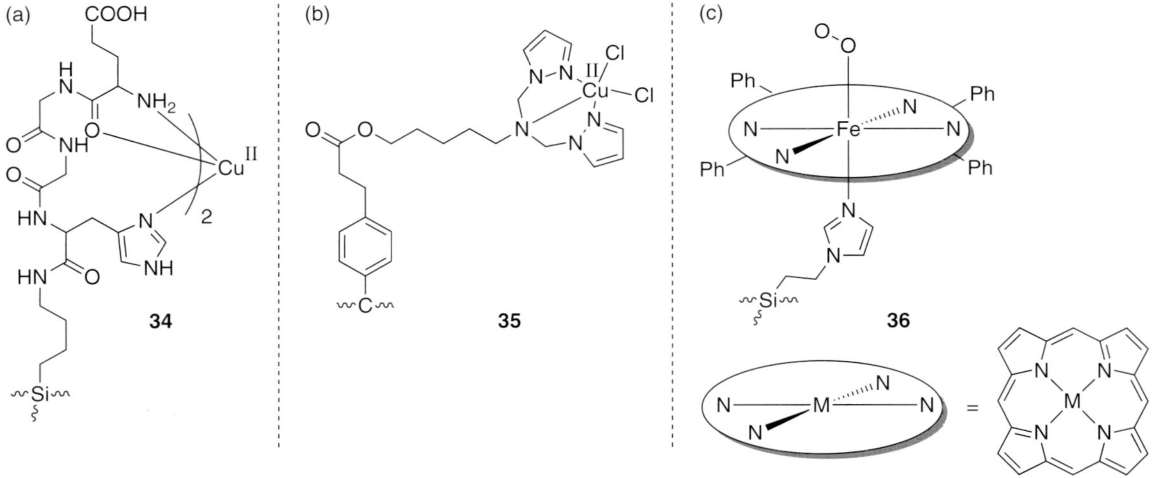

Figure 20.7 (a) Proposed assembly of Cu^{II} on a silica-immobilized His-Gly-Gly-Glu peptide **34**, (b) mononuclear Cu^{II} complex supported on graphite felt **35**, and (c) immobilized myoglobin model **36**.

applications. The prominent rationale for immobilization is to prevent bimolecular reactions to form the thermodynamically stable μ-oxido dimers, especially in iron porphyrin chemistry (Figure 20.1b) [13]. In the 1970s, myoglobin models **36** were developed by attaching iron(II) porphyrinato complexes onto surfaces that were able to reversibly bind dioxygen without deleterious formation of μ-oxido dimers (Figure 20.7c) [14]. Using homogeneous models, dimerization is prevented by the famous sterically demanding picket-fence porphyrins, while on the surface protection is guaranteed by the site isolation effect [15].

20.4
Elementary Reaction Steps Performed by Solid-Phase Supported Complexes

Elementary reaction steps generally relevant to (biomimetic) catalysis have been investigated in some detail on resin-bound complexes using IR, EPR, or NMR spectroscopic labels. A highly useful tool is the characterization of released complexes in solution by standard solution techniques after prosthetic group extraction (Figure 20.3). For instance, ligand coordination/decoordination has been investigated with axial pyridine (de-)coordination at square-planar Cu^{II} complexes **11** using EPR spectroscopy (Figure 20.2b) [20]. Ligand substitution reactions, for example, substitution of labile tetrahydrofuran (thf) by CO, PPh_3, or C≡N−R ligands (**37** → **38**, Figure 20.8a), and oxidative additions, for example, of allyl halides or $PhSnCl_3$ (**38** → **39**/**40**, Figure 20.8b), have been successfully studied by IR spectroscopy and prosthetic group extraction [22, 24]. Solid-phase NMR and EPR spectroscopic methods were helpful in probing atom transfer ($Mo{=}\underline{O}$ to PMe_3 **41** → **42**/$Me_2S\underline{O}$ to Mo^{IV} **42** → **41**, Figure 20.8c) [12] and outer-sphere single-electron transfer reactions (oxidation of **42** to Mo^{V} by ferrocenium ions and reduction of **41** to Mo^{V} by cobaltocene) [12]. Protonation and subsequent reduction (nitrido/imido $Mo^{VI/V}$ complexes **43** and **44**, Figure 20.9a) [53] and stability issues during catalytic turnover ($[V^{VI}O]^{2+}$ complex **45** of a polymer-anchored Schiff base derived from salicylaldehyde and cysteine, Figure 20.9b) were investigated using IR and EPR spectroscopic methods [54].

20.5
Immobilized Functional Active Site Models

The site isolation concept with simple *meso*-tetraaryl porphyrin ligands has found widespread application in oxidation catalysis with immobilized metalloporphyrins, for example, catechol and lignin oxidation with $Mn^{III}Cl(TPP)$ complexes **46** and H_2O_2 as oxidant [55, 56] or hydroquinone and thiol oxidation with $Co^{II}(TPP)$ complexes **47** and O_2 as oxidant [57, 58]. Although these reactions are of considerable synthetic utility [59], truly functional biomimetic models with a detailed knowledge of elementary reaction steps have not yet been developed.

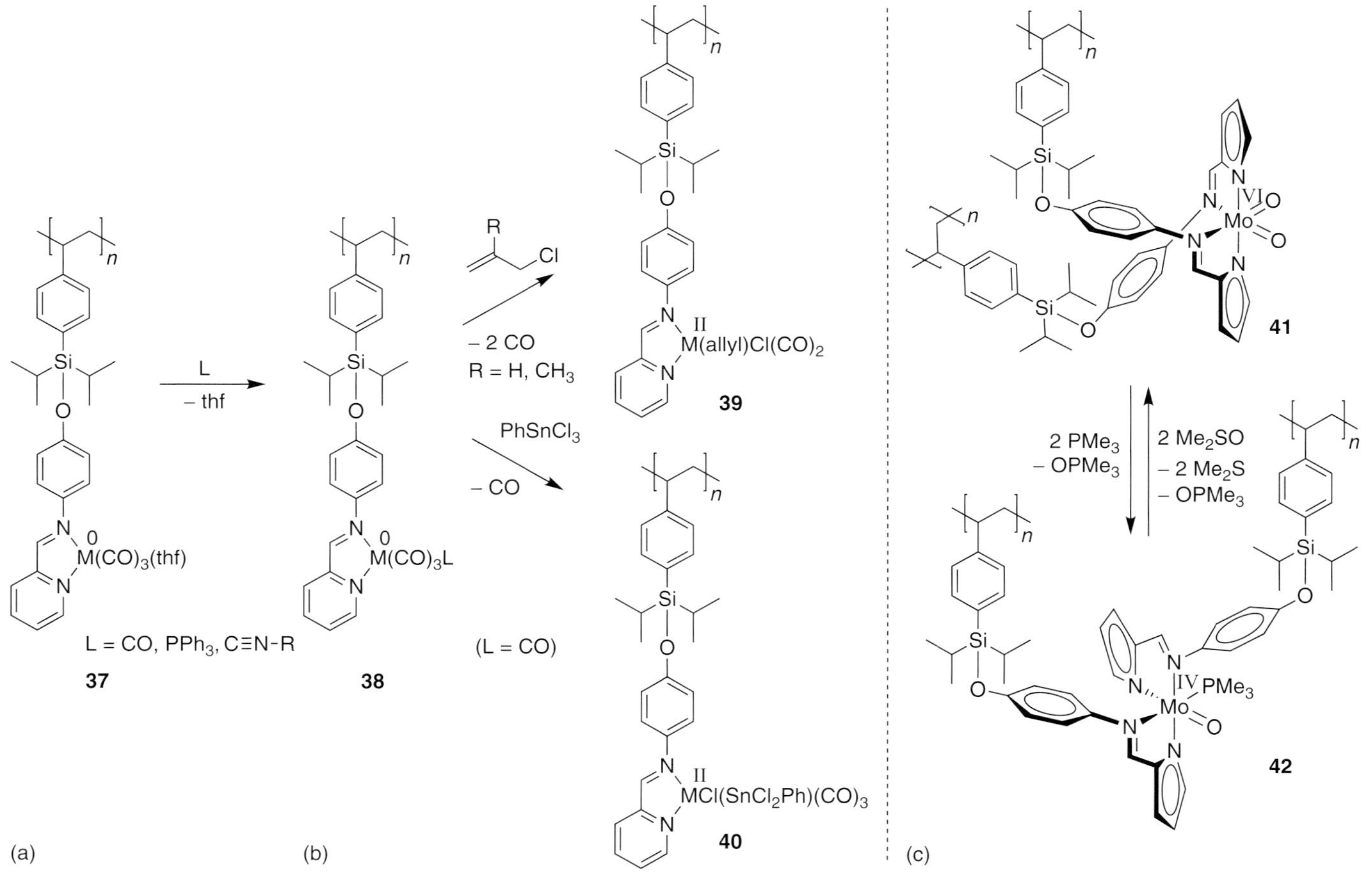

Figure 20.8 (a) Ligand substitution at immobilized carbonyl molybdenum(0) complexes **37** → **38**, (b) oxidative addition at immobilized carbonyl molybdenum(0) complexes **38** → **39**/**40**, and (c) oxygen atom transfer at dioxido molybdenum(VI) complexes **41** → **42**.

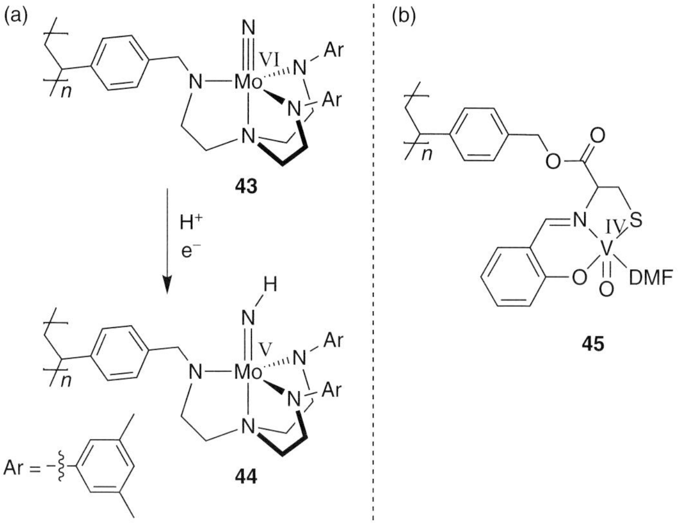

Figure 20.9 (a) Protonation and reduction of an immobilized molybdenum(VI) nitrido complex **43** to the imido molybdenum(V) complex **44** and (b) immobilized Schiff base vanadyl complex **45**.

A highly instructive example that has already been mentioned previously is the peroxidase-like activity of a hemin-containing MIP **22** (Figure 20.5a) [36]. Peroxidases and their models catalyze the oxidative dimerization of phenols, for example, homovanillic acid, using H_2O_2 as oxidant. The reported peroxidase MIP **22** displays intriguing features of enzyme-like behavior, namely Michaelis–Menten kinetics (saturation), substrate specificity (e.g., homovanillic acid is preferred over substrates lacking the methoxy group), and active site blocking by inhibitors (ferulic acid). Again, dimerization to μ-oxido bimetallic complexes [13] is prevented by the polymer matrix [36].

Preassembled cationic bimetallic structural models [60] **48** and **49** of the oxygen-evolving complex (OEC) of photosystem II (PS II) [61] have been immobilized onto clay surfaces (kaolinite, montmorillonite) and TiO_2 nanoparticles by ion exchange (Figure 20.10) [62–64]. Upon immobilization, the electronic properties (color) of the complexes are altered in such a way that they become competent to catalytically perform the four-electron water to dioxygen oxidation by proton-coupled electron transfer (PCET [65]) using the single-electron transfer reagent $(NH_4)_2[Ce^{IV}(NO_3)_6]$ as terminal oxidant. The soluble complex **48** evolves O_2 only with hypochlorite, likely according to a different mechanism [60]. Using Ce^{IV} as oxidant, soluble **48** is rapidly oxidatively degraded to MnO_4^- initiated by disproportionation of intermediate oxidation states. The disproportionation is suppressed on the clay mineral or TiO_2, which obviously protects the active sites from oxidative degradation by site isolation [63]. The detailed mechanism of this extraordinarily important reaction of immobilized biomimetic complexes remains unresolved up to now, although a density functional theory (DFT)

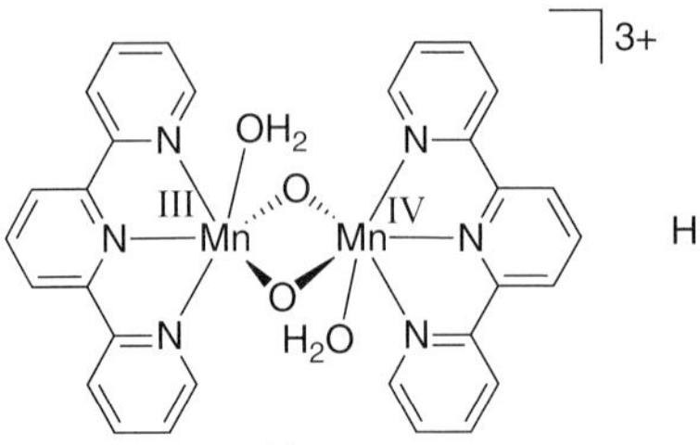

48

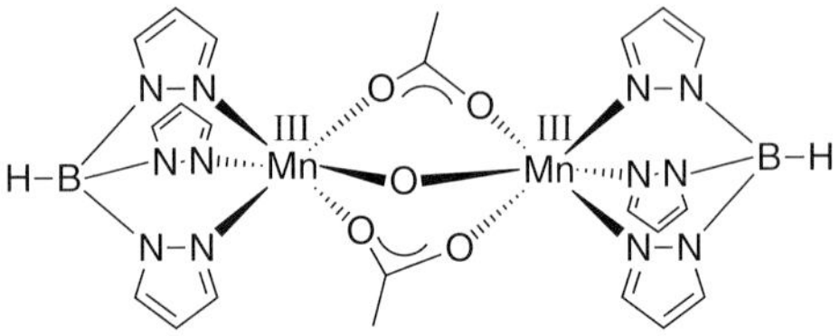

49

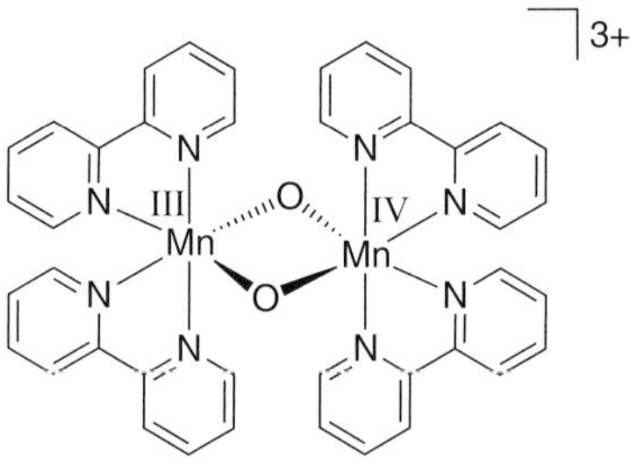

50

Figure 20.10 Binuclear manganese complexes **48–50** as OEC mimics before adsorption to phyllosilicates such as montmorillonite or kaolinite or to TiO_2.

study has been put forward [66]. Even the structural identity of the adsorbed manganese complexes remains unclear [67]. Clearly, more detailed mechanistic investigations concerning intermediates and elementary reaction steps occurring in the manganese complex/clay hybrid materials are critically needed.

The structural but nonfunctional OEC model complex $[(bpy)_2Mn^{III}(\mu\text{-}O)_2Mn^{IV}(bpy)_2]^{3+}$ (bpy, 2,2′-bipyridine) **50** (Figure 20.10) has been confined in the channels of the nanoporous oxide AlMCM-41 with Cr^{VI} charge-transfer chromophores embedded in the oxidic matrix. Excitation of the ligand-to-metal charge transfer (LMCT) band of the Cr–O unit in **50@**AlMCM-41 results in photoinduced electron transfer (PET) from the confined mixed-valent $Mn^{III}Mn^{IV}$ complex to the excited Cr^{VI}–O center to give the charge-separated state $Mn^{IV}Mn^{IV}/Cr^{V}$. This system obviously models the initial PET step in PS II (oxidative part of water splitting) [68].

A reverse PET, namely photoinduced charge injection from an electronically excited chromophore to a semiconductor surface, is typically the initial step in dye-sensitized solar cells (DSSCs) of the Grätzel-type, for example, with a trinuclear ruthenium dye **51** (Figure 20.11a) [69]. Such a PET step, reminiscent of the charge separation in DSSCs or in PS II, has been visualized by Kelvin probe force microscopy on a donor–chromophore–acceptor nanocomposite poly[TPA-$Ru^{II}(tpy)_2$]@ZnO (tpy, 2,2′;6′,2″-terpyridine) **52** prepared from a ZnO-immobilized $[Ru^{II}(tpy)_2]^{2+}$ chromophore (Figure 20.11b, TPA, triphenylamine) [70]. In **52**, both the electron acceptor (ZnO) and the electron donor (TPA) are assembled in a directional manner to assist the charge separation step

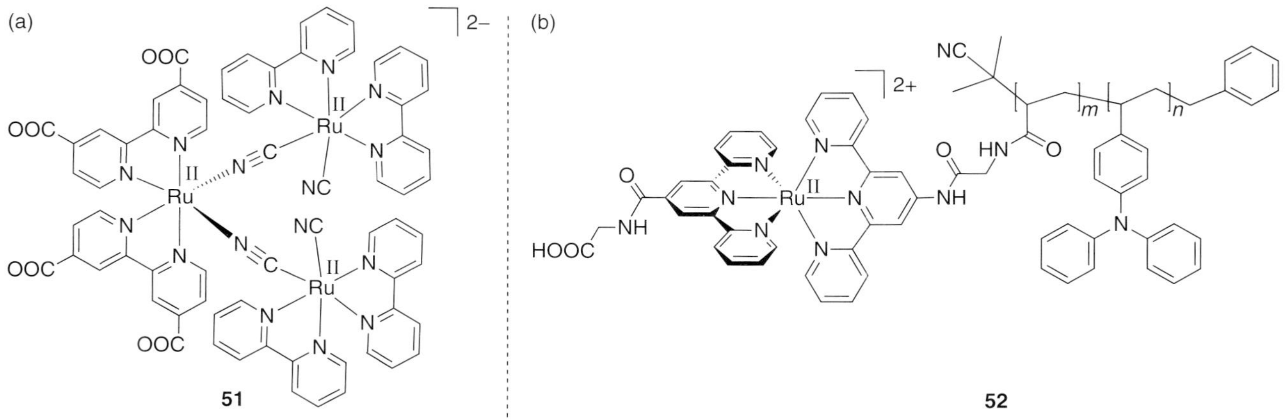

Figure 20.11 Photoinduced charge separation in immobilized systems relevant to the reductive part of water splitting (a) trinuclear ruthenium(II) chromophore **51** for adsorption onto TiO_2 and (b) polymeric ruthenium(II) chromophore including triarylamine electron acceptor units for adsorption onto ZnO.

that crudely mimics the charge separation in PS II with the special pair as the charge-separating chromophore, pheophytin as primary electron acceptor, and tyrosine Z as electron donor.

One of the most sophisticated active site modes are Collman's cytochrome c oxidase (CcO) [71] mimics **53**–**54** comprising three redox centers, namely an iron(II) picket-fence porphyrin, a copper(I) site with a tris(imidazole) ligation modeling the natural His_3 motif, and a phenol as tyrosine mimic (Figure 20.9) [72]. Together, the subunits are able to synergistically perform the four-electron reduction of dioxygen to water with suppression of superoxide or peroxide formation. This is formally the reverse reaction catalyzed by the OEC (see above). This elaborate system has been immobilized onto conducting gold surfaces using self-assembled monolayer (SAM) formation and "click" chemistry [72]. By adjusting the type of linker between the active site model and the gold surface (**53**: long aliphatic chain; **54**: short conjugated bridge in the SAM, Figure 20.12), it was possible to tune the rate of electron flow to the active FeCu site, with the aliphatic chain delivering the electrons more slowly (about $6\,s^{-1}$) than the conjugated bridge ($>10^4\,s^{-1}$). Interestingly, the electron transfer from cytochrome c in the natural CcO system is the rate-determining step of the enzymatic O_2 reduction, and thus the Au electrode with the aliphatic linker in the SAM **53** nicely emulates the slow electron transfer step within the catalytic cycle [72].

Immobilized dinitrogen complexes of molybdenum are clearly inspired by the FeMo cofactor of the natural enzyme nitrogenase (N_2ase) [73–75], which catalyzes the demanding six-electron reduction of N_2 to two NH_3 (6× PCET). The exact binding site of N_2 at the FeMo cofactor is still a matter of debate (iron–sulfur cluster periphery vs molybdenum center [74–76]), although the center of the iron–sulfur cluster is definitely excluded on the basis of recent spectroscopic and theoretical insights [77]. Interestingly, soluble complexes *trans*-$Mo^0(N_2)_2L_4$ **55** (L = phosphane) produce ammonia upon protonation with mineral acids [78], while analogous polymer-immobilized complexes (characterized by the IR signature of the end-on-coordinated N_2 ligands ($1946\,cm^{-1}$) and gel-phase ^{31}P NMR [22, 79]) fail to release ammonia [80]. This has been ascribed to the intermediate molybdenum hydrazido(2-) complexes $Mo^{III}X(NNH_2)L_4$ **56** which undergo disproportionation in solution to give NH_4^+ and N_2. Because of site isolation of the immobilized complexes, such a disproportionation is suppressed and, consequently, NH_4^+ is not formed. On the other hand, genuine immobilized $Mo^{III}F(NNH_2)L_4$ prepared from immobilized *trans*-$Mo^0(N_2)_2L_4$ **55** by treatment with HBF_4 and characterized by IR (3333, 3253, $3161\,cm^{-1}$) and gel-phase ^{31}P NMR spectroscopy releases hydrazine N_2H_4 upon acidification with HBr, demonstrating that the bimolecular disproportionation step is indeed prevented [80]. On the basis of these results, neither the homogeneous (different mechanism as compared to N_2ase featuring a site-isolated molybdenum center) nor the heterogenized system (different product) can be considered a true functional N_2ase mimic. Clearly, other active site models need to be considered and anchored to polymeric matrixes. Current work focuses on immobilized

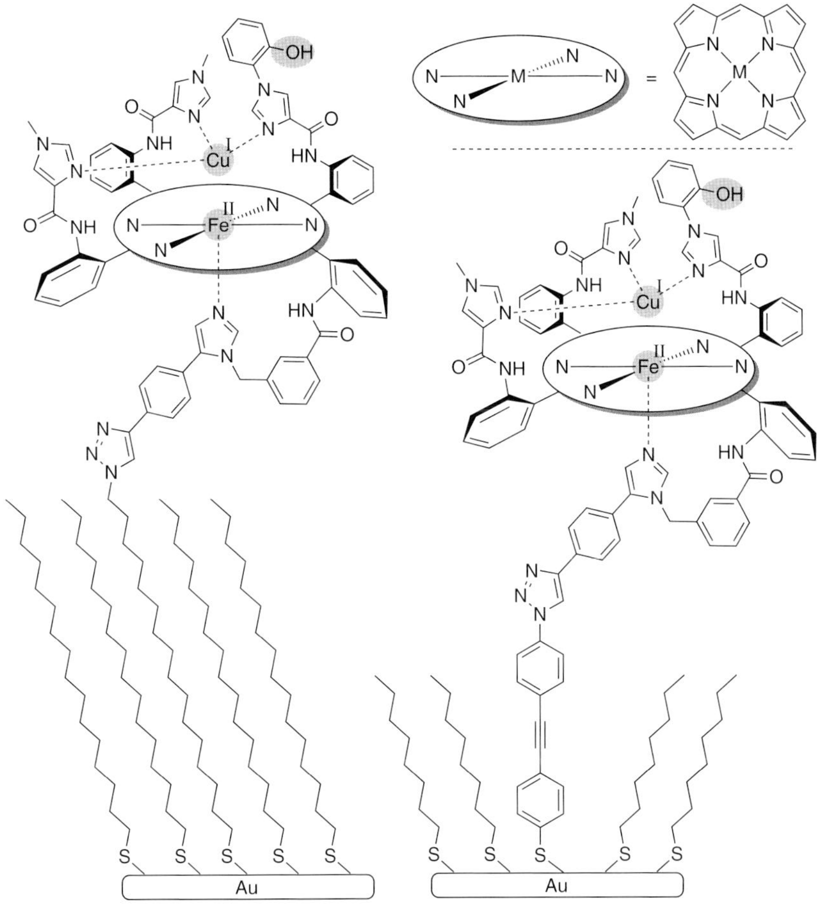

53 54

Figure 20.12 Immobilized CcO models **53** and **54** using two different linker types.

complexes **43** [53] derived from the Schrock–Yandulov N_2 reduction catalyst [74] (Figure 20.9a).

Heinze *et al.* have developed a truly functional model **41** of a molybdenum oxotransferase [81–84] by immobilizing anionic bidentate chelates N∩N′ onto polystyrene supports as ancillary ligands using the Si–O linker (Figure 20.8c) [12, 22]. Molybdenum oxytransferases catalyze the oxygen atom transfer (OAT) from water as oxygen source to substrates (e.g., sulfite oxidase), or vice versa (e.g., DMSO reductase). The initial reaction is an OAT elementary reaction step, a formal two-electron process reducing Mo^{VI} to Mo^{IV}, by electrophilic attack of a terminal oxido ligand onto the substrate. In aqueous molybdenum chemistry, the formation of unreactive oxido-bridged species is a thermodynamically favorable reaction (Figure 20.1a), which obviously has to be suppressed in biomimetic catalytic reactions. Regeneration of the enzyme's active site is achieved by two subsequent PCET reactions [56], which necessarily generate a Mo^{V} intermediate. In the biomimetic catalytic cycle, two challenging steps have to be mastered: first, the OAT step in which Mo^{IV} and Mo^{VI} oxido species are simultaneously present and easily form diamagnetic μ-oxido dimers of Mo^{V} by comproportionation, and, second, the one-electron redox chemistry of the $Mo^{IV/V}$ and $Mo^{V/VI}$ couples with the Mo^{V} species being susceptible to condensation in the presence of traces of water, again forming the stable μ-oxido dimers. The traditional way to overcome these problems is to employ sterically demanding ligands [83, 84]. However, a full biomimetic catalytic cycle (OAT, 2× PCET) has not yet been accomplished in homogeneous solution, although important intermediates have been identified [85]. By double attachment of $[Mo^{VI}O_2]$ fragments onto the support via the two N∩N′ chelate ligands, the biomimetic functional model **41** has been developed (Figure 20.8c). Compound **41** is capable of OAT to phosphanes as substrates [12], giving the molybdenum(IV) complex **42** (Figure 20.8c), and subsequent PCET reactions using diacetylferrocenium tetrafluoroborate as oxidant and water as the oxygen atom source [12] (Figure 20.13a). The correct formation of the immobilized bis(chelate) molybdenum(VI) complex $Mo^{VI}O_2(N{\cap}N')_2$ **41** was proven by extracting the whole prosthetic group by selective cleavage of Si–O bonds through fluoridolysis (prosthetic group extraction). The immobilized intermediate $Mo^{IV}O(PMe_3)(N{\cap}N')_2$ complex **42** has been characterized by ^{31}P cross-polarization magic angle spinning (CP-MAS) NMR spectroscopy. The mononuclear Mo^{V} intermediates **57** and **58** (Figure 20.13a) feature characteristic EPR signatures (g values below 2.0 and hfcs to $^{95/97}Mo$). For the diamagnetic Mo^{VI} and Mo^{IV} oxidation states **41** and **42**, homogeneous molecular models [12, 86, 87] fully substantiate the properties of the resin-bound systems. For the intermediate and reactive Mo^{V} species **57**, a sterically congested imido analog $[Mo^{V}(N^tBu)(PMe_3)(N{\cap}N')_2]^+$ **59** has been developed for spectroscopic, especially EPR, comparison (Figure 20.13b) [87]. Anionic immobilized Mo^{V} complexes **60** have been previously reported and characterized by EPR by Topich [88] (Figure 20.13c). These $[Mo^{V}Cl_2(O)(ONO)]^-$ complexes **60** with a tridentate dianionic ONO Schiff base ligand are also stable in homogeneous solution because of their negative charge preventing dimerization, and thus they are also EPR active ($g < 2.0$) both in solution and in the polymer-anchored state [88]. In the immobilized system **41**, formation of μ-oxido dimers $(N{\cap}N')_2Mo^{V}(O)$-O-$Mo^{V}(O)(N{\cap}N')_2$ of type **3** (Figure 20.1a) is obviously prevented by site isolation. All relevant intermediates of a biomimetic catalytic cycle in the oxidation states +VI, +V, and +IV have been fully characterized in the resin-bound version. Notably, only the heterogeneous system **41** is capable of performing the whole catalytic cycle of OAT and PCET [65] steps in a sustained manner and with significant turnovers using substituted ferrocenium ions as terminal oxidant and phosphazene base $P_1{}^tBu$ as proton acceptor (Figure 20.13a). The respective homogeneous oxido complexes are susceptible to formation of unreactive **3** during turnover, while the sterically shielded imido complex **59** is susceptible to hydrolysis and subsequent formation of μ-oxido dimers in the presence of water [87]. Using ^{18}O-labeled water as substrate, mass spectrometric analysis of

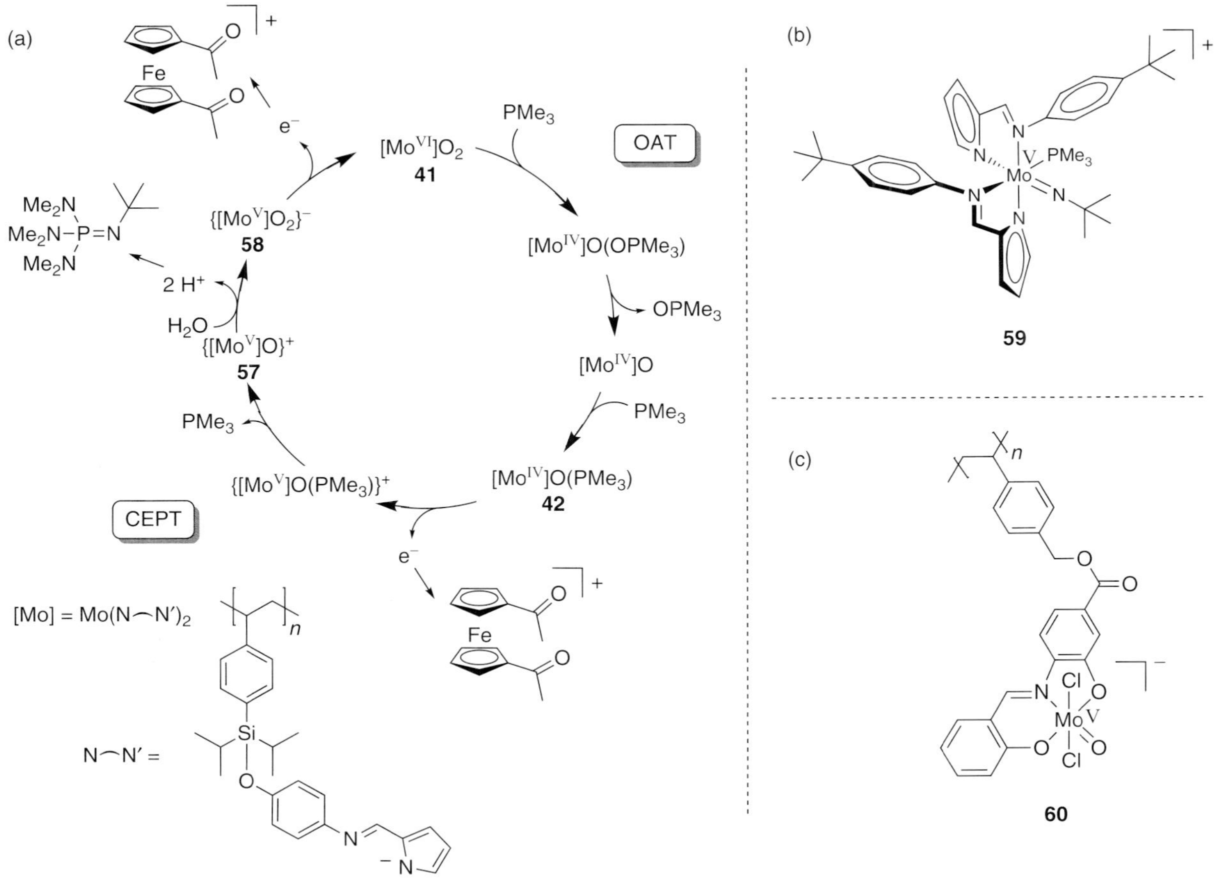

Figure 20.13 (a) Catalytic cycle of molybdenum oxotransferase mimic **41**, (b) cationic soluble sterically protected imido molybdenum(V) analog **59**, and (c) anionic immobilized molybdenum(V) complex **60**.

the catalysis products proves that the oxygen atom indeed originates from water (and not from adventitious dioxygen or other sources). ^{1}H NMR spectroscopic analysis of the supernatant catalysis solution shows that indeed 2 equiv of ferrocene are formed per product molecule $OPMe_3$, demonstrating the correct oxidant stoichiometry. Protonation of the $P_1{}^t$Bu base is corroborated by ^{31}P NMR spectroscopy of the supernatant solution. The combined data fully substantiate the required stoichiometry of the biomimetic catalytic cycle comprising OAT and two PCET reactions:

$$\mathrm{PMe_3 + 2Fc^+ + H_2O + 2P_1{}^tBu \rightarrow OPMe_3 + 2Fc + 2[HP_1{}^tBu]^+}$$
$$(\mathrm{Fc} = 1,1'\text{-diacetylferrocene})$$

The whole catalytic cycle of this system, including the mononuclear Mo^{VI}, Mo^{V}, and Mo^{IV} species as well as the detrimental binuclear μ-oxido species and relevant transition states for OAT, have also been viewed from a theoretical perspective [89]. This system hence represents a rare example in which only the heterogenized version is competent to perform the biomimetic catalysis while the homogeneous system is not and, furthermore, for which detailed information concerning the elementary reaction steps is available, being fully congruent with the biological prototype.

A functional Fe^{II} complex **61** for emulating 2-oxoglutarate-dependent iron oxygenases [34] has been devised using a thiol-appended tris(2-pyridylmethyl)amine N_4 ligand immobilized on gold nanoparticles [90]. The immobilized $[Fe^{II}(\kappa O,\kappa O$-benzoylformato$)(N_4)]^+$ complex **61** decarboxylates the benzoylformate under dioxygen to give benzoate and CO_2 eight times faster than the nonimmobilized congener. Again, site isolation prevents the formation of inactive dinuclear μ-carboxylato complexes on the gold surface.[90] The immobilized Fe^{II} complex **61** is even capable to catalytically decarboxylate benzoylformic acid in the presence of dioxygen and a cosubstrate, for example, thioethers, whereas the nonimmobilized congener is inactive. However, the Au–S bond of **61** is not perfectly inert under these conditions, and some adsorbed thiols detach from the nanoparticle and leach into the homogenous phase.[90] Hence, this example nicely illustrates the importance of all bonds between the solid phase and the catalytically active metal center

20.6 Final Remarks

First steps toward highly elaborate functional artificial metalloenzymes including active site and microenvironment and comprising fundamental aspects of structural, spectroscopic, and, most importantly, functional similarities toward natural systems have been highlighted in this review. Hydrolysis reactions (phosphor(di)esterases), C–C and C–X bond formations (ACS), X–O bond formation (molybdenum oxotransferase) and multielectron redox reactions (H_2ase, N_2ase,

OEC, CcO, catechol oxidase, peroxidase, oxotransferase), typically coupled to proton transfer reactions (PCET), have been exemplarily discussed on the basis of matrix-confined active sites.

Furthermore, attempts to include redox cofactors in the form of conducting matrixes and to model substrate and product selectivity and enzyme-like Michaelis–Menten kinetics in the form of shape and size selective matrixes have been demonstrated. Future visions of artificial metalloenzymes could combine protected functional active site models with electron supply chains and proton supply channels for sustained PCET reactions [65] – eventually coupled to suitable molecular photosensitizers (PS II analogs) – and more elaborate substrate recognition and activation sites, for example, by hydrogen-bonding units and further Lewis acidic or basic sites placed at strategic positions within the active site cavity for cooperative catalysis and allosteric regulation. A multitude of opportunities and challenges in the design and realization of true artificial metalloenzymes lies ahead to be tackled by synthetic bioinorganic chemists in cooperation with biochemists, polymer chemists, spectroscopists, and photochemists.

Abbreviations

acac	acetylacetonate
Ac-SCoA	acetyl coenzyme A
ACS	acetyl CoA synthetase
bpy	2,2′-bipyridine
CcO	cytochrome c oxidase
en	1,2-ethylenediamine
DSSC	dye-sensitized solar cell
EGDMA	ethylene glycol dimethacrylate
EXAFS	extended X-ray absorption fine structure
Fc	ferrocene
Fmoc	9-fluorenylmethoxycarbonyl
H_2ase	hydrogenase
hfc	hyperfine coupling
HOPG	highly ordered pyrolytic graphite
HRP	horseradish peroxidase
HSCoA	acetyl coenzyme A
LMCT	ligand-to-metal charge transfer
OAT	oxygen atom transfer
OEC	oxygen evolving complex
MIP	molecularly imprinted polymer
MMO	Methane mono-oxygenase
N_2ase	Nitrogenase
PCET	Proton-coupled electron transfer
PET	Photoinduced electron transfer

PEG	Polyethylene glycol
phen	9,10-Phenanthroline
PS II	Photosystem II
SAM	Self-assembled monolayer
SPPS	Solid-phase peptide synthesis
TPA	Triphenylamine
TPP	*meso*-Tetraphenylporphyrin
tpy	2,2′;6′,2″-Terpyridine

References

1. (a) Messerschmidt, A., Huber, R., Poulos, T., and Wieghardt, K. (eds) (2001) *Handbook of Metalloproteins*, John Wiley & Sons, Ltd, Chichester; (b) Kaim, W. and Schwederski, B. (2010) *Bioanorganische Chemie. Zur Funktion chemischer Elemente in Lebensprozessen*, 5th edn, Vieweg+Teubner Verlag, Wiesbaden.
2. Kraatz, H.-B. and Metzler-Nolte, N. (eds) (2006) *Concepts and Models in Bioinorganic Chemistry*, Wiley-VCH Verlag GmbH, Weinheim.
3. Xuereb, D., Dzierzak, J., and Raja, R. (2010) in *Heterogenized Homogeneous Catalysts for Fine Chemicals Production*, Catalysis by Metal Complexes, vol. **33**, Chapter 2 (eds P. Barbaro and F. Ligiari), Springer Science and Business Media B.V, 37–63.
4. Notenstein, J.M. and Katz, A. (2006) *Chem. Eur. J.*, **12**, 3954–3965.
5. Meurig Thomas, J., Raja, R., and Lewis, D.W. (2005) *Angew. Chem.*, **117**, 6614–6641; *Angew. Chem. Int. Ed.*, (2005), **44**, 6456–6482.
6. Welbers, L.L. and Borovik, A.S. (2005) *Acc. Chem. Res.*, **38**, 765–774.
7. De Vos, D.E. and Vankelecom, I.F.J. (2007) in *Chiral Catalyst Immobilization and Recycling* (ed P.A. Jacobs), Wiley-VCH Verlag GmbH, Weinheim.
8. Qu, Z., Giurgiu, L., and Roduner, E. (2006) *Chem. Commun.*, 2507–2509.
9. Tian, Y., Li, G.-D., and Chen, J.-S. (2003) *J. Am. Chem. Soc.*, **125**, 6622–6623.
10. Boscher, N.D., Duday, D., Heier, P., Heinze, K., Hilt, F., and Choquet, P. (2013) *Plasma Process. Polym.*, **10**, 336–344.
11. Boscher, N.D., Duday, D., Heier, P., Heinze, K., Hilt, F., and Choquet, P. (2013) *Surf. Coat. Technol.* doi: 10.1016/j.surfcoat.2012.12.034, in press, **234**, 48–52.
12. Heinze, K. and Fischer, A. (2007) *Eur. J. Inorg. Chem.*, **2007**, 1020–1026.
13. Jones, P., Mantle, D., and Wilson, I. (1983) *J. Chem. Soc., Dalton Trans.*, 161–164.
14. Leal, O., Anderson, D.L., Bowman, R.G., Basolo, F., and Burwell, R.L. Jr., (1975) *J. Am. Chem. Soc.*, **97**, 5125–5129.
15. (a) Collman, J.P., Gagne, R.R., Reed, C.A., Halbert, T.R., Lang, G., and Robinson, W.T. (1975) *J. Am. Chem. Soc.*, **97**, 1427–1439; (b) Collman, J.P. and Fu, L. (1999) *Acc. Chem. Res.*, **32**, 455–463.
16. Peters, L., Tepedino, M.-F., Haas, T., Heinemann, F.W., Wolf, M., and Burzlaff, N. (2011) *Inorg. Chim. Acta*, **374**, 392–405.
17. Türkoglu, G., Pubill Ulldemolins, C., Müller, R., Hübner, E., Heinemann, F.W., Wolf, M., and Burzlaff, N. (2010) *Eur. J. Inorg. Chem.*, **2010**, 2962–2974.
18. Hübner, E., Türkoglu, G., Wolf, M., Zenneck, U., and Burzlaff, N. (2008) *Eur. J. Inorg. Chem.*, **2008**, 1226–1235.
19. Hübner, E., Haas, T., and Burzlaff, N. (2006) *Eur. J. Inorg. Chem.*, **2006**, 4989–4997.
20. Heinze, K. and Reinhart, A. (2006) *Inorg. Chem.*, **45**, 2695–2703.
21. Terry, T.J., Dubois, G., Murphy, A., and Stack, T.D. (2007) *Angew. Chem.*, **119**, 963–965; *Angew. Chem. Int. Ed.*, (2007), **46**, 945–947.

22. Heinze, K. (2001) *Chem. Eur. J.*, **7**, 2922–2932.
23. Heinze, K., Jannack, T., and Winterhalter, U. (2000) *Chem. Eur. J.*, **6**, 4203–4210.
24. Heinze, K. and Bueno Toro, J.D. (2004) *Eur. J. Inorg. Chem.*, **2004**, 3498–3507.
25. Heinze, K. and Bueno Toro, J.D. (2003) *Angew. Chem.*, **115**, 4671–4674; *Angew. Chem. Int. Ed.*, (2003), **42**, 4533–4536.
26. Heinze, K., Jacob, V., and Feige, C. (2004) *Eur. J. Inorg. Chem.*, **2004**, 2053–2061.
27. Reinhardt, S. and Heinze, K. (2006) *Z. Anorg. Allg. Chem.*, **1465**, 1470.
28. Heinze, K. and Reinhardt, S. (2008) *Chem. Eur. J.*, **14**, 9482–9486.
29. Guillier, F., Orain, D., and Bradley, M. (2000) *Chem. Rev.*, **100**, 2091–2157.
30. Green, K.N., Jeffery, S.P., Reibenspiess, J.H., and Darensbourg, M.Y. (2006) *J. Am. Chem. Soc.*, **128**, 6493–6498.
31. Green, K.N., James, W.D., Cantillo, A.V., and Darensbourg, M.Y. (2007) *J. Organomet. Chem.*, **692**, 1392–1397.
32. Green, K.N., Hess, J.L., Thomas, C.M., and Darensbourg, M.Y. (2009) *Dalton Trans.*, 4344–4350.
33. Costas, M., Mehn, M.P., Jensen, M.P., and Que, L. Jr., (2004) *Chem. Rev.*, **104**, 939–986.
34. (a) Elkins, J.M., Hewitson, K.S., McNeill, L.A., Seibel, J.F., Schlemminger, I., Pugh, C.W., Ratcliffe, P.J., and Schofield, C.J. (2003) *J. Biol. Chem.*, **278**, 1802–1806; (b) Schofield, C.J. and Zhang, Z. (1999) *Curr. Opin. Struct. Biol.*, **9**, 722–731.
35. Bailey, D.C. and Langer, S.H. (1981) *Chem. Rev.*, **81**, 109–149.
36. Cheng, Z., Zhang, L., and Li, Y. (2004) *Chem. Eur. J.*, **10**, 3555–3561.
37. (a) Carlsson, G.H., Nicholls, P., Svistunenko, D., Berglund, G.I., and Hajdu, J. (2005) *Biochemistry*, **44**, 635–642; (b) Gajhede, M., Schuller, D.J., Henriksen, A., Smith, A.T., and Poulos, T.L. (1997) *Nat. Struct. Biol.*, **4**, 1032–1038.
38. Polborn, K. and Serverin, K. (2000) *Chem. Eur. J.*, **6**, 4604–4611.
39. (a) Krämer, R. (1999) *Coord. Chem. Rev.*, **182**, 243–261; (b) Hegg, E.L. and Burstyn, J.N. (1998) *Coord. Chem. Rev.*, **173**, 133–165.
40. Schiller, A., Scopelliti, R., Benmelouka, M., and Severin, K. (2005) *Inorg. Chem.*, **44**, 6482–6492.
41. Schiller, A., Scopelliti, R., and Severin, K. (2006) *Dalton Trans.*, 3858–3867.
42. Darnault, C., Volbeda, A., Kim, E.J., Legrand, P., Vernède, X., Lindahl, P.A., and Fontecilla-Camps, J.C. (2003) *Nat. Struct. Biol.*, **10**, 271–279.
43. (a) Chan, W.C. and White, P.D. (eds) (2000) *Fmoc Solid Phase Peptide Synthesis*, Oxford University Press, Oxford; (b) Heinze, K., Beckmann, M., and Hempel, K. (2008) *Chem. Eur. J.*, **14**, 9468–9480.
44. Rankin, J., Poole, A.D., Benyei, A.C., and Cole-Hamilton, D.J. (1997) *Chem. Commun.*, 1835–1836.
45. Horn, B., Limberg, C., Herwig, C., and Mebs, S. (2011) *Angew. Chem.*, **123**, 12829–12833; *Angew. Chem. Int. Ed.*, (2011), **50**, 12621–12625.
46. (a) Fontecilla-Camps, J.C., Volbeda, A., Cavazza, C., and Nicolet, Y. (2007) *Chem. Rev.*, **107**, 4273–4303; (b) Vignais, P.M. and Billoud, B. (2007) *Chem. Rev.*, **107**, 4206–4272; (c) Adams, M.W.W. and Stiefel, E.I. (1998) *Science*, **282**, 1842–1843.
47. Thomas, C.M., Rüdiger, O., Liu, T., Carson, C.E., Hall, M.B., and Darensbourg, M.Y. (2007) *Organometallics*, **26**, 3976–3984.
48. Ibrahim, S.K., Liu, X., Tard, C., and Pickett, C.J. (2007) *Chem. Commun.*, 1535–1537.
49. (a) Rosenzweig, A.C., Frederick, C.A., Lippard, S.J., and Nordlund, P. (1993) *Nature*, **366**, 537–543; (b) Siewert, I. and Limberg, C. (2009) *Chem. Eur. J.*, **15**, 10316–10328; (c) Friedle, S., Reisner, E., and Lippard, S.J. (2010) *Chem. Soc. Rev.*, **39**, 2768–2779.
50. Pirngruber, G.D., Frunz, L., and Lüchinger, M. (2009) *Phys. Chem. Chem. Phys.*, **11**, 2928–2938.
51. (a) Klabunde, T., Eicken, C., Sacchettini, J.C., and Krebs, B. (1998) *Nat. Struct. Biol.*, **5**, 1084–1090; (b) Mirica, L.M., Ottenwaelder, X., and Stack, T.D.P. (2004) *Chem. Rev.*, **104**, 1013–1045.
52. Marion, R., Muthusamy, G., and Geneste, F. (2012) *J. Catal.*, **286**, 266–272.

53. Wünsche von Leupoldt, A., Förster, C., Fiedler, T.J., Bings, N., and Heinze, K. (2013) *Eur. J. Inorg. Chem.*, **2013**, 6079–6090, submitted.
54. (a) Maurya, M.R. and Pessoa, J.C. (2011) *J. Organomet. Chem.*, **696**, 244–254; (b) Maurya, M.R., Kumar, U., Correia, I., Adão, P., and Pessoa, J.C. (2008) *Eur. J. Inorg. Chem.*, **2008**, 577–587.
55. Nuzzo, A. and Piccolo, A. (2013) *J. Mol. Catal. A*, **371**, 8–14.
56. Crestini, C., Pastorini, A., and Tagliatesta, P. (2004) *J. Mol. Catal. A*, **208**, 195–202.
57. Berner, S., Biela, S., Ledung, G., Gogoll, A., Bäckvall, J.-E., Puglia, C., and Oscarsson, S. (2006) *J. Catal.*, **244**, 86–91.
58. Rollmann, L.D. (1975) *J. Am. Chem. Soc.*, **97**, 2132–2136.
59. Landbeater, N.E. and Marco, M. (2002) *Chem. Rev.*, **102**, 3217–3274.
60. Limburg, J., Vrettos, J.S., Liable-Sands, L.M., Rheingold, A.L., Crabtree, R.H., and Brudvig, G.W. (1999) *Science*, **283**, 1524–1527.
61. (a) Umena, Y., Kawakami, K., Shen, J.-R., and Kamiya, N. (2011) *Nature*, **473**, 55–60; (b) Mullins, C.S. and Pecoraro, V.L. (2008) *Coord. Chem. Rev.*, **252**, 416–443.
62. Kurz, P. (2009) *Dalton Trans.*, 6103–6108.
63. (a) Yagi, M. and Narita, K. (2004) *J. Am. Chem. Soc.*, **126**, 8084–8095; (b) Yagi, M., Narita, K., Maruyama, S., Sone, K., Kuwabara, T., and Shimizu, K. (2007) *Biochim. Biophys. Acta*, **1767**, 660–665; (c) Yamazaki, H., Ueno, T., Aiso, K., Hirahara, M., Aoki, T., Nagata, T., Igarashi, S., and Yagi, M. (2013) *Polyhedron*, **52**, 455–460.
64. Li, G., Sproviero, E.M., Snoeberger, R.C. III,, Iguchi, N., Blakemore, J.D., Crabtree, R.H., Brudvig, G.W., and Batista, V.S. (2009) *Energy Environ. Sci.*, **2**, 230–238.
65. Weinberg, D.R., Gagliardi, C.J., Hull, J.F., Fecenko Murphy, C., Kent, C.A., Westlake, B.C., Paul, A., Ess, D.H., Granville McCafferty, D., and Meyer, T.J. (2012) *Chem. Rev.*, **112** (7), 4016–4093.
66. Zhou, T., Lin, X., Zheng, X., and Chem, J. (2013) *Theor. Comput.*, **9**, 1073–1080.
67. Artero, V. and Fontecave, M. (2013) *Chem. Soc. Rev.*, **42**, 2338–2356.
68. Weare, W.W., Pushkar, Y., Yachandra, V.K., and Frei, H. (2008) *J. Am. Chem. Soc.*, **130**, 11355–11363.
69. O'Regan, B. and Grätzel, M. (1991) *Nature*, **353**, 737–740.
70. zur Borg, L., Domanski, A.L., Breivogel, A., Bürger, M., Berger, R., Heinze, K., and Zentel, R. (2013) *J. Mater. Chem. C*, **1**, 1223–1230.
71. (a) Yoshikawa, S., Shinzawa-Itoh, K., Nakashima, R., Yaono, R., Yamashita, E., Inoue, N., Yao, M., Fei, M.J., Libeu, C.P., Mizushima, T., Yamaguchi, H., Tomizaki, T., and Tsukihara, T. (1998) *Science*, **280**, 1723–1729; (b) Ferguson-Miller, S. and Babcock, G.T. (1996) *Chem. Rev.*, **96**, 2889–2908; (c) Kim, E., Chufán, E.E., Kamaraj, K., and Karlin, K.D. (2004) *Chem. Rev.*, **104**, 1077–1134.
72. Collman, J.P. and Gosh, S. (2010) *Inorg. Chem.*, **49**, 5798–5810.
73. Einsle, O., Tezcan, F.A., Andrade, S.L., Schmid, B., Yoshida, M., Howard, J.B., and Rees, D.C. (2002) *Science*, **297**, 1696–1700.
74. (a) Yandulov, D.V. and Schrock, R.R. (2003) *Science*, **301**, 76–78; (b) Schrock, R.R. (2008) *Angew. Chem.*, **120**, 5594–5605; *Angew. Chem. Int. Ed.*, (2008), **47**, 5512–5522.
75. Hinrichsen, S., Broda, H., Gradert, C., Söncksen, L., and Tuczek, F. (2012) *Annu. Rep. Prog. Chem., Sect. A: Inorg. Chem.*, **108**, 17–47.
76. Seefeldt, L.C., Dance, I.G., and Dean, D.R. (2004) *Biochemistry*, **43**, 1401–1409.
77. (a) Spatzal, T., Aksoyoglu, M., Zhang, L., Andrade, S.A., Schleicher, E., Weber, S., Reese, D.C., and Einsle, O. (2011) *Science*, **334**, 940; (b) Lancaster, K.M., Roemelt, M., Ettenhuber, P., Hu, Y., Ribbe, M.W., Neese, F., Bergmann, U., and DeBeer, S. (2011) *Science*, **334**, 974–977.
78. Chatt, J., Pearman, A.J., and Richards, R.L. (1975) *Nature*, **253**, 39–40.
79. Robillard, M.S., Leith, J.S., van der Marel, G.A., van Boom, J.H., and Reedijk, J. (2003) *Eur. J. Inorg. Chem.*, **2003**, 1529–1532.

80. George, T.A. and Kaul, B.B. (1990) *Inorg. Chem.*, **29**, 4969–4974.
81. Hille, R. (1996) *Chem. Rev.*, **96**, 2757–2816.
82. Hille, R. (2002) *Trends Biochem. Sci.*, **27**, 360–367.
83. Enemark, J.H., Cooney, J.J.A., Wang, J.J., and Holm, R.H. (2004) *Chem. Rev.*, **104**, 1175–1200.
84. Schulzke, C. (2011) *Eur. J. Inorg. Chem.*, **2011**, 1189–1199.
85. Ng, V.W.L., Taylor, M.K., White, J.M., and Young, C.G. (2010) *Inorg. Chem.*, **49**, 9460–9469.
86. Heinze, K. and Fischer, A. (2010) *Eur. J. Inorg. Chem.*, **2010**, 1939–1947.
87. Hüttinger, K., Förster, C., Bund, T., Hinderberger, D., and Heinze, K. (2012) *Inorg. Chem.*, **51**, 4180–4192.
88. Topich, J. (1982) *Inorg. Chem.*, **21**, 2079–2082.
89. Heinze, K., Marano, G., and Fischer, A. (2008) *J. Inorg. Biochem.*, **102**, 1199–1211.
90. Sheet, D. Halder, P. (2013) *Angew. Chem.*, **125**, 13556–13560; Paine, T.K. *Angew. Chem. Int. Ed.*, **52**, 13314–13318.

21
High-Throughput Screening of Catalysts and Reactions

Oliver Trapp

21.1
Introduction

Finding new catalysts is of great importance for sustainable chemical processes in industrial applications as well as for expanding the spectrum of synthetic strategies in chemistry. Investigating reaction kinetics, particularly of catalyzed reactions, is of fundamental importance in chemistry, biochemistry, and related sciences because the understanding to control the transition state of a reaction allows the optimized design of new catalysts and benign processes. Therefore, the development of techniques for the comprehensive investigation of reaction kinetics is of central importance, and, especially, the investigation of individual pathways in stereoselective transformations requires techniques to directly measure enantioselectivities.

Despite major advances in the large and expanding field of high-throughput screening, typically batch reactors are employed to investigate reactions and catalytic processes. The central limiting factor of this approach is the fact that only one reaction per run can be investigated. This inevitably leads to indefinable reaction kinetics in case of competing reactions. Furthermore, this approach requires large quantities of substrates and catalysts because reaction, separation, and quantification of reactants and products to determine conversions have to be performed consecutively. Therefore, the integration of reaction and separation is not merely a desired goal for the determination of reaction rate constants. In particular, this is of great interest in chemical engineering because pure products are obtained in a single step [1, 2].

Flow-through micro reactors [3–7] represent an interesting alternative for carrying out chemical reactions in continuous flow and are increasingly used in synthetic organic chemistry [8–11], for example, in highly exothermal reactions, such as fluorinations, which can be safely performed. The striking advantage of such micro reactors is the large surface-to-volume ratio that can be achieved in these miniaturized systems.

Inspired by micro total analysis systems (μTASs), where sequential analytical operations can be automatically performed by computer control and which allow

Molecular Catalysts: Structure and Functional Design, First Edition. Edited by Lutz H. Gade and Peter Hofmann.

a very efficient analysis of biomolecules, microfluidic devices are also employed in synthetic organic chemistry as lab on a chip [12–16]. Even sequential reactions can be performed, and high-throughput screening of reactions is feasible [17–19].

Other advantages of microfluidic devices include precise temperature control, in particular for quartz chips and fused-silica capillaries with high thermal conductivity. This allows performing exact kinetic measurements even for highly exergonic reactions. Furthermore, such devices are highly attractive to study multiphase catalytic systems [20]. These play a predominant role in industrial processes. Typically, such processes are difficult to investigate because the interaction of the reactant with the catalyst is controlled by the mass transfer between the different phases and therefore the directly measured apparent reaction rate is composed of the inherent reaction rate and the diffusion rates. To reduce this effect, the interfacial area has to be increased. Microstructured reaction systems intrinsically have a high specific interfacial area per volume ($a_{\text{inter}} = 2/r$), which depends only on the radius for circular reaction channels; for example, for capillaries with inner diameters between 250 and 100 μm the specific interfacial area per volume ranges from 16 000 to 40 000 m^2/m^3. It can be expected that microfluidic systems will be revolutionizing chemical synthesis because chemical processes can be easily controlled. However, there are still many experimental challenges that need to be addressed; among these are the efficient control of mixing as well as interfacing with standard analytical instruments for advanced analysis.

A major issue is also the uniform modification of reaction channels with immobilized catalysts and the use of these devices for performing kinetic measurements. Conventional microfluidic devices typically use square or rectangular shaped reaction channels, and the system is often closed by a cover glass glued on top of the chip. Uniform coating of open channels with polymeric materials is difficult, and deviations in the thickness lead to nonlinear behavior, which makes kinetic analysis problematic. In contrast, micro capillaries can easily be modified uniformly by established coating procedures [21], allowing the creation of defined polymeric films with embedded catalysts. Micro capillaries are commonly used for performing separations in gas chromatography (GC), open-tubular liquid chromatography (OTLC), and capillary electrophoresis (CE). Therefore, it is obvious that the direct use of catalytically active capillaries as separation capillaries and at the same time performing reactions and analysis in the same analytical device will offer all the advantages offered by modern analytical instrumentation. Over the past 10 years, the investigation of degenerated interconversion processes such as the interconversion of stereoisomers (which is important to study the stereointegrity of drugs [22]) by dynamic chromatography [23–25] has attracted a lot of attention and stimulated the development of advanced reaction chromatographic techniques.

Despite the use of chromatographic separation methods as tools to investigate the course of reactions in detail, chromatographic data as well as reactions in the course of a chromatographic separation can also be used to perform physicochemical measurements. An example for this is the application of van

Deemter's equation, which allows the determination of diffusion coefficients of analytes [26, 27]. It has to be noted that the original derivation of van Deemter's equation described heat and mass transport in a fixed catalyst bed during regeneration [28]!

Integrating catalytic activity and separation selectivity into a single separation column offers many advantages. Reactants and products can be separated, and, in the case of reversible reactions, there is the possibility to shift the chemical equilibrium to the side of the reaction products. Reactions can be studied in great detail because the data collected in these experiments will reflect the reaction kinetics and chromatographic adsorption as well as parameters of diffusion [29]. According to the theoretical plate model of chromatography, each theoretical plate can be considered as a chemical reactor. If the reaction proceeds only in the catalytically active stationary phase, conversions in the gas phase can be neglected (cf. Figure 21.1). The equilibrium in the stationary phase can be separated into the contributions due to physisorption (distribution between the gas and stationary phase) and chemisorption (in the stationary liquid phase, the distribution of the reactants and products between the dissolved state and the state where reactant and product form a joint complex).

The first report on integrating catalysis and chromatographic separation was published by Tamaru in 1959 [30]. Palladium was deposited on Celite 545 and used as packing material in GC to study the decomposition of formic acid. Adsorption was investigated by using different carrier gases. It was found that the decomposition of formic acid on Pd^0 is of first order with respect to formic acid, which was in agreement with the Langmuir–Hinshelwood mechanism.

In 1960, Bassett and Habgood [31] investigated the isomerization of cyclopropane to propylene catalyzed by the Linde molecular sieve 13× exchanged with Ni^{2+} ions on a packed column. Cyclopropane was injected as a pulse onto the column, and the formation of the product was observed as an extremely broadened peak. Gil-Av and Herzberg-Minzly [32] investigated Diels–Alder

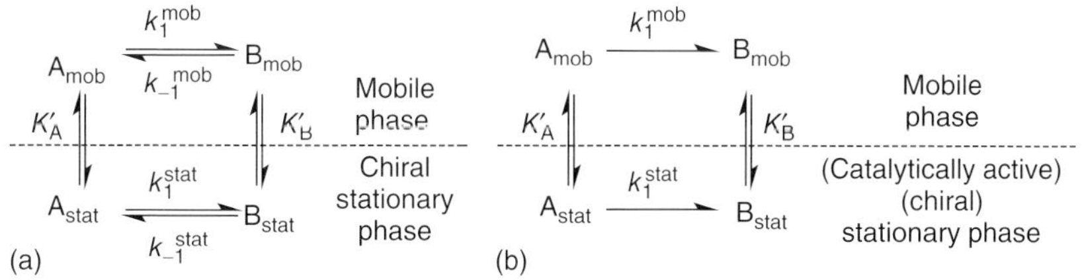

Figure 21.1 Equilibria in a chromatographic theoretical plate considering (a) reversible reactions (dynamic chromatography) and (b) irreversible reactions taking place in the mobile phase (mob) and stationary phase (stat). If the reaction proceeds only in presence of a catalyst, the reaction in the mobile phase can be neglected in scheme (b). k_1 and k_{-1} represent the forward and backward reaction rate constants in the mobile phase (mob) and stationary phase (stat), and K denotes the distribution constant.

reactions by impregnating the stationary phase of the GC column with chloromaleic anhydride and injecting various dienes to form the Diels–Alder product. However, in this approach they were not able to analyze the reaction product to study selectivities because the product remained on the column. By varying the contact time and observing the change in the peak area, the reaction rate constants were determined. Pratt and Langer [33] extended the concept of chromatographic reactors experimentally as well as theoretically. They investigated very carefully the kinetics of the dissociation of *endo*-dicyclopentadiene (retro Diels–Alder reaction). Later, Marriott and Lai [34] investigated the same reaction and compared packed and capillary columns. They found that packed columns with larger diameters were more efficient than coated capillary columns. In contrast, for catalyzed reactions high surface area is desirable. Therefore, using micro capillaries might be counterintuitive, but the high specific interfacial area per volume is dramatically improved compared to packed columns and the detection sensitivity is improved.

Matsen *et al.* [35] focused on the investigation of reactions of the type $A \rightleftharpoons B + C$ (decompositions) on a chromatographic column. As example reaction, the dehydrogenation of cyclohexane to benzene catalyzed by 0.6% Pt on an alumina support was chosen. The authors reported that, under ideal conditions, the conversion was 30% higher in the chromatographic reactor than under equilibrium conditions, attributing this difference to a shifting of the chemical equilibrium by the separation of hydrogen from the reaction product. Matsen *et al.* reported an attempt to modify or replace the stationary phase for the separation to improve the overall efficiency and resolution. They found that the isotherm for benzene adsorbed on fresh alumina was very nonlinear, causing the benzene peak to tail off slowly. Under such conditions, the reactor could be pulsed only infrequently, and conversions were no better than for a continuous reactor operating at the same average dilution of cyclohexane by the carrier gas. To overcome these problems, an alumina column was used that had been treated with a solution of 10% KOH in methanol. The treatment neutralized the very strongly acidic sites, which was attributed to the nonlinear isotherm. Skrdla [36] investigated the thermal decomposition of *tert*-butyl peroxide in the injector of a GC and on the separation column. Measurements were performed at different temperatures, and activation energies were calculated and compared with different kinetic models. It was found that the on-column reactor approach utilizing peak area measurements is by far superior in terms of speed, robustness, and accuracy.

This following section focuses on recent breakthroughs in combining separation and catalytic transformation in a single a chromatographic separation capillary. The focus will be on how selectivities and reaction contact times of the reactants with the catalyst can be tuned. The presented strategy overcomes the limitation of investigating only a single reaction per measurement.

21.2
Technical Requirements for On-Column Reaction Chromatography

21.2.1
Experimental Setups of On-Column Reaction Chromatography

Depending on the speed of a chemical transformation at a given temperature, two different approaches can be applied to perform kinetic investigations. The contact time is given by the migration time of a reactant through the catalytically active column. Therefore the catalytically active separation column should be short for fast reactions or, when highly active catalysts are employed, to prevent complete conversion, which is desired only in a preparative capillary reactor. A typical experimental setup consists of a preseparation column of about 5 m, followed by a catalytically active separation capillary with a length between 2 and 1000 cm, and finally a separation column of 25–50 m length that is coated with a suitable (chiral) stationary phase for qualitative and quantitative analysis (cf. Figure 21.2a). The function of the preseparation column is to (i) thermally equilibrate the reactants, after they are injection into a hot injector, for example, in GC; (ii) separate reactant libraries, so that no competing reactions occur during the migration through the reactive column; and (iii) remove impurities in the reactants, which is essential to avoid changes in the catalytic activity.

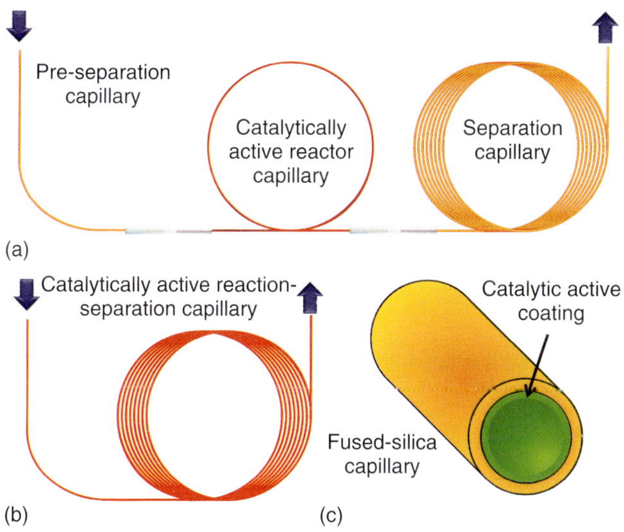

Figure 21.2 (a) On-column reaction chromatographic setup to investigate fast reactions, where catalytic transformation and separation are sequentially performed. (b) Complete integration of catalytic activity and separation selectivity by dissolving a catalyst in the stationary phase. (c) Schematic cross-section of a fused silica capillary with a homogeneous wall coating of a catalyst in a polymeric stationary phase.

The choice of the chromatographic method and the selection of suitable columns depend on the physical and chemical properties of reactants and products as well as on the desired reaction conditions. The use of internal standards is highly recommended for quantitative analysis of reactants, products, and by-products. It may serve as an important analytical tool to recognize unexpected side reactions such as on-column polymerizations of the reactants. Such reactions lead to a shift in the mass balance and can be detected with the help of an internal standard.

First-order reactions such as intramolecular rearrangements, degradation reactions, and so on, or pseudo-first order reactions that can be observed by using a reactive carrier gas in GC, that is, hydrogenations, or an additive in high-performance liquid chromatography (HPLC) or CE can be directly observed using the setup described above. Investigations of higher order reactions have to be planned with more consideration. In general, these reactions can be monitored by first injecting the slower migrating reactant and then the faster moving reactant. During the course of the separation, the second injected reactant will pass the first one at a certain point in the capillary. The time period when both reactants are in the same column section and can therefore react is called the *contact time*. Matching injection times and contact times in the presence of the catalyst requires a careful design of the experiments. For best results, temperature and flow-dependent measurements are first performed without the catalytically active separation column to determine the migration times of the reactants without any interactions with the catalyst and the stationary phase employed to dissolve or to immobilize the catalyst. Next, the same dataset is measured using the catalytically active capillary between the preseparation column and the separation column. The difference between the measurements without this capillary and the dataset employing the catalyst capillary yields the contact time Δt of the individual reactant with the catalyst. By integrating the reactant and product, the reaction rate constants are directly accessible.

The second approach integrates both the catalytic activity and separation efficiency by dissolving the catalyst in the stationary phase or immobilizing the catalyst on to the stationary phase and using columns that are uniformly coated with the catalytically active stationary phase over the complete capillary length (cf. Figure 21.2b). In this case, elution profiles are obtained that are characterized by plateau formation or a more complex convolution of reactant and product. Today, reaction rate constants can be easily determined with the unified equation of chromatography (see below) [37, 38].

Both experimental approaches can be coupled with additional detectors, for example, mass spectrometers (MS), to obtain additional structural information of the formed products and by-products.

21.2.2
Preparation of Capillary Reactors

Catalytically active capillaries for gas chromatography (GC), open tubular capillary electrophoresis (OTCE), or open tubular liquid chromatography (OTLC) measurements can be prepared by static coating [21], which is an established method to prepare GC separation columns. Here, the catalyst and a suitable polysiloxane are dissolved in a volatile solvent, for example, diethylether or dichloromethane (DCM), the capillary is filled and closed, and the solvent slowly removed by the application of a vacuum to one end of the capillary. The film thickness is defined by the concentration of the polymer and catalyst and the inner diameter of the capillary (cf. Figure 21.2c). After solvent removal, the capillaries are flushed with N_2 or any other inert gas for several hours. Depending on the stability of the catalyst and the activation protocol, the capillaries might have to be heated for several hours.

Catalysts immobilized to polysiloxanes can be prepared in a similar way. The use of immobilized ligands allows the creation of catalyst libraries, as short pieces of capillaries modified with immobilized ligands can be flushed with suitable metal precursors to produce catalysts inside the capillary. Compared to solid-phase chemistry, this approach offers the advantage of using an excess of reagents to achieve high conversions while at the same time allowing easy removal of unreacted reagents by flushing with solvents. This concept was introduced as "catalyst by the meter" [39].

It is recommended to choose a stationary phase that is compatible with both the reaction and the catalysts. Polysiloxanes allow continuous tuning of solvent properties. In this context, it is important to note that all reactions take place in the liquid stationary phase and therefore reaction rate constants are absolutely comparable to the reaction constants determined by reaction progress analysis, where samples are periodically taken from a reactor or vial.

21.2.3
High-Throughput Approach

Simultaneous injection of a library of reactants onto an on-column reaction chromatographic system with catalytically active capillaries allows high-throughput screening of potential reactions (cf. Figure 21.3). Performing such a screening within a chromatographic separation system has several advantages: (i) competing reactions are prevented; (ii) kinetic analysis of several reactions under the exactly same reaction conditions is easily accomplished; (iii) reactant consumption is minimal; and finally, (iv) reaction as well as analysis is carried out in a single step. The synchronous combination of catalysis and separation was demonstrated for hydrogenations over highly active Pd nanoparticles and the ring-closure metathesis over the Grubbs second-generation catalyst [40, 41]. In these studies, up to 147 reactions per hour could be performed and analyzed.

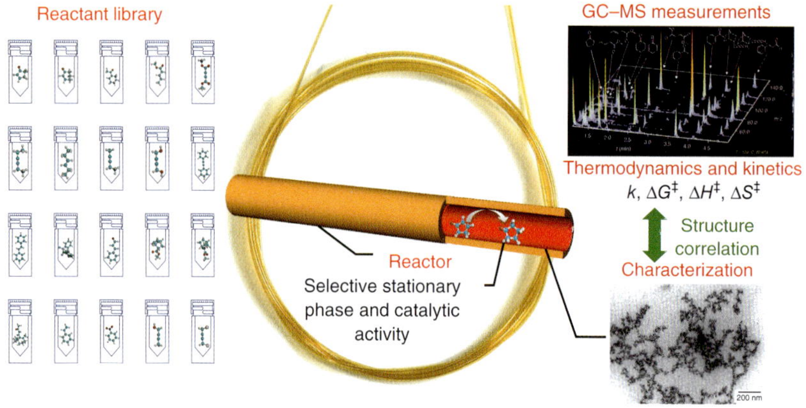

Figure 21.3 Schematic overview of on-column reaction chromatography. A reactant library is simultaneously injected onto a catalytically active separation column. From the conversions, the reaction kinetics can be determined.

21.3 Determination of Kinetic Data

21.3.1 Classical Reaction Kinetics for On-Column Reaction Chromatographic Experiments with Reaction and Consecutive Separation

In evaluation of experiments in which the catalyzed reaction and the separation are performed consecutively, the evaluation is quite straightforward, as classical approaches can be applied. The conversions are calculated from the integrated peak areas of the chromatograms, considering the response factors of the compounds, which are determined from the calibration data. The reaction time equals the contact time Δt determined by differential measurements with and without the catalytic capillary.

For a first-order reaction, where the reactant E is converted into the product P, the following equation can be applied to determine the reaction rate constant:

$$[E] = ([P] + [E])e^{-k\Delta t} \quad (21.1)$$

Any other reaction rate law can also be considered in such experiments, thus giving maximum flexibility for the investigation of catalytic reactions.

21.3.2 Evaluation of Conversion Profiles Obtained by On-Column Reaction Chromatography

The theoretical plate model, which was originally developed to describe a separation process based on partitioning, is the most graphic model. Here, the separation is described by a discontinuous process, assuming that all steps proceed repeatedly in separate uniform sections of a multicompartmentalized column consisting of N theoretical plates [42]. To integrate a reaction in this model, each theoretical plate is considered as a distinct chemical reactor (cf. Figure 21.1) [43, 44] and three steps are performed in each plate (cf. Figure 21.4): (i) distribution of the stereoisomers A and B between mobile phase (mob) and the stationary liquid phase (stat); (ii) reversible interconversion the stereoisomers during the residence time $\Delta t = t_M/N$ (hold-up time t_M); and (iii) shifting the content of the mobile phase to the subsequent theoretical plate while retaining the content of the stationary phase. In the case of irreversible reactions, that is, irreversible rearrangements, the reaction scheme is simplified (cf. Figure 21.1b), and, for catalyzed reactions, the contribution of the reaction in the mobile phase can be neglected.

The distribution of the reactant A and product B between mobile phase (mob) and the selective stationary phase (stat) is determined according to Equations 21.2a and 21.2b, respectively.

$$\begin{aligned} A_{mob} &= \frac{1}{1+k'_A}(\mathring{A}_{mob} + \mathring{A}_{stat}) \\ B_{mob} &= \frac{1}{1+k'_B}(\mathring{B}_{mob} + \mathring{B}_{stat}) \end{aligned} \tag{21.2a}$$

$$\begin{aligned} A_{stat} &= \frac{k'_A}{1+k'_A}(\mathring{A}_{mob} + \mathring{A}_{stat}) \\ B_{stat} &= \frac{k'_B}{1+k'_B}(\mathring{B}_{mob} + \mathring{B}_{stat}) \end{aligned} \tag{21.2b}$$

where A_{mob}, B_{mob}, A_{stat}, B_{stat} are the amounts of the reactant A and product B at equilibrium; $\mathring{A}_{mob}$, $\mathring{B}_{mob}$, $\mathring{A}_{stat}$, $\mathring{B}_{stat}$ are the amounts of A and B before the equilibrium (in the very general case); and k'_A and k'_B are the retention factors of A and B, calculated from the retention time t_R and the hold-up time t_0, determined by methane or regression of the retention times of n-alkanes, according to $k' = (t_R - t_0)/t_0$.

The converted part of the reactant during the residence time $\Delta t = t_0/N$ in the stationary phase and in the mobile phase in a theoretical plate is determined by the

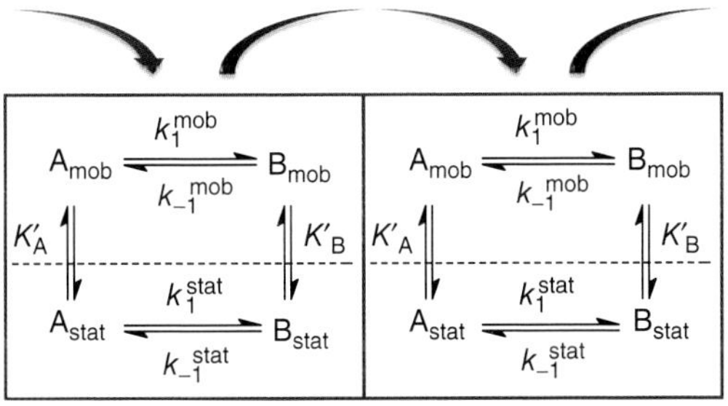

Figure 21.4 Application of the theoretical plate model of chromatography to evaluate on-column reactions in a chromatographic setup. Each theoretical plate is considered as chemical reactor.

respective rate constants. For reversible processes, the forward and backward rate constants k_1^{mob} and k_{-1}^{mob} in the mobile phase are equal, the equilibrium constant K^{mob} is unity, and the equilibrium constant in the chiral stationary phase (CSP) depends on the two-phase distribution constants (i.e., partition coefficients) K_{A} and K_{B} according to the principle of microscopic reversibility [45]:

$$K^{\text{stat}} = \frac{k_1^{\text{stat}}}{k_{-1}^{\text{stat}}} = \frac{K_{\text{B}}}{K_{\text{A}}} = \frac{k'_{\text{B}}}{k'_{\text{A}}} \tag{21.3}$$

This equation implies that the backward rate constant k_{-1}^{stat} is already determined for given values of k_1^{stat}, k'_{A}, and k'_{B} and that k_1^{stat} differs from k_{-1}^{stat} when and $K_{\text{B}} \neq K_{\text{A}}$. For the conversion of reactants, the principle of microscopic reversibility has to be neglected. Typically, only the apparent rate constants k_1^{app} and k_{-1}^{app}, which represent a weighted mean of the reaction rate constants in the mobile and stationary phase, are experimentally accessible (cf. Equation 21.4). By performing complementary experiments, these reaction rate constants can be accessed.

$$\begin{aligned} k_1^{\text{app}} &= \frac{1}{1+k'_{\text{A}}} k^{\text{mob}} + \frac{k'_{\text{A}}}{1+k'_{\text{A}}} k_1^{\text{stat}} \\ k_{-1}^{\text{app}} &= \frac{1}{1+k'_{\text{B}}} k^{\text{mob}} + \frac{k'_{\text{B}}}{1+k'_{\text{B}}} k_{-1}^{\text{stat}} \end{aligned} \tag{21.4}$$

The reversible first-order kinetics is described by

$$\frac{dx}{dt} = k_1^{\text{app}}([\text{A}_0] - [\text{X}]) - k_{-1}^{\text{app}}([\text{B}_0] + [\text{X}]) \tag{21.5}$$

where the amount [X] is the change of A and B. Equation 21.5 is solved by integration, using the initial condition

$$[\text{A}] = \frac{k_{-1}^{\text{app}}}{k_1^{\text{app}} + k_{-1}^{\text{app}}}([\text{A}_0] + [\text{B}_0]) + \frac{k_1^{\text{app}}[\text{A}_0] - k_{-1}^{\text{app}}[\text{B}_0]}{k_1^{\text{app}} + k_{-1}^{\text{app}}} \text{e}^{-(k_1^{\text{app}} + k_{-1}^{\text{app}})\Delta t} \tag{21.6}$$

The amount of [B] is calculated from the mass balance $[\text{A}_0] + [\text{B}_0] = [\text{A}] + [\text{B}]$. After these two steps (partitioning of the present species between mobile phase and stationary phase and undergoing (inter-)conversion), the content of the mobile phase is shifted to the subsequent theoretical plate, whereas the stationary phase is retained. While the given amount of the enantiomers is initially introduced in the first theoretical plate, the content of the mobile phase of the last theoretical plate is finally recorded as a chromatogram featuring an interconversion profile over the time t.

Until the late 1990s, the determination of reaction rate constants from such experiments was extremely time consuming and computationally expensive. Newer algorithms and software as well as the advances made in computer hardware have contributed to the increased interest this topic has received recently. Especially, the theoretical plate model can be easily modified to any type of reaction and allows quick adaptation to experimental requirements.

A statistical description of the theoretical plate model leads to the stochastic model [46], which is based on Gaussian distribution functions of the noninterconverted stereoisomers $\Phi(t)$ and uses a time-dependent probability density function

$\Psi(t)$ to describe the conversion profile. Based on the stochastic model, a unified equation was derived to directly access the reaction rate constants.

Another even more powerful approach is the application of the unified equation of chromatography, which allows determining the reaction rate constants of any first-order reaction directly from chromatographic elution profiles without the need for performing reaction progress analysis. This dramatically accelerates the evaluation of temperature-dependent kinetics, as the analysis time no longer limits the rate of measurements. Detailed kinetic data and activation parameters are of great importance to model and predict activities and selectivities by computational methods.

The unified equation of chromatography (Equations 21.7a and 21.7b) allows the direct calculation of reaction rate constants k_1 and k_{-1} and Gibbs activation energies $\Delta G^{\neq}$ for all types of first-order reactions taking place in chromatographic or electrophoretic systems, regardless of the initial concentrations of the reactants A and B and the equilibrium constant $K_{A/B}$.

A detailed description of the derivation is given in Ref. [37]. The input parameters of the unified equation are the chromatographic parameters obtained from the experiment (cf. Figure 21.5). If the first eluted peak of the reaction profile is higher than the second one, then Equation 21.7a has to be applied, and if the second peak is higher, Equation 21.7b has to be applied.

$$k_1^{\mathrm{ue}} = -\frac{1}{t_\mathrm{R}^\mathrm{A}}\left(\ln\left(\mathrm{B}_0 e^{-\frac{\mathrm{A}_\infty}{\mathrm{B}_\infty} k_1^{\mathrm{ue}} t_\mathrm{R}^i}\left(\frac{100 e^{-\frac{(t_\mathrm{R}^\mathrm{B}-t_\mathrm{R}^\mathrm{A})^2}{8\sigma_\mathrm{B}^2}} - h_\mathrm{p} e^{-\frac{(t_\mathrm{R}^\mathrm{B}-t_\mathrm{R}^\mathrm{A})^2}{2\sigma_\mathrm{B}^2}}}{\sigma_\mathrm{B}\sqrt{2\pi}} - \frac{100}{t_\mathrm{R}^\mathrm{B}-t_\mathrm{R}^\mathrm{A}}\right) + \frac{100\mathrm{B}_0 + \mathrm{A}_0\left(100 - h_\mathrm{p}\left(1+\sqrt{\frac{2}{\pi N}}\right)\right)}{t_\mathrm{R}^\mathrm{B}-t_\mathrm{R}^\mathrm{A}}\right) - \ln\left(\mathrm{A}_0\left(\frac{h_\mathrm{p} - 100 e^{-\frac{(t_\mathrm{R}^\mathrm{A}-t_\mathrm{R}^\mathrm{B})^2}{8\sigma_\mathrm{A}^2}}}{\sigma_\mathrm{A}\sqrt{2\pi}} + \frac{100 - h_\mathrm{p}\left(1+\sqrt{\frac{2}{\pi N}}\right)}{t_\mathrm{R}^\mathrm{B}-t_\mathrm{R}^\mathrm{A}}\right)\right)\right) \tag{21.7a}$$

$$k_1^{\mathrm{ue}} = -\frac{1}{t_\mathrm{R}^\mathrm{A}}\left(\ln\left(\mathrm{B}_0 e^{-\frac{\mathrm{A}_\infty}{\mathrm{B}_\infty} k_1^{\mathrm{ue}} t_\mathrm{R}^i}\left(\frac{100 e^{-\frac{(t_\mathrm{R}^\mathrm{A}-t_\mathrm{R}^\mathrm{B})^2}{8\sigma_\mathrm{B}^2}} - h_\mathrm{p}}{\sigma_\mathrm{B}\sqrt{2\pi}} + \frac{h_\mathrm{p}\left(1-\sqrt{\frac{2}{\pi N}}\right) - 100}{t_\mathrm{R}^\mathrm{B}-t_\mathrm{R}^\mathrm{A}}\right) + \frac{100\mathrm{A}_0 + \mathrm{B}_0\left(100 - h_\mathrm{p}\left(1-\sqrt{\frac{2}{\pi N}}\right)\right)}{t_\mathrm{R}^\mathrm{B}-t_\mathrm{R}^\mathrm{A}}\right) - \ln\left(\mathrm{A}_0\left(\frac{h_\mathrm{p} e^{-\frac{(t_\mathrm{R}^\mathrm{B}-t_\mathrm{R}^\mathrm{A})^2}{2\sigma_\mathrm{A}^2}} - 100 e^{-\frac{(t_\mathrm{R}^\mathrm{B}-t_\mathrm{R}^\mathrm{A})^2}{8\sigma_\mathrm{A}^2}}}{\sigma_\mathrm{A}\sqrt{2\pi}} + \frac{100}{t_\mathrm{R}^\mathrm{B}-t_\mathrm{R}^\mathrm{A}}\right)\right)\right) \tag{21.7b}$$

with $\sigma_i = w_i/8 \ln 2$and $i = \{\mathrm{A}, \mathrm{B}\}$.

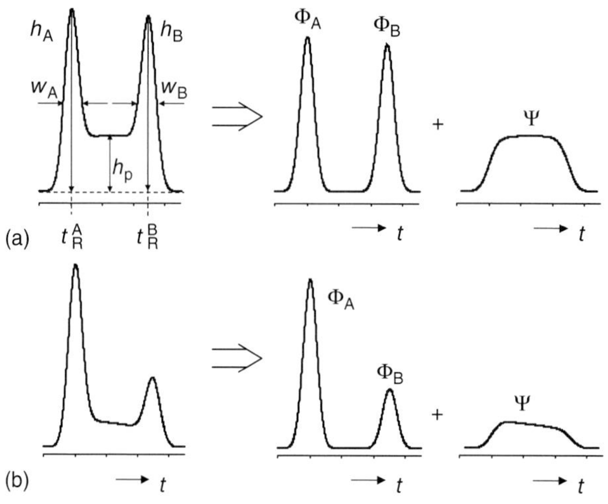

Figure 21.5 Mathematical separation of dynamic and on-column reaction chromatographic elution profiles into non-interconverted peaks, represented by time-dependent Gaussian distribution functions $\Phi_A(t)$ and $\Phi_B(t)$, and the interconverted part, represented by the stochastic distribution functions $\Psi(t)$. (a) Degenerated reversible (pseudo-) first-order reaction with an equilibrium constant $K_{A/B} = 1$ and (b) reversible (pseudo-) first-order reaction with $K_{A/B} > 1$.

Here, k is the reaction rate constant, h_p is the height of the plateau between the individual peaks at the mean retention time, A_0 and B_0 represent the concentrations of the reactant and product, t_R denotes the retention times of the individual peaks, N is the theoretical plate number, and w is the peak width (calculated from the outer parts of the individual peaks if the plateau height is greater than 50%).

21.4 Determination of Activation Parameters

The Gibbs free activation energy $\Delta G^{\neq}(T)$ can be calculated according to the Eyring equation (Equation 21.8) with k_B as the Boltzmann constant ($k_B = 1.380662 \times 10^{-23}$ J K^{-1}), T as the reaction temperature (K), h as Planck's constant ($h = 6.62617 \times 10^{-34}$ J s), and R as the gas constant ($R = 8.31441$ J K^{-1} mol^{-1}). The statistical factor κ is normally unknown and set to 1.0.

$$\Delta G^{\neq}(T) = -\mathrm{RT} \ln \left(\frac{k_1 h}{\kappa k_B T} \right) \tag{21.8}$$

The activation enthalpy $\Delta H^{\neq}$ is then obtained from the slope, and the activation entropy $\Delta S^{\neq}$ from the intercept of the Eyring plot ($\ln(k_1/T)$ as a function of T^{-1}).

21.5
On-Column Reaction Chromatography for the Investigation of Catalytic Reactions

21.5.1
Hydrogenations over Noble Metal Nanoparticles

Catalysts based on nonstabilized noble metal nanoparticles are very difficult to investigate because agglomeration and change of the catalytic activity can occur during reaction when these nanoparticles are dissolved in an organic solvent. On-column reaction gas chromatography (ocRGC), where the catalysis is performed in a fused-silica capillary, is a good method to investigate the catalytic activity of such materials. A prerequisite for such studies is a catalytically active stationary phase which is compatible with the chromatographic requirements. An example in which Pd nanoparticles embedded in an inert polydimethylsiloxane matrix without any interfering protecting shell, for example, tetraalkylammonium salts as surfactant, were prepared and used in hydrogenations was described by Trapp *et al.* [40, 41]. There, methylvinylsiloxanedimethylsiloxane copolymer (4.5% $Si(O)(CH_3)$ $(CH{=}CH_2)$ groups) was used to coordinate Pd ions to the vinyl groups in diethylether. To this mixture, hydridomethyldimethylpolysiloxane copolymer (25.7% $Si(O)(CH_3)H$ groups) was added for (i) the reduction of Pd^{2+} to Pd^0 and (ii) to crosslink with the methylvinylsiloxanedimethylsiloxane copolymer in a hydrosilylation reaction, catalyzed by Pd, to form a stabilizing matrix (cf. Figure 21.6).

The Pd nanoparticles were spherical and crystalline with a narrow size distribution of 3.2 ± 0.7 nm as determined by transmission electron microscopy (TEM). The viscous brownish-gray polymers were coated as a thin film of 0.25 μm onto

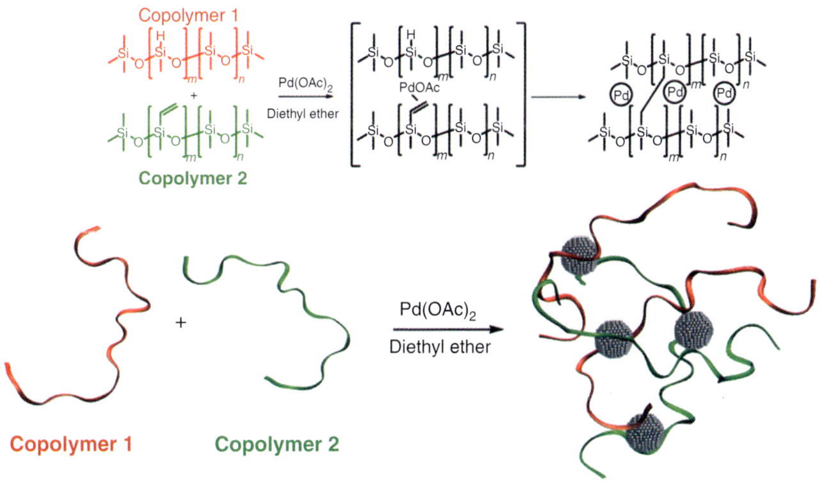

Figure 21.6 Preparation of highly active Pd nanoparticles embedded in a polysiloxane matrix.

the inner surface of fused-silica capillaries (i.d. 250 μm) and heated to 180 °C at the rate of 0.5 K min^{-1} under hydrogen flow, resulting in a permanently bonded polymer.

Figure 21.7 shows a scanning electron microscopy (SEM) image of the capillary. SEM and energy-dispersive X-ray (EDX) pattering of Si and Pd (16 h measurement) show the fused-silica micro capillary with the Pd/polysiloxane surface coating. The Pd loading is extremely low, only 0.73 ± 10^{-12} mol cm^{-1} capillary.

On-column catalysis was performed by coupling of this Pd nanoparticle capillary between a preseparation capillary (1 m) and a separation column (25 m). H_2 was used as the reactive carrier gas. Reactants and products were quantified by flame ionization detection (FID) and identified by MS. Reactant libraries consisting of 22 unsaturated compounds (alkenes, alkynes, aromatic hydrocarbons) and functionalized compounds (nitro compounds, aldehydes, ketones) to investigate the chemoselectivity could be simultaneously injected onto this column configuration at different temperatures and gas flows to vary the reaction time and to obtain temperature-dependent kinetic data. These Pd nanoparticles were extraordinarily active, and fast hydrogenations led to complete conversions even for capillaries of only 5 cm length and low temperatures of 60 °C. Therefore, to achieve incomplete conversions for the kinetic measurements all experiments had to be performed with a 2 cm capillary and reaction times in the range of 20 ms to 1 s. Still, for some compounds, for example, *trans*-cinnamaldehyde, methylbicyclo[2.2.1]hept-5-en-2-carboxylate, dimethylacetylenedicarboxylate, and styrene, 100% conversion for Pd nanoparticles with the highest activity was observed. Selected experimental data are summarized in Table 21.1.

For each compound, datasets were obtained from temperature and flow (contact time) dependent conversion measurements, which were then incorporated into kinetic models based on a Langmuir–Hinshelwood mechanism to determine the reaction rate constants k and activation parameters (Gibbs activation energy $\Delta G^{\neq}$, activation enthalpy $\Delta H^{\neq}$, and activation entropy $\Delta S^{\neq}$).

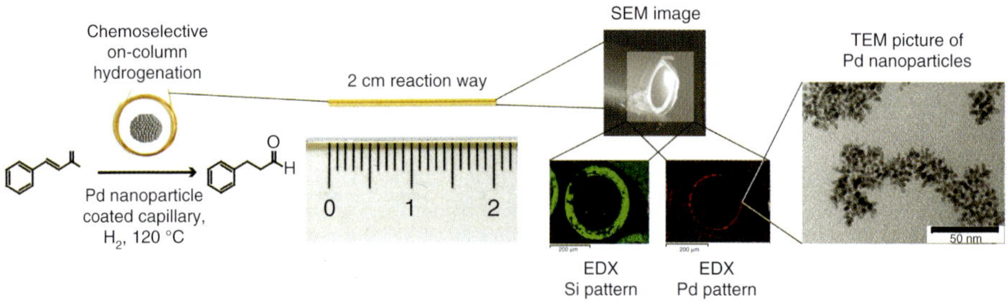

Figure 21.7 On-column hydrogenation over highly active Pd nanoparticles. Capillaries of only 2 cm coated with Pd nanoparticles stabilized in a polysiloxane matrix are used as the reactor. SEM and Si/Pd EDX measurements show the coating of the fused silica micro capillaries (i.d. 250, 0.25 μm film thickness).

Table 21.1 Selected results of the on-column hydrogenations over highly active Pd nanoparticles.

Substrate	Product	$k^{a)}$ (s^{-1})	$\Delta G^{\neq b)}$ ($kJ\,mol^{-1}$)	$\Delta H^{\neq}$	$\Delta S^{\neq}$ ($J\,K^{-1}\,mol^{-1}$)	$r^{c)}$ (s.d.)
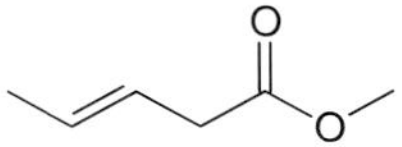	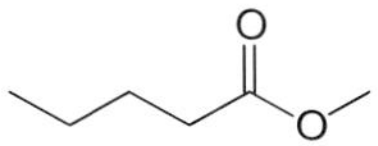	194	67.8	30.1 ± 0.5	−126 ± 3	0.997 (0.057)
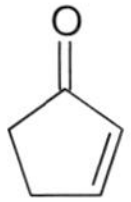		42	70.4	25.2 ± 0.5	−152 ± 4	0.996 (0.052)
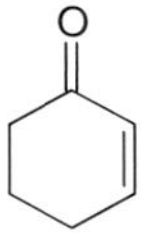	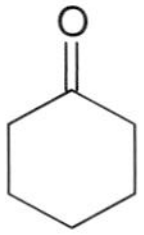	36	71.4	27.2 ± 0.7	−148 ± 6	0.996 (0.046)
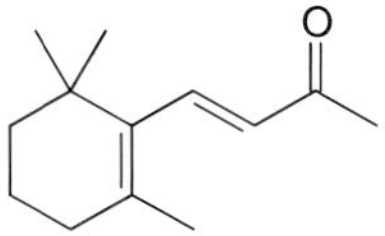	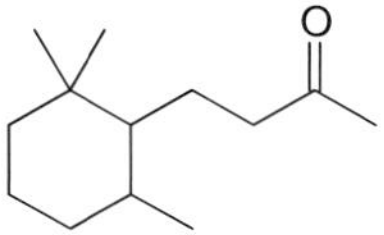	4	82.3	56.0 ± 1.0	−94 ± 2	0.998 (0.020)
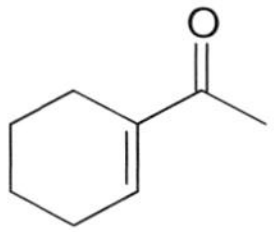	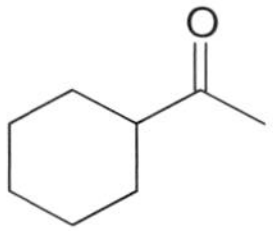	44	73.4	37.5 ± 0.6	−121 ± 3	0.999 (0.025)
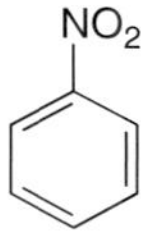	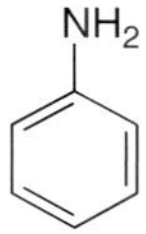	23	75.3	38.3 ± 1.5	−124 ± 7	0.985 (0.116)

a) Reaction rate constant at 120 °C.
b) Gibbs activation energy $\Delta G^{\neq}$ at 25 °C.
c) Correlation factor r and residual standard deviation (s.d.) of the linear regression of the Eyring plot.

21.5.2
Ring-Closing Metathesis

To prepare catalytically active micro columns to perform ring-closing metathesis [40], Grubbs second-generation catalysts were dissolved in dimethylpolysiloxane (GE SE 30). A capillary (10 m) with a film thickness of 1 μm was coated under strict exclusion of oxygen. It is very important that the catalyst loading is very low to avoid too fast reactions; therefore, only 1.6 μg catalyst per meter (1.9×10^{-9} mol m^{-1}) capillary was used. The on-column catalysis was performed by coupling this column with a preseparation column of 1 m. The eluted compounds were quantified and identified by FID and MS detection. Simultaneous injection of a reactant library consisting of 12 different compounds for ring-closure metathesis onto the catalytically active separation column allowed performing high-throughput kinetic screening of the catalyst. The obtained elution profiles were characterized by a plateau formation between the reactant and the product. The obtained elution profiles were analyzed by the unified equation to calculate reaction rate constants. The dissociation and chromatographic removal of the tricyclohexylphosphine ligand ($P(Cy)_3$) was not detected in these experiments, which was explained by the fact that the phosphine ligand is dissolved in the polysiloxane, thereby stabilizing the complex after the reaction. It is remarkable that the catalyst was stable over a wide temperature range of up to 150 °C without any detectable degradation of the catalytic activity or leaching. From temperature-dependent measurements, the activation parameters were obtained (cf. Figure 21.8) (for *N*,*N*-diallyltrifluoroacetamide $\Delta G^{\neq}(298\,\text{K}) = 84.1\,\text{kJ mol}^{-1}$, $\Delta H^{\neq} = 15.5 \pm 0.9\,\text{kJ mol}^{-1}$, $\Delta S^{\neq} = -230 \pm 8\,\text{J (K mol)}^{-1}$). By simultaneous injection of the reactant library, an extraordinary high throughput in the determination of reaction rate constants of 36 reaction rate constants per hour was achieved.

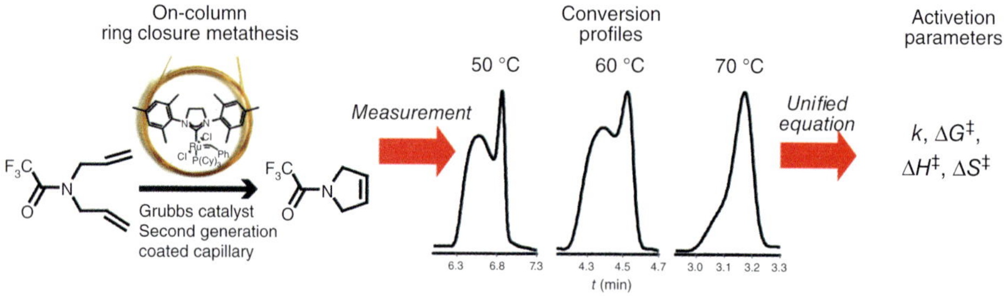

Figure 21.8 On-column metathesis over Grubbs second-generation catalyst. In these experiments, catalytic activity and separation selectivity are combined in a single 10-m capillary by dissolving the catalyst in the stationary separation phase. Elution profiles are obtained and characterized by a pronounced conversion profile from the reactant to the product. Kinetic analysis is directly performed with the unified equation.

21.5.3
Gosteli–Claisen Rearrangement

The feasibility of enantioselective catalysis in on-column reaction chromatography was exemplified by the Gosteli–Claisen rearrangement of allylvinyl ethers (cf. Scheme 21.1) [47].

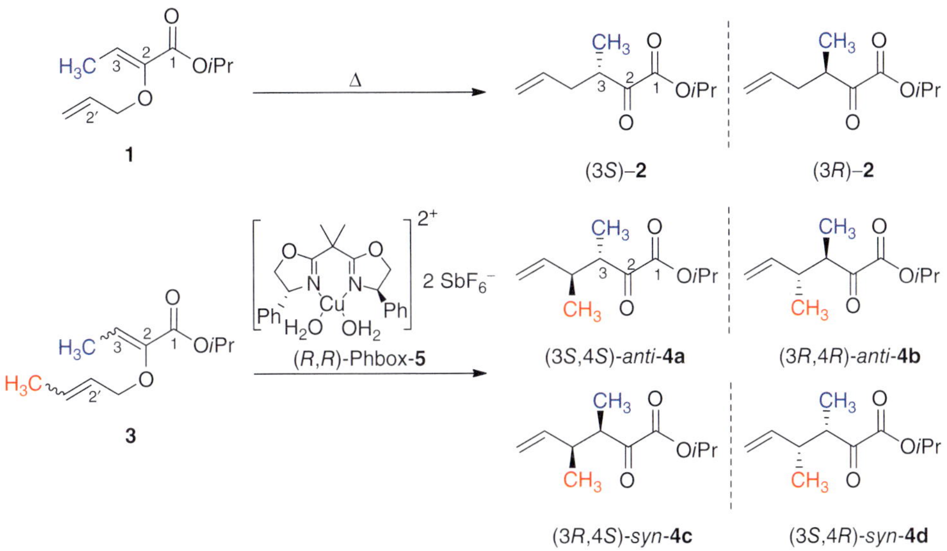

Scheme 21.1 Catalytic asymmetric Gosteli–Claisen rearrangement studied by enantioselective on-column reaction GC (ocRGC).

The enantiomers of the α-keto ester **2** (cf. Scheme 21.1) were separated with excellent resolution and selectivity by enantioselective GC in presence of the CSP Chirasil-β-Dex. It was observed that the allyl vinyl ether **1** already underwent the uncatalyzed Gosteli–Claisen rearrangement during enantioselective GC analysis at moderate temperatures, which was visible by a distinct plateau formation between the peaks of the enantiomers **2** and the starting substrate **1**. The identity of the peaks was proven by co-injection of substrate **1** and racemic compound **2**, and by GC–MS analysis, comparing fragmentation patterns of reference spectra. The presence of substrate **1** complicated the determination of enantiomeric excesses in conventional reaction progress analysis. This was due to the competing thermal background rearrangement during analysis, which diminished the true enantiomeric excess value, and furthermore influenced the enantiomer separation by decreasing the resolution *R*. With increasing separation temperature between 70 and 120 °C, the plateau formation clearly increased while the timescale of separation and enantioselectivity of the separation decreased (cf. Figure 21.9).

The depicted experiments can be evaluated using the unified equation (cf. Table 21.2). An irreversible first-order reaction was considered, because the

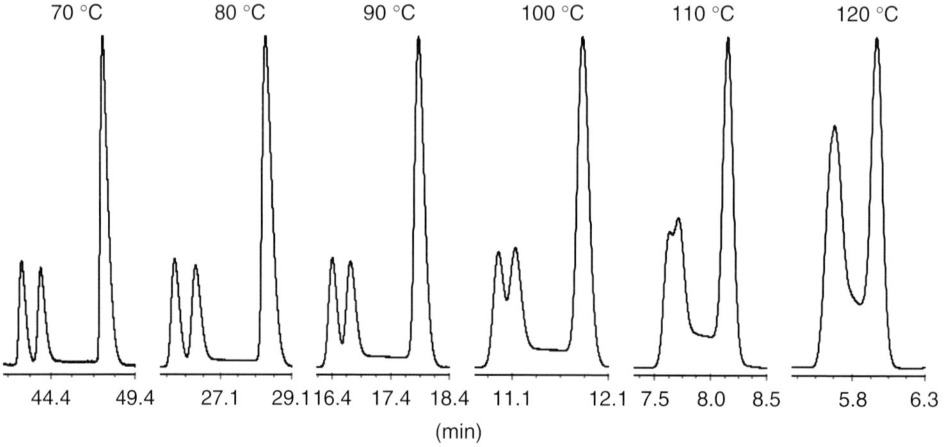

Figure 21.9 Elution profiles between 70 and 120 °C of enantioselective ocRGC of the uncatalyzed Gosteli–Claisen rearrangement of (*Z*)-**1** (last eluted peak) to (±)-**2** (first two eluted peaks) in presence of the CSP Chirasil-β-Dex (length 25 m; i.d. 250 μm; film thickness 500 nm). Helium was used as inert carrier gas at 120 kPa inlet pressure. The elution profiles are characterized by a distinct plateau formation, indicating the conversion of **1** to **2**.

Table 21.2 Summarized results of the uncatalyzed Gosteli–Claisen rearrangement of (*Z*)-**1** to (±)-**2** obtained by enantioselective ocRGC in presence of the CSP Chirasil-β-Dex.

T (°C)	h_p (%)	(*Z*)-**1** (%)	(±)-**2** (%)	N_1	N_2	k_1 (10^{-4} s^{-1})	$\Delta G^{\neq}(T)$ (kJ mol^{-1})
70.0	1.1	65.7	34.3	101 500	24 700	0.14	116.2
80.0	2.0	64.2	35.8	84 100	28 400	0.46	116.3
90.0	3.1	62.7	37.3	58 500	29 300	1.12	117.0
100.0	5.5	60.3	39.7	39 200	27 700	2.85	117.4
110.0	10.3	57.1	43.0	19 200	22 200	6.84	117.9
120.0	21.8	51.8	48.2	6 800	12 500	15.56	118.3

area under the plateau consisted only of the racemic compound **2**, which was corroborated by analysis of the GC–MS trace.

Activation parameters were calculated from temperature-dependent measurements between 70 and 120 °C. The activation enthalpy $\Delta H^{\neq}$ of the reaction was obtained from the slope, and the activation entropy $\Delta S^{\neq}$ from the intercept of the Eyring plot ($\ln(k_1/T)$ as a function of T^{-1}) (cf. Figure 21.10).

To extend ocRGC to the catalytic asymmetric Gosteli–Claisen rearrangement, the following experimental setup is described. Catalytically active separation columns were used as the micro-capillary reactor and are subsequently coupled with enantioselective separation columns to perform analysis of the reaction products. The inner surface of fused-silica capillaries (i.d. 250 μm) was coated with [Cu{(*R*,*R*)-Ph-box}]$(SbF_6)_2$ (*R*,*R*)-**5** dissolved in an inert polysiloxane matrix

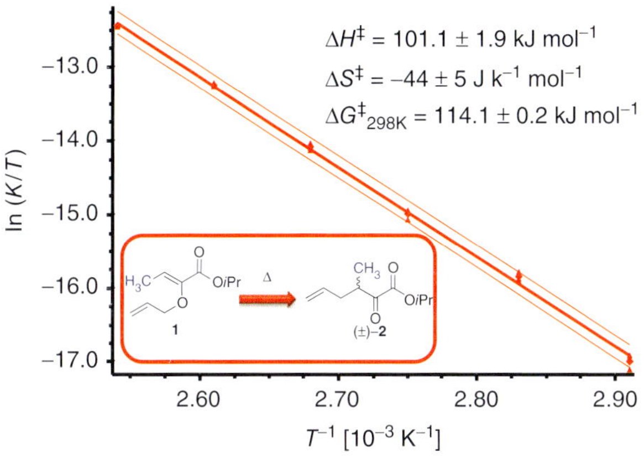

Figure 21.10 Eyring plot to determine activation parameters of the uncatalyzed Gosteli–Claisen rearrangement of (*Z*)-**1** to (±)-**2** obtained by temperature-dependent measurements using stereoselective ocRGC between 70 and 120 °C.

and these capillaries were used as the on-column micro-reactor in a GC, which allowed the control of the reaction temperatures with high precision. It was pointed out that the catalyst [Cu{(*R,R*)-Ph-box}]$(SbF_6)_2$ (*R,R*)-**5** was completely dissolved in the viscous polysiloxane, representing fully homogeneous reaction conditions. The experimental setup (cf. Figure 21.11) consisted of the catalytically active capillary coupled between a preseparation capillary (GE-SE-30; length 5 m; i.d. 250 μm; film thickness 500 nm) and an enantioselective separation capillary (CSP Chirasil-β-Dex; length 25 m; i.d. 250 μm; film thickness 500 nm). The enantioselective separation column separated the stereoisomeric reactants and products to determine conversions and selectivities. The suitable temperature range to perform the investigated Gosteli–Claisen rearrangement and achieve excellent stereoisomer separation of the reaction products was determined to be between 65 and 85 °C.

In order to study the catalytic asymmetric Gosteli–Claisen rearrangement of (*E,Z*)- and (*Z,E*)-**3** by ocRGC, these allyl vinyl ethers were injected into the described capillary arrangement with or without the catalytically active capillary column (cf. Figure 21.11). A priori, the uncatalyzed Gosteli–Claisen rearrangement of either (*E,Z*)- or (*Z,E*) **3** may provide a mixture of the four stereoisomers (±)-*anti*- and (±)-*syn*-**4**. Formation of (±)-*anti*-**4** from (*E,Z*)- or (*Z,E*)-**3** is energetically favored. In contrast to the case of substrate **1**, uncatalyzed formation of the rearrangement product from (*E,Z*)- or (*Z,E*)-**3** during the timescale of separation was not detectable by highly sensitive MS measurements and no characteristic plateau formation was observable in the applied temperature range.

The reactor capillary with the embedded catalyst (*R,R*)-**5** was coupled between the preseparation and the enantioselective separation capillary. Using this experimental setup, the diastereo- and enantioselective formation of **4a** from (*E,Z*)-**3** and **4b** from (*Z,E*)-**3** was detected. The advantage of this approach is that

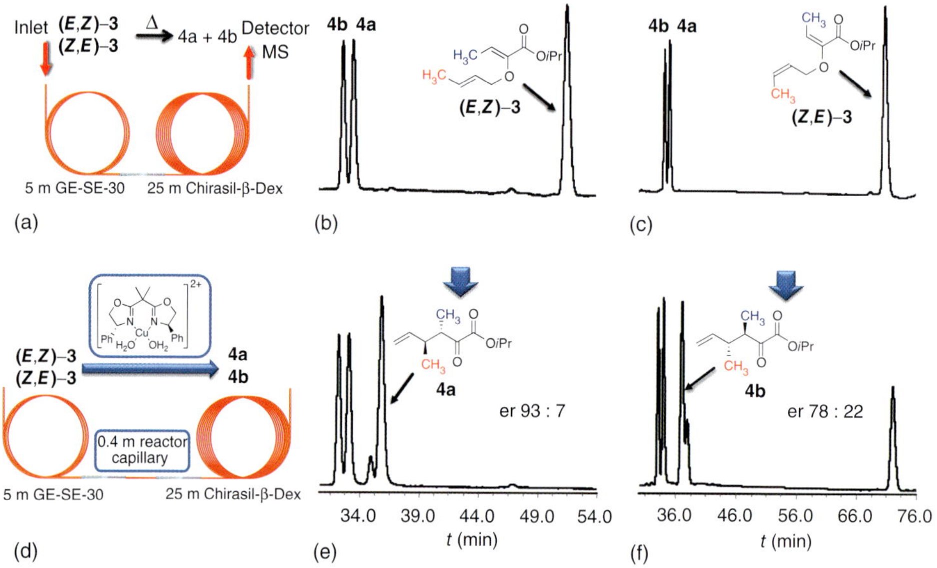

Figure 21.11 Experimental setup of the stereoselective on-column gas chromatographic experiment (a) to investigate the catalytic asymmetric Gosteli–Claisen rearrangement of (*E*,*Z*)- and (*Z*,*E*)-**3** in presence of the Lewis acid catalyst (*R*,*R*)-**5** dissolved in an inert polysiloxane matrix to form the corresponding α-keto isopropyl esters **4a** and **4b**, respectively. (b,c,e,f) Show representative chromatograms of the on-column experiment.

the enantiomeric rearrangement products **4a** and **4b**, formed by the uncatalyzed rearrangement in the injector, can be used as internal standard to determine the contact times Δ*t* of the reactants and the catalytically active capillary. The enantiomeric products from the catalyzed rearrangement were eluted with a time delay compared to the racemate formed through the uncatalyzed background reaction because of the separation of the reactants (*E*,*Z*)-**3** and (*Z*,*E*)-**3** from the noncatalytically formed product from the achiral preseparation column. Reaction products **4** showed less retention compared to the reactant in presence of the a CSP GE-SE-30 and Chirasil-β-Dex.

From these chromatographic data, conversions of the individual stereoisomers, enantiomeric ratios, and, consequently, the catalytic activity could be precisely determined.

The embedded catalyst (*R*,*R*)-**5** provided excellent conversions and enantiomeric ratio values for the catalytic asymmetric Gosteli–Claisen rearrangement of (*E*,*Z*)-**3** to **4a**. Notably, the catalyst maintained its activity and selectivity within the investigated temperature range between 65 and 85 °C.

To obtain the experimental activation parameters $\Delta H^{\neq}$ and $\Delta S^{\neq}$ of the catalytic asymmetric Gosteli–Claisen rearrangement catalyzed by (*R*,*R*)-**5**, the incomplete conversion of (*Z*,*E*)-**3** to **4a** and **4b** was investigated by temperature-dependent measurements between 65 and 85 °C. Best agreement in the calculation of reaction rate constants was achieved by the application of a pseudo-first-order reaction rate law for parallel reactions, where both pathways yielded the enantiomeric reaction products **4a** and **4b**. The Gibbs free activation energies $\Delta G^{\neq}(T)$ were calculated according to the Eyring equation. $\Delta G^{\neq}$ at 80 °C was $100.2 \pm 0.3\ \text{kJ mol}^{-1}$ for the rearrangement of (*Z*,*E*)-**3** to the major enantiomer **4b** and $104.2 \pm 0.3\ \text{kJ mol}^{-1}$ for the rearrangement of (*Z*,*E*)-**3** to the minor enantiomer **4a**. The activation enthalpy $\Delta H^{\neq}$ of the reaction was obtained from the slope, and the activation entropy $\Delta S^{\neq}$ from the intercept of the Eyring plot ($\ln(k/T)$ as a function of T^{-1}) (cf. Figure 21.12). Rearrangement of (*Z*,*E*)-**3** to **4b** in presence of (*R*,*R*)-**5** gave $\Delta H^{\neq} = 106.1 \pm 6.6\ \text{kJ mol}^{-1}$, and $\Delta S^{\neq} = 17 \pm 19\ \text{J (K mol)}^{-1}$, and (*Z*,*E*)-**3** to **4a** gave $\Delta H^{\neq} = 106.9 \pm 12.3\ \text{kJ mol}^{-1}$, and $\Delta S^{\neq} = 8 \pm 35\ \text{J (K mol)}^{-1}$. This approach made the investigation of single isomeric reactants possible because even traces of isomers or isomerized reactants were removed by the first separation dimension.

21.5.4
Combinatorial High-Throughput Screening: Catalyst by the Meter

N-Heterocyclic carbenes (NHCs) are highly versatile ligands and are of great importance in the development of highly active catalysts for a broad range of reactions with superior properties with regard to stability and applicability in various reaction media and conditions. Next, we present a strategy in which permanent immobilization of NHC pre-ligands on the inner surface of fused-silica capillaries (i.d. 250 μm) with a defined film thickness (between 50 and 1000 nm) of the polymeric matrix was used to rapidly prepare NHC-based metal catalysts for high-throughput screening of reactions.

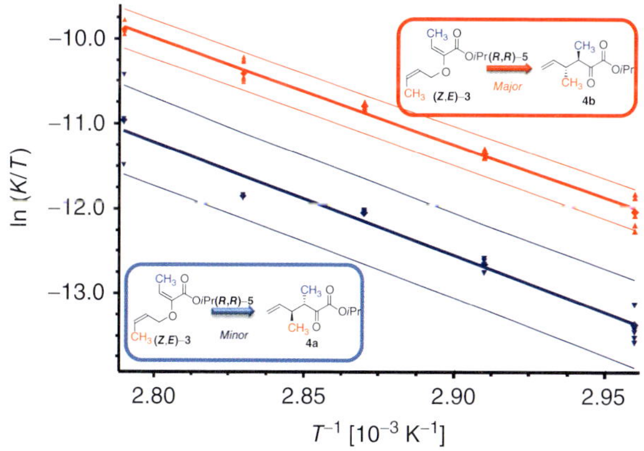

Figure 21.12 Eyring plot to determine activation parameters of the catalytic asymmetric Gosteli–Claisen rearrangement of (*Z*,*E*)-**3** to **4a** and **4b** obtained by temperature-dependent measurements using stereoselective ocRGC between 65 and 85 °C.

These NHC pre-ligand-modified fused-silica capillaries [39] can be converted to NHC ligands and, with suitable organometallic precursors, to precatalysts utilizing all advantages of solid-phase organic chemistry, that is, high reagent concentrations, simple removal of nonreacted reagents, and so on.

To immobilize a 1,3-bismesityl-2-imidazolidinylidene ligand, a strategy adapted from Blechert *et al.* [48] was used in the described study to modify the backbone of the 2-imidazolidinylidene moiety via an ether function. This allowed the introduction of terminal alkene chains, which can react under benign conditions in a hydrosilylation with hydridopolysiloxanes (cf. Scheme 21.2).

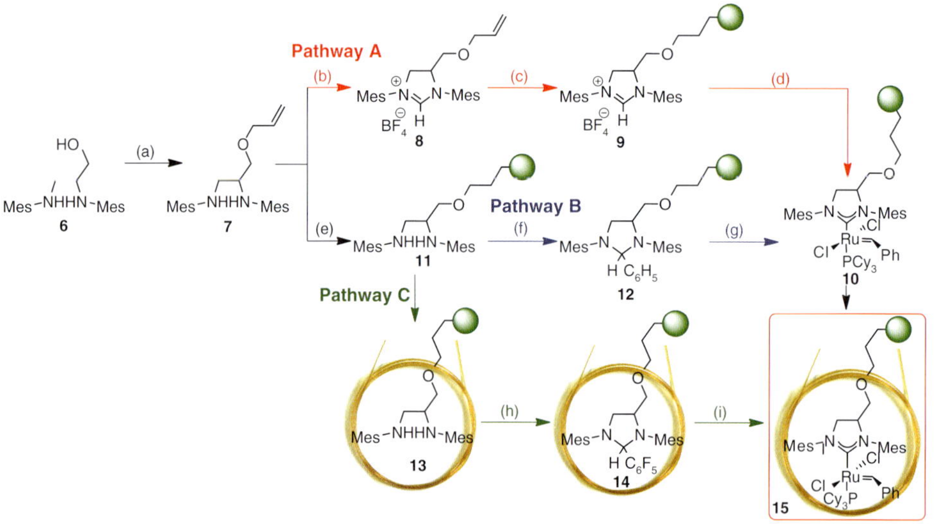

Scheme 21.2 (a) 3-Bromo-prop-1-ene, NaH, THF, 24 h, 64 °C. (b) NH_4BF_4, $HC(OEt)_3$. (c) Hydridomethyldimethylpolysiloxane, Karstedt catalyst, THF, ultrasonication for 3 h. (d) KOtBu, THF, rt, 1 h, $(Cy_3P)_2Cl_2Ru(=CHPh)$, toluene, 80 °C, 4 h. (e) Hydridomethyldimethylpolysiloxane, Karstedt catalyst, THF ultrasonication for 3 h. (f) Pentafluorobenzaldehyde, AcOH, 24 h. (g) $(Cy_3P)_2RuCl_2CHPh$, toluene, 80 °C, 4 h. (h) Pentafluorobenzaldehyde AcOH, *n*-pentane, 60 °C. (i) $(Cy_3P)_2RuCl_2CHPh$, *n*-pentane, 60 °C.

Various reaction pathways were studied on how to prepare capillaries with immobilized NHC ligands. The base-free routes in pathways B and C (cf. Scheme 21.1) are superior to route A. In pathway B, the diamine **7** is first immobilized via hydrosilylation to hydridomethyldimethylpolysiloxane to form the polymer **11**. The bonded diamine **11** is then cyclized with pentafluorobenzaldehyde to the immobilized 2-(pentafluorophenyl)imidazolidine **12**. It was found that the bonded carbene could be thermally generated and thus be immediately converted to form the bonded Grubbs second-generation catalyst **10** by ligand exchange against the tricyclohexylphosphine ligand. Polymeric diamine **11** was coated onto the inner surface of a fused-silica capillary (i.d. 250 μm) with a film thickness of

250 nm. The polymer film was then immobilized onto the glass surface by heating the capillary to 180 °C. This polymer was highly reactive and was permanently bonded to the glass surface via Si–O–Si bonds. The polymeric diamine **13** was then converted with pentafluorobenzaldehyde to the permanently bonded polymeric 2-(pentafluorophenyl)imidazolidine **14**. These surface-modified capillaries **14** show excellent long-term stability and they are easy to activate to release the carbene, to form together with organometallic precursors, NHC ligand based catalysts by the meter. Heating the capillaries modified by **14** in a water bath to 60 °C resulted in the formation of the immobilized free carbene which reacted with Grubbs first-generation catalyst by ligand exchange to give permanently bonded polymeric Grubbs second-generation catalyst capillary **15**. Permanently bonded polymeric Grubbs second-generation catalyst capillaries **15** showed some remarkable properties in comparison to Grubbs second-generation catalyst dissolved in poly(dimethylsiloxane) and coated onto the inner surface of fused-silica capillaries. These capillaries are moisture and air insensitive, can be installed several times over weeks without any loss of activity or leaching, and are stable even at elevated temperatures of 105 °C for several hours.

Figure 21.13 shows an on-column reaction chromatographic experiment using a 10 cm, permanently bonded, polymeric Grubbs second-generation catalyst capillary to achieve on-column ring-closing metathesis of *N,N*-diallyltrifluoroacetamide to *N*-trifluoroacetamide-3-pyrroline at 50 °C, and 10 cm permanently bonded polymeric 1,3-bis(2,4,6-trimethylphenyl)imidazol-2-ylidene gold(I) chloride to hydrogenate nitrobenzene to aniline at 95 °C.

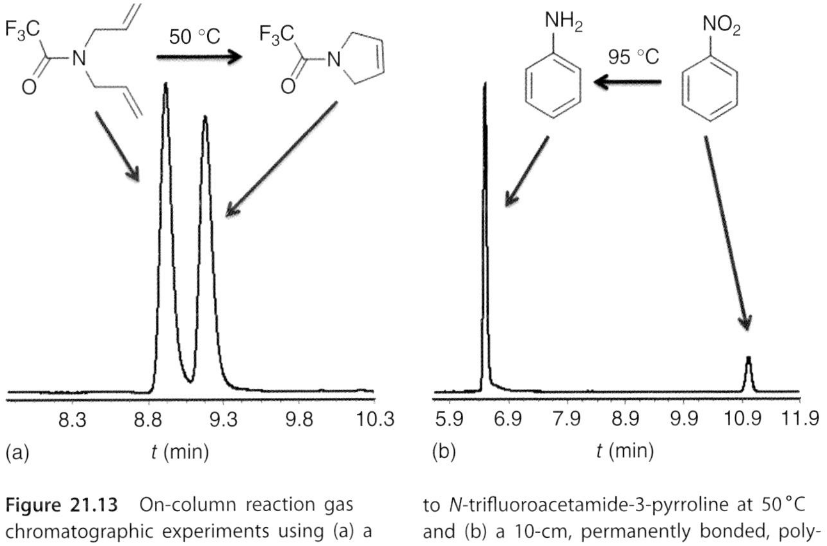

Figure 21.13 On-column reaction gas chromatographic experiments using (a) a 10-cm, permanently bonded, polymeric Grubbs second-generation catalyst capillary **15** to achieve on-column ring-closing metathesis of *N,N*-diallyltrifluoroacetamide to *N*-trifluoroacetamide-3-pyrroline at 50 °C and (b) a 10-cm, permanently bonded, polymeric 1,3-bis(2,4,6-trimethylphenyl)imidazol-2-ylidene gold(I) chloride to hydrogenate nitrobenzene to aniline at 95 °C.

21.6 Outlook

The general strategies to perform catalytic reactions in capillaries presented in this chapter can be utilized in the comprehensive kinetic characterization of catalysts and materials. Furthermore, catalytic capillaries may one day also prove to be useful tools for selective transformations in analytical applications or for structure elucidation. Also, preparative scale-up by stacking of reactor capillaries can easily be achieved. From there, the development of chemical micro plants for the production of fine chemicals can be envisioned as a natural step in terms of energy efficiency and environmental impact.

The presented recent approaches impressively illustrate the (r)evolution of chemist's toolkits from flasks and beakers to miniaturized reactors with highly efficient and selective catalysts, inherently combined with separation techniques.

References

1. Falk, T. and Seidel-Morgenstern, A. (1999) *Chem. Eng. Sci.*, **54**, 1479–1485.
2. Vu, T.D., Seidel-Morgenstern, A., Grüner, S., and Kienle, A. (2005) *Ind. Eng. Chem. Res.*, **44**, 9565–9574.
3. Haswell, S.J. and Watts, P. (2003) *Green Chem.*, **5**, 240–249.
4. Haswell, S.J. and Skelton, V. (2000) *Trends Anal. Chem.*, **19**, 389–395.
5. Jas, G. and Kirschning, A. (2003) *Chem. Eur. J.*, **9**, 5708–5723.
6. Watts, P. and Haswell, S.J. (2005) *Chem. Soc. Rev.*, **34**, 235–246.
7. Valera, F.E., Quaranta, M., Moran, A., Blacker, J., Armstrong, A., Cabral, J.T., and Blackmond, D.G. (2010) *Angew. Chem. Int. Ed.*, **49**, 2478–2485.
8. Jähnisch, K., Hessel, V., Löwe, H., and Baerns, M. (2004) *Angew. Chem. Int. Ed.*, **43**, 406–446.
9. Flögel, O., Codée, J.D.C., Seebach, D., and Seeberger, P.H. (2006) *Angew. Chem. Int. Ed.*, **45**, 7000–7003.
10. Nikbin, N., Ladlow, M., and Ley, S.V. (2007) *Org. Proc. Res. Dev.*, **11**, 458–462.
11. Bourne, S.L., Koos, P., O'Brien, M., Martin, B., Schenkel, B., Baxendale, I.R., and Ley, S.V. (2011) *Synlett*, **18**, 2643–2647.
12. Knight, J. (2002) *Nature*, **418**, 474–475.
13. Janasek, D., Franzke, J., and Manz, A. (2006) *Nature*, **442**, 374–380.
14. Haswell, S.J. (2006) *Nature*, **441**, 705.
15. Belder, D. (2006) *Anal. Bioanal. Chem.*, **385**, 416–418.
16. Fritzsche, S., Ohla, S., Glaser, P., Giera, D.S., Sickert, M., Schneider, C., and Belder, D. (2011) *Angew. Chem.*, **123**, 9639–9642.
17. de Bellefon, C., Tanchoux, N., Caravieilhes, S., Grenouillet, P., and Hessel, V. (2000) *Angew. Chem. Int. Ed.*, **39**, 3442–3445.
18. Sahoo, H.R., Kralj, J.G., and Jensen, K.F. (2007) *Angew. Chem. Int. Ed.*, **46**, 5704–5708.
19. Noel, T., Kuhn, S., Musacchio, A.J., Jensen, K.F., and Buchwald, S.L. (2011) *Angew. Chem.*, **123**, 6065–6068.
20. Kobayashi, J., Mori, Y., Okamoto, K., Akiyama, R., Ueno, M., Kitamori, T., and Kobayashi, S. (2004) *Science*, **304**, 1305–1308.
21. Grob, K. (1986) *Classical Split and Splitless Injection in Capillary Gas Chromatography with Some Remarks on PTV Injection*, Huethig, Heidelberg.
22. Schoetz, G., Trapp, O., and Schurig, V. (2001) *Electrophoresis*, **22**, 3185–3190.
23. Trapp, O., Schoetz, G., and Schurig, V. (2001) *Chirality*, **13**, 403–414.
24. Wolf, C. (2008) *Dynamic Stereochemistry of Chiral Compounds. Principles and Applications*, RSC Publishing, Cambridge.

25. Trapp, O. (2010) *Electrophoresis*, **31**, 786–813.
26. Purnell, J.H. (1955) *Nature*, **175**, 2009.
27. van Deemter, J.J., Zuiderweg, F.J., and Klinkenberg, A. (1956) *Chem. Eng. Sci.*, **5**, 271–289.
28. van Deemter, J.J. (1953) *J. Ind. Eng. Chem.*, **45**, 1227–1233.
29. Katsanos, N.A., Thede, R., and Roubani-Kalantzopoulou, F. (1998) *J. Chromatogr. A*, **795**, 133–184.
30. Tamaru, K. (1959) *Nature*, **183**, 319–320.
31. Bassett, D.W. and Habgood, H.W. (1960) *J. Phys. Chem.*, **64**, 769–773.
32. Gil-Av, E. and Herzberg-Minzly, Y. (1961) *Proc. Chem. Soc.*, 316.
33. Pratt, G.L. and Langer, S.H. (1969) *J. Phys. Chem.*, **73**, 2095–2097.
34. Marriott, P.J. and Lai, Y.-H. (1986) *Inorg. Chem.*, **25**, 3680–3683.
35. Matsen, J.M., Harding, J.W., and Magee, E.M. (1965) *J. Phys. Chem.*, **69**, 522–527.
36. Skrdla, P.J. (2004) *Int. J. Chem. Kinet.*, **36**, 386–393.
37. Trapp, O. (2006) *Anal. Chem.*, **78**, 189–198.
38. Trapp, O. (2006) *Electrophoresis*, **27**, 534–541.
39. Lang, C., Gärtner, U., and Trapp, O. (2011) *Chem. Commun.*, **47**, 391–393.
40. Trapp, O., Weber, S.K., Bauch, S., and Hofstadt, W. (2007) *Angew. Chem. Int. Ed.*, **46**, 7307–7310.
41. Trapp, O., Weber, S.K., Bauch, S., Bäcker, T., Hofstadt, W., and Spliethoff, B. (2008) *Chem. Eur. J.*, **14**, 4657–4666.
42. Craig, L.C. (1944) *J. Biol. Chem.*, **155**, 519–534.
43. Martin, A.J.P. and Synge, R.L.M. (1941) *Biochem. J.*, **35**, 1358–1368.
44. Kallen, J. and Heilbronner, E. (1960) *Helv. Chim. Acta*, **43**, 489–500.
45. Bürkle, W., Karfunkel, H., and Schurig, V. (1984) *J. Chromatogr.*, **288**, 1–14.
46. Trapp, O. and Schurig, V. (2001) *Comput. Chem.*, **25**, 187–195.
47. Troendlin, J., Rehbein, J., Hiersemann, M., and Trapp, O. (2011) *J. Am. Chem. Soc.*, **133**, 16444–16450.
48. Schürer, S.C., Gessler, S., Buschmann, N., and Blechert, S. (2000) *Angew. Chem. Int. Ed.*, **39**, 3898–3901.

Index

Molecular Catalysts: Structure and Functional Design, First Edition. Edited by Lutz H. Gade and Peter Hofmann.

d

l

m